Atmosphere, Earth, Ocean & Space

The series Atmosphere, Earth, Ocean & Space (AEONS) publishes state-of-art studies spanning all areas of Earth and Space Sciences. It aims to provide the academic communities with new theories, observations, analytical and experimental methods, and technical advances in related fields. The series includes monographs, edited volumes, lecture notes and professional books with high quality. The key topics in AEONS include but are not limited to: Aeronomy and ionospheric physics, Atmospheric sciences, Biogeosciences, Cryosphere sciences, Geochemistry, Geodesy, Geomagnetism, Environmental informatics, Hydrological sciences, Magnetospheric physics, Mineral physics, Natural hazards, Nonlinear geophysics, Ocean sciences, Seismology, Solar-terrestrial sciences, Tectonics and Volcanology.

More information about this series at http://www.springer.com/series/16015

De-Jin Wu · Ling Chen

Kinetic Alfvén Waves in Laboratory, Space, and Astrophysical Plasmas

南京大學出版社
NANJING UNIVERSITY PRESS

Springer

De-Jin Wu
Chinese Academy of Sciences
Purple Mountain Observatory
Nanjing, Jiangsu, China

Ling Chen
Chinese Academy of Sciences
Purple Mountain Observatory
Nanjing, Jiangsu, China

ISSN 2524-440X ISSN 2524-4418 (electronic)
Atmosphere, Earth, Ocean & Space
ISBN 978-981-13-7988-8 ISBN 978-981-13-7989-5 (eBook)
https://doi.org/10.1007/978-981-13-7989-5

Jointly published with Nanjing University Press, Nanjing, China
The print edition is not for sale in China Mainland. Customers from China Mainland please order the print book from: Nanjing University Press.

This Springer imprint is published by the registered company Springer Nature Singapore Pte Ltd.
The registered company address is: 152 Beach Road, #21-01/04 Gateway East, Singapore 189721, Singapore

Foreword

Kinetic Alfvén waves (KAWs) are small-scale dispersive Alfvén waves and can effectively exchange energy and momentum with plasma particles through the wave-particle interaction. Therefore, KAWs can play an important role in the heating of high-temperature plasmas, the acceleration of high-energy particles, and the anomalous transport of plasma particles, which occur commonly and frequently in various plasma environments from laboratory to space and astrophysics. In particular, a large amount of investigations based on in situ measurements by satellites in space plasmas has revealed that kinetic-scale fluctuations in the solar wind as well as in the magnetosphere can be identified well as the KAW-associated turbulence. By combining the Voyager 1 in-situ measurements (Lee & Lee, Nature Astronomy, 3, 154, 2019) with the earlier ground remote observations (Armstrong et al., Astrophys. J., 443, 209, 1995), the interstellar turbulence spectrum is found to extend from 50 m ($\sim$ the Debye length) to 10^{18} m ($\sim$30 pc), over 16 orders of magnitude. The turbulence spectrum consists of a Kolmogorov inertial range (10^6–10^{18} m) and a kinetic range (50–10^6 m). This implies that the KAW turbulence can ubiquitously exist in various cosmic plasma environments, just like the AW turbulence.

The monograph *Kinetic Alfvén Waves in Laboratory, Space, and Astrophysical Plasmas* by Drs. De-Jin Wu and Ling Chen presents a comprehensive description of the applications of KAWs in space and astrophysical plasmas. After introducing and summarizing the main physical properties of KAWs and their experimental demonstrations in Chaps. 1 and 2, they focus on the applications of KAWs to the crucial processes in the solar-terrestrial coupling system, including the auroral electron acceleration in the magnetosphere-ionosphere coupling (Chap. 3), the anomalous particle transport occurring at the magnetopause in the solar wind-magnetosphere interaction (Chap. 4), the turbulent cascading process of the large-scale AW turbulence towards the small-scale KAW turbulence in the solar wind turbulence (Chap. 5), and the solar corona and solar wind heating (Chap. 6). However, the authors do not stop at these well-known phenomena and proceed further across the extrasolar astrophysical plasmas. In Chap. 7, they discuss further possible applications of KAWs to extrasolar astrophysical plasmas, in particular, to the heating of cosmic hot gases such as stellar and accretion-disk coronae, galactic hot halos, and intracluster hot medium, as well as to the reacceleration of high-energy electrons and the collimation of radio jets in extragalactic extended radio sources.

KAWs and their applications are without doubt an increasingly interesting subject. The physics of KAWs can sensitively depend on the local plasma parameters. This indicates that they may play very different roles in various cosmic environments. This important and interesting characteristic of KAWs also is fully displayed in this monograph, which collects a number of KAW-related applications scattered in various journals and proceedings, but not always easily accessible. In a word, this is a particularly useful monograph for graduate students and young scientists joining the fields of plasma physics, space physics, and astrophysics. The materials collected in this monograph also are interesting for the most advanced researchers working actively in these fields.

October 2019

Prof. Lou-Chuang Lee
Institute of Earth Sciences
Academia Sinica, Taipei, Taiwan

Preface

Alfvén waves (AWs) are low-frequency and long-wavelength electromagnetic fluctuations in conducting fluids and ubiquitous in various space and astrophysical magnetic plasma environments. The most important property of AWs is that they can efficiently transport energy and momentum between distant plasma regions connected to each other by steady magnetic fields because the group velocity of AWs propagates exactly along the steady magnetic fields, independent of the orientation of the wave front. Therefore, AW can play an important role in the electrodynamical coupling of active and passive magnetic plasmas and has been a subject of intense study since it was found by Hannes Alfvén in 1942. In particular, it is the discovery of AWs that led to a whole new field of magnetohydrodynamics (MHD), which combines electromagnetic theory and fluid dynamics and creates fruitful applications in different parts of plasma physics.

Kinetic Alfvén waves (KAWs) are dispersive AWs with a short wavelength comparable to the kinetic scales of plasma particles, such as the ion (or ion-acoustic) gyroradius or the electron inertial length. Although KAWs retain some basic properties of MHD AWs, such as the quasi-parallel propagation of the wave group velocity, they have a lot of the novel important characteristics the MHD AWs do not possess. These new characteristics can be attributed to the so-called kinetic effects of plasma particles due to the wave scale matching with that of the particle motion. This implies that KAWs may exchange effectively energy and momentum with plasma particles and can be responsible for the acceleration or heating of plasma particles, which are very common occurrence from laboratory to space and astrophysical plasmas. Since the pioneer work of Chen and Hasegawa in the early 1970s, KAWs have been attracting a lot of attention of researchers from different fields of laboratory, space, and astrophysical plasmas. In particular, observational identifications of KAWs in space plasmas and experimental measurements of KAWs in laboratory plasmas in the 1990s resulted in the reevaluation of the importance of KAWs in the dynamics of various magnetic plasmas and remotivated the interest of KAW studies. Since 2000s, the study of KAWs is further extended and generalized to solar physics and other astrophysics, in particular, one finds that KAWs can play important roles in the dynamics of solar wind AW turbulence as well as in the inhomogeneous heating of the solar atmosphere.

Why are KAWs an important and interesting subject? First, the most important reason is that their characteristic scales are comparable to the kinetic scales of plasma particles. These scales are that between macroscopic phenomena and microscopic processes of plasmas, and hence are the most typical scales of plasma collective phenomena and fine structures, which are the most distinctive feature of plasma physics. It is in these scales that plasma waves and particles can most effectively exchange their energy and momentum, or i.e., the plasma wave-particle interaction can occur most efficiently. This makes KAWs capable of playing an important role in both the particle energization and fine structure phenomena of magnetic plasmas, which can be observed universally in space and solar plasmas.

Secondly, one of the important reasons is their ubiquitous existence in various plasma environments. It is well-known that AWs are low-frequency electromagnetic fluctuations observed most commonly in extensive plasma environments, such as in the magnetosphere, solar wind and solar atmosphere, and interstellar and intergalactic media. The standard Goldreich-Sridhar theory of interstellar turbulence implies that KAWs are natural and inevitable results of AWs cascading toward smaller scales. Recent observation studies indeed show that KAWs exist ubiquitously in space and solar plasmas and play important roles in the auroral electron acceleration, magnetospheric particle transport, solar wind turbulent cascade, and solar atmospheric heating.

The third reason for KAWs to be important and interesting is the richness and diversity of their wave-particle interaction. In space, solar, and other astrophysical plasma environments, existing plasmas almost all are complex plasmas, which consist of electrons and multiple ion species, including electrons, protons, alpha particles, other heavy ions, and sometimes even charged dust particles. In general, different species of particles with different charges and masses have different kinetic scales and the wave-particle interaction of KAWs sensitively depends on the kinetic scales of the species. In consequence, a significant difference can appear among different species components in plasmas, which often presents in observations of space and solar plasmas.

Finally, the sensitive dependence of the KAW physics on local plasma parameters also is an important reason for KAWs to be an important and interesting subject. The physics of KAWs, especially their electromagnetic polarization senses, sensitively depends on the plasma kinetic scales and hence on the local plasma parameters, such as the static magnetic field, the unperturbed plasma density and temperatures, the plasma kinetic to magnetic pressure ratio β, and so on. These plasma parameters can vary remarkably in the order of magnitude in the magnetosphere, the solar wind, the solar atmosphere, and other astrophysical plasma environments. This results in the distinctness of the wave-particle interaction of KAWs in various environments as shown in observations.

Seven years ago, I wrote a monograph on KAWs, *Kinetic Alfvén Wave: Theory, Experiment, and Application* (2012, Science Press, Beijing), which is the first monograph focusing on KAWs and provides a systematic and broad introduction to the KAW physics with the emphasis on the basic theories of KAWs. KAWs have been an actively and increasingly interesting subject. In these years, a lot of observations and applications of KAWs from laboratory and space plasmas have greatly enriched and advanced our understanding of KAWs. Without a doubt, the solar-terrestrial coupling system provides us a unique natural laboratory for the comprehensive and thorough study of KAWs. The present book is a companion volume of the monograph above and is finished cooperatively by me with my colleague Dr. Ling Chen. In this companion, we will show you an interesting journey of KAWs from laboratory plasmas to space and astrophysical plasmas, especially in the solar-terrestrial coupling system. You can find the different physics properties of KAWs in various plasma environments and their diverse roles played in these different environments due to the sensitively parametric dependence of the physics properties of KAWs on local plasma parameters.

This companion volume focuses on the applications of KAWs in space and astrophysical plasmas, especially on some crucial phenomena in the solar-terrestrial coupling system, including the magnetosphere-ionosphere coupling and the auroral electron acceleration (Chap. 3), the solar wind-magnetosphere interaction and the anomalous transport of particles at the magnetopause (Chap. 4), the solar wind turbulence (Chap. 5), and the solar corona and solar wind heating (Chap. 6). In addition, for the sake of consistency, this volume also covers a concise review of the main physics properties of KAWs (Chap. 1) and the experimental demonstrations of these properties in laboratory plasmas (Chap. 2). Finally, in closing, some possible applications of KAWs to extrasolar astrophysical plasmas are discussed in Chap. 7.

It is a very difficult, arduous, and never ending task to make the text error-free although we have made all efforts. We would always appreciate deeply it if the readers of this book will kindly inform us about any errors and misprints they find, preferentially by e-mail to djwu@pmo.ac.cn.

Nanjing, China
October 2019

De-Jin Wu

Contents

Chapter 1
Basic Physical Properties of KAWs

In the companion volume, *Kinetic Alfvén Wave: Theory, Experiment, and Application*, we have presented the systematic theories and basic principles of kinetic Alfvén waves (KAWs) based on both kinetic and fluid descriptions of magnetic plasmas. The present volume builds upon that and proceeds into the in-depth analyses and discussions of observation, generation, and application of KAWs in various plasma environments in the solar-terrestrial coupling system, which is a unique natural laboratory for the comprehensive and thorough study of KAWs. However, first of all, in this introductory chapter we concisely review main physical properties of KAWs without proving, which are from the companion volume and set up the basis for the following chapters.

1.1 From AWs to KAWs

In large scales much larger than the microscopic kinetic scales of charged particles a plasma (i.e., a partially or fully ionized gas) can be treated as a conducting fluid. When a conducting fluid carrying a current density $\mathbf{j}$ moves in the presence of a magnetic field $\mathbf{B}$, the interaction between the magnetic field and the motion of the fluid can be described by the Lorentz force,

$$\mathbf{j} \times \mathbf{B} = \frac{1}{\mu_0} (\nabla \times \mathbf{B}) \times \mathbf{B} = \nabla \cdot \left(\frac{B^2}{\mu_0} \hat{\mathbf{b}}\hat{\mathbf{b}} - \frac{B^2}{2\mu_0} \mathcal{I} \right), \tag{1.1}$$

which is equivalent to an isotropic pressure $P_B = B^2/2\mu_0$, combined with a tension $T_B = B^2/\mu_0$ along the magnetic field, where $\hat{\mathbf{b}} \equiv \mathbf{B}/B$ is the unit vector along the magnetic field and $\mathcal{I}$ the unit tensor. Analogous to a stretched string, it is this tension T_B that leads to an incompressible electromagnetic-hydrodynamic wave propagating along the magnetic field at the velocity

D.-J. Wu and L. Chen, *Kinetic Alfvén Waves in Laboratory, Space, and Astrophysical Plasmas*, Atmosphere, Earth, Ocean & Space,
https://doi.org/10.1007/978-981-13-7989-5_1

$$v_A = \sqrt{T_B/\rho} = B/\sqrt{\mu_0 \rho}, \tag{1.2}$$

called the Alfvén velocity, where ρ is the mass density of the fluid. This electromagnetic-hydrodynamic wave, which was first predicted to exist by Alfvén (1942) and ten years later was experimentally demonstrated in laboratory by Bostick and Levine (1952), now has been well known as Alfvén wave (AW) to honor its discoverer, Professor Hannes Alfvén. In literature, however, it also commonly called the shear AW in order to distinguish it from other one of electromagnetic-hydrodynamic waves, the compressive AW, which is caused by the isotropic magnetic pressure P_B, in ways similar to the gaseous kinetic pressure p, and was first discovered by Herlofson (1950). Taking account of the kinetic pressure, the coupling of the magnetic and kinetic pressures results in the compressive AW to separate into two magnetosonic waves, that is, the fast and slow magnetosonic waves.

It is the discovery of AW that led to a new subject, magnetohydrodynamics (MHD), which combines fluid dynamics and electrodynamics. Since its discovery, AW has been having extensive and fruitful applications in various fields of plasma physics, fluid and gas dynamics, and astrophysics. Up to now, AW itself still is an increasingly interesting topic in laboratory, space, and astrophysical plasmas.

Without loss of generality, let us take the z-axis of the Cartesian coordinate system along the static magnetic field $\mathbf{B}_0$ and the wave vector $\mathbf{k}$ in the x–z plane. Thus we have

$$\hat{\mathbf{z}} = \hat{\mathbf{b}} \text{ and } \hat{\mathbf{k}} = \sin\theta\hat{\mathbf{x}} + \cos\theta\hat{\mathbf{z}}, \tag{1.3}$$

where θ is the angle between $\mathbf{k}$ and $\mathbf{B}_0$, $\hat{\mathbf{x}}$ and $\hat{\mathbf{z}}$ denote the unit vectors along the x and z axes, respectively. In this coordinate system, the dispersion relation for shear AW reads as follows,

$$\omega = v_A k \cos\theta = v_A k_z, \tag{1.4}$$

where $k_z = k\cos\theta$ is the parallel wave number. The dispersion relations of the fast and slow magnetosonic waves are

$$\frac{\omega^2}{k^2} = \frac{v_M^2}{2}\left(1 \pm \sqrt{1 - 4\frac{v_A^2 v_S^2}{v_M^4}\cos^2\theta}\right), \tag{1.5}$$

where $v_S^2 \equiv \partial p/\partial\rho = \gamma p_0/\rho_0$ is the ordinary sound velocity, p_0 and ρ_0 are the unperturbed kinetic presure and mass density of the plasma, respectively, γ is the adiabatic index of the plasma, $v_M^2 \equiv v_A^2 + v_S^2$ is called the magnetic sound velocity, and "+" and "−" correspond to the fast and slow waves, respectively. It is obvious that, when $v_S^2 \ll v_A^2$ (i.e., the plasma kinetic pressure much smaller than the magnetic pressure) the slow wave vanishes and the fast wave reduces to the compressive AW that has an isotropic dispersion relation:

$$\omega = v_A k. \tag{1.6}$$

This implies the fast and slow wave modes practically are coupling modes of the compressive AW mode driven by the magnetic pressure and the ordinary acoustic wave mode driven by the kinetic pressure.

The shear AW and the fast and slow magnetosonic waves are three basic modes of low-frequency MHD waves in plasmas. They all are nondispersive waves and can be described well by the MHD theory of plasmas. However, the MHD description is valid only in the low-frequency and long-wavelength limit, in which the wave frequency is much lower than the ion gyrofrequency and the wavelength is much longer than the kinetic scales of particles, such as the ion or ion-acoustic gyroradius, the electron inertial length, and so on. When the wave scales approach these kinetic scales, the individual motion of charged particles can play an important role in the physics of the waves. Stefant (1970) first noted that AWs with a short perpendicular wavelength comparable to the ion or ion-acoustic gyroradius can couple to the ion acoustic mode because of the charge separation produced by the unmatched motions between electrons and ions, in which ions no longer follow the static magnetic field lines but electrons are still attached to the field lines. By use of kinetic theory, Chen and Hasegawa (1974a, b) and Hasegawa and Chen (1975, 1976a) further studied these coupled AWs in detail and found that when an AW with fixed frequency ω and parallel wave number $k_\parallel$ propagates in an inhomogeneous plasma, it would be absorbed in the resonant plane where the parallel phase speed, $\omega/k_\parallel$, matches up to the local Alfvén velocity, $v_A(x)$, and resonantly mode converts to a dispersive wave, which includes the effect of finite ion gyroradius, called kinetic Alfvén wave (KAW). In particular, they pointed out that KAWs can efficiently energize plasma particles (Hasegawa and Chen 1976a).

The particle energization, including heating and acceleration, is very common plasma active phenomena in laboratory, space, and astrophysical plasmas. Since the pioneer work by Chen and Hasegawa, many aspects of KAWs have been widely investigated not only in their theories but also in their experiments including experimental measurements in laboratory plasmas and observational identifications in space plasmas. The applications of KAWs in space and solar plasmas have been discussed extensively by many authors, such as in the magnetosphere-ionosphere coupling and the auroral electron acceleration (Hasegawa 1976; Goertz and Boswell 1979; Lysak and Carlson 1981; Lysak and Dum 1983; Goertz 1984; Kivelson and Southwood 1986; Chmyrev et al. 1988; Kletzing 1994; Wygant et al. 2002; Wu 2003b, c; Wu and Chao 2003, 2004a; Bespalov et al. 2006; Chaston et al. 2007a; Sharma et al. 2009), in the solar wind-magnetosphere coupling and the anomalous particle transport in the magnetopause (Hasegawa and Mima 1978; Lee et al. 1994; Johnson and Cheng 1997, 2001; Johnson et al. 2001; Wu et al. 2001; Chaston et al. 2007c), in the dissipation dynamics of the solar wind turbulence (Leamon et al. 1999; Malik and Sharma 2005; Sharma and Malik 2006; Brodin et al. 2006; Malik et al. 2007; Singh 2007; Gary and Smith 2009; Sahraoui et al. 2010; Podesta et al. 2010; Voitenko and De Keyser 2011; Rudakov et al. 2011; Smith et al. 2012; Salem et al. 2012; Boldyrev and Perez 2012; Podesta and TenBarge 2012; Podesta 2013), and in the inhomogeneous heating of the solar atmosphere (Voitenko 1998; Wu and Fang 1999, 2003, 2007; Voitenko and Goossens 2002, 2004, 2005a; Voitenko et al. 2003;

Wu 2005; Stasiewicz 2006; Wu and Yang 2006, 2007; Wu et al. 2007; Singh and Sharma 2007; Wang et al. 2009).

In particular, great progresses in experimental studies of KAWs are achieved in both laboratory and space plasmas. For example, in the measurements of auroral region crossing by the Freja and FAST satellites, some low-frequency electromagnetic fluctuations with strong electric spikes and strong density fluctuations were identified well as KAWs (Louarn et al. 1994; Wahlund et al. 1994a; Volwerk et al. 1996; Wu et al. 1996a, b, 1997; Huang et al. 1997; Stasiewicz et al. 1997, 1998, 2000a, b; Bellan and Stasiewicz 1998; Chaston et al. 1999). On the other hand, many physical properties of KAWs predicted by the theory are demonstrated by a series of experimental investigations of KAWs in laboratory plasmas, which were carried out on LAPD (LArge Plasma Device) at UCLA (University of California, Los Angeles) (Gekelman et al. 1994, 1997, 2000; Maggs and Morales 1996; Maggs et al. 1997; Leneman et al. 1999; Burke et al. 2000a, b; Mitchell et al. 2001; Kletzing et al. 2003a; Vincena et al. 2004), where the technology of both plasma sources and diagnoses as well as a large chamber were developed. Moreover, the phenomena observed in the laboratory experiments show striking similarities to what has been observed by satellites in space plasmas. These experimental studies from both the laboratory and space plasmas motivated greatly the interest in KAWs, in particular, in their application to various dynamical phenomena of magneto-plasmas from laboratory to space and astrophysical plasmas. As pointed out by Gekelman (1999): “in the past few years the quantum jump in data collection on the Freja and FAST missions has led to the reevaluation of the importance of these waves in the highly structured plasma that was probed”.

KAWs have been becoming increasingly interest and their applications have been discussed extensively in fields of solar and space plasmas. In this introductory chapter we present a concise summary of basic physical properties of KAWs and their experimental vertifications in laboratory plasmas will be introduced in the next chapter.

1.2 Dispersion and Propagation of KAWs

The dispersion of a wave traveling through a plasma can be caused by the kinematic coupling between the electromagnetic fields of the wave and the individual motions of plasma particles in the kinetic scales. In principle, in the low-frequency limit of the wave frequency much less than the ion gyrofrequency (i.e., $\omega \ll \omega_{ci}$), such as for the ideal MHD waves, the propagation of the wave is nondispersive, that is, the phase speed and group velocity of the wave both are independent of the wave vector, because the fast periodic motion of individual plasma particles in the nearly uniform wave fields does not effectively cause net energy and momentum exchanges between the wave and particles. Similarly, in the high-frequency limit of the wave frequency much larger than the electron hybrid resonant frequency (i.e., $\omega \gg \omega_{UH} \equiv \sqrt{\omega_{ce}^2 + \omega_{pe}^2}$, also called the upper hybrid resonant frequency), such as for the radiation electromagnetic waves far away their emission source regions, the propagation of the wave also is nondispersive, because the fast periodicity of the wave fields in the nearly steady motion of particles leads to no net energy and momentum exchanges between the wave and particles. However, when the wavelength is comparable to the particle kinetic scales the periodicity of individual particle motions will be broken due to the particles can feel evidently the variation of the wave fields. In result, the wave-particle interaction will lead to effectively net energy and momentum exchanges between the wave and particles, and moreover the energy and momentum exchanging rates sensitively depend on the ratio of the wavelength to the particle kinetic scales. In consequence, the propagation properties of the wave (i.e., the phase speed and the group velocity) sensitively depend on the wavelength (or i.e., the wave vector), that is, the wave becomes dispersive.

For KAWs of their scales approaching the kinetic scales of plasma particles, the motion of individual particles can play an important role in the physics of the wave and the related kinetic effects can significantly change not only the dispersion relation and propagation property of the wave but also the electromagnetic characteristics and polarization sense of the wave. To take account of the so-called kinetic effects, in general, it is necessary to employ the kinetic theory for the description of plasmas. In the low-frequency limit, however, the two-fluid description, which is much simpler than the exact kinetic description, can well give almost all kinetic effects except very few exceptions such as the Landau damping effect (Lysak and Lotko 1996). For the case of KAWs, in particular, Yang et al. (2014) compared in detailed the results from the two-fluid model to those from the gyrokinetic model. They found that they can agree well with each other not only in the long wavelength regime ($\lambda = 2\pi/k > \rho_i \equiv v_{T_i}/\omega_{ci}$ the ion gyroradius) for any the ion-electron temperature ratio $\tau \equiv T_i/T_e$ but also in the intermediate short wavelength regime ($\rho_i > \lambda > \rho_e \equiv v_{T_e}/\omega_{ce}$ the electron gyroradius) for the moderate or higher ion-electron temperature $\tau > 1$. For the case of $\tau < 1$, the electron gyroradius effect may cause some considerable deviation of the description of KAWs with the intermediate short wavelength in the two-fluid model from that in the gyrokinetic model.

To avoid obscuring the physics with the complex mathematics in the full kinetic theory, in principle, we will employ two-fluid model of plasmas to describe the physics of KAWs through this book, which takes account of the kinetic effects and the kinetic model is used only when it is necessary, for instance, when to discuss the Landau or inverse Landau damping processes caused by their wave-particle interaction, which can not be described properly in the two-fluid model. In addition, in space and solar plasma environments the kinetic scales of particles usually all are much smaller than the mean collision free path of particles, λ_c. For example, for typical local plasma parameters one has $\rho_i/\lambda_c \sim 10^{-9}$ in the magnetosphere, $\sim 10^{-5}$ in the solar wind, and $\sim 10^{-6}$ in the solar corona, all much less than the unit. This indicates that the collisionless plasma is an enough well approximation for the case of space and solar plasmas, in which the effect of the classic Coulomb collisions on KAWs is reasonably ignorable.

On the other hand, if we further confine to the nonrelativistic cases, in which the Alfvén velocity (v_A), the thermal velocity ($v_{T_s} \equiv \sqrt{T_s/m_s}$ for the $s = e,\ i$ species), and the wave phase speed ($v_p \equiv \omega/k$) all are much less than the light velocity (c), the kinetic scales of plasma particles will be much larger than the plasma Debye length on which plasma fluctuations may obviously deviate from the charge neutrality. In fact, for the Debye length, λ_D, we have

$$\frac{\lambda_D^2}{\lambda_e^2} = \frac{2\tau}{1+\tau}\frac{v_{T_e}^2}{c^2} < 2\frac{v_{T_e}^2}{c^2} \ll 1$$
$$\frac{\lambda_D^2}{\rho_{is}^2} = \frac{2\tau}{(1+\tau)^2}\frac{v_A^2}{c^2} < \frac{v_A^2}{c^2} \ll 1, \tag{1.7}$$

where $\lambda_e \equiv c/\omega_{pe}$ is the electron inertial length, ω_{pe} is the electron plasma frequency (also called the Langmuir frequency), $\rho_{is} \equiv v_{is}/\omega_{ci} = \sqrt{1+\tau}\rho_s$ and $\rho_s \equiv \rho_i/\sqrt{\tau}$ are the effective ion gyroradius and the ion-acoustic gyroradius, respectively, and $v_{is} \equiv \sqrt{1+\tau}v_s$ and $v_s \equiv v_{T_i}/\sqrt{\tau}$ are the effective ion-acoustic velocity and the ion-acoustic velocity, respectively. The inequation (1.7) implies that plasma fluctuations in the kinetic scales, especially in the case of KAWs, can well satisfy the quasi-neutral condition in the low-frequency approximation of $\omega^2 \ll \omega_{ci}^2$ and the nonrelativistic condition of v_A^2, $v_{T_s}^2 \ll c^2$.

Let us consider one dimensional plane wave with wave vector $\mathbf{k} = k_x\hat{\mathbf{x}} + k_z\hat{\mathbf{z}}$ and frequency ω that propagates in a collisionless plasma with the two species of ions and electrons, which is magnetized by a uniform external magnetic field along the z-axis, $\mathbf{B}_0 = B_0\hat{\mathbf{z}}$. The low-frequency dispersion equation can be obtained as follows (Chen and Wu 2011a; Wu 2012)

$$AM_z^6 - BM_z^4 + CM_z^2 - D = 0, \tag{1.8}$$

where the coefficients

$$A = (1+Q)\left(1+Q+\lambda_e^2k_x^2\right)\left(1+Q+\frac{\beta}{2}\lambda_e^2k_x^2\right)\cos^2\theta;$$
$$B = \left[(1+Q)^2 + \rho_{is}^2k_x^2 + Q^2\frac{\beta}{2}\lambda_e^2k_x^2\right]\cos^2\theta$$
$$+(1+Q)\left(1+Q+\lambda_e^2k_x^2\right) + \left(1+Q+\lambda_e^2k_x^2\right)^2\frac{\beta}{2};$$
$$C = (1+Q)(1+\beta) + \left(1+Q^2\right)\rho_{is}^2k_x^2;$$
$$D = \frac{\beta}{2}, \tag{1.9}$$

where $Q \equiv m_e/m_i$ is the electron-ion mass ratio, usually an ignorable small quantity comparing with one, $\beta_{e(i)}$ is the electron (ion) kinetic to magnetic pressure ratio, and $\beta \equiv (1+\tau)\beta_e = 2v_{is}^2/v_A^2$ is the total plasma kinetic-magnetic pressure ratio.

In the long-wavelength limit of both $k_x^2\lambda_e^2$ and $k_x^2\rho_{is}^2 \ll 1$ the low-frequency dispersion equation (1.8) reduces to

$$\left(M_z^2 - 1\right)\left[M^4 - \left(1 + \frac{\beta}{2}\right)M^2 + \frac{\beta}{2}\cos^2\theta\right] = 0, \tag{1.10}$$

where $M \equiv M_z \cos\theta$ is the phase speed in the unit of the Alfvén velocity v_A along the wave-vector direction and $M_z \equiv \omega/k_z v_A$ is the parallel phase speed in v_A. As expected, this reduced dispersion equation leads to the well-known three ideal MHD modes, the AW mode by the first factor and the fast and slow magnetosonic wave modes by the second factor. From the expressions for the coefficients in Eq. (1.9) it is clear that the dispersion of the waves is caused by the short-wavelength effect of the finite perpendicular wavenumber in the forms of $\rho_{is}^2 k_x^2$ and $\lambda_e^2 k_x^2$, that is, when the perpendicular wavelength becomes comparable to the effective ion gyroradius ρ_{is} or to the electron inertial length λ_e.

In the low-frequency and nonrelativistic approximations the three solutions of the two-fluid dispersion equation (1.8) can be obtained as following:

$$\begin{aligned} M_{zF}^2 &= \frac{B}{3A} + \frac{2}{3}\sqrt{\frac{B^2}{A^2} - \frac{3C}{A}}\cos\frac{\phi}{3} \\ M_{zS}^2 &= \frac{B}{3A} + \frac{2}{3}\sqrt{\frac{B^2}{A^2} - \frac{3C}{A}}\cos\frac{\phi + 2\pi}{3} \\ M_{zA}^2 &= \frac{B}{3A} + \frac{2}{3}\sqrt{\frac{B^2}{A^2} - \frac{3C}{A}}\cos\frac{\phi + 4\pi}{3}, \end{aligned} \tag{1.11}$$

where

$$\cos\phi = \frac{2B^3 - 9ABC + 27A^2 D}{2\left(B^2 - 3AC\right)^{3/2}}. \tag{1.12}$$

They are the kinetic modified versions of the three MHD modes by taking account of the short-wavelength effect, that is, the kinetic fast magnetosonic wave (KFW), the kinetic slow magnetosonic wave (KSW), and the KAW.

In the parallel propagation limit of $\theta = 0°$ the dispersion equation (1.8) reduces to

$$\left(M^2 - 1\right)^2\left(M^2 - \frac{\beta}{2}\right) = 0, \tag{1.13}$$

which two solutions, $M^2 = 1$ and $M^2 = \beta/2$, represent the dispersion relations of parallel-propagation AWs and ordinary sonic waves, respectively, implying that the perpendicular short-wavelength effect does not influence the phase speeds of these kinetic waves in the parallel-propagation case. On the other hand, in the perpendicular propagation limit of $\theta = 90°$, the dispersion equation (1.8) reduces to

$$M^2 = \frac{1 + \beta/2 + Q\rho_{is}^2 k^2}{1 + Q\rho_{is}^2 k^2} \simeq 1 + \frac{\beta}{2}, \tag{1.14}$$

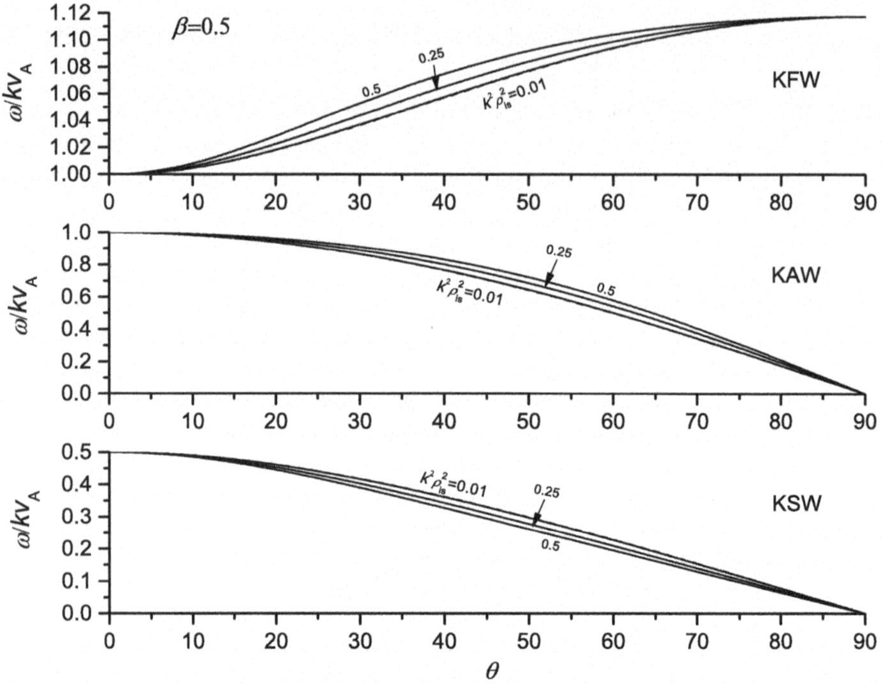

Fig. 1.1 The phase speed versus the propagation angle: Panels, from the top down, present of KFWs, KAWs, and KSWs, respectively, where a fixed plasma pressure parameter $\beta = 0.5$ has been used, the solid lines are given by fixed wave numbers $k^2\rho_{is}^2 = 0.01, 0.25$, and 0.5, respectively, and the dashed lines plot the corresponding MHD solutions

which is the KFW dispersion relation and other one solution of $M = 0$ indicates that, in similar to their MHD versions, KAWs and KSWs can not propagate at the direction exactly perpendicular to the ambient magnetic field.

Figure 1.1 plots the phase speed ($M \equiv \omega/kv_A = M_z \cos\theta$) of KFWs (top panel), KAWs (middle panel), and KSWs (bottom panel), presented by the dispersion relations of Eq. (1.11), versus the propagation angle θ (i.e., the angle between the wave vector $\mathbf{k}$ and the ambient magnetic field $\mathbf{B}_0$) for the fixed plasma pressure parameter $\beta = 0.5$, in which the solid lines represent the three different cases with the normalized wave numbers $k^2\rho_{is}^2 = 0.01, 0.25$, and 0.5, respectively, and for the sake of comparison, the corresponding MHD solutions with $k^2\rho_{is}^2 \to 0$ are presented by the dashed lines.

From Fig. 1.1, the phase speeds of the three kinetic modes (i.e., KFWs, KAWs, and KSWs) all are almost coincident with their MHD limits (given by the dashed lines) for the small wave number case of $k^2\rho_{is}^2 = 0.01$, but evidently deviate from the MHD limits for the finite wave number cases of $k^2\rho_{is}^2 = 0.25$ and 0.5, implying that the modifications in the phase speed are caused by the finite short-wavelength kinetic effect. Figure 1.1 also shows that the phase speeds approach to the MHD limits in the parallel and perpendicular propagation limits as pointed out above. In addition, from Fig. 1.1 it can be found that the phase speeds of KFWs and KAWs increase with the finite wave number $k\rho_{is}$ but the phase speed of KSWs decreases with the finite wave number $k\rho_{is}$. Although the phase speed modifications due to the short-wavelength effect are evident, they are only quantitative differences without qualitative changes in the high-β case shown in Fig. 1.1. On the other hand, for the low-β case of $\beta \ll 1$ the propagation property of KAWs can be changed qualitatively by the short-wavelength effect.

As well-known, the plasma pressure parameter β can significantly influence the propagation of compressive AWs, but the propagation of shear AWs is entirely independent of the plasma β parameter in their MHD modes. In the kinetic modes with the short-wavelength effects, however, the plasma β parameter can significantly affect the propagation of KAWs but has little effects on that of KFWs and KSWs. Figure 1.2 shows the dependence of the phase speeds of KFWs (top panel), KAWs (middle panel), and KSWs (bottom panel) on the plasma β parameter for the fixed propagation angle $\theta = 60°$ and the three cases (solid lines) with wave numbers $k^2\rho_{is}^2 = 0.01$, 0.25, and 0.5, respectively. In similar to Fig. 1.1, the corresponding MHD solutions are presented by the dashed lines for the sake of comparison.

From Fig. 1.2 it can be found that, for the two compressive modes, KFWs (top panel) and KSWs (bottom panel), the short-wavelength effect due to finite wave numbers (i.e., $k\rho_{is}$) causes only small quantitative modifications of their phase speeds in a very wide range of the plasma β, for instance, the phase speeds of KFWs with the wave numbers $k^2\rho_{is}^2 = 0.01, 0.25$, and 0.5 almost all coincide with its MHD solution and the phase speeds of KSWs with finite $k^2\rho_{is}^2 = 0.25$ and 0.5 slightly decrease by small quantities in the high-β case of $\beta > 0.1$. The propagation property of KAWs (middle panel), however, has been changed significantly and qualitatively by the short-wavelength effect. As shown in the middle panel of Fig. 1.2 that the KAW phase speeds of the three cases with finite wave numbers $k^2\rho_{is}^2 = 0.01$, 0.25, and 0.5 all evidently deviate from the MHD solution presented by the dashed line (i.e., $v_A \cos\theta = 0.5v_A$). In particular, the phase speed modification of KAWs due to the short-wavelength effect changes from increasing with $k\rho_{is}$ in the so-called kinetic regime of $\beta > 2Q \sim 10^{-3}$) to decreasing with $k\rho_{is}$ in the so-called inertial regime of $\beta < 2Q \sim 10^{-3}$. Accordingly the propagation of KAWs also changes from a super-Alfvénic propagation with the phase speed higher than the corresponding MHD phase speed (i.e., $v_A \cos\theta$) in the kinetic regime to a sub-Alfvénic propagation with the phase speed lower than the corresponding MHD phase speed in the inertial regime.

In general, in a low-β plasma of $\beta < 1$ the magnetic field can play an important role in the dynamics of the plasma and AWs can transport the information of plasma fluctuations more quickly than sound waves. On the other hand, as shown above, the important kinetic effects of the KAW propagation due to the short-wavelength modification also occur in the low-β regime. Below we focus discusses on the case of KAWs in low-β plasmas, which have more important and extensive interests in space and astrophysical plasmas.

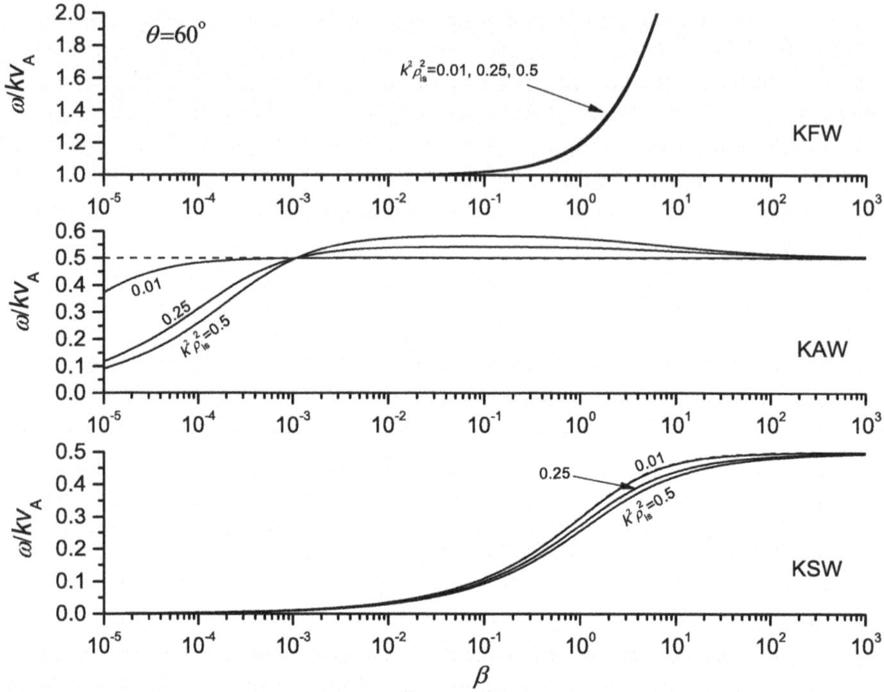

Fig. 1.2 The phase speed versus the plasma β parameter: Panels, from the top down, present those of KFWs, KAWs, and KSWs, respectively, where a fixed propagation angle $\theta = 60°$ has been used, the solid lines are given by fixed wave numbers $k^2\rho_{is}^2 = 0.01, 0.25$, and 0.5, respectively, and the dashed lines plot the corresponding MHD solutions

KAWs in low-β plasmas

An important and interesting case is KAWs in low-β plasmas, which are often encountered in space and astrophysical plasma environments, such as the Earth's magnetospheric plasma, the solar coronal plasma, and astrophysical radio plasmas. In a low-β plasma environment the magnetic field usually can accumulate and preserve abundant free energies into the local plasma electric currents, which can supply energy sources to drive energetic active phenomena that occur ubiquitously in various space and astrophysical plasma environments.

In the low-β approximation of $\beta \ll 1$, the coefficients in Eq. (1.9) reduce to

$$
\begin{aligned}
A &\simeq \left(1 + \lambda_e^2 k_x^2\right)\cos^2\theta;\\
B &\simeq 1 + \lambda_e^2 k_x^2 + \left(1 + \rho_{is}^2 k_x^2\right)\cos^2\theta;\\
C &\simeq 1 + \rho_{is}^2 k_x^2;\\
D &= \frac{\beta}{2} \ll 1
\end{aligned}
\tag{1.15}
$$

and hence the dispersion equation (1.8) reduces to

$$M_z^2 \left(M^2 - 1\right) \left(M_z^2 - K_A\right) \simeq \frac{\beta/2}{1 + \lambda_e^2 k_x^2} \ll 1, \tag{1.16}$$

where

$$K_A \equiv \frac{1 + \rho_{is}^2 k_x^2}{1 + \lambda_e^2 k_x^2} = \frac{1 + (1 + \tau)\, \alpha_e^2 \lambda_e^2 k_x^2}{1 + \lambda_e^2 k_x^2} \tag{1.17}$$

with the electron thermal resonant parameter

$$\alpha_e \equiv \frac{\rho_s}{\lambda_e} = \frac{v_{T_e}}{v_A}. \tag{1.18}$$

In the derivation of the above equation the small amounts of order Q or β compared with one have been neglected.

For the case of $M_z^2 \ll 1$, the solution of the reduced dispersion equation (1.16) gives the kinetic dispersion relation of the slow mode wave,

$$M_{zS}^2 \simeq \frac{\beta/2}{1 + \lambda_e^2 k_x^2} K_A^{-1} \Rightarrow \frac{\omega^2}{k_z^2} = \frac{v_{is}^2}{1 + \rho_{is}^2 k_x^2}, \tag{1.19}$$

or i.e., the dispersion relation of ion acoustic waves in a magnetic plasma. This implies that the ion acoustic wave is the slow mode wave modified by the short perpendicular wavelength effect. For the case of $M^2 \sim 1$, one has $M_z^2 > 1$ and hence the solution of the dispersion equation (1.16) leads to the dispersion relation of the fast mode wave, $M_F^2 \simeq 1$. For the case of $M_z^2 \sim 1$ one has $M^2 > 1$ and hence the solution of the dispersion equation (1.16) leads to the dispersion relation of KAWs,

$$M_{zA}^2 \simeq K_A \Rightarrow \frac{\omega^2}{k_z^2 v_A^2} = \frac{1 + \rho_{is}^2 k_x^2}{1 + \lambda_e^2 k_x^2} = \frac{1 + (1 + \tau)\, \alpha_e^2 \lambda_e^2 k_x^2}{1 + \lambda_e^2 k_x^2}, \tag{1.20}$$

which is the short-wavelength modified version of the AW dispersion relation.

From the dispersion relation of Eq. (1.20) one can find that, similar to AWs, KAWs have a field-aligned group velocity v_{gz} identical to its parallel phase speed v_{pz}, that is,

$$v_{gz} \equiv \frac{\partial \omega}{\partial k_z} = \sqrt{K_A} v_A = \sqrt{\frac{1 + \rho_{is}^2 k_x^2}{1 + \lambda_e^2 k_x^2}} v_A = v_{pz}, \tag{1.21}$$

which is independent of the parallel wavenumber k_z. This indicates that KAWs can efficiently transport energies at the velocity $\sim v_A$ along the static magnetic field like AWs.

Unlike AWs with a zero perpendicular group velocity, however, KAWs have a non-zero perpendicular group velocity v_{gx} as follows:

$$v_{gx} \equiv \frac{\partial \omega}{\partial k_x} = \frac{\left(\rho_{is}^2 - \lambda_e^2\right) k_x^2 \cot\theta}{\left(1+\rho_{is}^2 k_x^2\right)\left(1+\lambda_e^2 k_x^2\right)} v_{gz} \ll v_{gz}, \tag{1.22}$$

where $\cot\theta = k_z/k_x \ll 1$ for the quasi-perpendicular propagation. Although the perpendicular component of the group velocity, in general, is much less than its parallel component, it shall lead to some important changes of KAWs different from AWs. For instance, when KAWs propagate in the kinetic regime where $\rho_{is} > \lambda_e$, the wave energy is forward propagating with respect to the perpendicular wave vector k_x, that is, $v_{gx} > 0$, while in the inertial regime where $\rho_{is} < \lambda_e$, the wave energy is backward propagating, that is, $v_{gx} < 0$. In particular, the perpendicular group velocity goes to zero (i.e., $v_{gx} = 0$) at the transition point of $\rho_{is} = \lambda_e$. This change of the perpendicular group velocity in sign was invoked by Streltsov and Lotko (1995) to explain trapping of KAWs on auroral magnetic field lines between the magnetosphere and ionosphere. Recently, this lingering of KAWs in the transition regime of $\rho_{is} \sim \lambda_e$ has been investigated experimentally in laboratory plasmas (Vincena et al. 2004) and the result shows that the wave energy tends to pile up in the transition regime, as expected in the discussion here.

For the two parametric regimes of KAWs, the kinetic and inertial regimes, which are distinctly separated by the plasma pressure parameter $\beta > 2Q$ and $< 2Q$, respectively, we have $\rho_{is} > \lambda_e$ and $< \lambda_e$, respectively. Therefore, the dispersion of KAWs is dominated by the effective ion gyroradius via the term in the numerator $\rho_{is}^2 k_x^2$ in the kinetic regime and by the electron inertial length via the term in the denominator $\lambda_e^2 k_x^2$ in the inertial regime, respectively. In particular, one has $M_{zA} = \sqrt{K_A} > 1$ in the kinetic regime, implying that KAWs propagate at a supper-Alfvén velocity of $v_{gz} = v_{pz} > v_A$ and have the positive dispersion, that is, the phase speed increasing with the wavenumber k_x. While in the inertial regime $M_{zA} = \sqrt{K_A} < 1$, implying that KAWs propagate at a sub-Alfvén velocity of $v_{gz} = v_{pz} < v_A$ and have the negative dispersion, that is, the phase speed decreasing with the wavenumber k_x.

On the other hand, in the transition regime between the kinetic and inertial regimes we have

$$\rho_{is} \simeq \lambda_e \Rightarrow v_{T_e} \simeq \frac{v_A}{\sqrt{1+\tau}}. \tag{1.23}$$

For the cold ion case of $T_i \ll T_e$ (i.e., $\tau \ll 1$), this leads to $v_{T_e} \simeq v_A$ and $M_{zA} \simeq 1$, that is, $v_{T_e} \simeq v_{pz}$. This indicates that in this transition regime the Landau damping possibly plays an important role due to the thermal resonant effect of electrons, for which the two-fluid description, in general, is invalid and the kinetic theory is necessary. In fact, as pointed out above, in this transition regime the perpendicular group velocity of KAWs $v_{gx} \simeq 0$, this results in KAWs to pile up in the transition regime. The experimental studies of the dispersion relation for this transition regime also showed that the kinetic version can give more precise dispersion relation in comparison with measurements of the KAWs (Leneman et al. 1999; Kletzing et al. 2003a).

The results above show that the propagation property of KAWs sensitively depends on the plasma pressure parameter β. In the low but finite β approximation of $1 \gg$

$\beta \gg 2Q$, Chen and Hasegawa (1974a, b) and Hasegawa and Chen (1975, 1976a) firstly presented the reduced version of the dispersion relation of Eq. (1.20) in the kinetic regime,

$$\omega^2 = \left(1 + \frac{4}{3}\rho_i^2 k_x^2\right) k_z^2 v_A^2, \tag{1.24}$$

where the factor $4/3$ is due to the kinetic modification, and first called it as kinetic Alfvén wave (KAW). Goertz and Boswell (1979) noticed that KAWs can play an important role in the magnetosphere-ionosphere coupling, where $\beta \ll 2Q$, and gave other reduced dispersion relation of KAWs

$$\omega^2 = \frac{k_z^2 v_A^2}{1 + \lambda_e^2 k_x^2} \tag{1.25}$$

for the inertial regime. Based on the two-fluid description of plasmas, Streltsov and Lotko (1995) extended their results to include both the kinetic and inertial regimes and obtained the dispersion relation of KAWs in the form of Eq. (1.20). Lysak and Lotko (1996) compared the two-fluid dispersion relation of KAWs and its Vlasov kinetic version and found that it is often a very good approximation of the full kinetic dispersion relation, at least, in the low frequency limit. Moreover, the measurement of the parallel phase speed directly verifies the two-fluid dispersion relation (Gekelman et al. 1994). Including the ion inertial effect (i.e., the Hall effect with the finite frequency modification), Wu (2003a) presented the dispersion relation of KAWs with the finite frequency modification as following form:

$$\omega^2 = \frac{1 + \rho_{is}^2 k_x^2 + \rho_i^2 k_z^2}{1 + \lambda_e^2 k_x^2 + \lambda_i^2 k_z^2} k_z^2 v_A^2. \tag{1.26}$$

Hollweg (1999) further studied KAWs in the high-β regime of $\beta \sim 1$, where $\rho_{is} \sim \lambda_i \gg \lambda_e$, and found KAWs coupling to KFWs and KSWs. In this case the finite frequency modification, in general, is important unless the parallel wavelength $\lambda_z \gg \lambda_i \sim \rho_{is}$. The latter implies $k_z\lambda_i \sim k_z\rho_{is} \ll k_x\rho_{is}$, that is, the quasi-perpendicular propagation of $k_z \ll k_x$ (Wu 2003a).

Quasi-perpendicular propagating KAWs

Other one important and interesting limit case is quasi-perpendicular propagating KAWs. The theory of MHD turbulence in a magneto-plasma (Sridhar and Goldreich 1994; Goldreich and Sridhar 1995, 1997) predicts that the wave energy in the inertial regime of MHD wave spectrums cascades primarily by preferentially developing small scales perpendicular to the unperturbed magnetic field. This implies that in the small-scale kinetic regime the turbulence develops preferentially into the anisotropic KAWs with $k_\perp \gg k_\parallel$, that is, the quasi-perpendicular propagating KAWs. Numerical simulations of magnetized turbulence (Maron and Goldreich 2001; Cho et al. 2002) also support the idea that such a turbulence is strongly anisotropic. Meanwhile, in situ measurements of turbulence in the solar wind (Belcher and Davis 1971;

Matthaeus et al. 1990; Luo and Wu 2010; Luo et al. 2011) and observations of interstellar scintillation (Wilkinson et al. 1994; Trotter et al. 1998; Rickett et al. 2002; Dennett-Thorpe and de Bruyn 2003) provide evidences for significant anisotropy. In consequence, at least when approaching to the dissipation regime of wave spectrums, the ultimate waves on the small scales are expected to be highly anisotropic in the way of quasi-perpendicular propagation with $k_\perp^2/k_\parallel^2 \gg 1$.

For the quasi-perpendicular propagating case of $\cos^2\theta = k_z^2/k^2 \ll 1$, the coefficients of Eq. (1.9) reduce to

$$
\begin{aligned}
A &\simeq 0;\\
B &\simeq \left(1+\lambda_e^2k_x^2\right) + \left(1+\lambda_e^2k_x^2\right)^2\frac{\beta}{2};\\
C &\simeq 1+\beta+(1+\tau)\,\alpha_e^2\lambda_e^2k_x^2;\\
D &= \frac{\beta}{2}
\end{aligned}
\tag{1.27}
$$

and the dispersion equation (1.8) reduces to the following quadratic equation of M_z^2:

$$
\left[1+\frac{\beta}{2}\left(1+\lambda_e^2k_x^2\right)\right]M_z^4 - \frac{1+\beta+(1+\tau)\,\alpha_e^2\lambda_e^2k_x^2}{1+\lambda_e^2k_x^2}M_z^2 + \frac{1}{1+\lambda_e^2k_x^2}\frac{\beta}{2} = 0,
\tag{1.28}
$$

which two roots, that is, the two dispersion relations are

$$
\begin{aligned}
M_{zA(S)}^2 = &\frac{1+\beta+(1+\tau)\,\alpha_e^2\lambda_e^2k_x^2}{\left(1+\lambda_e^2k_x^2\right)\left(2+\beta+\beta\lambda_e^2k_x^2\right)}\\
&\times\left\{1\pm\sqrt{1-\frac{\left(1+\lambda_e^2k_x^2\right)\left(2+\beta+\beta\lambda_e^2k_x^2\right)}{\left[1+\beta+(1+\tau)\,\alpha_e^2\lambda_e^2k_x^2\right]^2}\beta}\right\}.
\end{aligned}
\tag{1.29}
$$

They represent the coupling modes of KAWs and KSWs (Wu et al. 1996c; Wu and Wang 1996). In the low-β case of $\beta \ll 1$, the KAW dispersion relation of Eq. (1.20) is recovered by the solution with the sign "+" and other solution with sign "−" leads to the ion acoustic wave. This indicates that these two coupling modes are decoupled in the low-β limit of $\beta \ll 1$.

On the other hand, the quasi-perpendicular propagation of KAWs also indicates the remarkable and outstanding anisotropy of KAWs in the spatial structure, that is, they have the parallel wavelength $\lambda_\parallel$ much larger than the perpendicular wavelength $\lambda_\perp$. Thus, their spatial structures look as filaments or density striations along the magnetic field lines, in which the field-aligned gradient is much lower than the cross-field gradient. In fact, this anisotropy of KAWs also is the manifestation of the anisotropic nature of the kinetic effect of KAWs that is caused by the Larmor cyclotron motion of charged particles in the plane perpendicular to the static magnetic field $\mathbf{B}_0$. The gyromotion of charged particles due to the Lorentz force by the static field can effectively inhibit the cross-field motions of the particles in the plane perpendicular to the static field, while along the static field lines the particles may move freely. In consequence, the plasma can maintain large density and temperature gradients across the magnetic field, while the freely, fast field-aligned motion may quickly smooth out the gradients along the field lines. This leads to the formation of field-aligned filamentous structures, which are characterized by the parallel length scales of the inhomogeneity much larger than its perpendicular scales.

High-resolution observations have now given an impressive image of the solar corona as a rapidly evolving dynamic plasma where energetic phenomena occur mainly on very small scales that are characterized by filaments, called coronal plasma loops (Golub et al. 1990). In addition, similar filamentous structures also are very common in observations of terrestrial auroral plasmas, such as field-aligned density striations and auroral fine structures in the forms of narrow discrete arcs, auroral curls and folds (rays), and flickering aurora, which have feature scales comparable to those of ion cyclotron motions. We believe that the anisotropic nature of KAWs has potential importance for the formation of these fine filamentous structures.

1.3 Polarization and Wave-Particle Interaction of KAWs

In the physics of the wave-particle interaction, the basic role of the wave electric field is to energize charged particles and that of the wave magnetic field is to scatter the orbits of the particles but not to energize the particles because the Lorentz force by the magnetic field is always perpendicular to the moving velocity of the charged particles and hence has a zero working rate. Moreover, for low-frequency waves with frequency well below the ion gyrofrequency, such as KAWs, the parallel and perpendicular components of the wave electric fields also will result in remarkably different energization phenomena in the wave-particle interaction of different species. For instance, in the energization of charged particles by the perpendicular electric field of KAWs the ions have the energy exchanging efficiency much higher than that of electrons because the transverse scales of KAWs are comparable to the ion gyroradius but much larger than the electron gyroradius (Voitenko and Goossens 2004, 2006; Wu and Yang 2006, 2007). However, in the field-aligned acceleration or heating of charged particles by the parallel electric field of KAWs the electrons are much more efficient than the ions because the electrons usually have an instantaneous moving velocity much higher than that of the ions (Wygant et al. 2002; Wu 2003a, b; Wu and Chao 2003, 2004a, b; Bespalov et al. 2006; Varma et al. 2007; Chaston et al. 2007a, b). In consequence, the energization of charged particles by KAWs is characterized by a strong anisotropy and mass-charge dependence on the particle species, as shown in observations. Therefore, the electromagnetic polarization states of KAWs also are very important for us comprehensively to understand the physics of the wave-particle interaction of KAWs.

The electromagnetic polarization of KAWs, in general, is much more complex than that of AWs. For the low-β and quasi-perpendicular propagation case that has major interest in space and astrophysical plasmas, however, the electromagnetic polarization states of KAWs can be approximated well by Wu (2003a), Chen and Wu (2011b):

$$\frac{E_\parallel}{E_\perp} \simeq \left|\frac{\left(\alpha_e^2 - M_{zA}^2\right)\left(M_{zA}^2 - 1\right)}{M_{zA}^2 - \left(1 + \tau M_{zA}^2\right)\alpha_e^2}\right| \cot\theta$$
$$\frac{B_\parallel}{B_\perp} \simeq \left|\frac{M_{zA}^2 - 1}{(1+\tau)\,\alpha_e^2 - M_{zA}^2}\right| \frac{\sqrt{Q}\,(1+\tau)\,\alpha_e^2}{M_{zA}k_x\lambda_e}$$
$$\frac{E_\perp}{B_\perp} \simeq \left|\frac{M_{zA}^2 - \left(1 + \tau M_{zA}^2\right)\alpha_e^2}{(1+\tau)\,\alpha_e^2 - M_{zA}^2}\right| \frac{v_A}{M_{zA}}, \tag{1.30}$$

where $E_{\parallel(\perp)}$ and $B_{\parallel(\perp)}$ are the parallel (perpendicular) components of the electric and magnetic fields of KAWs, respectively, the parallel phase speed of KAWs in the Alfvén velocity v_A, M_{zA}, may be given by Eq. (1.29), the electron thermal speed in v_A, $\alpha_e = v_{T_e}/v_A$, can be related to the plasma pressure parameter β by

$$\beta = 2Q\,(1+\tau)\,\alpha_e^2, \tag{1.31}$$

and $\tau = T_i/T_e$ is the ion-electron temperature ratio.

Figure 1.3 shows the electromagnetic polarization properties of KAWs given by Eq. (1.30), the parallel to perpendicular electric field ratio ($E_\parallel/E_\perp$ in the top panel), the parallel to perpendicular magnetic field ratio ($B_\parallel/B_\perp$ in the middle panel), and the perpendicular electric to magnetic field ratio in the Alfvén velocity ($E_\perp/B_\perp v_A$ in the bottom panel), where the solid lines present the cases with the fixed parameters $\alpha_e = 0.3$, 3, and 30. As well known, for the polarization properties of ideal MHD AWs we have $E_\parallel/E_\perp = 0$, $B_\parallel/B_\perp = 0$, and $E_\perp/B_\perp v_A = 1$. From Fig. 1.3 it is very clear that for all cases the deviations of the polarization properties of KAWs from that of AWs all increase evidently with the perpendicular wavenumber $k_x^2\rho_{is}^2$, that is, the kinetic effect in the KAW polarization is attributed very clearly to the modification of the finite perpendicular wavelength of KAWs.

The most important kinetic modification effect of KAWs different from AWs is the generation of the parallel electric and magnetic fields of KAWs. The results presented in Fig. 1.3 indicate that the parameter α_e also can significantly influence the polarization properties of KAWs. For example, the parallel to perpendicular electric field ratio of KAWs ($E_\parallel/E_\perp$) with smaller $\alpha_e = 0.3$ is evidently higher than that with larger $\alpha_e = 3$ and 30, as shown in the top panel of Fig. 1.3, while in the middle panel of Fig. 1.3 it is found that the parallel to perpendicular magnetic field ratio of KAWs ($B_\parallel/B_\perp$) with larger $\alpha_e = 30$ is higher than that with smaller $\alpha_e = 3$ and 0.3. In similar to the parallel to perpendicular electric field ratio, the perpendicular electric to magnetic field ratio of KAWs ($E_\perp/B_\perp v_A$) also has the tendency of increasing with the α_e decreasing.

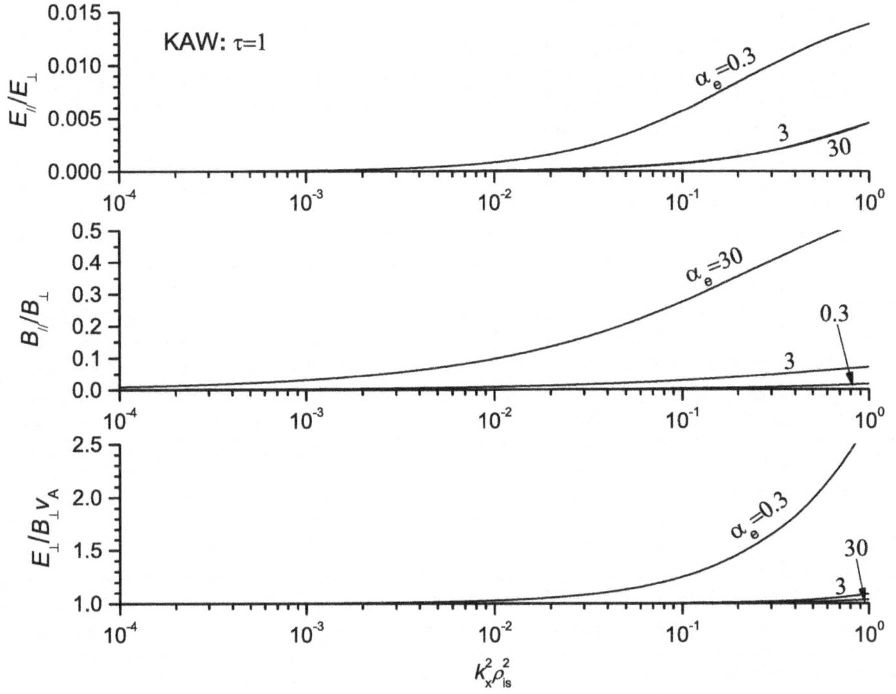

Fig. 1.3 The electromagnetic polarization states versus the perpendicular wave number $k_x^2\rho_{is}^2$: The solid lines are calculated by the fixed parameters $\alpha_e = 0.3$, 3, and 30, respectively, where other parameters $\theta = 89°$ and $\tau = 1$ have been used for the quasi-perpendicular propagation and isothermal plasma case

For the fixed τ, the plasma pressure parameter β can be determined directly by the parameter α_e in Eq. (1.31). The plasma pressure parameter β can strongly influence not only the propagation properties of KAWs but also their electromagnetic polarization states. Figure 1.4 presents the dependency of the electromagnetic polarization properties of KAWs on the parameter α_e, where the solid lines present the cases with the fixed temperature ratio $\tau = 0.1$, 1, and 10. Without loss of generality the fixed perpendicular wave number $k_x^2\rho_{is}^2 = 0.1$ has been used in the calculation of Fig. 1.4, and the propagation angle $\theta = 89°$ is the same to Fig. 1.3.

From Fig. 1.4, one can clearly find that the dependency of the KAW polarization on the parameter α_e (or i.e., β) has very distinctive features. The influence on the parallel to perpendicular electric field ratio $E_\parallel/E_\perp$ (the top panel) presents mainly in the parametric regime of $\alpha_e < 1$, that is, the low-β inertial regime of $\beta < 2Q$, implying that the parallel electric field of KAWs is generated mainly by the inertia motion of electrons. On the other hand, contrarily, the parallel to perpendicular magnetic field ratio of KAWs $B_\parallel/B_\perp$ (the middle panel) rises in the parametric regime of $\alpha_e > 1$, that is, the kinetic regime of $\beta > 2Q$. This is because the parallel component of the KAW magnetic field represents the compressible fluctuation of the

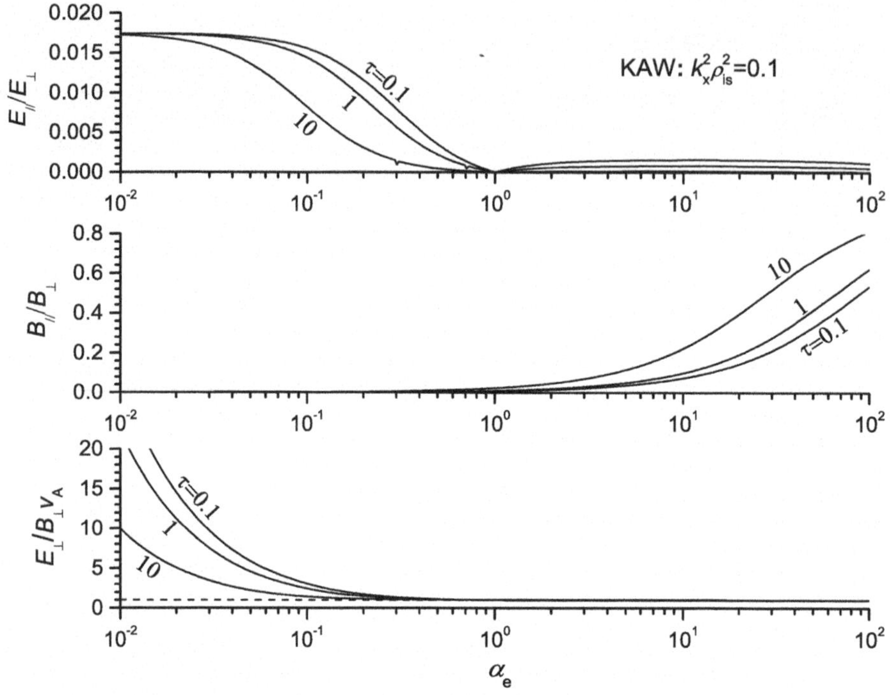

Fig. 1.4 The electromagnetic polarization states versus the parameter α_e: The solid lines are calculated by the fixed parameters $\tau = 0.1$, 1, and 10, respectively, where other parameters $\theta = 89°$ and the perpendicular wave number $k_x^2\rho_{is}^2 = 0.1$ have been used

KAW magnetic field and hence can be produced by the coupling of the plasma kinetic pressure with the magnetic pressure. The tendency of the ratio $B_{\parallel}/B_{\perp}$ increasing with the temperature ratio τ also confirms further this inference, as shown in the middle panel of Fig. 1.4. In addition, from the bottom panel of Fig. 1.4 also it is very clear that the deviation of the ratio $E_{\perp}/B_{\perp}v_A$ from one (i.e., its MHD solution) increases remarkably with the α_e decreasing in the inertial regime of $\alpha_e < 1$.

It is worth noticing that the parallel electric field of KAWs $E_{\parallel} = 0$ at $\alpha_e \simeq 1$, that is, at the inertial-kinetic transition regime, as shown in the top panel of Fig. 1.4. In fact, the parallel electric field of KAWs in the kinetic regime of $\alpha_e > 1$, which is associated with the plasma kinetic pressure fluctuation, has an opposite direction to that in the inertial regime. While at the transition regime $\alpha_e \simeq 1$ the electron inertia and the kinetic pressure may reach some balance and it is this balance that results in the disappearance of the parallel electric field of KAWs. The fact of $E_{\parallel} = 0$ at $\alpha_e \simeq 1$ also indicates that the x-component of the Poynting flux of KAWs approaches to zero. This is consistent with the result of the perpendicular group velocity of KAWs $v_{gx} = 0$ at the transition regime, which can be obtained by Eq. (1.22).

Figure 1.4 also shows clearly the dependency of the ratios $E_{\parallel}/E_{\perp}$, $B_{\parallel}/B_{\perp}$, and $E_{\perp}/B_{\perp}v_A$ on the temperature ratio $\tau = T_i/T_e$. Figure 1.5 further displays the polar-

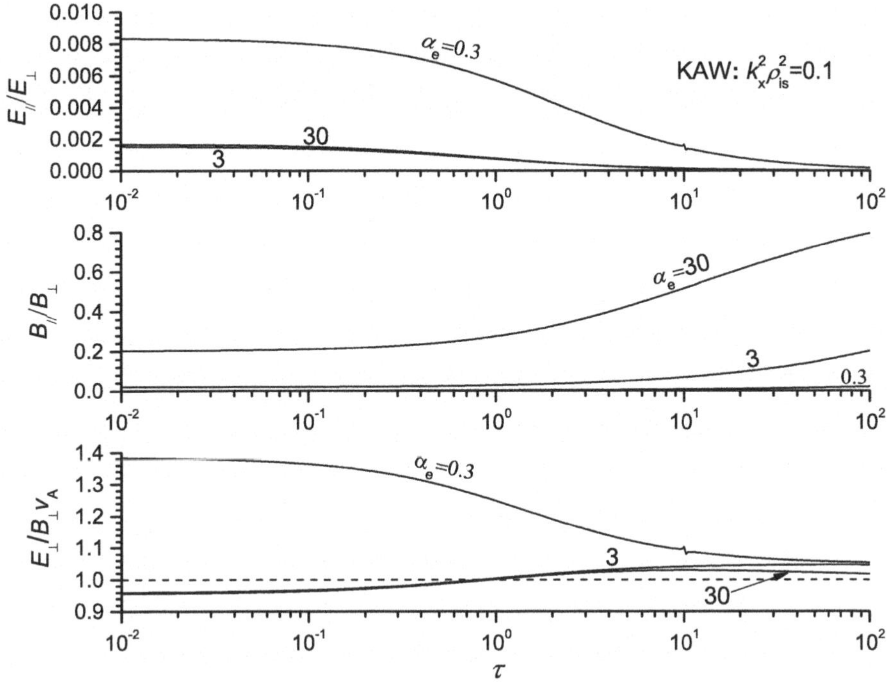

Fig. 1.5 The electromagnetic polarization states versus the parameter τ: The solid lines are calculated by the fixed parameters $\alpha_e = 0.3$, 3, and 30, respectively, where other parameters $\theta = 89°$ and $k_x^2\rho_{is}^2 = 0.1$ have been used

ization properties of KAWs versus the temperature ratio τ, where the solid lines present the cases with the fixed $\alpha_e = 0.3, 3$, and 30, and other parameters $k_x^2\rho_{is}^2 = 0.1$ and $\theta = 89°$ are the same to Fig. 1.4.

From the top panel of Fig. 1.5, the parallel electric field of KAWs, which is mainly attributed to the electron inertia in the inertial regime of $\alpha_e < 1$, decreases as the τ (i.e., the ion temperature) increases, implying that the ion kinetic pressure may counteract partly the effect of the electron inertia. On the other hand, the ion kinetic pressure can considerably affect the parallel magnetic field of KAWs, that is, the compressive magnetic fluctuation as shown in the middle panel of Fig. 1.5. In the bottom panel of Fig. 1.5, it can be found that as the parameter τ (or i.e., the ion temperature) increases, the perpendicular electric to magnetic ratio of KAWs, $E_\perp/B_\perp v_A$, decreases in the inertial regime case of $\alpha_e = 0.3 < 1$ but increases in the kinetic regime cases of $\alpha_e = 3$ and $30 > 1$. In particular, for the kinetic regime cases of $\alpha_e = 3$ and 30 the ratio $E_\perp/B_\perp v_A$ is slightly lower and higher than one in the parametric regimes of $\tau < 1$ (the electron temperature domination) and $\tau > 1$ (the ion temperature domination), respectively.

The sensitive dependency of the polarization properties of KAWs on the local plasma parameters can have important applications in space and astrophysical plasmas. Since the wave-particle interaction has directly close relation with the polarization states of KAWs the parametric dependency of the KAW polarization can produce different energization phenomena as well as fine structures of local plasmas in various cosmic plasma environments due to widely varying ranges of these local plasma parameters. For example, in the solar atomsphere, the plasma β can change from larger than one in the photosphere to much less than one in the corona, and even well below the electron-ion mass ratio (i.e., the inertial regime of $\beta < 2Q$) in local tenuous magnetic flux tubes.

On the other hand, in the near-earth space it is generally believed that the plasma temperature changes from below 1 eV in the ionosphere up to several hundreds of eV in the magnetosphere with altitudes $\sim$3–4R_E ($R_E \simeq 6370$ km is the Earth radius) where the plasma is dominated by the plasma sheet with temperature of $\sim$300–600 eV (Kletzing et al. 1998, 2003b). Often electrons and ions have evidently different temperatures. In the central plasma sheet, observations have established that the ion temperature ranges from 1 to 10 keV (Frank et al. 1976; Eastman et al. 1984; Takahashi and Hones 1988). While the analytic works by Kletzing et al. (1998, 2003b) based on measurements of temperature up to the highest altitudes measured by S3-3 show that the electron temperature does not exceed 10 eV at altitudes below $\sim 1.5R_E$.

In the near-Earth magnetotail (<20 R_E), a comprehensive statistical study of the plasma sheet using the AMPTE satellite data showed that more than 80% of the particle measurements fell within the range $5.5 < \tau < 11$ (Baumjohann et al. 1989). Slavin et al. (1985) and Schriver et al. (1998) performed statistical survey using ISEE 3 satellite data and found that the parameter τ varied between 4.8 and 7.8 in the deep magnetotail from 30 to 220 R_E downtail. Also, there appears to be significant deviations of the ion temperature toward values consistently less than the electron temperature, that is, $\tau < 1$ (Banks 1967; Bilitza 1991).

For different local plasma environments some reduced expressions of the KAW polarizations are very useful and can more clearly depict the physical scenario of the wave-particle interaction associated with KAWs. For the electric to magnetic field ratio ($E_\perp/B_\perp$) and the electric field ratio ($E_\parallel/E_\perp$) of KAWs, the following reduced expressions can be well approximations (Wu 2003a)

$$\frac{E_\perp}{B_\perp} = \frac{\left(1+\lambda_e^2 k_x^2\right)\left(1+\rho_i^2 k_x^2\right)}{1+\rho_{is}^2 k_x^2} M_{zA} v_A = \frac{1+\rho_i^2 k_x^2}{M_{zA}} v_A, \tag{1.32}$$

and

$$\frac{E_\parallel}{E_\perp} = \frac{\left(1-\alpha_e^2\right)\lambda_e^2 k_x^2 + \lambda_e^2 \rho_i^2 k_x^4}{\left(1+\lambda_e^2 k_x^2\right)\left(1+\rho_i^2 k_x^2\right)} \cot\theta. \tag{1.33}$$

For the cold ion case of $\tau \ll 1$, one has $\rho_i \ll \rho_s$ and hence the above expressions can further reduce to

$$\frac{E_\perp}{B_\perp} \simeq \frac{v_A}{M_{zA}} \tag{1.34}$$

and

$$\frac{E_\parallel}{E_\perp} \simeq \frac{\left(1-\alpha_e^2\right)\lambda_e^2 k_x^2}{1+\lambda_e^2 k_x^2} \cot\theta. \tag{1.35}$$

In contrast, for the warm ion case of $\tau \gg 1$, one has $\rho_s \ll \rho_i$ and the above polarization relations can reduce to

$$\frac{E_\perp}{B_\perp} \simeq \left(1+\lambda_e^2 k_x^2\right) M_{zA} v_A = \sqrt{\left(1+\lambda_e^2 k_x^2\right)\left(1+\rho_i^2 k_x^2\right)} v_A > v_A. \tag{1.36}$$

and

$$\frac{E_\parallel}{E_\perp} \simeq \frac{\lambda_e^2 k_x^2}{1+\lambda_e^2 k_x^2} \cot\theta = \frac{\lambda_e^2 k_x k_z}{1+\lambda_e^2 k_x^2}. \tag{1.37}$$

These polarization relations have been identified by in situ observations of satellites in the auroral plasmas (Chaston et al. 1999; Stasiewicz et al. 2000b).

1.4 Excitation of KAWs in Homogeneous Plasmas

KAWs can be generated due to the excitation of various plasma instabilities, including macroscopic instabilities such as electric currents, shear flows, density and temperature gradients, temperature anisotropy, and so on, as well as microscopic instabilities due to plasma wave-particle resonances such as beam particles, or other reverse Landau damping. Here we focus our concerns on the growth rates of KAWs excited by linear macroscopic instabilities. Usually, macroscopic instabilities in plasmas may be treated well in the two-fluid model. However, it is necessary to employ the kinetic theory in order to discuss more clearly the kinetic effects of KAWs caused by short wavelengths comparable to the kinetic scales of plasma particles. Neglecting the collision effect, the kinetic description of plasmas can be given by the Vlasov equation of the velocity distribution function of plasma particles, combining the Maxwell equations of fluctuating electromagnetic fields.

In this section, we discuss the excitation of KAWs by plasma instabilities driven by the temperature anisotropy and the field-aligned drift electric current in a homogeneous plasma and the excitation of KAWs driven by the density gradient in an inhomogeneous plasma is discussed in the next section. In a homogeneous plasma, the free energy of driving KAW instabilities is supplied mainly by the temperature anisotropy and the electric current due to the relative drift between the electron and ion components of the plasma. When a static ambient magnetic field presents, the temperature anisotropy is an intrinsic characteristic for a magnetized plasma because the kinematic difference of the plasma particles in the plane perpendicular to and along the magnetic field, especially for the cases of collisionless plasmas such as astrophysical and space plasmas. On the other hand, field-aligned currents can often present in a magnetic plasma, such as Birkeland currents in the polar magnetosphere and current-carrying loops in the solar atmosphere. These field-aligned currents are carried usually by electrons drift motion along magnetic field lines.

In a homogeneous plasma the velocity distributions of plasma particles with anisotropic temperatures and field-aligned currents can be fitted by the so-called bi-Maxwell distribution function below:

$$f_{s0}(\mathbf{v}) = f_{s0}(v_\parallel,\ v_\perp) = \frac{n_{s0}}{(2\pi)^{3/2}\, v_{T_{s\parallel}} v_{T_{s\perp}}^2} e^{-\left[\left(v_\parallel - v_{s\parallel}\right)^2/2v_{T_{s\parallel}}^2 + v_\perp^2/2v_{T_{s\perp}}^2\right]}, \tag{1.38}$$

where $v_{T_{s\perp(\parallel)}} \equiv \sqrt{T_{s\perp(\parallel)}/m_s}$ and $T_{s\perp(\parallel)}$ are the perpendicular (parallel) thermal speed and the perpendicular (parallel) temperature for species s, respectively, and n_{s0} and $v_{s\parallel}$ are the number density and field-aligned drift velocity for species s, respectively.

Without loss of generality, let us denote $\mathbf{B}_0 = B_0\hat{\mathbf{z}}$ and $\mathbf{k} = k_\perp\hat{\mathbf{x}} + k_\parallel\hat{\mathbf{z}}$. In the low-frequency approximation,

$$\omega^2,\ k_\parallel^2 v_{s\parallel}^2,\ \text{and}\ k_\parallel^2 v_{T_{s\parallel}}^2 \ll \omega_{cs}^2\ \ (s = i, e), \tag{1.39}$$

Chen et al. (2011) and Wu (2012) obtained the kinetic dispersion equation of above the current-carrying bi-Maxwell plasma as follows:

$$\left(\frac{\omega}{\omega_{ci}}\right)^4 - A\left(\frac{\omega}{\omega_{ci}}\right)^2 + B\left(\frac{\omega}{\omega_{ci}}\right) + C = 0 \tag{1.40}$$

with the coefficients

$$\begin{aligned} A &= \frac{1}{c_{i1} - 2c_{i3}}\left(c_3 + \frac{c_4^2}{c_{i1}} - \frac{c_5^2}{c_1}\right) + \left(\frac{1}{c_{i1}} - \frac{1}{c_1}\right)c_2, \\ B &= \frac{2k_\parallel\lambda_i}{c_{i1} - 2c_{i3}}\left(\frac{c_4}{c_{i1}} - \frac{c_5}{c_1}\right)\frac{v_D}{v_A}, \\ C &= \frac{c_2}{c_{i1} - 2c_{i3}}\left[\left(\frac{1}{c_{i1}} - \frac{1}{c_1}\right)c_3 - \frac{(c_4 - c_5)^2}{c_{i1}c_1}\right] \\ &\quad - \frac{k_\parallel^2\lambda_i^2}{c_{i1} - 2c_{i3}}\left(\frac{1}{c_{i1}} - \frac{1}{c_1}\right)\frac{v_D^2}{v_A^2}, \end{aligned} \tag{1.41}$$

where the parameters $c_{1,2,3,4,5}$ are given by

$$
\begin{aligned}
c_1 &= \frac{\beta_{i\perp}}{2\chi_i}\left(\frac{Z'(\zeta_0^e)}{\beta_{e\parallel}} + I_0 e^{-\chi_i}\frac{Z'(\zeta_0^i)}{\beta_{i\parallel}}\right),\\
c_2 &= \frac{k_\parallel^2\lambda_i^2}{2}\left(2 - \beta_{e\parallel}\delta_{T_e} - c_{i1}\beta_{i\parallel}\delta_{T_i}\right),\\
c_3 &= \frac{k_\parallel^2\lambda_i^2}{2}\left[2 - \beta_{e\parallel}\delta_{T_e} - (c_{i1} - 2c_{i3})\,\beta_{i\parallel}\delta_{T_i}\right] + \\
&\quad \frac{2\chi_i}{\beta_{i\perp}}\left[1 + \beta_{e\perp}\left(1 + \frac{1-\delta_{T_e}}{2}Z'(\zeta_0^e)\right) + \beta_{i\perp}c_{i2}\left(1 + \frac{1-\delta_{T_i}}{2}Z'(\zeta_0^i)\right)\right],\\
c_4 &= 1 - c_{i2} - \frac{k_\parallel^2\lambda_i^2}{2}\beta_{i\parallel}\left(1 + 2\delta_{T_i}\right)\frac{c_{i3}}{\chi_i},\\
c_5 &= \frac{1-\delta_{T_i}}{2}c_{i2}Z'(\zeta_0^i) - \frac{1-\delta_{T_e}}{2}Z'(\zeta_0^e) - \frac{k_\parallel^2\lambda_i^2}{2}\beta_{i\parallel}\left(1 + \delta_{T_i}\right)\frac{c_{i3}}{\chi_i},
\end{aligned}
\tag{1.42}
$$

and the other parameters $c_{i1,2,3,4}$ are following functions of χ_i:

$$
\begin{aligned}
c_{i1} &\equiv \frac{1 - I_0(\chi_i)e^{-\chi_i}}{\chi_i},\\
c_{i2} &\equiv \left(I_0(\chi_i) - I_1(\chi_i)\right)e^{-\chi_i},\\
c_{i3} &\equiv 2\chi_i\sum_{l=1}^{\infty}\frac{\left(I_l(\chi_i)e^{-\chi_i}\right)'}{l^2},\\
c_{i4} &\equiv 2\sum_{l=1}^{\infty}\frac{I_l(\chi_i)e^{-\chi_i}}{l^2}.
\end{aligned}
\tag{1.43}
$$

In these expressions above,

$$
Z(y) \equiv \frac{1}{\sqrt{\pi}}\int_{-\infty}^{\infty}\frac{e^{-x^2}}{x-y}dx \tag{1.44}
$$

is the so-called plasma dispersion function with the argument

$$
\zeta_0^e \equiv \frac{\omega - k_\parallel v_D}{\sqrt{2}k_\parallel v_{T_{e\parallel}}} \text{ and } \zeta_0^i \equiv \frac{\omega}{\sqrt{2}k_\parallel v_{T_{i\parallel}}}, \tag{1.45}
$$

and the prime denotes the first derivative with respect to the argument,

$$
I_l(\chi_i) = i^{-l}J_l(i\chi_i) \tag{1.46}
$$

is the modified Bessel function of order l with the argument

$$
\chi_i \equiv \rho_i^2 k_\perp^2 = \frac{k_\perp^2 v_{T_{i\perp}}^2}{\omega_{ci}^2}, \tag{1.47}
$$

which is responsible for the effect of the ion gyroradius, that is, the short-wavelength effect, and another two parameters are the electron field-aligned drift velocity,

$$v_D \equiv v_{e\parallel}, \tag{1.48}$$

and the temperature anisotropy,

$$\delta_{T_s} \equiv \frac{T_{s\parallel} - T_{s\perp}}{T_{s\parallel}} = 1 - \frac{T_{s\perp}}{T_{s\parallel}} = 1 - \frac{\beta_{s\perp}}{\beta_{s\parallel}}, \tag{1.49}$$

for the electrons ($s = e$) and ions ($s = i$). In addition, in the derivation of the above expressions the approaching approximation $\chi_e \to 0$ and $v_{i\parallel} = 0$ have been used because $\chi_e \ll \chi_i$ and the field-aligned current is carried usually by the electron field-aligned drift.

The above results can be used as the basis to discuss the kinetic effects of the short-wavelength modification (e.g., a finite χ_i) in the low-frequency approximation of Eq. (1.39). Below let us discuss the two special cases for the excitation of KAWs, which are driven by the temperature anisotropy (i.e., $\delta_{T_s} \neq 0$) and the field-aligned current (i.e., $v_D \neq 0$), respectively.

Excitation of KAWs by anisotropic temperatures

It is well known that the so-called fire-hose instability driven by the temperature (or the kinetic pressure) anisotropy is one of the most important mechanisms for the generation of excited AWs and has been extensively discussed in the literature on the basis of the bi-adiabatic hydromagnetics in the long-wavelength limit (Chew et al. 1956). The short-wavelength modified version of this fire-hose instability by the temperature anisotropy, which takes account of the finite ion gyroradius effect, can be responsible for the generation of KAWs. For the case without the field-aligned current, that is, $v_D = 0$, the dispersion equation of Eq. (1.40) can reduce to

$$\left(\frac{\omega}{\omega_{ci}}\right)^4 - A\left(\frac{\omega}{\omega_{ci}}\right)^2 + C = 0 \tag{1.50}$$

with the coefficients

$$\begin{aligned} A &= \frac{1}{c_{i1} - 2c_{i3}}\left(c_3 + \frac{c_4^2}{c_{i1}} - \frac{c_5^2}{c_1}\right) + c_2\left(\frac{1}{c_{i1}} - \frac{1}{c_1}\right), \\ C &= \frac{c_2}{c_{i1} - 2c_{i3}}\left[c_3\left(\frac{1}{c_{i1}} - \frac{1}{c_1}\right) - \frac{(c_4 - c_5)^2}{c_{i1}c_1}\right], \end{aligned} \tag{1.51}$$

where the plasma dispersion function $Z(\zeta_0^e)$ has the argument $\zeta_0^e = \omega/\sqrt{2}k_\parallel v_{T_{e\parallel}}$.

For the fluid limit with the zero ion gyroradius (i.e., $\chi_i \to 0$) the dispersion equation (1.50) further reduces to

$$\left(\frac{\omega}{\omega_{ci}}\right)^4 - (2c_2 + c_4^2)\left(\frac{\omega}{\omega_{ci}}\right)^2 + c_2^2 = 0. \tag{1.52}$$

which has dispersion relations as follows:

$$\frac{\omega}{\omega_{ci}} = \pm\frac{c_4}{2}\left(1 \pm \sqrt{1 + 4\frac{c_2}{c_4^2}}\right). \tag{1.53}$$

In particular, in the low-frequency and long-wavelength MHD limit one has $\omega/\omega_{ci} \sim k_\parallel \lambda_i \to 0$ and hence $c_4^2 \ll |c_2|$. In result, the dispersion equation (1.52) further reduces to

$$\frac{\omega^2}{\omega_{ci}^2} \simeq c_2 \simeq \left(1 - \frac{1}{2}\delta_T \beta_\parallel\right) k_\parallel^2 \lambda_i^2 \Rightarrow \omega^2 \simeq k_\parallel^2 v_A^2 \tag{1.54}$$

for thc isotropic case of $\delta_T = 0$, that is, the well-known dispersion relation for the ideal AWs. For the anisotropic case of $\delta_T \neq 0$, on the other hand, this leads to the well-known fire-horse instability of AWs (Rosenbluth 1956; Parker 1958) when

$$c_2 < 0 \Rightarrow \delta_T \beta_\parallel = \beta_\parallel - \beta_\perp > 2 \text{ or i.e., } \delta_T > 2/\beta_\parallel, \tag{1.55}$$

where $\beta_{\perp(\parallel)} = \beta_{i\perp(\parallel)} + \beta_{e\perp(\parallel)}$ is the total plasma perpendicular (parallel) pressure parameter and $\delta_T = 1 - \beta_\perp/\beta_\parallel$ is the total temperature anisotropy.

Including the ion gyroradius effect of a small but finite χ_i the dispersion equation (1.50) reduces to

$$\left(\frac{\omega}{\omega_{ci}}\right)^4 - (2c_2 + c_4^2 + b\chi_i)\left(\frac{\omega}{\omega_{ci}}\right)^2 + (c_2^2 + c_2 b \chi_i) = 0, \tag{1.56}$$

where the coefficient of the finite-χ_i modification term

$$b = \frac{2}{\beta_{i\perp}}\left(1 + \beta_{i\perp}\delta_{T_i} + \beta_{e\perp}\delta_{T_e}\right) + \frac{\beta_{e\parallel}}{\beta_\parallel}\frac{\left(\delta_{T_e} - \delta_{T_i}\right)^2}{1 - \delta_{T_i}}. \tag{1.57}$$

Assuming an isotropic electron (i.e., $\delta_{T_e} = 0$) and ion temperature (i.e., $\delta_{T_i} = 0$), Yoon et al. (1993) and Chen and Wu (2010) discussed the kinetic fire-hose instability driven by the ion and electron temperature anisotropy, respectively, which is associated with KAWs in high-β plasmas. Figure 1.6 displays the growth rate of KAWs excited by the electron (solid lines) and ion (dashed lines) temperature anisotropy versus the finite perpendicular wavenumber $\chi_i = \rho_i^2 k_\perp^2$, where the solid and dashed lines present the cases of the electron temperature anisotropic parameter $\delta_{T_e} = 0.6$, 0.7, and 0.8 with the ion temperature isotropy $\delta_{T_i} = 0$ and the ion temperature anisotropic parameter $\delta_{T_i} = 0.6$, 0.7, and 0.8 with the electron temperature isotropy $\delta_{T_e} = 0$, respectively, and the fixed parameters $\beta_{e\perp} = \beta_{i\perp} = 2$ and $k_z \lambda_i = 0.05$ have been used for all cases.

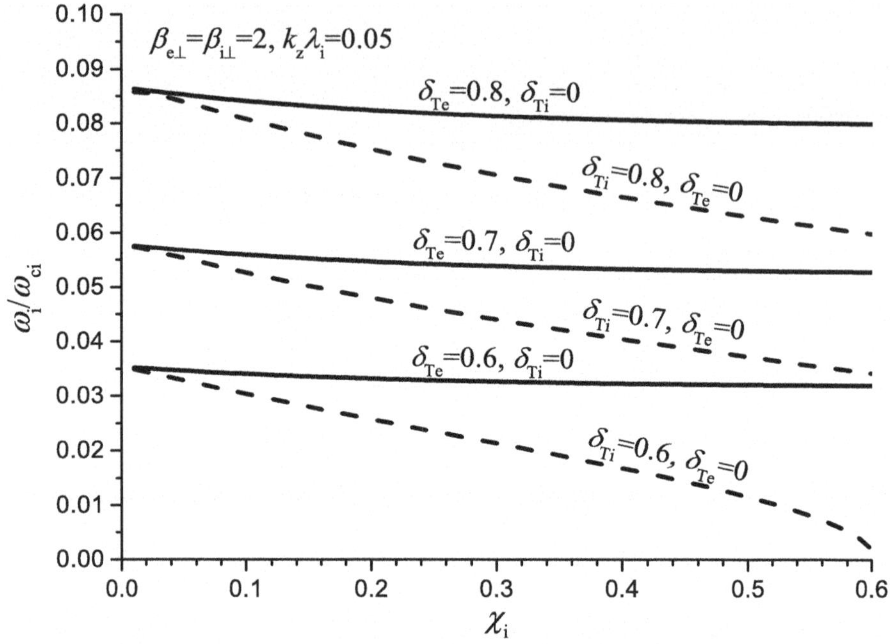

Fig. 1.6 The normalized growth rate ω_i/ω_{ci} versus the normalized perpendicular wavenumber $\chi_i = \rho_i^2 k_\perp^2$ for the fixed parameters $\beta_{e\perp} = \beta_{i\perp} = 2$, and $k_z\lambda_i = 0.05$, where the solid (dashed) lines represent the electron (ion) temperature anisotropy with parameters $\delta_{T_{e(i)}} = 0.6, 0.7, 0.8$, and $\delta_{T_{i(e)}} = 0$

The results presented in Fig. 1.6 show that in the long-wavelength limit (i.e., the fluid limit) of $\chi_i \to 0$, the ion and electron temperature anisotropy have the same KAW growth rate. As the wave number χ_i increases, however, the growth rate of KAWs excited by the ion temperature anisotropy evidently decreases, but that by electron temperature anisotropy almost is invariant. This indicates that the short-wavelength modification can evidently depress the growth rate of KAWs by the ion temperature anisotropy, but has little effect on the growth rate of KAWs by the electron anisotropy. In consequence, the electron temperature anisotropy can excite more effectively KAWs with finite perpendicular wave number χ_i than the ion temperature anisotropy.

The short-wavelength modification due to the finite perpendicular wavenumber χ_i can evidently influence not only the growth rate but also the critical instability condition for the ion temperature anisotropy. Figure 1.7 shows the dependency of the growth rate of KAWs by the ion (left) and electron (right) temperature anisotropic parameter δ_{T_i} and δ_{T_e}, where the fixed parameters $\beta_{e(i)\perp}$ and $k_z\lambda_i$ are the same with Fig. 1.6, the parameters $\delta_{T_e} = 0$ (the electron temperature isotropy) and $\delta_{T_i} = 0$ (the ion temperature isotropy) have been used in the left and right panels, respectively, and the solid lines represent the cases of $\chi_i = 0.01, 0.1$, and 0.5, respectively.

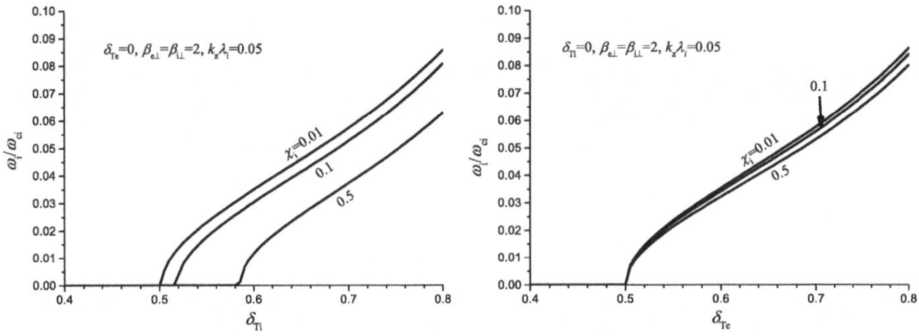

Fig. 1.7 The normalized growth rates ω_i/ω_{ci} versus the ion or electron temperature anisotropic parameter δ_{T_i} (left) or δ_{T_e} (right), where the parameters $\delta_{T_e} = 0$ and $\delta_{T_i} = 0$ have been used in the left and right panels, respectively, and other fixed parameters are the same with Fig. 1.6

From Fig. 1.7, as expected, for the two cases of the ion (left) and electron (right) temperature anisotropy the growth rates of KAWs both increase remarkably with the anisotropic parameter δ_{T_i} (left) and δ_{T_e} (right) once $\delta_{T_{i(e)}}$ exceeds the corresponding threshold $\delta^c_{T_{i(e)}}$ for the KAW instability. For the case driven by the ion temperature anisotropy, however, the instability threshold $\delta^c_{T_i}$ has the increasing tendency with the finite wave number χ_i, for instance, from $\delta^c_{T_i} \simeq 0.5$ for $\chi_i = 0.01$ increasing to $\delta^c_{T_i} \simeq 0.52$ and 0.56 for $\chi_i = 0.1$ and 0.5, respectively, as shown in the left panel of Fig. 1.7. While for the case driven by the electron temperature anisotropy presented in the right panel of Fig. 1.7 the instability threshold $\delta^c_{T_e}$ for the three cases of $\chi_i = 0.01$, 0.1, and 0.5 are almost invariant and all are $\delta^c_{T_e} \simeq 0.5$.

In fact, for the fixed parameters $\beta_{e\perp} = \beta_{i\perp} = 2$ and $\delta_{T_{e(i)}} = 0$ here it can be found that the instability threshold $\delta^c_{T_{i(e)}} = 0.5$ is the corresponding MHD threshold of the fire-horse instability of AWs given by Eq. (1.55). In the derivation of the dispersion equation (1.40), on the other hand, the approximation of $\chi_e \to 0$ has been used because $\chi_e \ll \chi_i$. It is because of this approximation that the short-wavelength modification by the finite ion perpendicular wave number χ_i has little effect on the KAW instability driven by the electron temperature anisotropy but can evidently depress that driven by the ion temperature anisotropy. In consequence, as the finite perpendicular wave number χ_i increases the KAW instability driven by the ion temperature anisotropy has an increasing (and hence higher) instability threshold and a decreasing (and hence lower) growth rate than that driven by the electron temperature anisotropy. In other words, the electron temperature anisotropy can more effectively excite the instability growth of KAWs with finite perpendicular wave numbers χ_i than the ion temperature anisotropy.

Excitation of KAWs by field-aligned currents

The fire-horse instability by the temperature anisotropy required the threshold condition $\beta_\parallel - \beta_\perp > 2$. Therefore in a low-$\beta$ plasma the fire-horse instability can not work. In similar, the KAW instability of field-aligned currents considerably depends on the plasma β parameter. Chen et al. (2013) investigated the excitation of KAWs by the field-aligned current instability in a low-β plasma with $\beta < 2Q \ll 1$. Taking $\delta_{T_{e(i)}} = 0$ for the plasma with the isotropic temperatures, the dispersion equation (1.40) can reduce to

$$\frac{\omega^2}{k_\parallel^2 v_A^2} - \frac{2k_\perp^2\lambda_e^2}{1+k_\perp^2\lambda_e^2}\frac{v_D}{v_A}\frac{\omega}{k_\parallel v_A} + \frac{k_\perp^2\lambda_e^2}{1+k_\perp^2\lambda_e^2}\frac{v_D^2}{v_A^2} - \frac{1+(3/4)\tau\alpha_e^2 k_\perp^2\lambda_e^2}{1+k_\perp^2\lambda_e^2} = 0, \quad (1.58)$$

where the large-argument expansion of the plasma dispersion function $Z(\zeta_0^{e(i)})$ has been used because

$$|\zeta_0^i| \sim v_A/v_{T_i} \sim \sqrt{2/\beta_i} \gg |\zeta_0^e| \sim v_A/v_{T_e} \sim \sqrt{2Q/\beta_e} \gg 1. \quad (1.59)$$

From the dispersion equation (1.58) KAWs can be excited with a growth rate ω_i

$$\omega_i = \frac{k_\parallel v_A}{1+k_\perp^2\lambda_e^2}\sqrt{k_\perp^2\lambda_e^2\frac{v_D^2}{v_A^2} - (1+k_\perp^2\lambda_e^2)\left(1+\frac{3\tau\alpha_e^2}{4}k_\perp^2\lambda_e^2\right)} \quad (1.60)$$

when the electron field-aligned drift velocity v_D exceeds the threshold velocity v_D^c, that is,

$$\frac{v_D}{v_A} > \frac{v_D^c}{v_A} = \frac{\sqrt{(1+k_\perp^2\lambda_e^2)\left(1+3\tau\alpha_e^2 k_\perp^2\lambda_e^2/4\right)}}{k_\perp\lambda_e} > 1. \quad (1.61)$$

Usually, this is not easy to be satisfied because there is a high Alfvén velocity $v_A > v_{T_e}$ in the low-β plasma of $\beta < 2Q$ and the instability condition by Eq. (1.61) requires that the electrons have a field-aligned drift velocity larger than their thermal velocity, that is, $v_D > v_{T_e}$.

In the intermediate-β regime of $2Q < \beta \ll 1$, however, the instability condition above can be reduced. In fact, for the case of $2Q < \beta \ll 1$ one has $|\zeta_0^i| \sim v_A/v_{T_i} \sim \sqrt{2/\beta_i} \gg 1$ and $|\zeta_0^e| \sim v_A/v_{T_e} \sim \sqrt{2Q/\beta_e} \ll 1$, implying that the plasma dispersion functions $Z(\zeta_0^e)$ and $Z(\zeta_0^i)$ are suitable for the small- and large-argument expansion, respectively. In consequence, the dispersion equation (1.40) reduces to (Chen and Wu 2012)

$$0 = \frac{\omega^2}{k_\parallel^2 v_A^2} - \frac{2(1-c_{i2})\tau + 2c_{i1}\chi_i}{\left(1+2\beta_e^{-1}+2c_{i2}\tau\right)c_{i1}\chi_i + (1-c_{i2})^2\tau}\frac{v_D}{v_A}\frac{\omega}{k_\parallel v_A}$$
$$+\frac{(\tau + c_{i1}\chi_i)\left(v_D^2/v_A^2 - 2\beta_i^{-1} - 2c_{i2}\right) - c_{i2}^2}{\left(1+2\beta_e^{-1}+2c_{i2}\tau\right)c_{i1}\chi_i + (1-c_{i2})^2\tau}\chi_i, \quad (1.62)$$

which can have a lower threshold drift velocity $v_D^c < v_A$ for the instability of KAWs than that given by Eq. (1.61).

In the high-β case of $\beta \geq 1$, one has $|\zeta_0^e| \ll |\zeta_0^i| \leq 1$, implying that the plasma dispersion functions $Z(\zeta_0^e)$ and $Z(\zeta_0^i)$ both are suitable for the small-argument expansion. In particular, for the case of plasma temperature isotropy with $\delta_{T_i} = \delta_{T_e} = 0$,

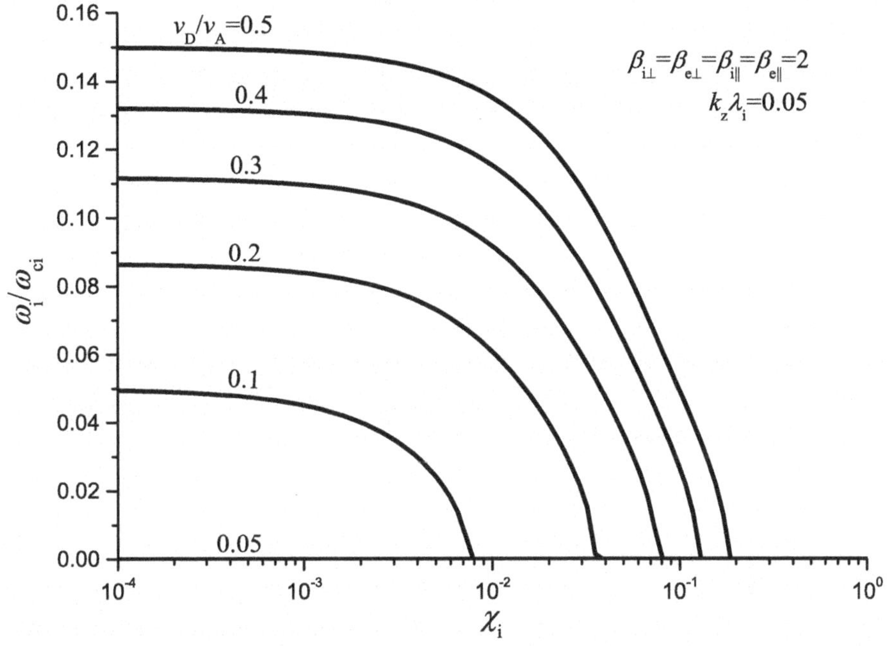

Fig. 1.8 The growth rate of KAWs ω_i/ω_{ci} versus the perpendicular wave number χ_i, where solid lines correspond to the electron drift velocity $v_D/v_A = 0.05, 0.1, 0.2, 0.3, 0.4$, and 0.5, respectively, and the parameters $\beta_{e(i)\|(\perp)} = 2$ and $k_z\lambda_i = 0.05$ have been used

the parameters $c_{1,2,3,4,5}$ given by Eq. (1.42), which presented in the coefficients A, B, and C of the dispersion equation (1.40), can reduce to

$$c_1 \simeq c_{i1} - \frac{\beta_i + \beta_e}{\beta_e \chi_i}, \quad c_2 \simeq k_\|^2 \lambda_i^2, \quad c_3 \simeq c_2 + \frac{2\chi_i}{\beta_i}, \quad c_4 \simeq 1 - c_{i2} - \frac{c_2}{2}\frac{\beta_i}{\chi_i} c_{i3}, \quad c_5 \simeq c_4, \tag{1.63}$$

where $k_\|\lambda_i = k_\| v_A/\omega_{ci} \sim \omega/\omega_{ci} \ll 1$ is a small quantity for low-frequency KAWs and the approximation $Z'(\zeta_0^e) \simeq Z'(\zeta_0^i) \simeq -2$ has been used for sufficiently small arguments $\zeta_0^e \ll 1$ and $\zeta_0^i \ll 1$.

Figure 1.8 shows the growth rate of KAWs excited by the field-aligned drift velocity v_D of the electrons in a high-β plasma versus the finite perpendicular wave number χ_i, where the fixed parameters $\beta_{e(i)\|(\perp)} = 2$ and $k_z\lambda_i = 0.05$ have been used for the high-β and temperature isotropy case, and the solid lines present the growth rates for the cases $v_D/v_A = 0.5, 0.4, 0.3, 0.2, 0.1$, and 0.05, respectively. From Fig. 1.8, it is clear that the growth rate of KAWs increases with the drift velocity v_D/v_A as expected, but decreases with the finite perpendicular wave number χ_i. In particular, for fixed drift velocity v_D/v_A there is a threshold wave number χ_i^c that the growth rate decreases to zero when $\chi_i \geq \chi_i^c$, and the threshold wave number χ_i^c increases as the drift velocity v_D/v_A increases.

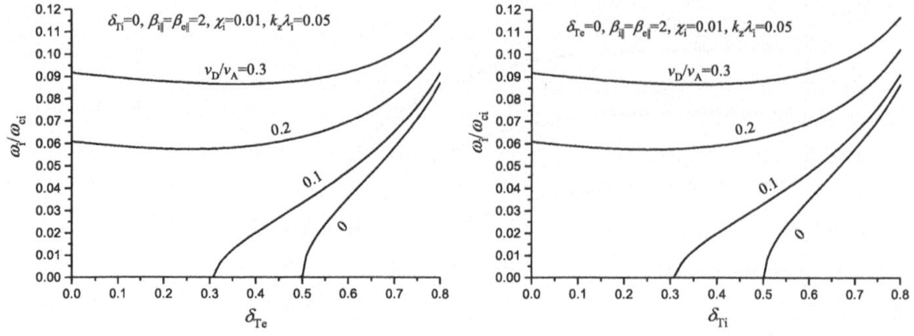

Fig. 1.9 The growth rate of KAWs ω_i/ω_{ci} versus the anisotropy factor $\delta_{T_{e(i)}}$, where the left and right panels present the cases of $\delta_{T_i}=0$ and $\delta_{T_e}=0$, respectively, the solid lines represent the results when $v_D/v_A=0, 0.1, 0.2$, and 0.3, respectively, and the parameters $\beta_{e\parallel}=\beta_{i\parallel}=2$, $\chi_i=0.01$, and $k_z\lambda_i=0.05$ have been used for all cases

In a high-β plasma, on the other hand, the temperature anisotropy also can work effectively to excite KAWs by the kinetic fire-horse instability. However, the critical condition for the fire-hose instability will be changed considerably due to the presence of field-aligned currents. Figure 1.9 shows the normalized growth rate of KAWs driven by the temperature anisotropy when the presence of a nonzero field-aligned current (i.e., $v_D\neq 0$), where the left and right panels present the cases driven by the electron (with $\delta_{T_i}=0$) and ion (with $\delta_{T_e}=0$) temperature anisotropy, respectively, the solid lines represent the results when $v_D/v_A=0, 0.1, 0.2$, and 0.3, respectively, and other parameters $\beta_{e\parallel}=\beta_{i\parallel}=2$, $\chi_i=0.01$, and $k_z\lambda_i=0.05$ have been used.

For the fixed parameter $\chi_i=0.01$ used in Fig. 1.9, the drift velocities of $v_D/v_A=0$ and 0.1 both are well below the threshold of the current instability (see Fig. 1.8), and hence the growth rates of KAWs are mainly contributed by the electron (left) and ion (right) temperature anisotropy. However, the threshold of the temperature anisotropy instability still is influenced evidently by the nonzero drift velocity $v_D/v_A=0.1$ and is decreased to $\delta^c_{T_{e(i)}}\sim 0.3$ from $\sim$0.5 at $v_D/v_A=0$. In the two cases of the drift velocity exceeding the threshold of the current instability (i.e., $v_D/v_A=0.2$ and 0.3), on the other hand, the growth rates of KAWs are almost invariant when $\delta_{T_{e(i)}}<\delta^c_{T_{e(i)}}$ and slightly increase with $\delta_{T_{e(i)}}$ when $\delta_{T_{e(i)}}>\delta^c_{T_{e(i)}}$ due to the contribution from the temperature anisotropy instability. In addition, as shown in Figs. 1.6 and 1.7, KAWs excited by the electron and ion temperature anisotropy have almost same growth rates and the instability thresholds in the low wave number limit of $\chi_i\ll 1$. Here, the almost same result presented in the left and right panels of Fig. 1.9 is because a small parameter $\chi_i=0.01\ll 1$ has been used here.

1.5 Excitation of KAWs in Inhomogeneous Plasmas

Field-aligned density striation is one of the most common inhomogeneity phenomena in magnetic plasmas, such as in the solar coronal plasma and in the terrestrial auroral plasma, where KAWs can play an important role in the inhomogeneous heating of solar coronal plasmas as well as in the local acceleration of auroral energetic electrons. In fact, in a magnetic plasma charged particles of the plasma respond to the Lorentz force of the magnetic field by freely streaming along the ambient magnetic **B**, whilst executing circular Larmor orbits, in the plane perpendicular to **B**, and hence the plasma particles move along helical orbits. Moreover, as the magnetic field strengthens, the resulting helical orbits become more tightly winded, and more effectively tying the particles to the field lines. This inhibits the plasma particles transversely to cross the field lines in scales larger than the particle gyroradius, while along the field lines the particles can move freely. In consequence, a magnetized plasma can maintain large density and/or temperature gradients across the magnetic field, while the fast parallel motion can quickly smooth out the gradients along the field lines. Therefore, one of the most important characteristics of magnetized plasmas is the anisotropy in the density distribution, which leads to the formation of field-aligned filamentous structures characterized by the parallel length scales much larger than the perpendicular scales. In particular, the most common perpendicular scales of these field-aligned filamentous structures, also called density striations, are in the orders of the kinetic scales of particles, that is, the ion gyroradius or the electron inertial length (Wu 2012).

By use of the two-fluid mode of plasmas, Wu and Chen (2013) investigated the excitation of KAWs by density striation in an inhomogeneous plasma magnetized by a homogeneous magnetic field $\mathbf{B}_0 = B_0\hat{z}$. Assuming that the unperturbed density $n_0(x)$ is inhomogeneous along the x-direction and has characteristic spatial scale L_x as follows:

$$L_x \equiv \left(\frac{\partial \ln n_0(x)}{\partial x}\right)^{-1}, \tag{1.64}$$

the low-frequency dispersion equation with the frequency $\omega < \omega_{ci}$ and wave vector $\mathbf{k} = k_x\hat{x} + k_z\hat{z}$ can be obtained as follows (Wu and Chen 2013):

$$\frac{\omega^2}{k_z^2 v_A^2} = (c_r + ic_i)\, K_A, \tag{1.65}$$

where

$$c_r = \frac{\rho_s^2 k_x^2}{1+\rho_s^2 k_x^2} + \frac{l_x^2}{\left(1+l_x^2\right)\left(1+\rho_s^2 k_x^2\right)},$$
$$c_i = \frac{l_x}{\left(1+l_x^2\right)\left(1+\rho_s^2 k_x^2\right)}, \tag{1.66}$$

and $l_x \equiv k_x L_x = 2\pi L_x/\lambda_x$ is the dimensionless inhomogeneity scale.

The complex solution of the dispersion equation (1.65), that is, the dispersion relation

$$\omega(k_x, k_z) = \omega_r(k_x, k_z) + i\omega_i(k_x, k_z) \tag{1.67}$$

can be given by

$$\begin{aligned} \frac{\omega_r^2}{k_z^2 v_A^2} &= \left(\sqrt{1+\frac{c_i^2}{c_r^2}}+1\right)\frac{c_r}{2}K_A \\ \frac{\omega_i^2}{k_z^2 v_A^2} &= \left(\sqrt{1+\frac{c_i^2}{c_r^2}}-1\right)\frac{c_r}{2}K_A. \end{aligned} \tag{1.68}$$

In the limit case of $l_x \to \infty$, that is, a homogeneous plasma, one has $c_r \to 1$ and $c_i \to 0$ from Eq. (1.66). The dispersion relation (1.68) reduces to

$$\begin{aligned} \frac{\omega_r^2}{k_z^2 v_A^2} &= K_A = \frac{1+\rho_s^2 k_x^2}{1+\lambda_e^2 k_x^2} \\ \frac{\omega_i^2}{k_z^2 v_A^2} &= 0, \end{aligned} \tag{1.69}$$

which is the standard dispersion relation of KAWs without growth (Streltsov and Lotko 1995; Lysak and Lotko 1996; Wu 2003a; Chen and Wu 2011a). On the other hand, for the strongly inhomogeneous limit of $l_x \to 0$ the dispersion relation (1.68) leads to

$$\begin{aligned} \frac{\omega_r^2}{k_z^2 v_A^2} &= \frac{\rho_s^2 k_x^2}{1+\rho_s^2 k_x^2} K_A = \frac{\rho_s^2 k_x^2}{1+\lambda_e^2 k_x^2} \\ \frac{\omega_i^2}{k_z^2 v_A^2} &= 0, \end{aligned} \tag{1.70}$$

which represents electrostatic waves in the ion-acoustic mode.

For the case of a finite inhomogeneity, that is, a finite $l_x = 2\pi L_x/\lambda_x$, the growth rate of KAWs can be given by ω_i in Eq. (1.68). Figure 1.10 shows the normalized growth rate $\omega_i/k_z v_A$ versus the perpendicular wave number $\lambda_e k_x$, where the solid lines represent the cases of $\alpha_e \equiv \rho_s/\lambda_e = 0.1$, 1, and 10, respectively, and the fixed parameter $l_x = 1$ has been used.

From Fig. 1.10, the growth rate approaches to a constant independent of α_e in the long-wavelength limit of $\lambda_e k_x \ll 1$ but goes to zero in the short-wavelength limit of $\lambda_e k_x \gg 1$. In the case of finite wave numbers ($\lambda_e k_x$) the growth rate considerably depends on the parameter $\alpha_e \equiv \rho_s/\lambda_e = v_{T_e}/v_A$. Figure 1.11 presents the normalized growth rate $\omega_i/k_z v_A$ versus the inhomogeneity parameter l_x, where the solid lines

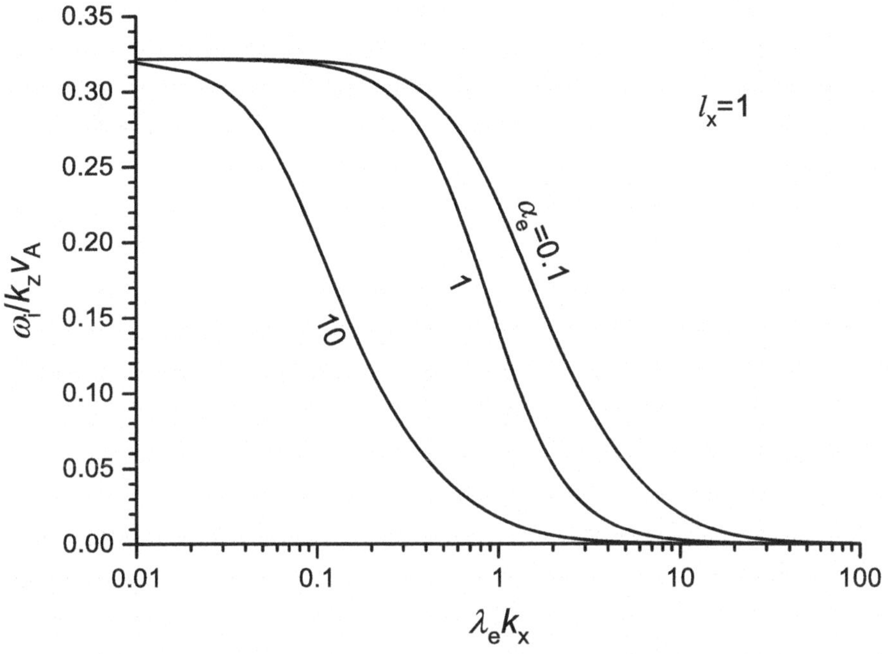

Fig. 1.10 The normalized growth rate $\omega_i/k_z v_A$ versus the normalized wave number $\lambda_e k_x$, where the solid lines represent the cases of $\alpha_e = 0.1$, 1, and 10, respectively, and the fixed parameter $l_x = 1$ has been used

represent the cases of $\lambda_e k_x = 0.1$, 0.3, and 1, respectively, and the fixed parameter $\alpha_e^2 = 10$ has been used.

Figure 1.11 shows that the growth rate approaches to zero both in the strong and weak inhomogeneity limit of $l_x \ll 1$ and $l_x \gg 1$, as mentioned above. In the case of the finite inhomogeneity of $l_x \sim 1$, the growth rate remarkably increases as the perpendicular wave number $\lambda_e k_x$ decreases.

Taking the inhomogeneity with periodically oscillating density striation

$$n(x) = n_0 + n_1 \sin(\kappa_0 x) \Rightarrow$$
$$l_x \equiv k_x \left(\frac{\partial \ln n(x)}{\partial x}\right)^{-1} = \frac{k_x}{\kappa_0} \frac{n_0/n_1 + \sin(\kappa_0 x)}{\cos(\kappa_0 x)}, \tag{1.71}$$

Chen et al. (2015) investigated the KAW instability by the density gradient and found that both the real frequency and the growth rate of the excited KAWs are evidently dependent on the spatial position x due to the presence of an inhomogeneous density gradient. The results also show that the real frequency increases with the spatial scale of inhomogeneity κ_0^{-1} for the fixed wave number k_x, while the growth rate has a maximum similar to that excited by the homogeneous density gradient above.

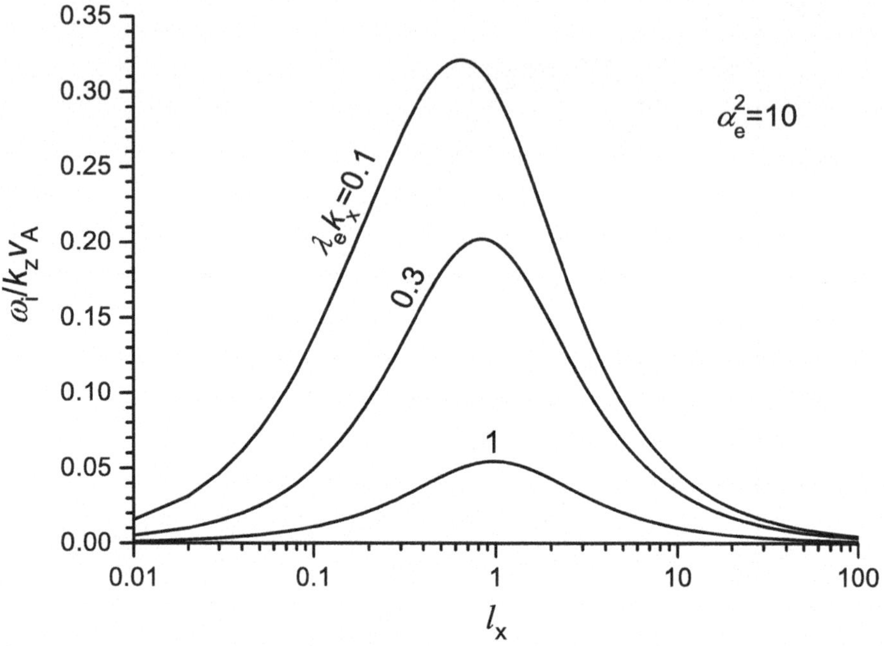

Fig. 1.11 The normalized growth rate $\omega_i/k_z v_A$ versus the inhomogeneity parameter l_x, where the solid lines represent the cases of $\lambda_e k_x = 0.1, 0.3$, and 1, respectively, and the fixed parameter $\alpha_e^2 = 10$ has been used

Recent particle-in-cell simulation by Tsiklauri (2011, 2012) showed that KAWs can be effectively excited by the transverse density inhomogeneity and that the parallel and perpendicular electric fields efficiently lead to the field-aligned acceleration of electrons and the cross-field heating of ions, respectively. The results by these simulation further confirmed the close relationship between the generation of KAWs and the formation of the transverse density inhomogeneity, that is, the field-aligned density striation.

In a weakly inhomogeneous plasma, however, the generation of KAWs by the instability above is not an effective mechanism. Other one more effective mechanism for generating KAWs is the resonant mode conversion of AWs, which was proposed by Hasegawa and Chen (1976a). When externally applied AWs at a fixed frequency ω propagate into an inhomogeneous plasma, the mode conversion occurs in the Alfvén resonant surface determined by $v_A(x_0) = \omega/k_\parallel$, where $v_A(x_0)$ is the local Alfvén velocity in the inhomogeneous plasma with a density gradient along x direction. After the resonant conversion, KAW with the wave vector $\mathbf{k} = k_y\hat{y} + k_\parallel\hat{z}$ can be generated in the higher density side (i.e., lower local Alfvén velocity) of the vicinity of the resonant surface when $v_{T_e} > v_A$ (i.e., $\alpha_e > 1$) in the resonant surface at $x_0 = 0$, where

$$v_A(x) = \frac{1}{\sqrt{1+k_\perp^2\rho_{is}^2}}\frac{\omega}{k_\parallel} < \frac{\omega}{k_\parallel} = v_A(x_0), \tag{1.72}$$

where $\rho_{is}^2 = (3/4 + T_e/T_i)\,\rho_i^2$. However, if $v_{T_e} < v_A$ (i.e., $\alpha_e < 1$) in the resonant surface at $x_0 = 0$, the generated KAW after the mode conversion propagates toward the lower density side (i.e., higher local Alfvén velocity), where

$$v_A(x) = \sqrt{1+k_\parallel^2\lambda_e^2}\frac{\omega}{k_\parallel} > \frac{\omega}{k_\parallel} = v_A(x_0). \tag{1.73}$$

The classic example of the resonant mode conversion is the Langmuir oscillation in an inhomogeneous plasma (Barston 1964). For sake of simplicity, assuming a cold plasma with a sharp density gradient:

$$n(x) = n_{01} + (n_{02} - n_{01})\,H(x), \tag{1.74}$$

where $H(x)$ is the Heaviside unit jump function, n_{01} and n_{02} ($> n_{01}$) are the plasma densities in the two sides of the plane $x = 0$. The wave equation of the electrostatic potential $\phi_0(x)\exp[-i(\omega t - k_y y)]$ can be written as

$$\nabla \cdot (\epsilon(x)\nabla\phi) = 0, \tag{1.75}$$

where

$$\epsilon(x) = 1 - \frac{\omega_{p1}^2 + \left(\omega_{p2}^2 - \omega_{p1}^2\right)H(x)}{\omega^2} \tag{1.76}$$

is the permittivity. One has the constant permittivity $\epsilon = 1 - \omega_{p1}^2/\omega^2$ for $x < 0$ and $\epsilon = 1 - \omega_{p2}^2/\omega^2$ for $x > 0$ and $\nabla\epsilon = 0$ for $x \neq 0$. Therefore, the wave equation (1.75) at $x \neq 0$ can be rewritten as

$$\epsilon(x)\nabla^2\phi = 0, \tag{1.77}$$

in which the factor $\epsilon(x) = 0$ gives the ordinary Langmuir oscillations $\omega = \omega_{p1}$ and $\omega = \omega_{p2}$ for $x < 0$ and $x > 0$, respectively, which are the intrinsic electrostatic oscillations of the bulk plasmas in the two sides of the discontinuous plane $x = 0$. The other factor, $\nabla^2\phi = 0$, leads to

$$\left(\partial_x^2 - k_y^2\right)\phi_0(x) = 0, \tag{1.78}$$

which nontrivial solution represents the surface wave that is constrained to propagate nearby the discontinuous surface $x = 0$ and has the form

$$\phi = \phi_0(0)\exp\left[-k_y|x| - i\left(\omega t - k_y y\right)\right], \tag{1.79}$$

where $k_y > 0$ has been assumed without loss of generality. The dispersion relation of the surface wave can be obtained by the boundary condition that ϕ and $\epsilon\partial_x\phi$ both are continuous at $x = 0$, and the result is

$$-\epsilon(x)|_{0^+} = \epsilon(x)|_{0^-} \Rightarrow \frac{\omega_{p2}^2}{\omega^2} - 1 = 1 - \frac{\omega_{p1}^2}{\omega^2} \Rightarrow \omega^2 = \frac{\omega_{p1}^2 + \omega_{p2}^2}{2}, \quad (1.80)$$

that is, the eigenfrequency of the surface wave $\omega_s = \sqrt{\left(\omega_{p1}^2 + \omega_{p2}^2\right)/2}$.

However, if the plasma density $n(x)$ is smoothly varying in x, the wave equation (1.75) can be reduced to

$$\partial_x\left[\epsilon(x)\partial_x\phi_0(x)\right] - k_y^2\epsilon(x)\phi_0(x) = 0, \quad (1.81)$$

where

$$\epsilon(x) = 1 - \frac{\omega_p^2(x)}{\omega^2} = 1 - \frac{\omega_{p0}^2}{\omega^2}g(x), \quad (1.82)$$

$g(x) \equiv n(x)/n_0$ is the normalized density profile and ω_{p0}^2 is the plasma frequency at the density n_0. Although the plasma density and hence $\epsilon(x)$ change continuously, Eq. (1.81) has a singularity at the point of $x = x_s$ when $\epsilon(x_s) = 0$, that is, $g(x_s) = \omega^2/\omega_{p0}^2$, implying that the local plasma frequency $\omega_{ps} = \sqrt{g(x_s)}\omega_{p0} = \omega \equiv \omega_s$ (the frequency of the applied wave). In fact, near $x = x_s$ Eq. (1.81) is dominated by the first term and its approximate solution can be given by (Barston 1964; Sedlácek 1971a, b)

$$\phi_0(x) = \ln|x - x_s| \pm i\pi H(x_s - x) \quad (1.83)$$

for a linear profile of $g(x)$ near $x = x_s$, where the sign of the imaginary part is determined by the causality relation. The plane $x = x_s$ is called the resonant surface and the approximate solution also represents the "surface wave", which is constrained to propagate near the resonant surface $x = x_s$. In particular, this surface wave solution has a logarithmic singularity at the resonant surface $x = x_s$ and its imaginary part implies the absorption of the wave, called the resonant absorption, resulted from the phase mixing of the continuous eigenmode in the continuous plasma medium (Baldwin and Ignat 1969; Golant and Piliya 1972).

In physics, however, this logarithmic singularity can be attributed to the "cold plasma" assumption that ignores the fact that the microscopic "thermal" oscillation of charged particles responding on the applied electrostatic surface wave occurs on a finite spatial scale, that is, the Debye length λ_D to be proportional to the "thermal velocity". Therefore, in the thin layer of $|x - x_s| < \lambda_D$ the approximate solution is meaningless, and this eliminates the logarithmic singularity in the approximate solution of Eq. (1.83). In fact, taking account of the effect of the finite Debye length (i.e., the "thermal" plasma with a finite temperature), the dispersive term, $k^2v_{T_e}^2$, in the dispersion relation for the Langmuir oscillation, $\omega^2 = \omega_p^2 + 3k^2v_{T_e}^2$, can lead to two orders higher derivatives in Eq. (1.81) or i.e. Eq. (1.75). In result, the wave equation (1.81) becomes fourth order equation, in which the two orders represent the surface wave with the dispersion relation of $\omega_s = \omega_{ps}$ and the other two orders do the Langmuir wave with the dispersion relation of $\omega^2 = \omega_p^2 + 3k^2 v_{T_e}^2$. While at the resonant surface $x = x_s$ the solution is no longer singular but represents the mode conversion of the surface wave to the Langmuir wave, implying that the so-called resonant absorption of the surface wave in the cold plasma approximation actually is the surface wave energy to be converted into the Langmuir wave and carried away from the resonant surface through the Langmuir wave propagating at the x direction (Nickel et al. 1963).

As shown by Hasegawa and Chen (1975), an analogical mode conversion can occur between a surface AW and KAW, in which the kinetic modified effect is caused by the finite ion gyroradius (ρ_i) instead of the Debye length (λ_D). The wave equation of the electric potential ϕ can be written as

$$0 = \left[M_z^2 \frac{3\rho_i^2}{4} \partial_x^3 + \partial_x^2 \left(\frac{\rho_s^2}{g(x)} \partial_x \right) \right] (g(x)\partial_x \phi) + \\ \left[\partial_x \left(M_z^2 g(x) - 1 \right) \partial_x - k_y^2 \left(M_z^2 g(x) - 1 \right) \right] \phi = 0, \tag{1.84}$$

where $M_z = \omega / k_z v_A$, v_A is the Alfvén velocity at $x = x_0$ where $g(x_0) = 1$, and $g(x)$ is the density profile. Ignoring the kinetic effects due to the finite ρ_i and ρ_s, the wave equation (1.84) has the same form as Eq. (1.81) with $\epsilon(x) = 1 - M_z^2 g(x)$ and hence describes the bulk AW and the surface AW at the resonant surface $x = x_s$ with $\epsilon(x_s) = 0$, where the solution of the wave equation has a logarithmic singularity and the applied AW with the resonant frequency $\omega_s = k_z v_A / \sqrt{g(x_s)}$ is resonantly absorbed.

In the case of a homogeneous plasma with $g(x) = 1$, however, Eq. (1.84) reduces to

$$\left(M_z^2 \frac{3\rho_i^2}{4} + \rho_s^2 \right) \partial_x^4 \phi + \left(M_z^2 - 1 \right) \partial_x^2 \phi - \left(M_z^2 - 1 \right) k_y^2 \phi = 0, \tag{1.85}$$

which describes the coupling between AW with the perpendicular wave number k_y and dispersion relation $M_z^2 = 1$ (represented by the second and third terms) and KAW with the perpendicular wave number $k_x \sim \partial_x$ and dispersion relation $M_z^2 \simeq 1 + \left(3\rho_i^2/4 + \rho_s^2 \right) k_x^2$ (represented by the first and second terms). Similarly, in an inhomogeneous plasma the resonant mode conversion of the surface AW with a frequency $\omega_s = k_z v_A / \sqrt{g(x_s)}$ to KAW can occur at the resonant surface $x = x_s$, where the singularity of the solution of the wave equation (1.84) of including the kinetic effect of the finite ion and ion-acoustic gyroradius disappears and the surface AW energy is converted resonantly into KAW and carried away from the resonant surface through the KAW propagating at the x direction.

Without loss of generality, taking $x_s = x_0 = 0$, that is, the $g(x) \simeq 1 + \kappa x$ near $x = 0$ and $\omega_s = k_z v_A$. Integrating near $x = 0$, the wave equation (1.84) can be reduced to

$$\bar{\rho}_{is}^2 \partial_x^2 E_x + \kappa x E_x = E_0, \tag{1.86}$$

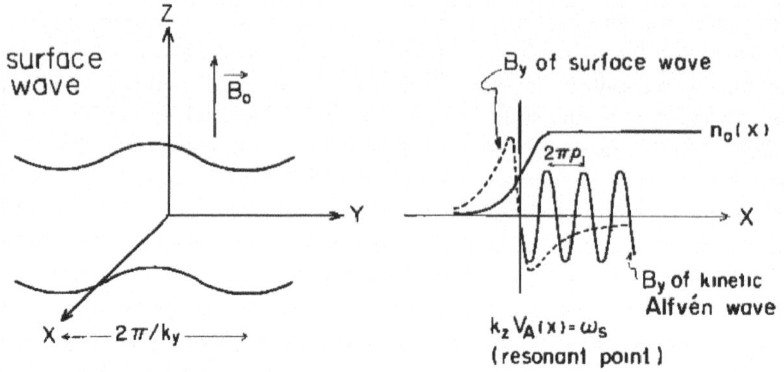

Fig. 1.12 A sketch of the resonant mode conversion of the surface AW (left) to KAW at the resonant surface in a density gradient plasma (right) (from Hasegawa 1976)

where $E_x = -\partial_x \phi$ is the x component of the wave electric field, E_0 is an integration constant representing the boundary condition, that is, the far-field value of E_x at the sufficiently large negative x ($x \sim -\kappa^{-1}$), and the feature kinetic length $\bar{\rho}_{is}$ is given by

$$\bar{\rho}_{is}^2 \equiv \frac{3}{4}\rho_i^2 + M_z^{-2}\rho_s^2. \tag{1.87}$$

In addition, the WKB approximation condition of $\kappa \ll |\partial_x \ln \phi| \sim |k_x|$ has been used in the derivation of Eq. (1.86). The solution of the wave equation (1.86) can be expressed in the terms of the Airy functions (Abramowitz and Stegun 1968). In particular, its asymptotic solution for $|x/\Delta| \gg 1$ can be given by

$$E_x = -\frac{\pi^{1/2} E_0}{(\kappa\bar{\rho}_{is})^{2/3}} \left(\frac{\Delta}{x}\right)^{1/4} H(x) \exp\left\{ i \left[\frac{2}{3}\left(\frac{x}{\Delta}\right)^{3/2} + \frac{\pi}{4} \right] \right\} + \frac{E_0}{\kappa x}, \tag{1.88}$$

where the jump function $H(x)$ indicates the KAW propagating towards the higher density region of $x > 0$ and the feature scale

$$\Delta \equiv \frac{\bar{\rho}_{is}}{(\kappa\bar{\rho}_{is})^{1/3}} = (\kappa\bar{\rho}_{is})^{2/3}\kappa^{-1}. \tag{1.89}$$

The term $E_0/\kappa x$ in the far-field asymptotic solution (1.88) represents the surface wave field and the oscillating term represents the generated KAW due to the resonant mode conversion. In particular, from Eq. (1.89) the mode conversion KAW has a feature perpendicular wavelength $\lambda_\perp \sim \Delta$ much larger than the kinetic scale $\bar{\rho}_{is}$ but much less than the inhomogeneity scale κ^{-1}. Figure 1.12 shows their qualitative features.

From Eq. (1.88) or Fig. 1.12, the first few peak amplitudes of the KAW are given by $E_0/(\kappa\bar{\rho})^{2/3} \gg E_0$, whereas away from the resonant surface, say at $x \sim 1/\kappa$, the amplitude becomes $E_0/\sqrt{\kappa\rho}$. In particular, it is worth noting that the amplitude of

KAW not only is much larger than that of the surface AW but also decreases much more slowly than the surface AW as x increases, that is, away from the resonant surface.

1.6 Generation of KAWs by Nonlinear Processes

Besides the temperature anisotropy, the field-aligned drift current, and the density inhomogeneity presented in the previous sections, there are other many free-energy sources, which can directly excite the growth of MHD large-scale AWs and kinetic small-scale KAWs through linear plasma instabilities. One of them is the Kelvin-Helmholtz instability when the presence of a velocity shear in plasma flows, which can excite AWs in the MHD scales. In kinetic scales, the small-scale modified version of the Kelvin-Helmholtz instability is often considered as an effective exciting source of KAWs in the magnetopause (Pu and Zhou 1986), in the polar cusp (Duan et al. 2005), in the magnetotail (Duan et al. 2012), and in the plasma sheet boundary layer (Tiwari et al. 2006, 2008). In this section, however, we focus our attention to another kind of generation mechanisms for KAWs, in which the driving sources are nonlinear wave-wave coupling processes.

Other kind of nonlinear generation mechanisms for KAWs is the parametric decay due to wave-wave coupling processes. The first proposed pump wave is the whistler wave, which can in parametric decay into a daughter whistler wave and a KAW (Chen 1977). Soon, using the two-fluid model and Vlasov and Maxwell equations, (Shukla and Mamedow 1978) studied the nonlinear decay of a lower hybrid pump wave (LHW), which can be generated by cross-field drift currents in space plasmas, into a KAW and a whistler wave which has a frequency below the LHW frequency in an intermediate-β plasma of $2Q \ll \beta \ll 1$. They proposed that, in the ionosphere, the pump LHW could excite the KAW and the whistler wave, and influence the dynamics of space plasmas.

Patel et al. (1985) studied the nonlinear process of the parametric decay of a large amplitude electrostatic ion cyclotron wave (ESICW) into a KAW and a daughter ESICW in an inhomogeneous plasma. The threshold value for the parametric decay process is determined by the inhomogeneity scale of the plasma density or the magnetic field. In the magnetosphere, the energy density of the pump ESICW of a few percent of the thermal energy is enough to drive this nonlinear instability, which is often less than observed ion electrostatic turbulence field (Anderson and Gurnett 1973; Hudson et al. 1978). So in most parts of the magnetosphere, the observed turbulent electric field can well satisfy the required threshold value of driving the nonlinear decay process, which probably may be applied to low-frequency pulsations from Pc 2 to Pc 5 in the magnetosphere.

In a low-β magnetized plasma of $\beta < 2Q$, an upper hybrid wave (UHW) is proposed as a pump wave that decays into a daughter UHW and a KAW due to nonlinear wave-wave interaction process and causes the spreading of the UHW spectrum. When the decay process takes place, synchronism conditions for the effective wave-wave interaction should hold

$$\omega_1 = \omega_2 + \omega_3, \quad \boldsymbol{k}_1 = \boldsymbol{k}_2 + \boldsymbol{k}_3, \tag{1.90}$$

where ω and $\boldsymbol{k}$ are the wave frequency and vector, respectively; the subscripts 1, 2, and 3 represent the pump UHW, another UHW, and KAW. Based on the two-fluid equations, the nonlinear growth rate of KAWs in this decay process can be obtained as follows (Yukhimuk et al. 1998):

$$\gamma = \frac{1}{2\lambda_e}\sqrt{\frac{\mathcal{E}_E}{2}\frac{T_i + T_e}{m_e}\frac{\omega_{ce}}{\omega_{pe}}}\bar{\gamma}(\theta_1, \theta_2), \tag{1.91}$$

where $\mathcal{E}_E = (\epsilon_0/2)E_0^2/(n_0T_e)$ is the wave energy density of the pump UHW in the unit of the thermal energy density of electrons n_0T_e, $\bar{\gamma}(\theta_1, \theta_2)$ is an angle-correlation factor with order of the unit, and θ_1 and θ_2 are the propagating angles of the pump and daughter UHWs with respect to the direction of the ambient magnetic field, respectively. Taking typical parameters for the lower magnetosphere as: $T_e \sim T_i \sim$ 1 eV, $\omega_{pe} \sim 10\omega_{ce} \sim 10^6$–$10^7$ s^{-1}, $\omega_{ci} \sim 10^2$–10^3 s^{-1}, $n_0 \sim 10$ cm^{-3}, and $\mathcal{E}_E \sim 10^{-5}$, one has the growth rate by Eq. (1.78) $\gamma \sim 1$–10 s^{-1}.

In addition, many other wave modes, such as ion Bernstein waves (Sharma and Tripathi 1988), high-frequency electromagnetic waves (Papadopoulos et al. 1982; Tripathi and Sharma 1988), Langmuir waves (Voitenko et al. 2003), and low-frequency MHD waves (Voitenko and Goossens 2002, 2005a, b; Fedun et al. 2004; Zhao et al. 2011), all had been proposed to act as the pump wave in parametric to excite KAWs.

When waves travel, they interact with the ambient medium they are traveling and the interaction ultimately leads to dissipation of waves. Large-scale AWs traveling in a homogeneous plasma can not exchange efficiently energy directly with the medium because of the mismatch between wave scales and microscopic kinetic scales of particles in the local medium. Nonlinear wave-wave interaction, however, can cause the energy in large-scale AWs cascades towards small-scale waves in a manner that ultimately leads to the wave energy dissipation due to the wave-particle interaction when the scales of small-scale waves match to the kinetic scales of particles in the local environment.

In fact, KAWs are dissipative as well as dispersive AWs because their perpendicular wavelengths have scales comparable to the microscopic kinetic scales of particles, such as the ion gyroradius and the electron inertial length. The turbulent cascade process of AWs leads to the energy transferring from large-scale AWs to small-scale KAWs, and then the wave-particle interaction of KAWs leads to the wave energy effectively transferring from the waves into plasma particles since their scales match the kinetic scales of particles. Thus, on the one hand KAWs dissipate the AW energy through the turbulent cascade process and on the other hand they energize plasma particles due to the wave-particle interaction. In other words, KAWs can play an important role of a bridge in transferring the energy of AWs to the kinetic energy of plasma particles. So, another important mechanism for the generation of KAWs is the anisotropic cascade process toward small-scale waves in AW turbulence.

Turbulent AWs exist ubiquitously in space and astrophysical plasmas. Transferring energy from large scales to small scales is very important process for the energization of plasma particles as well as the dissipation of the wave energy. This leads to the dynamic spectrum of turbulent AWs in the solar wind and other interplanetary and interstellar media. In a series study, Goldreich and Sridhar have established theories of MHD turbulence in a homogeneous magneto-plasma (Sridhar and Goldreich 1994; Goldreich and Sridhar 1995, 1997), now called the Goldreich-Sridhar theory. One of the most important conclusions of the Goldreich-Sridhar theory is to predict an anisotropic cascade process of the AWs energy in the inertial regime of the turbulent spectrum that develops primarily small scales perpendicular to the background magnetic field, that is, $k_\perp \gg k_\parallel$ in small-scale regimes. This anisotropy in small scales is one of the typical characteristics of KAWs. In other words, the Goldreich-Sridhar theory predicts that the AW's turbulence has an intense tendency that cascades the wave energy towards KAWs.

Numerical simulations of magnetized turbulence (Maron and Goldreich 2001; Cho et al. 2002) also support the idea that such a turbulence is strongly anisotropic. Meanwhile, in situ measurements of turbulence in the solar wind (Belcher and Davis 1971; Matthaeus et al. 1990; Luo and Wu 2010; Luo et al. 2011) and observations of interstellar scintillation (Wilkinson et al. 1994; Trotter et al. 1998; Rickett et al. 2002; Dennett-Thorpe and de Bruyn 2003) provide evidence for this significant anisotropy.

In the most recent work, Voitenko and De Keyser (2011) identified and investigated a weakly dispersive range (WDR) of KAW turbulence in the MHD-kinetic turbulence transition. They found perpendicular wavenumber spectra $\propto k_\perp^{-3}$ and $\propto k_\perp^{-4}$ formed in WDR by strong and weak turbulence of KAWs, respectively. These steep WDR spectra connect shallower spectra in the MHD and strongly dispersive KAW ranges, which results in a specific double-kink (2-k) pattern often seen in observed turbulent spectra. The first kink occurs where MHD turbulence transforms into weakly dispersive KAW turbulence; the second one is between weakly and strongly dispersive KAW ranges. Their analysis suggests that partial turbulence dissipation due to amplitude-dependent non-adiabatic ion heating may occur in the vicinity of the first spectral kink. The threshold-like nature of this process results in a conditional selective dissipation that affects only the largest over-threshold amplitudes and that decreases the intermittency in the range below the first spectral kink. Several recent counter-intuitive observational findings can be explained by the coupling between such a selective dissipation and the nonlinear interaction among weakly dispersive KAWs. In consequence, as least when approaching to the dissipation regime of wave spectrums, the ultimate waves on the small-scales are expected to be highly anisotropic in the way of quasi-perpendicular propagation with $k_\perp^2/k_\parallel^2 \gg 1$.

References

Abramowitz, M., & Stegun, I. A. (1968). *Handbook of mathematical functions with formulas, graphs and mathematical tables*. New York: Dover Publications.

Alfvén, H. (1942). Existence of electromagnetic-hydrodynamic waves. *Nature*, *150*, 405–406.

Anderson, R. R., & Gurnett, D. A. (1973). Plasma wave observations near the plasmapause with the S3-A satellite. *Journal of Geophysical Research*, *78*, 4756–4764.

Baldwin, D. E., & Ignat, D. W. (1969). Resonant absorption in zero-temperature nonuniform plasma. *Physical Fluids*, *12*, 697.

Banks, P. M. (1967). Ion temperature in the upper atmosphere. *Journal of Geophysical Research*, *72*, 3365–3385.

Barston, E. M. (1964). Electrostatic oscillations in inhomogeneous cold plasmas. *Annals of Physics*, *29*, 282–303.

Baumjohann, W., Paschmann, G., & Cattell, C. A. (1989). Average plasma properties in the central plasma sheet. *Journal of Geophysical Research*, *94*, 6597–6606.

Belcher, J. W., & Davis, L. (1971). Large-amplitude Alfvén wave in the interplanetary medium. *Journal of Geophysical Research*, *76*, 3534–3563.

Bellan, P. M., & Stasiewicz, K. (1998). Fine-scale cavitation of ionospheric plasma caused by inertial Alfvén wave ponderomoive force. *Physical Review Letters*, *80*, 3523–3526.

Bespalov, P. A., Misonova, V. G., & Cowley, S. W. (2006). Field-aligned particle acceleration on auroral field lines by interaction with transient density cavities stimulated by kinetic Alfvén waves. *Annales Geophysicae*, *24*, 2313–2329.

Bilitza, D. (1991). The use of transition heights for the representation of ion composition. *Advances in Space Research*, *11*, 183–186.

Boldyrev, S., & Perez, J. C. (2012). Spectrum of kinetic-Alfvn turbulence. *The Astrophysical Journal Letters*, *758*, L44.

Bostick, W. H., & Levine, M. (1952). Experimental demonstration in the laboratory of the existence of magneto-hydrodynamic waves in ionized helium. *Physical Review*, *94*, 671–671.

Brodin, G., Stenflo, L., & Shukla, P. K. (2006). Nonlinear interactions between kinetic Alfvén and ion-sound waves. *Solar Physics*, *236*, 285–291.

Burke, A. T., Maggs, J. E., & Morales, G. J. (2000a). Spontaneous fluctuations of a temperature filament in a magnetized plasma. *Physical Review Letters*, *84*, 1451–1454.

Burke, A. T., Maggs, J. E., & Morales, G. J. (2000b). Experimental study of fluctuations excited by a narrow temperature filament in a magnetized plasma. *Physical Plasmas*, *7*, 1397–1407.

Chaston, C. C., Carlson, C. W., McFadden, J. P., et al. (2007a). How important are dispersive Alfvén waves for auroral electron acceleration? *Geophysical Research Letters*, *34*, L07101.

Chaston, C. C., Hull, A. J., Bonnell, J. W., et al. (2007b). Large parallel electric fields, currents, and density cavities in dispersive Alfvén waves above the aurora. *Journal of Geophysical Research*, *112*, A05215.

Chaston, C. C., Wilber, M., Mozer, F. S., et al. (2007c). Mode conversion and anomalous transport in kelvin-helmholtz vortices and kinetic Alfvén waves at the earth's magnetopause. *Physical Review Letters*, *99*, 175004.

Chaston, C. C., Carlson, C. W., Peria, W. J., et al. (1999). FAST observations of inertial Alfvén waves in the dayside aurora. *Geophysical Research Letters*, *26*, 647–650.

Chen, L. (1977). Parametric excitation of kinetic Alfvén waves by whistler waves. *Plasma Physics*, *19*, 47–51.

Chen, L., & Hasegawa, A. (1974a). Plasma heating by spatial resonance of Alfvén wave. *Physical Fluids*, *17*, 1399–1403.

Chen, L., & Hasegawa, A. (1974b). A theory of long-period magnetic pulsations, 1. Steady state excitation of filed line resonance. *Journal of Geophysical Research*, *79*, 1024–1032.

Chen, L., & Wu, D. J. (2010). Kinetic Alfvén wave instability driven by electron temperature anisotropy in high-β plasmas. *Physics of Plasmas*, *17*, 062107.

Chen, L., & Wu, D. J. (2011a). Exact solutions of dispersion equation for MHD waves with short-wavelength modification. *Chinese Science Bulletin, 56*, 955–961.
Chen, L., & Wu, D. J. (2011b). Polarizations of coupling kinetic Alfvén and slow waves. *Physics of Plasmas, 18*, 072110.
Chen, L., & Wu, D. J. (2012). Kinetic Alfvén wave instability driven by field-aligned currents in solar coronal loops. *The Astrophysical Journal, 754*, 123.
Chen, L., Wu, D. J., & Hua, Y. P. (2011). Kinetic Alfvén wave instability driven by a field-aligned current in high-β plasmas. *Physical Review E, 84*, 046406.
Chen, L., Wu, D. J., & Huang, J. (2013). Kinetic Alfvén wave instability driven by field-aligned currents in a low-β plasma. *Journal of Geophysical Research, 118*, 2951.
Chen, L., Wu, D. J., Zhao, G. Q., et al. (2015). A possible mechanism for the formation of filamentous structures in magnetoplasmas by kinetic Alfvén waves. *Journal of Geophysical Research, 120*, 61–69.
Chew, G. L., Goldberger, M. L., & Low, F. E. (1956). The Boltzmann equation and the one-fluid hydromagnetic equations in the absence of particle collisions. *Proceedings of the Royal Society of London A, 236*, 112–118.
Chmyrev, V. M., Bilichenko, S. V., Pokhotelov, O. A., et al. (1988). Alfvén vortices and related phenomena in the ionosphere and the magnetosphere. *Search Results, 38*, 841–854.
Cho, J., Lazarian, A., & Vishniac, E. T. (2002). Simulation of magnetohydrodynamic turbulence in a strong magnetized medium. *The Astrophysical Journal, 564*, 291–301.
Dennett-Thorpe, J., & de Bruyn, A. G. (2003). Annual modulation in the scattering of J1819+3845: Peculiar plasma velocity and anisotropy. *Astronomy and Astrophysics, 404*, 113–132.
Duan, S. P., Li, Z. Y., & Liu, Z. X. (2005). Kinetic Alfvén wave driven by the density inhomogeneity in the presence of loss-cone distribution function-particle aspect analysis. *Planetary and Space Science, 53*, 1167–1173.
Duan, S. P., Liu, Z. X., & Angelopoulos, V. (2012). Observations of kinetic Alfvén waves by THEMIS near a substorm onset. *Chinese Science Bulletin, 57*, 1429–1435.
Eastman, T. E., Frank, L. A., Peterson, W. K., & Lennartsson, W. (1984). The plasma sheet boundary layer. *Journal of Geophysical Research, 89*, 1553–1572.
Fedun, V. N., Yukhimuk, A. K., & Voitsekhovskaya, A. D. (2004). The transformation of MHD Alfvén waves in space plasma. *Journal of Plasma Physics, 70*, 699–707.
Frank, L. A., Ackerson, K. L., & Lepping, R. P. (1976). On hot tenuous plasmas, fireballs, and boundary layers in the earth's magnetotail. *Journal of Geophysical Research, 81*, 5859–5881.
Gary, S. P., & Smith, C. W. (2009). Short-wavelength turbulence in the solar wind: linear theory of whistler and kinetic Alfvén fluctuations. *Journal of Geophysical Research, 114*, A12105.
Gekelman, W. (1999). Review of laboratory experiments on Alfvén waves and their relationship to space observations. *Journal of Geophysical Research, 104*, 14417–14435.
Gekelman, W., Leneman, D., Maggs, J. E., & Vincena, S. (1994). Experimental observation of Alfvén cones. *Physics of Plasmas, 1*, 3775–3783.
Gekelman, W., Vincena, S., Leneman, D., & Maggs, J. (1997). Laboratory experiments on shear Alfvén waves and their relationship to space plasmas. *Physics of Plasmas, 102*, 7225–7236.
Gekelman, W., Vincena, S., Palmer, N., et al. (2000). Experimental measurements of the propagation of large amplitude shear Alfvén waves. *Plasma Physics and Controlled Fusion, 42*, B15–B26.
Goertz, C. K. (1984). Kinetic Alfvén waves on auroral field lines. *Planetary and Space Science, 32*, 1387–1392.
Goertz, C. K., & Boswell, R. W. (1979). Magnetosphere-ionosphere coupling. *Journal of Geophysical Research, 84*, 7239–7246.
Golant, V. E., & Piliya, A. D. (1972). Reviews of topical problems: linear transformation and absorption of waves in a plasma, soviet phys. *Uspekhi, 14*, 413–437.
Goldreich, P., & Sridhar, S. (1995). Towards a theory of interstellar turbulence. *II. strong Alfvénic turbulence. Astrophys. J., 438*, 763–775.
Goldreich, P., & Sridhar, S. (1997). Magnetohydrodynamic turbulence revisited. *Astrophys. J., 485*, 680–688.

Golub, L., Herant, M., Kalata, K., et al. (1990). Sub-arcsecond observations of the solar X-ray corona. *Nature*, *344*, 842–844.
Hasegawa, A. (1976). Particle acceleration by MHD surface wave and formation of aurora. *Journal of Geophysical Research*, *81*, 5083–5090.
Hasegawa, A., & Chen, L. (1975). Kinetic process of plasma heating due to Alfvén wave excitation. *Physical Review Letters*, *35*, 370–373.
Hasegawa, A., & Chen, L. (1976a). Kinetic process of plasma heating by resonant mode conversion of Alfvén wave. *Physics of Fluids*, *19*, 1924–1934.
Hasegawa, A., & Mima, K. (1978). Anomalous transport produce by kinetic Alfvén wave turbulence. *Journal of Geophysical Research*, *83*, 1117–1123.
Herlofson, N. (1950). Magnetohydrodynamic waves in a compressible fluid conductor. *Nature*, *165*, 1020–1021.
Hollweg, J. V. (1999). Kinetic Alfvén waves revisited. *Journal of Geophysical Research*, *104*, 14811–14819.
Huang, G. L., Wang, D. Y., Wu, D. J., et al. (1997). The eigenmode of solitary kinetic Alfvén waves by Freja satellite. *Journal of Geophysical Research*, *102*, 7217–7224.
Hudson, M. K., Lysak, R. L., & Mozer, F. S. (1978). Magnetic field-aligned potential drops due to electrostatic ion cyclotron turbulence. *Geophysical Research Letters*, *5*, 143–146.
Johnson, J. R., & Cheng, C. Z. (1997). Kinetic Alfvén waves and plasma transport at the magnetopause. *Geophysical Research Letters*, *24*, 1423–1426.
Johnson, J. R., & Cheng, C. Z. (2001). Stochastic ion heating at the magnetopause due to kinetic Alfvén waves. *Geophysical Research Letters*, *28*, 4421–4424.
Johnson, J. R., Cheng, C. Z., & Song, P. (2001). Signatures of mode conversion and kinetic Alfvén waves at the magnetopause. *Geophysical Research Letters*, *28*, 227–230.
Kivelson, M. G., & Southwood, D. J. (1986). Coupling of global magnetospheric MHD eigenmodes to field line resonance. *Journal of Geophysical Research*, *91*, 4345–4351.
Kletzing, C. A. (1994). Electron acceleration by kinetic Alfvén waves. *Journal of Geophysical Research*, *99*, 11095–11103.
Kletzing, C. A., Bounds, S. R., Martin-Hiner, J., et al. (2003a). Measurements of the shear Alfvén wave dispersion for finite perpendicular wave number. *Physical Review Letters*, *90*, 035004.
Kletzing, C. A., Mozer, F. S., & Torbert, R. B. (1998). Electron temperature and density at high latitude. *Journal of Geophysical Research*, *103*, 14837–14845.
Kletzing, C. A., Scudder, J. D., Dors, E. E., & Curto, C. (2003b). Auroral source region: Plasma properties of the high-latitude plasma sheet. *Journal of Geophysical Research*, *108*, 1360–1375.
Leamon, R. J., Smith, C. W., Ness, N. F., et al. (1999). Dissipation range dynamics: Kinetic Alfvén waves and the importance of β_e. *Journal of Geophysical Research*, *104*, 22331–22344.
Lee, L. C., Johnson, J. R., & Ma, Z. W. (1994). Kinetic Alfvén waves as a source of plasma transport at the dayside magnetopause. *Journal of Geophysical Research*, *99*, 17405–17411.
Leneman, D., Gekelman, W., & Maggs, J. E. (1999). Laboratory observations of shear Alfvén waves launched from a small source. *Physical Review Letters*, *82*, 2673–2676.
Louarn, P., Wahlund, J. E., Chust, T., et al. (1994). Observations of kinetic Alfvén waves by the Freja spacecraft. *Physical Review Letters*, *21*, 1847–1850.
Luo, Q. Y., & Wu, D. J. (2010). Observations of anisotropic scaling of solar wind turbulence. *The Astrophysical Journal Letters*, *714*, L138.
Luo, Q. Y., Wu, D. J., & Yang, L. (2011). Measurement of intermittency of anisotropic magnetohydrodynamic turbulence in high-speed solar wind. *The Astrophysical Journal Letters*, *733*, L22.
Lysak, R. L., & Carlson, C. W. (1981). Effect of microscopic turbulence on magnetosphere-ionosphere coupling. *Geophys. Res. Lett.*, *8*, 269–272.
Lysak, R. L., & Dum, C. T. (1983). Dynamics of magnetosphere-ionosphere coupling including turbulent transport. *Journal of Geophysical Research*, *88*, 365–380.
Lysak, R. L., & Lotko, W. (1996). On the kinetic dispersion relation for shear Alfvén waves. *Journal of Geophysical Research*, *101*, 5085–5094.

Maggs, J. E., & Morales, G. J. (1996). Magnetic fluctuations associated field-aligned striations. *Geophysical Research Letters, 23*, 633–636.

Maggs, J. E., Morales, G. J., & Gekelman, W. (1997). Laboratory studies of field-aligned density striations and their relationship to auroral processes. *Physics of Plasmas, 4*, 1881–1888.

Malik, M., & Sharma, R. (2005). Nonlinear evolution of kinetic Alfvén waves and filament formation. *Solar Physics, 229*, 287–304.

Malik, M., Sharma, R. P., & Singh, H. D. (2007). Ion-acoustic wave generation by two kinetic Alfvén waves and particle heating. *Solar Physics, 241*, 317–328.

Maron, J., & Goldreich, P. (2001). Simulations of incompressible magnetohydrodynamic turbulence. *The Astrophysical Journal, 554*, 1175–1196.

Matthaeus, W. H., Goldstein, M. L., & Roberts, D. A. (1990). Evidence for the presence of quasi-two-dimensional nearly incompressible fluctuations in the solar wind. *Journal of Geophysical Research, 95*, 20673–20683.

Mitchell, C., Vincena, S., Maggs, J. E., & Gekelman, W. (2001). Laboratory observation of Alfvén resonance. *Geophysical Research Letters, 28*, 923–926.

Nickel, J. C., Parker, J. V., & Gould, R. W. (1963). Resonance Oscillations in a Hot Nonuniform Plasma Column. *Physical Review Letters, 11*, 183.

Papadopoulos, K., Sharma, R. R., & Tripathi, V. K. (1982). Parametric excitation of Alfvén waves in the ionosphere. *Journal of Geophysical Research, 87*, 1491–1494.

Parker, E. N. (1958). Dynamic instability of an anisotropic ionized gas of low density. *Physical Review, 109*, 1874–1876.

Patel, V. L., Tripathi, V. K., & Sharma, O. P. (1985). Parametric excitation of shear Alfvén modes by electrostatic ion cyclotron waves in the magnetosphere. *Journal of Geophysical Research, 90*, 9590–9594.

Podesta, J. J. (2013). Evidence of Kinetic Alfvén waves in the solar wind at 1 AU. *Solar Physics, 286*, 529–548.

Podesta, J. J., & TenBarge, J. M. (2012). Scale dependence of the variance anisotropy near the proton gyroradius scale: Additional evidence for kinetic Alfvén waves in the solar wind at 1 AU. *Journal of Geophysical Research, 117*, A10106.

Podesta, J. J., Borovsky, J. E., & Gary, S. P. (2010). A Kinetic Alfvén wave cascade subject to collisionless damping cannot reach electron scales in the solar wind at 1 AU. *The Astrophysical Journal, 712*, 685–691.

Pu, Z. Y., & Zhou, Y. (1986). The kinetic Alfvén wave instability driven by a sheared plasma flow and the associated anomalous transport. *Scientia Sinica A, 29*, 301–311.

Rickett, B. J., Kedziora-Chudczer, L., & Jauncey, D. L. (2002). Interstellar scintillation of the polarized flux density in quasar PKS 0405–385. *The Astrophysical Journal, 581*, 103–126.

Rosenbluth, M. N. (1956). Stability of the pinch, *Los Alamos Science Lab*, LA-2030.

Rudakov, L., Mithaiwala, M., Ganguli, G., et al. (2011). Linear and nonlinear landau resonance of kinetic Alfvén waves: Consequences for electron distribution and wave spectrum in the solar wind. *Physics of Plasmas, 18*, 012307.

Sahraoui, F., Goldstein, M. L., Belmont, G., et al. (2010). Three dimensional anisotropic k-spectra of turbulence at sub-proton scales in the solar wind. *Physical Review Letters, 105*, 131101.

Salem, C. S., Howes, G. G., Sundkvist, D., et al. (2012). Identification of kinetic Alfvén wave turbulence in the solar wind. *The Astrophysical Journal Letters, 745*, L9.

Schriver, D., Ashour-Abdalla, M., & Richard, R. L. (1998). On the origin of the ion-electron temperature difference in the plasma sheet. *Journal of Geophysical Research, 103*, 14879–14895.

Sedlácek, Z. (1971a). Electrostatic oscillations in cold inhomogeneous plasma I: Differential equation approach. *Journal of Plasma Physics, 5*, 239–263.

Sedlácek, Z. (1971b). Electrostatic oscillations in cold inhomogeneous plasma Part 2: Integral equation approach. *Journal of Plasma Physics, 6*, 189–199.

Sharma, A., & Tripathi, V. K. (1988). Excitation of kinetic Alfvén waves in the ion-Bernstein wave heating of a plasma. *Physics of Fluids, 31*, 3697–3701.

Sharma, R. P., & Malik, M. (2006). Nonlinear interaction of the kinetic Alfvén waves and the filamentation process in the solar wind plasma. *Astronomy and Astrophysics, 457*, 675–680.

Sharma, R. P., & Singh, H. D. (2009). Density cavities associated with inertial Alfvén waves in the auroral plasma. *Journal of Geophysical Research, 114*, A03109.

Shukla, P. K., & Mamedow, M. A. (1978). Nonlinear decay of a propagating lower-hybrid wave in a plasma. *Journal of Plasma Physics, 19*, 87–96.

Singh, H. D. (2007). Interpretation of solar wind reconnection exhaust in terms of kinetic Alfveń wave group-velocity cones. *Geophysical Research Letters, 34*, L13106.

Singh, H. D., & Sharma, R. P. (2007). Generation of coherent wave packets of kinetic Alfvén waves in solar plasmas. *Physics of Plasmas, 14*, 102304.

Slavin, J. A., Smith, E. J., Sibeck, D. G., et al. (1985). An ISEE 3 study of average and substorm conditions in the distant magnetotail. *Journal of Geophysical Research, 90*, 10875–10895.

Smith, C. W., Vasquez, B. J., & Hollweg, J. V. (2012). Observational constraints on the role of cyclotron damping and kinetic Alfvén waves in the solar wind. *The Astrophysical Journal, 745*, 8.

Sridhar, S., & Goldreich, P. (1994). Towards a theory of interstellar turbulence I. weak Alfvénic turbulence. *The Astrophysical Journal, 432*, 612–621.

Stasiewicz, K. (2006). Heating of the solar corona by dissipative Alfvén solitons. *Physical Review Letters, 96*, 175003.

Stasiewicz, K., Bellan, P., Chaston, C., et al. (2000a). Small scale Alfvénic structure in the aurora. *Space Science Reviews, 92*, 423–533.

Stasiewicz, K., Gustafsson, G., Marklund, G., et al. (1997). Cavity resonators and Alfvén resonance cones observed on Freja. *Journal of Geophysical Research, 102*, 2565–2575.

Stasiewicz, K., Holmgren, G., & Zanetti, L. (1998). Density depletions and current singularities observed by Freja. *Journal of Geophysical Research, 103*, 4251–4260.

Stasiewicz, K., Khotyaintsev, Y., Berthomier, M., & Wahlund, J. E. (2000b). Identification of widespread turbulence of dispersive Alfvén waves. *Geophysical Research Letters, 27*, 173–176.

Stefant, R. J. (1970). Alfvén wave damping from finite gyroradius coupling to the ion acoustic mode. *Physics of Fluids, 13*, 440–450.

Streltsov, A., & Lotko, W. (1995). Dispersive field line resonances on auroral field lines. *Journal of Geophysical Research, 100*, 19457–19472.

Takahashi, K., & Hones, E. W, Jr. (1988). ISEE 1 and 2 observations of ion distributions at the plasma sheet-tail lobe boundary. *Journal of Geophysical Research, 93*, 8558–8582.

Tiwari, B. V., Mishra, R., Varma, P., & Tiwari, M. S. (2006). Generation of kinetic Alfvén wave by velocity shear in the plasma sheet boundary layer during substorm. *Indian Journal of Pure & Applied Physics, 44*, 917–926.

Tiwari, B. V., Mishra, R., Varma, P., & Tiwari, M. S. (2008). Shear driven kinetic Alfvén wave with general loss-cone distribution function in the plasma sheet boundary layer. *Earth Moon Planet, 103*, 43–63.

Tripathi, Y. K., & Sharma, R. P. (1988). Some parametric instabilities of an ordinary electromagnetic wave in magnetized plasmas. *Physical Review A, 38*, 2991–2995.

Trotter, A. S., Moran, J. M., & Rodriguez, L. F. (1998). Anistropic radio scattering of NGC 6334B. *The Astrophysical Journal, 493*, 666–679.

Tsiklauri, D. (2011). Particle acceleration by circularly and elliptically polarised dispersive Alfvén waves in a transversely inhomogeneous plasma in the inertial and kinetic regimes. *Physics of Plasmas, 18*, 092903.

Tsiklauri, D. (2012). Three dimensional particle-in-cell simulation of particle acceleration by circularly polarised inertial Alfvén waves in a transversely inhomogeneous plasma. *Physics of Plasmas, 19*, 082903.

Varma, P., Mishra, S. P., Ahirwar, G., & Tiwari, M. S. (2007). Effect of parallel electric field on Alfvén wave in thermal magnetoplasma. *Planetary and Space Science, 55*, 174–180.

Vincena, S., Gekelman, W., & Maggs, J. E. (2004). Shear Alfvén wave perpendicular propagation from the kinetic to the inertial regime. *Physical Review Letters, 93*, 105003.

Voitenko, Y. (1998). Excitation of kinetic Alfvén waves in a flaring loop. *Solar Physics*, *182*, 411–430.

Voitenko, Y., & De Keyser, J. (2011). Turbulent spectra and spectral kinks in the transition range from MHD to kinetic Alfvén turbulence. *Nonlinear Processes in Geophysics*, *18*, 587–597.

Voitenko, Y., & Goossens, M. (2002). Nonlinear excitation of small-scale Alfvén waves by fast waves and plasma heating in the solar atmosphere. *Solar Physics*, *209*, 37–60.

Voitenko, Y., & Goossens, M. (2004). Cross-field heating of coronal ions by low-frequency kinetic Alfvén waves. *The Astrophysical Journal*, *605*, L149–L152.

Voitenko, Y., & Goossens, M. (2005a). Cross-scale nonlinear coupling and plasma energization by Alfvén waves. *Physical Review Letters*, *94*, 135003.

Voitenko, Y. & Goossens, M. (2005b). Nonlinear coupling of Alfvén waves with widely different cross-field wavelengths in space plasmas. *Journal of Geophysical Research*, 110 A10S01.

Voitenko, Y., & Goossens, M. (2006). Energization of plasma species by intermittent kinetic Alfvén waves. *Space Science Reviews*, *122*, 255–270.

Voitenko, Y., Goossens, M., Sirenko, O., & Chian, A. (2003). Nonlinear excitation of kinetic Alfvén waves and whistler waves by electron beam-driven Langmuir waves in the solar corona. *Astronomy & Astrophysics*, *409*, 331–345.

Volwerk, M., Louarn, P., Chust, T., et al. (1996). Solitary kinetic Alfvén waves-A study of the Poynting flux. *Journal of Geophysical Research*, *101*, 13335–13343.

Wahlund, J. E., Louarn, P., Chust, T., et al. (1994a). On ion-acoustic turbulence and the nonlinear evolution of kinetic Alfvén waves in aurora. *Geophysical Research Letters*, *21*, 1831–1834.

Wang, X. G., Ren, L. W., Wang, J. Q., & Xiao, C. J. (2009). Synthetic solar coronal heating on current sheets. *The Astrophysical Journal*, *694*, 1595–1601.

Wilkinson, P. N., Narayan, R., & Spencer, R. E. (1994). The scatter-broadened image of Cygnus X-3. *Monthly Notices of the Royal Astronomical Society*, *269*, 67–88.

Wu, D. J. (2003a). Effects of ion temperature and inertia on kinetic Alfvén waves. *Communications in Theoretical Physics*, *39*, 457–464.

Wu, D. J. (2003b). Model of nonlinear kinetic Alfvén waves with dissipation and acceleration of energetic electrons. *Physical Review E*, *67*, 027402.

Wu, D. J. (2003c). Dissipative solitary kinetic Alfvén waves and electron acceleration. *Physics of Plasmas*, *10*, 1364–1370.

Wu, D. J. (2005). Dissipative solitary kinetic Alfvén waves and electron acceleration in the solar corona. *Space Science Reviews*, *121*, 333–342.

Wu, D. J. (2012). *Kinetic Alfvén Wave: Theory*. Experiment and Application: Science Press, Beijing.

Wu, D. J., & Chao, J. K. (2003). Auroral electron acceleration by dissipative solitary kinetic Alfvén waves. *Physics of Plasmas*, *10*, 3787–3789.

Wu, D. J., & Chao, J. K. (2004a). Model of auroral electron acceleration by dissipative solitary kinetic Alfvén wave. *Journal of Geophysical Research*, *109*, A06211.

Wu, D. J., & Chao, J. K. (2004b). Recent progress in nonlinear kinetic Alfvén waves. *Nonlinear Processes in Geophysics*, *11*, 631–645.

Wu, D. J., & Chen, L. (2013). Excitation of kinetic Alfvén waves by density striation in magneto-plasmas. *The Astrophysical Journal*, *771*, 3.

Wu, D. J., & Fang, C. (1999). Two-fluid motion of plasma in Alfvén waves and heating of solar coronal loops. *The Astrophysical Journal*, *511*, 958–964.

Wu, D. J., & Fang, C. (2003). Coronal plume heating and kinetic dissipation of kinetic Alfvén waves. *The Astrophysical Journal*, *596*, 656–662.

Wu, D. J., & Fang, C. (2007). Sunspot chromospheric heating by kinetic Alfvén waves. *The Astrophysical Journal*, *659*, L181–L184.

Wu, D. J., & Wang, D. Y. (1996). Solitary kinetic Alfvén waves on the ion-acoustic velocity branch in a low-β plasma. *Physics of Plasmas*, *3*, 4304–4306.

Wu, D. J., & Yang, L. (2006). Anisotropic and mass-dependent energization of heavy ions by kinetic Alfvén waves. *Astronomy & Astrophysics*, *452*, L7–L10.

Wu, D. J., & Yang, L. (2007). Nonlinear interaction of minor heavy ions and kinetic Alfvén waves and their anisotropic energization in coronal holes. *The Astrophysical Journal, 659*, 1693–1701.
Wu, B. H., Wang, J. M., & Lee, L. C. (2001). Generation of kinetic Alfvén waves by mirror instability. *Geophysical Research Letters, 28*, 3051–3054.
Wu, D. J., Huang, G. L., Wang, D. Y., & Fälthammar, C. G. (1996a). Solitary kinetic Alfvén waves in the two-fluid model. *Physics of Plasmas, 3*, 2879–2884.
Wu, D. J., Huang, G. L., & Wang, D. Y. (1996b). Dipole density solitons and solitary dipole vortices in an inhomogeneous space plasma. *Physical Review Letters, 77*, 4346–4349.
Wu, D. J., Huang, J., Tang, J. F., & Yan, Y. H. (2007). Solar microwave drifting spikes and solitary kinetic Alfvén waves. *The Astrophysical Journal, 665*, L171–L174.
Wu, D. J., Wang, D. Y., & Fälthammar, C. G. (1996c). Coulpling Alfvénic and ion-acoustic solitons. *Chinese Physics Letters, 13*, 594–597.
Wu, D. J., Wang, D. Y., & Huang, G. L. (1997). Two dimensional solitary kinetic Alfvén waves and dipole vortex structures. *Physics of Plasmas, 4*, 611–617.
Wygant, J. R., Keiling, A., Cattell, C. A., et al. (2002). Evidence for kinetic Alfvén waves and parallel electron energization at 4–6 R_E altitudes in the plasma sheet boundary layer. *Journal of Geophysical Research, 107*, 1201–1215.
Yang, L., Wu, D. J., Wang, S. J., & Lee, L. C. (2014). Comparison of two-fluid and gyrokinetic models for kinetic Alfvén waves in solar and space plasmas. *The Astrophysical Journal, 792*, 36.
Yoon, P. H., Wu, C. S., & de Assis, A. S. (1993). Effect of finite ion gyroradius on the fire-hose instability in a high beta plasma. *Physics Fluids B, 5*, 1971–1979.
Yukhimuk, V., Voitenko, Y., Fedun, V., & Yukhimuk, A. (1998). Generation of kinetic Alfvén waves by upper-hybrid pump waves. *Journal of Plasma Physics, 60*, 485–495.
Zhao, J. S., Wu, D. J., & Lu, J. Y. (2011). Kinetic Alfvén waves excited by oblique MHD Alfvén waves in coronal holes. *The Astrophysical Journal, 735*, 114.

Chapter 2
Laboratory Experiments of KAWs

The existence of AWs, as electromagnetic hydrodynamic waves, was predicted first by Alfvén (1942), however, the experimental study in laboratory had not been made until 1949. One of important reasons for this delay is that the long wavelength and low frequency characteristics of AWs make them very difficult to study experimentally in laboratory. The similar case happened on KAWs. When KAWs were discovered in 1970's, their ability to heat plasma particles attracted attention of researches. Unfortunately heating experiments were often disappointing due to the poor energy confinement of the plasma sources in laboratory. The situation had not been improved evidently until the large plasma device (LAPD) was built in 1989 and started to perform at UCLA (University of California, Los Angeles). Since that, a series of experimental investigations of KAWs in laboratory plasmas have been performed successfully on LAPD and great progresses in experimental studies of KAWs have been achieved in laboratory plasmas. Moreover, the phenomena that have been observed by the satellites in space plasmas since 1990's show striking similarities to the results observed in the laboratory experiments.

In this chapter, we first briefly introduce the early endeavor experimentally to measure KAWs in laboratory plasmas. Then we review the laboratory investigations of KAWs, which were performed mainly on LAPD and well reproduced the theoretically expected results presented in Chap. 1. The descriptions of laboratory experimental device and plasma diagnostic tools are referred to, for example, the book by Cross (1988).

2.1 Early Experiments of Plasma Heating by KAWs

In 1970's, Hasegawa and Chen (1974; 1975; 1976a) first proposed that when propagating in an inhomogeneous plasma, AWs may directly convert into KAWs by the resonant mode conversion and, in particular, the dissipation of KAWs due to the

D.-J. Wu and L. Chen, *Kinetic Alfvén Waves in Laboratory, Space, and Astrophysical Plasmas*, Atmosphere, Earth, Ocean & Space,
https://doi.org/10.1007/978-981-13-7989-5_2

collision damping, the Landau damping, or other nonlinear effects can be responsible for plasma heating. In an inhomogeneous plasma in general, AWs have a continuous spectrum within a frequency range of $\omega_A(x)_{min} \le \omega \le \omega_A(x)_{max}$, where $\omega_A(x) \equiv v_A(x)k_z$, called the local Alfvén frequency. When a monochromatic electromagnetic wave with frequency ω_0 is applied externally to the inhomogeneous plasma, the wave energy can be absorbed by the plasma at the so-called resonant surface $x = x_0$ that satisfies the resonant condition (Hasegawa and Uberoi 1982)

$$\omega_0 = \omega_A(x)|_{x=x_0} = v_A(x_0)k_z(x_0). \tag{2.1}$$

In the resonant surface, the energy flux of the absorbed AW (or i.e., the converted rate) can be written as (Chen and Hasegawa 1974a)

$$fc \simeq \frac{\omega_0}{k_\perp}\frac{B_x^2(x_0)}{\mu_0}. \tag{2.2}$$

The ion and electron heating rates depend on the dissipation mechanisms. In a collisional plasma, the ion heating is dominated by the viscosity of the transverse wave field and the heating rate is

$$n_0\frac{dT_i}{dt} = \frac{1}{2}\text{Re}\left(\mathbf{j}\cdot\mathbf{E}^*\right) \simeq 0.7\nu_i\chi_i\frac{\epsilon_0|E_x|^2}{2}\frac{\omega_{pi}^2}{\omega_{ci}^2} \simeq \nu_i\frac{|B_y(\kappa^{-1})|^2}{2\mu_0}, \tag{2.3}$$

where $B_y(\kappa^{-1})$ is the value of the wave magnetic field at $x = \kappa^{-1}$ and κ is the inhomogeneity length scale. The electron heating in a collisional plasma is governed by the Ohmic dissipation of the field-aligned current, that is,

$$n_0\frac{dT_e}{dt} = \frac{1}{2}\text{Re}\left(j_zE_z^*\right) \simeq \nu_e\frac{T_e}{T_i}\frac{\left(1 - I_0e^{-\chi_i}\right)^2}{\chi_i}\frac{\epsilon_0|E_x|^2}{2}\frac{\omega_{pi}^2}{\omega_{ci}^2}\frac{\omega_0^2}{k_z^2v_{T_e}^2}. \tag{2.4}$$

To compare the ion and electron heating rates in a collision plasma, we have

$$\frac{dT_e/dt}{dT_i/dt} \simeq \frac{\nu_e}{\nu_i}\frac{T_e}{T_i}\frac{\omega_0^2}{k_z^2v_{T_e}^2} \simeq \sqrt{\frac{m_e}{m_i}}\left(\frac{T_i}{T_e}\right)^{3/2}\frac{1}{\beta_i}. \tag{2.5}$$

In a collisionless plasma, on the other hand, the ion and electron heating rates can be attributed to the linear Landau damping for small amplitude applied AWs. Therefore, we have

$$n_0\frac{dT_i}{dt} = \frac{\omega_0}{e^{1/\beta_i}}\sqrt{\frac{\pi}{\beta_i}\frac{T_e^2}{T_i^2}}\frac{I_0e^{-\chi_i}\left(1 - I_0e^{-\chi_i}\right)^2}{\chi_i}\frac{\epsilon_0|E_x|^2}{2}\frac{\omega_{pi}^2}{\omega_{ci}^2} \tag{2.6}$$

and

$$n_0 \frac{dT_e}{dt} = \omega_0 \sqrt{\frac{\pi}{\beta_i}} \sqrt{\frac{T_e}{T_i}} \sqrt{\frac{m_e}{m_i}} \frac{\left(1 - I_0 e^{-\chi_i}\right)^2}{\chi_i} \frac{\epsilon_0 |E_x|^2}{2} \frac{\omega_{pi}^2}{\omega_{ci}^2}. \tag{2.7}$$

The ratio of the electron to ion heating rates is

$$\frac{dT_e/dt}{dT_i/dt} \simeq \sqrt{\frac{m_e}{m_i}} \left(\frac{T_i}{T_e}\right)^{3/2} \frac{e^{1/\beta_i}}{\sqrt{\pi}}. \tag{2.8}$$

When large amplitude AWs are applied externally a plasma, however, nonlinear processes are expected to occur as a result of the resonant mode conversion, such as the parametric decay of KAW into another KAW and an ion-acoustic wave for the case of $\tau \ll 1$ or into another KAW through the nonlinear ion Landau damping for the case of $\tau \sim 1$ (Hasegawa and Uberoi 1982). The corresponding ion heating rate can be given approximately by

$$\frac{dT_i}{dt} \simeq \frac{\gamma B_y^2}{2\mu_0}, \tag{2.9}$$

where

$$\gamma \simeq \frac{\omega_{ci}}{4} \frac{k_\perp^2 \rho_s^2}{\beta^{1/4}} \left|\frac{B_y}{B_0}\right| \quad \text{for } T_e \gg T_i,$$
$$\gamma \simeq \frac{\omega_{ci}}{8} \frac{k_\perp^2 \rho_s^2}{\beta} \left|\frac{B_y}{B_0}\right|^2 \frac{\omega_{ci}}{\omega_A} \quad \text{for } T_e \simeq T_i. \tag{2.10}$$

In 1970's, the possibility to apply the plasma heating by KAWs on thermonuclear fusion experiments was extensively investigated in laboratory plasmas. The fusion reaction requires a plasma temperature $\sim$10 keV, while the saturation temperature by the classic Ohmic heating is only $\sim$1 keV. Thus, a supplementary heating is needed to heat the plasma in the reactor to the fusion temperature. AWs with radio frequency range of orders of MHz are one of attractive candidates because strong power sources of hundreds MW had become available in 1970's (Hasegawa and Uberoi 1982). Below, we introduce four of the KAW heating experiments in 1970's based on the description presented in Hasegawa and Uberoi (1982).

One of the four is the AW heating experiment on Proto-Cleo plasma at Wisconsin, America (Golovato et al. 1976). The Proto-Cleo device is a stellarator with a major radius of 40 cm and an average minor radius of 5 cm. The typical plasma parameters are $n_0 \sim 10^{12}$ cm^{-3}, $T_e \sim$ 10–20 eV, $B_p \sim 3$ kG (the poloidal magnetic field), and $t_c \sim 1$ ms (the energy confinement time). A generator operating up to 200 kW at the pulse length of 1 ms was used to launch the radio frequency (rf) wave. The experimental results show that the plasma temperature increases immediately after the rf was cut off. Moreover, the peak temperature appears roughly in the resonant surface of AWs, implying the plasma heating is caused by the resonant mode conversion of AWs into KAWs in the resonant surface. Another important result of this experiment is to find enhanced anomalous transport due to the applied rf wave. The anomalous transport varies approximately linearly with respect to the rf amplitude and is believed to be caused by magnetic island formation.

The AW heating experiment on R-02 stellarator at Sukhumi, Russia is interesting one because of the high-intensity rf wave (Demirkhanov et al. 1977). In this experiment it was found that efficient heating occurs when the plasma density is larger than 5×10^{13} cm^{-3}, at which density AW can be coupled into the plasma. In particular, the results indicate that the heating has the clear nonlinear feature with respect to the rf amplitude and a much higher efficiency. The temperature increases to almost 20 times of the initial temperature. The nonlinear threshold amplitude of the rf wave is about 50 G. Moreover, a remarkable and interesting phenomenon was found in this experiment, that is, the fluctuating field that leads to the anomalous transport has disappeared when AW is coupled in the higher density discharge phase, implying that the better confinement of the plasma is achieved in the presence of the large-amplitude AW. This can be understood in terms of the negative ponderomotive effect because the magnetic ponderomotive force, unlike the electrostatic ponderomotive force, tends to attract plasma into the high-intensity region of the rf wave. Therefore, the nonlinear AW heating has a very attractive feature in that the applied rf wave works to heat as well as to confine the plasma.

Another AW heating experiment was done on the Heliotron-D in Kyoto, Japan (Obiki et al. 1977). The machine has a major radius of 110 cm, a minor radius of 10 cm, a toroidal field of 2 to 3 kG, and an electron density ranging between 3×10^{12} and 3×10^{13} cm^{-3}. In consideration of that the KAW excited at the resonant surface propagates toward the plasma center only when $v_{T_e} > v_A$, in order to heat the bulk plasma in this experiment, the plasma is ohmically heated to satisfy the condition of $v_{T_e} > v_A$ before the AW heating is applied. Otherwise, the KAW propagates toward the plasma edge and heats only the plasma surface. One remarkable result of this experiment is to demonstrate this threshold condition of efficient heating as the function of initial electron temperature. Therefore, an enough high initially electron temperature to satisfy $v_{T_e} > v_A$ is an essential condition for the bulk heating of the plasma.

Besides these three toroidal plasma devices, the AW heating experiment also was done on the linear theta-pinch plasma device in Lausanne, Switzerland (Keller and Pochelon 1978). The theta coil is 142 cm long and 9 cm in inner diameter. The main magnetic field reaches 16 kG in 3.8 μs. The quartz discharge tube has an inner diameter of 5.2 cm. Two electrodes, 1.2 cm in diameter, are set at the two ends, 142 cm apart. The typical plasma parameters are 75% in ionization, 0.76 cm in mean plasma radius, 2×10^{16} cm^{-3} in mean electron density, and 40 eV in maximum plasma temperature. The temperature and heating power are measured by diamagnetic probe. The results show that the heating efficiency reaches approximately 50% in this experiment. The resonant absorption is also confirmed by this experiment.

These AW heating experiments in 1970's demonstrated not only the effectivity of the AW heating for fusion plasmas, but also the resonant conversion of AWs to KAWs at the resonant surface, which was first proposed by Chen and Hasegawa

(1974a, b, c). In particular, they illustrated indirectly the existence of KAWs. Direct experimental measurements of KAWs in laboratory started in 1980's.

The first experimental study on properties of KAWs was performed in a linear cylindrical device by Cross (1983) in Australia. The device is 260 cm long and 15 cm in diameter, with an axial background magnetic field 7–8 kG. The plasma (hydrogen and argon discharge) has typical density $n_0 \sim 10^{15}$ cm^{-3} and temperature $T_e \simeq T_i \simeq 2$ eV, where the equality of the electron and ion temperatures is a result of the very high electron-ion collision frequency ($\nu_{ei} \sim 5$ GHz much larger than the ion gyrofrequency $\omega_{ci}/2\pi \sim 11 - 12$ MHz and hence than the wave frequency $\omega/2\pi$). The experiment was carried out in the afterglow of the plasma, a short time after the end of the discharge. KAW was launched using a coaxial cage antenna, which produced only a B_θ field, a very localized KAW with a small perpendicular scale length on the order of the size of the antenna. The experimental results showed that KAW primarily propagates along the magnetic field as theoretically predicted. In addition, the fact that the wave remains localized near to the antenna indicates that the wave appears to be highly damped. This is attributed to the wave having a finite $k_\perp$ hence a radially spreading across field lines as well as the high collisional damping in the collisional plasma.

A similar experiment was also done on a toroidal tokamak device (Borg et al. 1985). The plasma density $n_0 \sim 10^{13}$ cm^{-3} and the initially electron temperature $T_e \simeq 10$ eV (but probably lower when the wave was measured after the main tokamak discharge). The analysis of the experimental data indicates that the ratio of the collision frequency to the wave frequency, or i.e., the damping coefficient is about $\Gamma \equiv \nu_{ei}/\omega \simeq 100 \gg 1$. They concluded that in a collisional plasma AWs may consist of an arbitrary spectrum of $k_\perp$ up to a maximum at $k_\perp \lambda_e = 1$.

In the tokamak experiment by Weisen et al. (1989), they set up the wave pattern of resonant mode conversion into KAW at the resonant surfaces within the device. The measured density fluctuations are well consistent with that caused by the radial propagation and damping of KAWs based on the predictions of the KAW theory.

These KAW experiments, however, all almost were highly collisional and hence KAWs were strongly damped. Further experimental investigations of KAWs, especially the researches of the kinetic effects of KAWs in weakly damping or collisionless plasmas were performed at in LAPD of UCLA. In next sections, we turn our attention to those experiments of KAWs carried out in the LAPD, which well demonstrated many aspects of physical properties of KAWs predicted by previous theories of KAWs.

2.2 Measurements of KAW Dispersion Relations

The LAPD is a large linear plasma research device at UCLA, which was designed to study space plasma processes (Gekelman et al. 1991). Its construction began in 1985 and was completed in 1989. The original LAPD has a 0.5×0.5 m^2 oxide-coated cathode as a source which produces a ~10 m long plasma column with density up to

5×10^{12} cm^{-3} and temperature up to 15 eV in a linear vacuum chamber (1 m diameter by 10 m long). There are 128 radial ports on the chamber to ensure excellent access for probes and antennas and an internal probe drive can move a set of probes to any position within the plasma column. A set of 68 magnet coils surrounding the chamber can generate an axial magnetic field up to 3000 G. Therefore, the characteristic kinetic scales of plasma particles in the LAPD have typical orders as follows: the Debye length $\lambda_D \sim 10^{-4}$ m, the electron and ion gyroradii $\rho_e \sim 5 \times 10^{-5}$ m and $\rho_i \sim \times 10^{-3}$ m, respectively, the electron and ion inertial lengths $\lambda_e \sim 5 \times 10^{-3}$ m and $\lambda_i \sim 0.5$ m, respectively, the Alfvén velocity $v_A \sim 10^6$ m/s, and the kinetic to magnetic pressures ratio $\beta \sim 10^{-4}$ (or i.e., $\alpha_e \sim 1$).

About ten years later, the original LAPD was further upgraded to a larger device longer than 20 m (Leneman et al. 2006). Its plasma source has been upgraded too. The new source incorporates a 1 m square heater, which can heat the barium oxide coated cathode to 750–1000 °C, and can produce plasma column up to 0.9 m in diameter (depending on the magnetic field configuration). The LAPD is the finest basic plasma research device in the world because of not only its largest dimensions but also its source to be able to produce a quiescent, highly ionized, vast plasma column with millions Debye lengths in long and hundreds ion gyroradii in diameter, in which the ions also can be strongly magnetized.

The LAPD is especially suitable for studying of KAWs because of its large chamber dimensions much larger than the ion inertial length in long and the electron inertial length or the ion gyroradius in diameter. When the wave exciters are set as frequencies lower than the ion gyrofrequency and sizes in order of the electron inertial length and the ion or ion-acoustic gyroradius (i.e., the characteristically perpendicular scales of KAWs), KAWs can be launched directly by the exciter and propagate in the plasma column within the chamber. In particular, it has become possible to achieve an almost collisionless plasma with the damping coefficient $\Gamma \ll 1$ and to study simultaneously both the inertial and kinetic regimes of KAWs in the LAPD.

The first experimental measurements of KAWs on the LAPD was reported by Gekelman et al. (1994), in which the plasma was operated under the conditions appropriate to the inertial regime. KAWs in the inertial regime were immediately launched in the discharge afterglow using wire mesh disk exciters of 4 mm and 8 mm radius into a helium plasma of density $\sim 10^{12}$ cm^{-3} and ionization $\sim$90%, an axial uniform magnetic field $\sim$1.1 kG, and hence the electron skin depth $\lambda_e \simeq$ 5 mm. The low collisionality due to the low density and the high ionization in the LAPD, hence the low dissipation allows a larger axial propagation distance of the wave from the exciter and a better measurement of the wave propagation properties. In this experiment, the inertial KAW was observed to spread with distance away from the exciter and the spreading follows a cone-like pattern whose angle $\theta = \tan^{-1}\left(k_\parallel/k_\perp\right) \sim \tan^{-1}\left(k_\parallel \lambda_e\right)$ is of order one degree (i.e., a quasi-perpendicular propagation angle of $\sim$89°) in the experiment because the perpendicular dispersion causes the wave energy to spread perpendicular to the magnetic field.

The wave field radiated by the disk exciter, oriented with its normal along the magnetic field, was measured at various axial locations of 150 to 350 m from the emitter, much larger than the parallel wavelengths of order a few meters. The frequency of

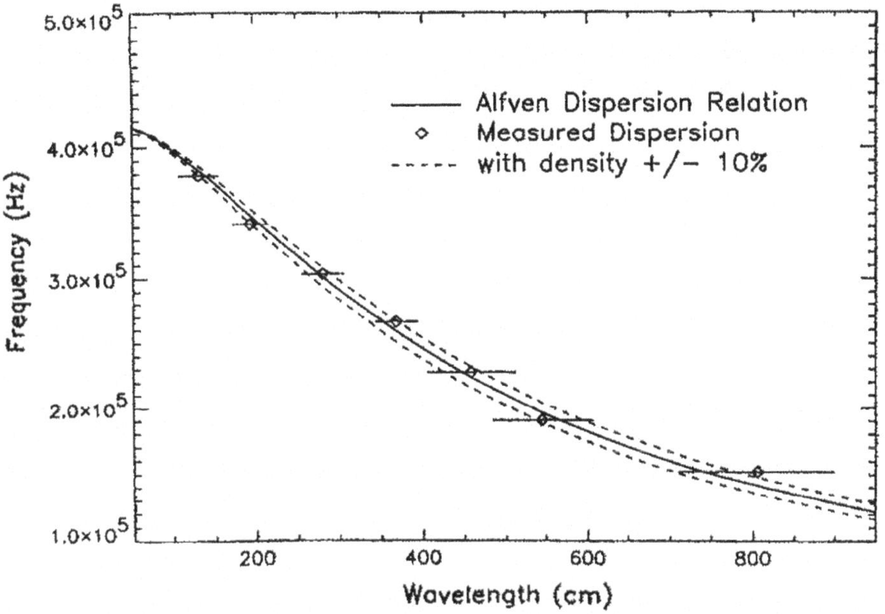

Fig. 2.1 The parallel dispersion relation of KAWs in the inertial regime, that is, the wave frequency versus parallel wavelength: The theoretical curves are calculated by Eq. (2.11) for $k_\perp = 0$ and $n_e = 0.9$ (lower dashed curve), 1.0 (middle solid curve), and 1.1×10^{12} cm^{-3} (upper dashed curve), respectively (reprinted from Gekelman et al., Phys. Plasmas, 1, 3775–3783, 1994, with the permission of AIP Publishing)

the emitter $f = \omega/2\pi$ varies from $f \sim 150$ kHz through 380 kHz, lower than the ion gyrofrequency $f_{ci} = \omega_{ci}/2\pi \simeq 420$ kHz. Taking account of the finite frequency modification the dispersion relation of KAWs in the inertial regime can read as

$$M_z^2 \equiv \frac{\omega^2}{k_\parallel^2 v_A^2} = \frac{1 - \omega^2/\omega_{ci}^2}{1 + k_\perp^2 \lambda_e^2}. \tag{2.11}$$

The measured dispersion indicates that the wave field is launched from the disk exciter and has a perpendicular wavelength $\lambda_\perp \simeq \lambda_\parallel \tan\theta \gg \lambda_e$ for $\theta = 1°$. Figure 2.1 shows the parallel dispersion relation of the measured waves, that is, the wave frequency versus the parallel wavelength, where the solid line is the theoretical curve given by Eq. (2.11) with density $n_e = 1.0 \times 10^{12}$ cm^{-3} and $k_\perp^2 \lambda_e^2 \ll 1$, and the two dashed curves with densities $n_e = 0.9$ and 1.1×10^{12} cm^{-3}, respectively. The measured result can very closely match to the solid curve predicted by the KAW theory.

In addition, the collision frequency is estimated as $\nu_c \sim 350$ kHz in this experiment, which indicates a lower damping coefficient by the collision dissipation, for example, $\Gamma \simeq 0.2$ for the wave at frequency of 280 kHz. The measurement of the wave damping, however, shows a higher damping coefficient required, implying that the Landau damping possibly has an important contribution to the dissipation of the wave energy (Morales et al. 1994).

The experiment was extended to include the kinetic regime of KAWs by Leneman et al. (1999, 2000). In the LAPD experiments the accelerated electrons, from an electron beam source sited on the chamber end, ionize He gas producing a highly ionized plasma. The LAPD discharge plasma usually has a lower ion-electron temperature ratio $\tau \ll 1$. Therefore, the dispersion relation of KAWs in the kinetic regime, including the finite frequency modification, can reduce to

$$M_z^2 \equiv \frac{\omega^2}{k_\parallel^2 v_A^2} = 1 - \frac{\omega^2}{\omega_{ci}^2} + k_\perp^2 \rho_s^2. \tag{2.12}$$

In the experiment by Leneman et al. (1999, 2000) KAWs were launched from a disk antenna with radius $a = 0.5$ cm and normal oriented parallel to the axial magnetic field. The launched wave frequency and parallel wavelength are typically $\omega \sim 0.64\,\omega_{ci}$ and $\lambda_\parallel \sim 2.5$ m, respectively. The produced He plasma column is 36 cm in diameter ($\sim 150\rho_i$) and 9.4 m long ($\sim 4\lambda_\parallel$). In order to produce changes of the parameter α_e from $\alpha_e < 1$ (the inertial regime) to > 1 (the kinetic regime), the plasma parameters, B_0, n_e, and T_e, are changeable in the ranges $0.7 \leq B_0 \leq 2.2$ kG, $1.4 \leq n_e \leq 2.4 \times 10^{12}$ cm^{-3}, and $1.3 \leq T_e \leq 11$ eV, respectively. In particular, the electron inertial length ranges from $\lambda_e \simeq 0.34$ to 0.45 cm and the ion cyclotron frequency does from $\omega_{ci} \simeq 266$ to 836 kHz.

The measured radial structure of wave fields are consistent with the theoretical predictions of KAWs presented by Morales and Maggs (1997). The measurements of the wave phase and axial phase velocity further confirm that the wave fields observed in the experiments are KAWs, which satisfy the KAW dispersion relation with the finite frequency modification. In particular, the perpendicular propagation of waves are observed both in the kinetic and inertial regimes. The results show that the wave propagates forward (i.e., $v_{g\perp} > 0$) in the kinetic regime and backward (i.e., $v_{g\perp} < 0$) in the inertial regime, as expected by the theories of KAWs. In addition, the wave currents also are calculated based on magnetic field measurements and are found to flow in toroidal vortices, which are similar to the observations by the Freja satellite in the auroral plasma (Volwerk et al. 1996).

These elementary experimental researches have demonstrated definitely the measured perturbations launched from the disk antenna have some essential characteristics of KAWs, for instance, the transverse profiles of KAW fields, the opposite directions of their perpendicular propagation in the kinetic and inertial regimes, and the parallel dispersion relation with the finite frequency modification of their axial propagation (i.e., the $\omega - k_\parallel$ relation). However, the most important kinetic characteristic of KAWs is their perpendicular dispersion relation, that is, the $\omega - k_\perp$ relation, which first was measured by Kletzing et al. (2003a) in the LAPD.

2.3 Propagation and Dissipation of KAWs

It is the perpendicular dispersion relation (i.e., the $\omega - k_\perp$ relation) of KAWs that contains the kinetic effects of KAWs different from AWs due to the short-wavelength (i.e., the finite $k_\perp \rho_{is}$ or $k_\perp \lambda_e$) modification. Kletzing et al. (2003a) first measured the perpendicular dispersion relation of KAWs for the kinetic-inertial transition regime of $v_{T_e} \sim v_A$ in the upgraded LAPD. The experiment was conducted in a fully ionized helium plasma column of 10 m length and 40 cm diameter with the electron density of $n_0 \simeq 1.6 \times 10^{12}$ cm^{-3} and temperature $T_e \simeq 3.05$ eV. The ambient magnetic field was set as $B_0 = 700$ and 1000 G, which yields the velocity ratio $v_{T_e}/v_A \simeq 1.25$ and $v_{T_e}/v_A \simeq 0.88$, respectively. The ion temperature was estimated as $T_i \simeq 1$ eV. Other plasma parameters $\lambda_e \simeq 0.42$ cm and $\rho_s \simeq 0.53$ cm for $B_0 = 700$ G and 0.37 cm for $B_0 = 1000$ G. For both cases, the waves were launched at a frequency $\omega \simeq 0.525\,\omega_{ci}$.

They measured two different time series of a signal with the same component of $k_\perp$ taken from the spatial Fourier transform at two different distances from the antenna for the same input signal. Then the later signal at the larger distance (hence weaker) was shifted with respect to the earlier signal at the smaller distance (hence stronger) by a time given by the greatest cross correlation coefficient between the two signals. The parallel phase velocity $v_{p\parallel} = \omega/k_\parallel$ of the wave can be obtained by combining the time delay corresponding to the greatest correlation coefficient with the distance between the two sets of measured signals. By repeating this procedure for different $k_\perp$, they obtained a set of measurements of $v_{p\parallel}$ as a function of $k_\perp$, that is, the so-called perpendicular dispersion relation. For all measurements, the linear correlation coefficients determined by this procedure exceed $r = 0.98$, implying that these signals at the two measurement locations come from the same wave which has propagated along the magnetic field.

For the kinetic-inertial transition regime of $\alpha_e = v_{T_e}/v_A \sim 1$, taking account of the finite frequency modification, the kinetic dispersion equation of KAWs in the low-frequency approximation ($\omega^2 < \omega_{ci}^2$) can reduce to (Kletzing et al. 2003a; Wu 2012):

$$\frac{\omega^2}{k_\parallel^2 v_A^2} = \left(1 - \frac{\omega^2}{\omega_{ci}^2}\right)\frac{\chi_i}{1-\Gamma_0(\chi_i)} + \frac{k_\perp^2 \rho_s^2}{1+\zeta_0^e Z(\zeta_0^e)}, \tag{2.13}$$

where $\Gamma_0(\chi_i) = e^{-\chi_i} I_0(\chi_i) \simeq 1 - \chi_i + (3/4)\chi_i^2$ for small χ_i and $\zeta_0^e = \omega/\sqrt{2}k_\parallel v_{T_e} \simeq \alpha_e^{-1}\sqrt{\left(1-\omega^2/\omega_{ci}^2\right)/2}$.

Figure 2.2 shows the measured results of two sets of the parallel phase velocity $v_{p\parallel}$ versus the perpendicular wave number $k_\perp$, where the upper panel shows measurements for $v_{T_e}/v_A = 1.25$ and the lower panel for $v_{T_e}/v_A = 0.88$. In both cases all plasma parameters were maintained at essentially the same values except for the background magnetic field which was changed to control v_A. For the sake of comparison, in Fig. 2.2 the solid line plots the theoretical prediction by the numerical result of the kinetic dispersion equation (2.13), which includes the finite frequency modification, and two additional dash-dot-dot lines represent the variation in the analytical

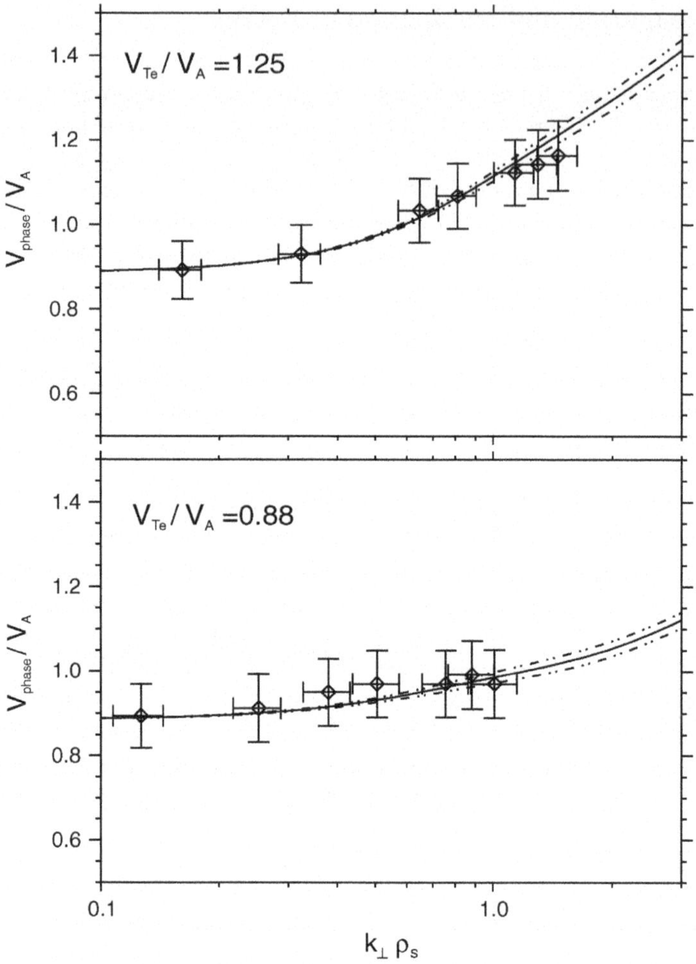

Fig. 2.2 The perpendicular dispersion relation of KAWs: the solid line represents the numerical solution of the theoretical dispersion relation by Eq. (2.13) and two dash-dot-dot lines are obtained by varying the electron temperature T_e by 1 standard deviation (i.e., 0.2 eV) (reprinted figure with permission from Kletzing et al., Phys. Rev. Lett., 90, 035004, 2003a, copyright 2003 by the American Physical Society)

result due to varying the electron temperature by 1 standard deviation (i.e., 0.2 eV). From Fig. 2.2 their experimental results show that the measurements can very well agree with the dispersion relations with the finite frequency modification of KAWs both in the kinetic and inertial regimes.

By use of the radial profile of the plasma density, Vincena et al. (2004) further compared the positions of the zero perpendicular group velocity ($v_{g\perp} = 0$) and the kinetic-inertial transition point ($\alpha_e = 1$). In their experiment the radial distribution of the parameter α_e can be fitted well by

$$\alpha_e^2 = 1 - \tanh\left(\frac{r - r_1}{r_2}\right), \tag{2.14}$$

where $r_1 = 18.4$ cm is the transition position at which $\alpha_e = 1$ and $r_2 = 4.3$ cm is a fitting parameter. The position of $v_{g\perp} = 0$, r_0, may be determined experimentally by the peak of the radial distribution of the normalized wave energy density (i.e., the ring energy density). The wave is launched in the kinetic regime at the center of the plasma and propagates radially until it passes into the plasma edge region, where the plasma density and temperature drop and transit into the inertial regime. Theoretically, on the other hand, r_0 can be calculated numerically by the derivative of Eq. (2.13) with respect to $k_\perp$, that is, $v_{g\perp} = \partial\omega/\partial k_\perp = 0$.

For the LAPD plasma, however, the approximation of $T_i = 0$ (and hence $\chi_i = 0$) is valid. From the dispersion equation (2.13), therefore, the perpendicular group velocity can reduce to

$$v_{g\perp} = -\frac{2k_\perp\rho_s^2\omega}{Z'(\zeta_0^e)} = -\frac{2k_\perp\rho_s^2\omega Z'_R(\zeta_0^e)}{|Z'(\zeta_0^e)|^2}, \tag{2.15}$$

where $Z'(\zeta_0^e) = -2\left[1 + \zeta_0^e Z(\zeta_0^e)\right]$ is the derivative of the plasma dispersion function Z with respect to its argument, the subscript "R" denotes the real part, and the latter equality is because we usually interpret the real part of the solution as being the physically measurable quantity. In the kinetic and inertial limits one has $Z' = -2$ and $\left(\zeta_0^e\right)^{-2}$ and hence $v_{g\perp} = k_\perp\rho_s^2\omega$ and $-k_\perp\lambda_e^2\omega\left(1 - \omega^2/\omega_{ci}^2\right)$, respectively. In general, we have $v_{g\perp} = 0 \Rightarrow Z'_R(\zeta_{0R}^e) = 0$ at $\zeta_{0R}^e = \pm\zeta_0 \simeq 0.924$, or i.e., $\zeta_0^2 \simeq \left(1 - \omega^2/\omega_{ci}^2\right)/2\alpha_e^2 \Rightarrow \alpha_e^2 \simeq \left(1 - \omega^2/\omega_{ci}^2\right)/2\zeta_0^2$. By use of Eq. (2.14), we have approximately

$$r_0 = r_1 + r_2 \tanh^{-1}\left(1 - \frac{1 - \omega^2/\omega_{ci}^2}{2\zeta_0^2}\right). \tag{2.16}$$

Figure 2.3 displays the dependence of the measured peak position of the normalized ring energy density (r_0) on the wave frequency (ω) in (a) and that of the zero perpendicular group velocity predicted by Eqs. (2.16) and (2.13) in curves (b) and (c), respectively, where curve (d) shows the low-frequency approximate solution given by $\alpha_e = 1$. From Fig. 2.3 the measured peak-energy position (a) and the zero crossing positions (b) and (c) have good agreement for high-frequency waves of $\omega/\omega_{ci} \geq 0.6$, but not as good at low frequencies. In fact, the measured results of the low-frequency waves are approaching to the low-frequency approximate solution given by $\alpha_e = 1$, which is independent of the wave frequency. In addition, from Fig. 2.3 also it can be found that the approximate solution (b) is nearly coincident with the numerical solution (c), implying that the justification of the assumptions used to obtain Eq. (2.15) is confirmed further.

For the plasma parametric regime of $v_{T_e} \sim v_A$ (hence $\sim v_{p\parallel}$), on the other hand, it is important to take account of the Landau damping effects. In fact, the solutions of the dispersion equation (2.13) are complex roots for the wave frequency ω. Figure 2.4

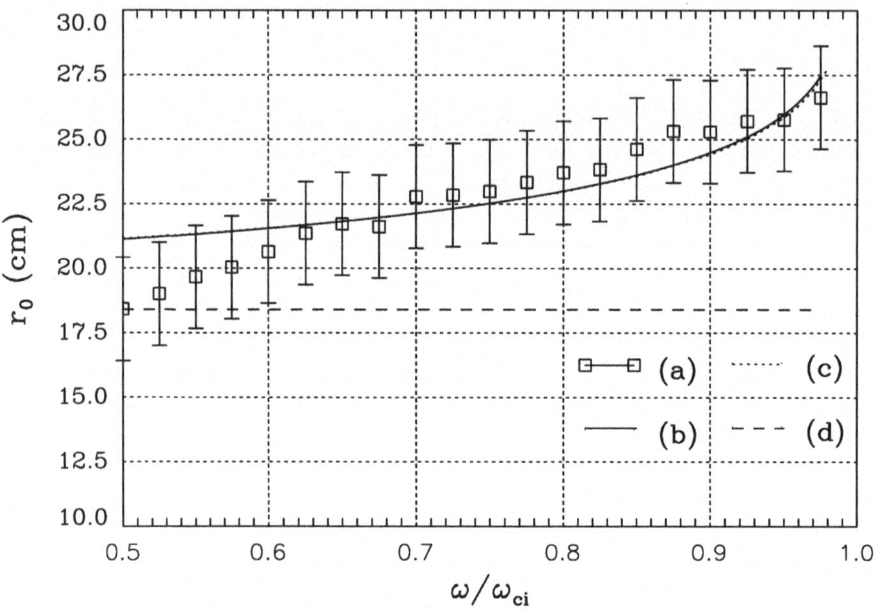

Fig. 2.3 Radial position of zero perpendicular group velocity r_0 versus wave frequency ω: **a** measured peak position of ring energy density; **b** prediction by the approximate expression of Eq. (2.16); **c** numerical solution by Eq. (2.13), which nearly overlaps the approximate solution in (**b**); **d** the position of the kinetic-inertial transition point, that is, the low-frequency approximate solution given by $\alpha_e = 1$ (reprinted figure with permission from Vincena et al., Phys. Rev. Lett. 93, 105003, 2004, copyright 2004 by the American Physical Society)

shows measured and theoretical damping rate as a function of $k_\perp$ in the experiment by Kletzing et al. (2003a), where the experimental values are determined by using the amplitude decrease between locations and the propagation time to compute a damping rate and the theoretical damping includes a nonconserving Krook collision term (Gross 1951) as well as Landau damping. The results presented in Fig. 2.4 show that the measured and theoretical damping rates are good agreement for small $k_\perp \rho_s < 0.5$. For $k_\perp \rho_s > 0.5$, the agreement is not as good, but the basic trend is correct.

The elementary experiment of KAWs by Gekelman et al. (1994) has shown that the collisionless Landau damping possibly plays an important role in the dissipation of KAWs. In order to test further the effects of collisional and collisionless damping on the dissipation of KAWs, Thuecks et al. (2009) compared experimental results with various collisional models, including the nonconserving Krook collision operator (Gross 1951)

$$\left(\frac{\partial f_e}{\partial t}\right)_c = -\nu_e \left(f_e - f_{e0}\right) = -\nu_e f_{e1}, \tag{2.17}$$

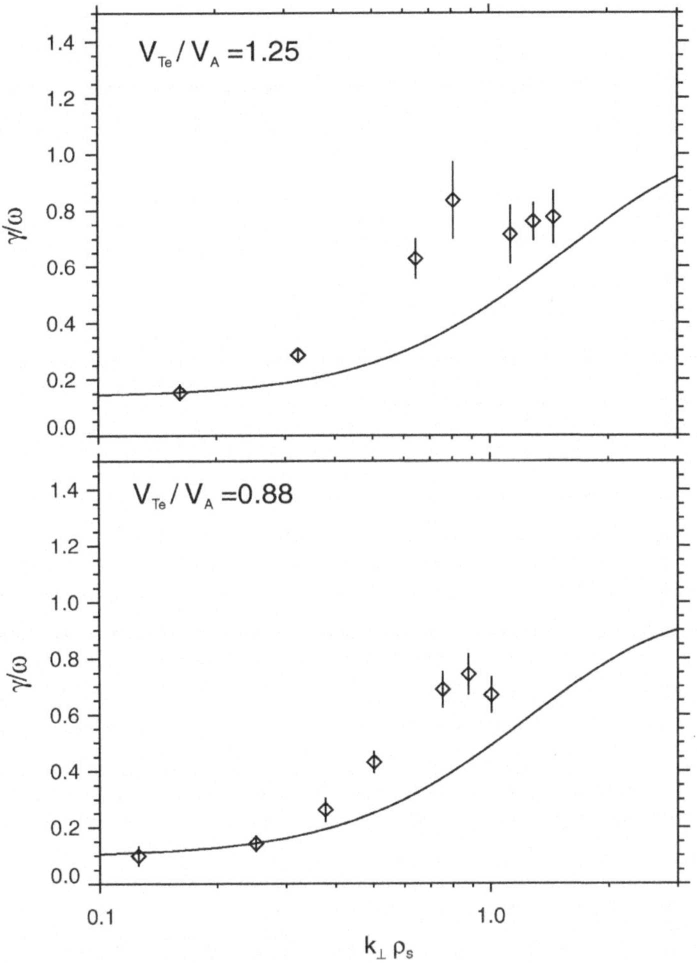

Fig. 2.4 The damping rate of KAWs: solid line is given by the image part of the numerical solution of the theoretical dispersion relation by Eq. (2.13) (reprinted figure with permission from Kletzing et al., Phys. Rev. Lett., 90, 035004, 2003a, copyright 2003 by the American Physical Society)

where f_{e0} and f_{e1} are the equilibrium and fluctuation electron distribution functions, respectively, the particle number conserving Krook operator (Bhatnagar et al. 1954)

$$\left(\frac{\partial f_e}{\partial t}\right)_c = -\nu_e\left(f_e - \frac{n}{n_0}f_{e0}\right) = -\nu_e\left(f_{e1} - \frac{n_1}{n_0}f_{e0}\right), \tag{2.18}$$

where $n_1 = \int f_{e1}d^3\mathbf{v}$ is the perturbed electron density, and the particle number-energy conserving Krook operator (Bhatnagar et al. 1954)

$$\left(\frac{\partial f_e}{\partial t}\right)_c = -\frac{n}{n_0}\nu_e + \frac{n^2}{n_0}\nu_e F_{e0}$$
$$= -\nu_e \left[f_{e1} - \frac{n_1}{n_0} f_{e0} - \frac{f_{e0}}{2n_0}\left(\frac{u^2}{3v_{T_e}^2} - 1\right) \int \left(\frac{v^2}{v_{T_e}^2} - 3\right) f_{e1} d^3\mathbf{v} \right], \tag{2.19}$$

where F_{e0} has the same form as f_{e0}, but with a variable temperature $T_e = T_{e0} + T_{e1}$, as well as the Lorentz collision operator (Koch and Horton 1975)

$$\left(\frac{\partial f_e}{\partial t}\right)_c = \nu_{ei} \frac{2\sqrt{2}v_{T_e}^3}{v^3} \frac{\partial}{\partial \mu}\left[\left(1 - \mu^2\right) \frac{\partial f_{e1}}{\partial \mu}\right] \tag{2.20}$$

with a velocity-dependent collision frequency, where $\mu \equiv v_z/v = \cos\theta$ and θ is a pitch angle.

Thuecks et al. (2009) experimentally measured the dispersion relation of KAWs (including parallel phase velocities and damping rates) in the kinetic regime with $v_{T_e}/v_A \simeq 1.8$ and in the inertial regime with $v_{T_e}/v_A \simeq 0.18$. The axial magnetic field was set as 600 G (the ion gyrofrequency 229 kHz for a helium plasma) and 2300 G (the ion gyrofrequency 876 kHz for a helium plasma) for the kinetic and inertial regimes, respectively. The wave frequency was set near 0.25 and 0.50 of the ion cyclotron frequency, namely, the wave frequencies of 70 and 130 kHz in the kinetic regime and of 250 and 380 kHz in the inertial regime. The comparison with the theoretically collisional models shows that, in the kinetic regime, the damping rates predicted by the collisional models all are obviously higher than the measured damping rates. Moreover, for the dispersion relation the best agreement between the theoretical models and the experimental measurements is achieved by completely ignoring the electron collisions. This indicates that the dissipation of KAWs is dominated by the collisionless Landau damping in the kinetic regime.

In contrast to the case in the kinetic regime, in the inertial regime the comparison between the experimental measurements and theoretical models indicates that it is necessary and essential to include the electron collisional effects to describe KAW behavior. The Landau damping alone predicts much less damping rate than the experimentally measured that. In addition, also higher phase velocities are predicted when the electron collisional effects are neglected. Moreover, the phase velocity and damping rates predicted by the various Krook type models nearly overlap each other (as well as the experimentally measured results) over the range of $k_\perp \rho_s$ from 0.1 through $\gtrsim 1$. However, they also noted that the prediction by the Lorentz collision operator agrees with neither the experimental measurement nor the Krook type predictions in the inertial regime. The difference of the electron collisional effects on the dissipation of KAWs is possibly related to the difference of the electron physics in the kinetic and inertial regimes, and maybe, true scenario is probably much more complex and far from revealed (Thuecks et al. 2009).

Although the kinetic theory with collisions gives the more precise dispersion relation for comparison with the measurements, it provides little intuitive form for understanding the wave physics. Employing a simplified dispersion relation for cold electrons ($v_{T_e} \ll v_A$) with the finite frequency and the electron collision effects:

$$\frac{\omega^2}{k_\parallel^2 v_A^2} = \frac{1 - \omega^2/\omega_{ci}^2}{1 + (1 + i\nu_{\rm eff}/\omega)\, k_\perp^2 \lambda_e^2}, \tag{2.21}$$

where the collisional effect is introduced by taking $m_e \to (1 + i\nu_{\rm eff}/\omega)\, m_e$ only in the electron momentum equation, Kletzing et al. (2010) further analyzed the comparison between experimental measurements and theoretical models of dispersion relation of KAWs in the inertial regime. In the upgraded LAPD a longer plasma column of ~16.5 m length and ~40 diameter was produced, in which the electron density and temperature were estimated as $n_e \simeq 7.7 \times 10^{11}$ cm^{-3} and $T_e \simeq 1.9$ eV, respectively, and a lower ion temperature $T_i \simeq 1.25$ eV was assumed. The axial magnetic field was set as 2300 G. For these plasma parameters, one has $\alpha_e = v_{T_e}/v_A \simeq 0.2$ implying the inertial regime condition, $\lambda_e \simeq 0.61$ cm, and $\rho_s \simeq 0.13$ cm. In addition, the waves were launched at two frequencies of 250 and 380 kHz, and hence the relative ratio frequencies of $\omega/\omega_{ci} \simeq 0.29$ and 0.43, respectively. Their result showed that the better agreement between the experimental measurement and the theoretical prediction by Eq. (2.21) can be obtained by the parameter $\nu_{\rm eff}/\omega \simeq 2.7$ for the 380 kHz wave and 4.1 for the 250 kHz wave, that is, by the effective collision frequency $\nu_{\rm eff} \sim 1$ MHz.

2.4 Excitation and Evolution of KAWs by Plasma Striations

Plasma striations are magnetic field-aligned structures in density, temperature, or current, also called filamentous structures, which are very easy to form and hence are very ubiquitous inhomogeneity phenomena in various magnetic plasma environments of space physics and astrophysics, such as in the terrestrial auroral plasma and in the solar coronal plasmas. These field-aligned filamentous structures can be characterized by the parallel length scales much larger than the perpendicular scales due to the anisotropy of charged particle motions in magnetic fields. It is evident that the most common as well as smallest perpendicular scales of inhomogeneity in these plasma striations are the particle kinetic scales, that is, the typical scales of KAWs, such as the ion gyroradius or the electron inertial length (Wu 2012). Therefore, these plasma striations can play an important role in the dynamics of KAWs, especially in their excitation and evolution, as shown in Sect. 1.5.

One of main scientific goals of the LAPD at UCLA is to model microphysical processes in space and astrophysical plasmas, including plasma waves and their wave-particle interactions. In particular, low-frequency wave activities associated with field-aligned plasma striations are observed commonly in auroral magnetospheric and ionospheric plasmas (Chmyrev et al. 1988; Boehm et al. 1990; Louarn et al. 1994; Lundin et al. 1994a). It is believed extensively that these low-frequency wave activities could result in local ion heating (Erlandson et al. 1994) as well as non-local electron acceleration as the waves propagate over long distances (Hasegawa 1976; Wu and Chao 2004b).

Motivated by the need to understand the interaction dynamics of low-frequency KAWs with field-aligned plasma striations, Maggs and Morales (1996, 1997) designed and performed an experimental study in the LAPD, in which a controlled field-aligned density depletion as a plasma striation was formed along the central axis of a plasma column in the LAPD. Figure 2.5 shows a schematic diagram illustrating the elements for the formation of a controlled plasma striation in the center of the plasma column in the chamber of the LAPD (upper part) and a typical radial profile of the plasma density in the vicinity of the density striation (lower part). As shown in Fig. 2.5, the striation extends along the magnetic field to the entire length of the plasma column in the chamber of the LAPD ($\sim$10 m) but has a thin radius ($\sim$2 cm, a few the electron inertial length and one-tenth of the plasma column radius), and hence can be considered to be embedded in an essentially infinite plasma.

The background column plasma is generated by electrons emitted from a heated barium oxide coated cathode and subsequently accelerated by a semitransparent grid anode located 60 cm from the cathode. The electrons are accelerated into the vacuum chamber and impact neutral helium gas to generate a helium plasma with a 40 cm diameter and $\sim$10 m in long, a degree of ionization higher than 75%, densities in the range $n_e \simeq 1 - 4 \times 10^{12}$ cm^{-3} (hence $\lambda_e \simeq 0.53$–0.27 cm), electron temperature $T_e \simeq 5$–15 eV, and ion temperature $T_i \simeq 0.1$–1 eV, much lower than the electron temperature T_e. The axial ambient magnetic field is regulated in the range of 0.5 to 2 kG to attain different plasma β values.

In the center of the plasma striation, the electron density and temperature are about $n_e \sim 0.5 \times 10^{12}$ cm^{-3} and $T_e \sim 6$ eV, respectively, both obviously lower than those in the surrounding plasma column. Discharge currents, which are carried by the primarily emitted electrons, are controlled in the range of 0.3 to 3 kA to obtain different radial density gradients in the density striation edge. The plasma density outside the density striation nearly linearly increases when the discharge current increases, while the density in the center of the striation much more slowly rises. In consequence, the density gradient at the edge of the striation considerably increases as the discharge current increases. During the discharge the density and temperature gradient scales across the magnetic field in the density striation are, typically, $L_n \sim Ł_T \sim 1$–2 cm, and the ion and ion-acoustic gyroradii are $\rho_i \sim 0.05$–0.2 cm and $\rho_s \sim 0.2$–0.5 cm, respectively.

When a weak current system flows around the striation both density and magnetic fluctuations are observed spontaneously to develop in the plasma density striation. The spontaneously generated magnetic fluctuations are measured by a small induction-loop probe consisting of mutually orthogonal coils (58 turns of 2.5 mm radius) in a triaxial arrangement. The signals corresponding on density fluctuations are collected by small Langmuir probes. The detected signal intensity is growing during the discharge current turns on and eventually achieves a steady state consisting

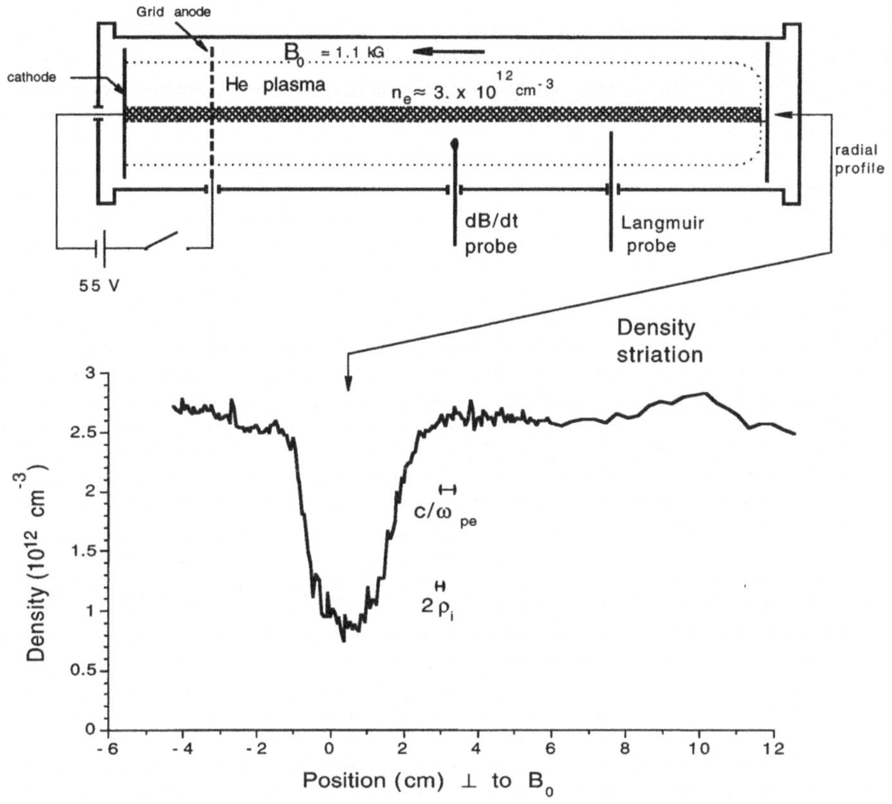

Fig. 2.5 A schematic of plasma striation experiments in the LAPD: A thin density striation extends to the entire length of the plasma column in the chamber of the LAPD (upper part) and a typical radial profile of the plasma density in the vicinity of the density striation (lower part) (reprinted from Maggs & Morales, Phys. Plasmas, 4, 290–299, 1997, with the permission of AIP Publishing)

of highly coherent low-frequency oscillations. After the discharge current shut off (i.e., in the afterglow), the measured fluctuations start to decay while simultaneously decreasing their oscillation frequency.

However, this does not imply that the field-aligned discharge currents are responsible for the wave growth. In fact, the axial currents are carried by two distinct populations. One is primary electrons emitted from the cathode which carry an electrical current directed towards the cathode. Other one is carried by a drift of the bulk electron population in the main body of the plasma column, which cancels the discharge current carried by the primary fast electrons. While in this region the total current nearly is zero when integrated over the entire cross-section of the plasma column. The radial distributions of the axial currents also indicate that in both the discharge and afterglow the current is minimum at the center of the density striation and largest in the bulk plasma. Although there are clear differences in the axial currents of the plasma with and without the density striation, the implication obtained by examining the radial profiles of the axial current is that the axial current is not the driving source of the fluctuations. The driving source for the fluctuations is the cross-field density and temperature gradients in the plasma striation.

The spectral analysis by the Fourier transform of the fluctuations shows that the power spectra of the magnetic and density fluctuations, $|\delta B(\omega)|$ and $|\delta n(\omega)|$, both reach their peak amplitudes at a frequency of 37 kHz ($\sim$0.085 $(\omega_{ci}/2\pi)$), implying the presence of an eigenmode. The measured radial distributions of the magnetic and density fluctuations indicate that they are well trapped within the narrow density striation and have a rough reflection symmetry about the center of the density striation. In particular, the magnetic and density fluctuation amplitudes peaks near the maximum density gradient, implying that the density gradient indeed is responsible for the primarily driving source for the fluctuations.

The polarization analysis of the magnetic fluctuations gives the amplitude ratio of the perpendicular to parallel components of the fluctuation magnetic field, $|\delta B_\perp(\omega)| / |\delta B_\parallel(\omega)| \sim 100 \gg 1$, which clearly reveals that the magnetic fluctuations are shear AWs with magnetic incomprssibility. In fact, as inferred by the authors, much of the small parallel component probably results from a slight misalignment of the magnetic probe which results in a portion of the large perpendicular component appearing in the coil when measuring the parallel component.

In the experiments, also it is noted that although the radially trapped eigenmodes driven by the gradients can simultaneously exhibit large fluctuations in density ($|\delta n/n_0| \sim 0.2$) and magnetic field ($|\delta B/B_0| \sim 2 \times 10^{-3}$), the correspondence between density and magnetic eigenmodes does not necessarily hold for all frequencies as the plasma β varies. The nature of the fluctuations can be identified as the drift-Alfvén wave and the broadband magnetic shear AW turbulence in the kinetic regime of $\alpha_e > 1$ and the inertial regime of $\alpha_e < 1$, respectively.

Figure 2.6 shows frequency spectra of density and magnetic field fluctuations for $\alpha_e > 1$ (a) and $\alpha_e < 1$ (b). From Fig. 2.6, for $\alpha_e^2 \sim 1.8 > 1$ the measured results show that there is an excellent correspondence between the peaks in the density and magnetic spectra, particularly in the lower frequency range of $\omega/\omega_{ci} < 0.1$, where the magnetic incompressibility of the fluctuations indicates that the eigenmode at $\omega/\omega_{ci} \simeq 0.085$ can be identified as the drift-KAW mode (Maggs and Morales 1997). For a lower $\alpha_e^2 \sim 0.5 < 1$, the correspondence between peaks in the density and magnetic spectra is limited to a lower frequency range $\omega/\omega_{ci} < 0.04$, where a clear peak near $\omega/\omega_{ci} \simeq 0.03$ can be identified as the eigenmode of pure electrostatic drift waves. In the frequency region $\omega/\omega_{ci} > 0.04$, on the other hand, the evident peaks present only in the magnetic fluctuations and can be identified as the eigenmodes of KAWs, while in the density fluctuations there is not any eigenmode structure to appear but an exponential frequency spectrum, which is a universal feature for the fluctuations of the plasma column driven by edge plasma gradients near a conducting wall in the laboratory.

In a similar experiment, Burke et al. (2000a, b) further investigated the spontaneously developing fluctuations in magnetic field and density associated with a thin field-aligned temperature striation embedded in a large magnetized plasma column.

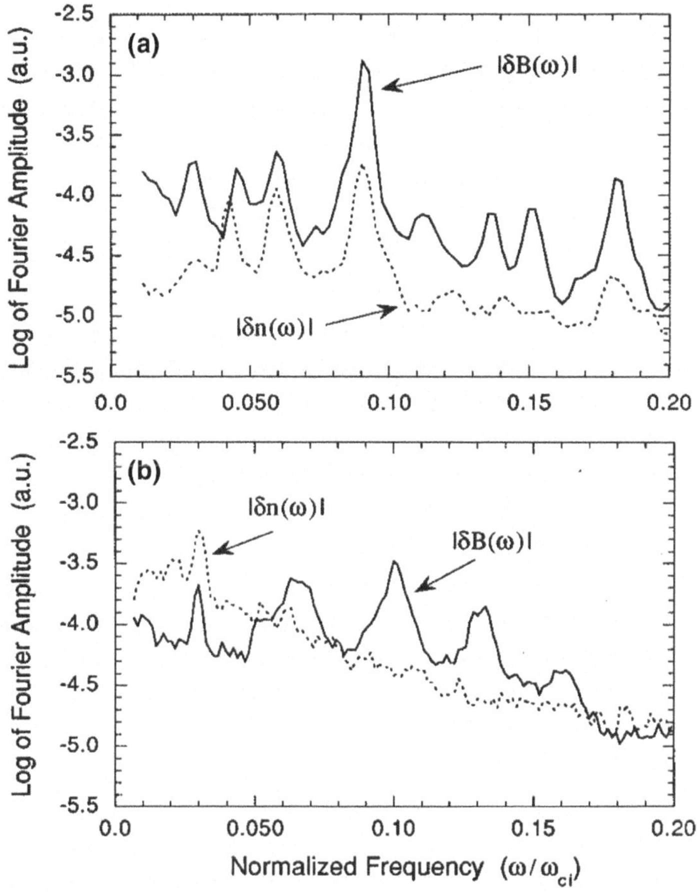

Fig. 2.6 Frequency spectra of magnetic field and density fluctuations for $\alpha_e > 1$ (**a**) and $\alpha_e < 1$ (**b**) (reprinted from Maggs & Morales, Phys. Plasmas, 4, 290–299, 1997, with the permission of AIP Publishing)

They found that the drift-KAWs, which were observed and identified in the experiments by Maggs and Morales (1996, 1997), exhibited some interesting nonlinear features. For instance, some strongly coherent nonlinear structures of drift-KAWs with the form of two dimensional dipole vortex, which are observed commonly by satellites in space plasmas as two dimensional solitary KAWs (Wu et al. 1996a, 1997), were observed in their experiment to progress in the azimuthal mode number $m = 1$ and to couple to the low frequency mode with subsequently generating sideband structures in the frequency spectrum, developing harmonic structures, and eventually transiting to broadband turbulence.

Their experiment was performed under similar plasma conditions with that by Maggs and Morales (1996, 1997). Discharge currents of 1–5 kA are used to generate an axially and radially nearly uniform cylindrical plasma column with densities $n_e \simeq 1\text{–}4\times10^{12}$ cm^{-3}, electron and ion temperatures $T_e \simeq 6\text{–}8$ eV and $T_i \sim 1$ eV, respectively. The axially confining magnetic field varies in the range of 0.5–1.5 kG. The spontaneously developing fluctuations are measured in the so-called plasma afterglow phase after the discharge pulse is terminated. In the afterglow plasma T_e decays rapidly (on a time scale of 0.1 ms) due to classical axial transport to the ends of the device and cooling due to energy transfer to the ions.

Instead of the density depletion in the experiment by Maggs and Morales (1996, 1997), a narrow electron beam with low energies (15–20 eV) and small transverse extension (3 mm diameter) is field-aligned injected into the large magnetized plasma column. The injected beam electrons are slowed down and thermalized in a fairly short distance ($\sim$70 cm for 20 eV electrons) due to elastic collisions with the slow bulk plasma electrons and ions. The role of the beam is simply to produce a narrow field-aligned temperature filament of a few millimeters in diameter and about a meter in length. When this temperature filament, as a heating source, presents in the afterglow plasma with a lowered ambient electron temperature of $T_e \lesssim 1$ eV and a density $n \simeq 2 \times 10^{12}$ cm^{-3}, it is possible to locally increase T_e by 5–10 times the ambient temperature. This leads to the presence of steep temperature gradients between the central plasma temperature striation and the surrounding plasma column, which can be responsible for primarily driving source for spontaneously developing fluctuations in magnetic field, density, and temperature.

The electron beam is injected at $t = 0.5$ ms after the pulse discharge current that produced the large cylindrical plasma column is turned off, i.e., during the afterglow phase. The measurement of the ion saturation current $I_s \propto n_e\sqrt{T_e}$ at the center axis distance 285 cm from the beam injector shows that the plasma temperature striation formed by the injected beam effectively heats the afterglow plasma in an interval $\sim$2 ms as expected. High-frequency fluctuations are found to develop spontaneously starting at $t \simeq 2.3$ ms and are accompanied by an increase in the average temperature. At $t \simeq 3.5$ ms low-frequency oscillations start to appear and, simultaneously, the average temperature undergoes a drop, which implies an enhanced rate of heat transport (see Fig. 3 in Burke et al. 2000b). Detailed correlation analyses of I_s versus n_e and T_e indicate that the high- and low-frequency fluctuations in the saturation current I_s are primarily dominated by the density and temperature fluctuations, respectively.

The mapping dimensional structure of the high-frequency fluctuations near the largest temperature gradient displays a clear dipole vortex structure in density and magnetic field (i.e., corresponding to an azimuthal mode number $m = 1$) with magnetic shear polarization and a transverse scale on the order of the electron inertial length (see Fig. 4 in Burke et al. 2000b). In fact, similar structures in space plasmas first were found by Wu et al. (1996a, 1997) by use of the Freja satellite data during it passed the auroral plasma at the altitude of $\sim$1700 km and they identified these structures as two dimensional solitary KAWs with a dipole density soliton and vortex electromagnetic structure. While in the laboratory experiment by Burke et al. (2000a, b) the magnetic vortex structures were clearly observed to be associated with two microscopic axial currents embedded within the temperature striation. Their experimental measurements in the LAPD, however, further found that the vortex structure exhibits a very striking topological change corresponding to a steady progression of the azimuthal mode number, m, from an initial dipole eigenmode of $m = 1$ to the nearest mode of $m = 2$, eventually reaching mode numbers in excess of $m = 6$ (see Fig. 11 in Burke et al. 2000b).

In addition, the frequency-spectrum analysis for the measured data from $t = 3.5$ to 4.5 ms shows that the high- and low-frequency modes have peak frequencies at $\omega_H \simeq 50$ and $\omega_L \simeq 10$ kHz (the ion gyrofrequency $f_{ci} = 380$ kHz), respectively. In particular, the high-frequency mode and its harmonic at about 100 kHz both show evident sideband structures at about 40 and 60 kHz around the fundamental frequency and 90 and 110 kHz around the harmonic frequency. These sideband modes should result from nonlinear coupling between the high- and low-frequency modes.

Another common nonlinearity arising from a coherent eigenmode is the generation of multiplying harmonic structures in the frequency spectrum. A typical example of the development of multiplying harmonic structures in the frequency spectrum is presented in Fig. 2.7, which shows the power spectrum in log scale of the ion saturation current measured at a radial position $r = 3$ mm from the center of the temperature striation and an axial distance 285 cm from the beam injector. In Fig. 2.7, the three panels describe the harmonic structures at three different times after the temperature striation is formed at $t = 0$, that is, at $t = 2$, 3, and 4 ms. From Fig. 2.7, it can be found that in the early stage $t = 2$ ms, there is only a single frequency peak (i.e., the fundamental frequency of the eigenmode) when the fluctuations have a lower amplitude. Then, in the intermediate stage $t = 3$ ms, a larger amplitude is achieved and three harmonic peaks in addition to the fundamental clearly appear in the frequency spectrum. While at the later time $t = 4$ ms, there are six additional harmonic peaks to be discernable in the spectrum. Although the interesting nonlinear coherent structures and their multiplying harmonic development may last quite long time, it is expectable that this long-time lasting to develop eventually results in the presence of a broad band turbulent spectrum as experimentally measured at $t = 8$ ms (see Fig. 12 in Burke et al. 2000b).

In the upgrated LAPD, Pace et al. (2008a, b) more in detail study the transition from the coherent structures to the broadband turbulence in a plasma temperature striation experiment. The upgraded LAPD has a longer vacuum chamber lengthened to 20 m and hence can produce a longer plasma column, in which fluctuations are allowed to propagate into farther distance and to sustain a longer developing and evolving time. The plasma source of the upgraded LAPD, which consists of a hot-cathode (a pure nickel sheet coated with a thin layer of barium oxide heated to an emission temperature of 800 °C) and a 55-cm apart anode (a semitransparent molybdenum mesh), can create repeatedly at a 1 Hz rate ms-duration pulsed beams of electrons having primary energies in the 50 eV range and carrying discharge currents of 1–3 kA. These primary energetic electrons that pass through the mesh anode ionize an initial neutral He gas in the vacuum chamber to produce a 70-cm diameter He plasma column with typical electron density $n_e \sim 1\text{–}3\times10^{12}$ cm^{-3}, electron temperature $T_e \sim 6$ eV, and ion temperature $T_i \sim 1$ eV.

A plasma temperature striation, as a hot electron channel, can be established in the plasma column by injecting a thin small electron beam with a diameter of 3-mm,

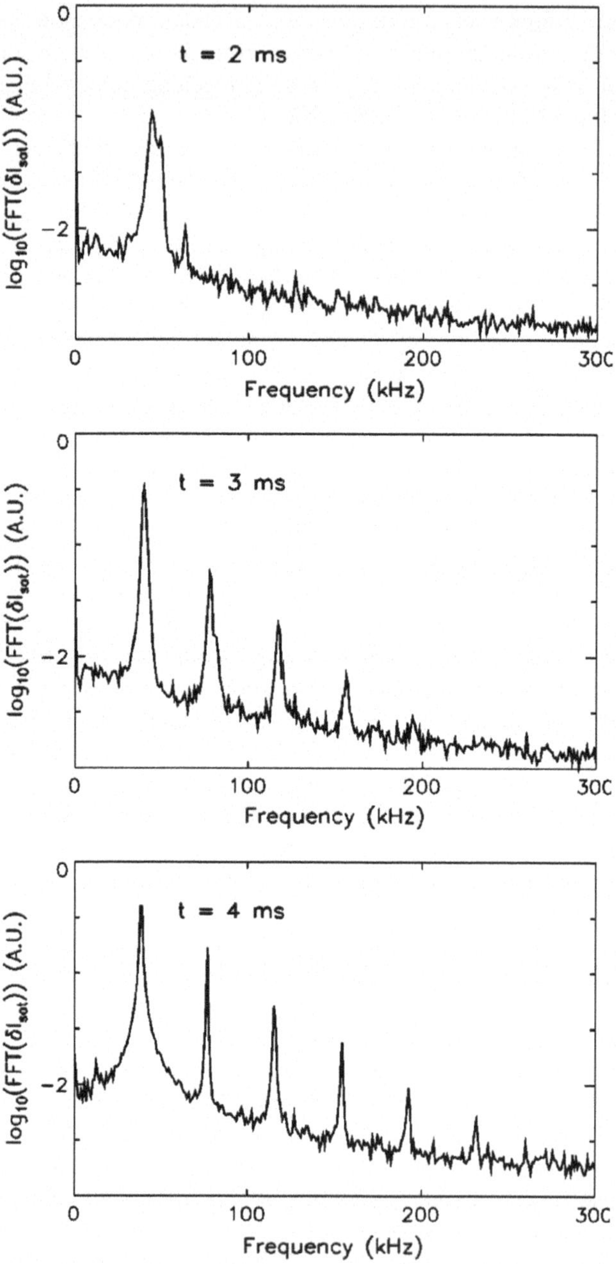

Fig. 2.7 Development of harmonic structures in frequency spectrum of ion saturation current in the evolution of the drift-KAW or two-dimensional solitary KAW (SKAW), where panels, from the top down, correspond to the harmonic structures at $t = 2$, 3, and 4 ms after the temperature filament is formed ($t = 0$) (reprinted from Burke et al., Phys. Plasmas, 7, 1397–1407, 2000b, with the permission of AIP Publishing)

energy ~20 eV, and carrying current ~250 mA into the afterglow-phase plasma that has a cooled temperature $T_e \sim 1$ eV due to the electron heat conduction, about 0.5 ms after the main discharge pulse is turned off. The heat conduction parallel to the magnetic field results in the formation of a plasma temperature striation with 8 m in length, 5 mm in radial extent, and 5 eV in electron temperature in the center of the plasma column. This plasma temperature striation has an evident temperature gradient but a nearly uniform density (because of the time scale of the ambipolar flow for the density transport much slower than that of the heat conduction) across the magnetic field and is embedded in an approximately infinite and colder plasma column.

The experiment by Pace et al. (2008a) found that the temporal evolution of temperature fluctuations caused by the plasma temperature striation exhibits three stages. The first is the initial classical transport stage lasting an interval ~2 ms after the beam injection, in which a quiescent temperature evolution is observed. The second is the nonlinear growing stage approximately starting at 2 ms after the beam injection, in which high-frequency oscillations of corresponding the coherent drift KAWs excited by the temperature gradient become apparent, as observed in the experiment by Burke et al. (2000a, b). The third is the incoherent broadband turbulence stage appearing after 5 ms of the beam injection, in which incoherent high-frequency oscillations mixed with low-frequency oscillations start to develop into a broadband turbulence.

In particular, their experiment further found that the coherent structure and turbulent spectrum are dominated by the classical and anomalous transports, respectively, as expected theoretically. The experiment also demonstrates that the broadband turbulence displays an ensemble-averaged exponential frequency spectrum, which could be traced to the generation of coherent pulse structures with a Lorentz temporal signature caused by nonlinear interactions of drift-KAWs driven by the electron temperature gradient in a microscopic scale length (6 mm). A normalized, temporal Lorentz pulse centered at time t_c and lasting interval Δt can be fitted by so-called the Lorentz function as follows:

$$g(t) = \frac{\Delta t^2}{(t - t_c)^2 + \Delta t^2}. \tag{2.22}$$

Its Fourier transform

$$\tilde{g}(\omega) = \pi \Delta t e^{-\omega \Delta t} e^{i\omega t_c} \tag{2.23}$$

has the amplitude exponentially dependent on the frequency.

For the sake of comparison, they also performed a limiter-edge experiment, in which a macroscopic density gradient (3.5 cm scale length) is established by inserting a metallic plate at the edge of the plasma column. In this case there is only a steady-state turbulent regime with a pure exponential frequency spectrum but without mixture of coherent peaks, unlike the case for the plasma temperature striation above. The broadband turbulence in this experiment has the same intrinsically turbulent nature as that in the temperature striation experiment. Moreover, a numerical model consisting of an ensemble of random Lorentz pulses is found to reproduce the exponential spectrum observed in both experiments (Shi et al. 2009). This strongly suggests that the exponential frequency spectrum is not only a signature of Lorentz pulses generated by drift-KAWs but also a possibly universal feature of pressure-gradient driven turbulence in magnetized plasmas.

2.5 Generation of KAWs by Other Plasma Phenomena

KAWs and their MHD counterparts AWs in large scales are ubiquitous in space plasmas and can be generated in various magnetic plasma phenomena and play important roles in their both dynamics and kinetics. Here we briefly introduce a few laboratory experiments of the generation and interaction of KAWs, which were carried out in the LAPD to model a few other space plasma phenomena.

Generation in exploding laser-produced plasmas

In space and astrophysical plasmas there are many situations in which a dense plasma expands into a background magnetized plasma that can support AWs and KAWs, for example, bursty bulk flows in the terrestrial magnetotail (Cao et al. 2006), coronal mass ejections in the solar corona, supernovas in collapsing stars, and astrophysical jets in active galaxies. To model such ubiquitously natural phenomena, a class of experiments which involve the expansion of a dense laser-produced plasma (LPP) into a background highly magnetized plasma has been carried out on the LAPD (Van Zeeland et al. 2001, 2003; Gekelman et al. 2003; Van Zeeland and Gekelman 2004). The LPP is generated by the irradiation of a metallic (aluminium) target with an axial incidence laser. Without a background plasma, the LPP expansion leads to a radially inward directed ambipolar electric field because the ions are nearly unmagnetized whereas the electrons are strongly magnetized. This electric field pulls outward the electrons across the magnetic field to neutralize the ions and produces an electron $\mathbf{E} \times \mathbf{B}$ drift current. The drift current is diamagnetic and completely expels the magnetic field to form a LPP diamagnetic cavity. This diamagnetic cavity then recoils and pulls the LPP to expand along the magnetic field.

When a background plasma presents, however, the electron cross-field neutralization process becomes more complex because background electrons can also flow down the magnetic field lines, which leads to the drift current reducing. Moreover, the parallel current due to the background electrons can excite KAWs, which have been identified as standing waves formed along the background He plasma column (Van Zeeland et al. 2001; Van Zeeland and Gekelman 2004). In fact, the Fourier spectrum of these standing waves displays a pronounced cutoff at the He ion cyclotron frequency as well as distinct peaks below. These peaks are similar to those observed in previous LAPD experiments on KAWs with $k_\perp \rho_s \lesssim 0.1$, which are launched using a small-scale antenna (0.635 cm in diameter) driven by a current pulse (Mitchell et al. 2001, 2002). They also found that there are many standing eigenmodes in the shear Alfvén frequency range.

Later on, similar experiments were renewedly performed in the upgraded LAPD, but by a laser beam aligned perpendicular to the background magnetic field (Van Zeeland et al. 2003; Gekelman et al. 2003). Several different stages in the progression of a LPP expanding across an ambient magnetic field were investigated more in detail.

In the initial perpendicular expansion, the initial laser impact results in immediate ionization of atoms on the target surface and a blast of fast electrons. These fast electrons rip ions from the target surface due to a large ambipolar electric field. The more massive ions hold back the electrons and eventually overshoot these strongly magnetized electrons due to their relative unmagnetized state. This leads to the presence of a radially inward directed ambipolar electric field, which causes a current due to the $\mathbf{E} \times \mathbf{B}$ drift of electrons. This electron drift current, in conjunction with the diamagnetic drift current due to the $\nabla p \times \mathbf{B}$ drift, causes the LPP diamagnetism. When a background plasma is present, the background electrons can shuttle along field lines and partially short out the initial ambipolar field allowing laser-plasma electrons to escape. The current distribution is approximately coaxial in nature with the central current channel corresponding to escaping electrons coming from the target plasma and the outer return-current channel associated with electrons from the background plasma (see Fig. 3 in Van Zeeland et al. 2003). The resulting return and escaping electron current densities depend on the electron thermal and Alfvén velocities, and the specifical relation can be fitted approximately by

$$j = c_0 j_{T_e} \left(\frac{v_{T_e}}{v_A} \right)^{c_1} \tag{2.24}$$

for the range $0.2 < v_{T_e}/v_A < 2$, where $j_{T_e} \equiv e n_0 v_{T_e}$ is the background thermal current density, and the fitting parameters $c_{0e} = 0.3 \pm 0.01$, $c_{1e} = -0.62 \pm 0.04$ for the escaping current density and $c_{0r} = 0.16 \pm 0.01$, $c_{1r} = -0.68 \pm 0.04$ for the return current density. Currently, this dependence is not understood.

In the following further perpendicular expansion, as the unmagnetized ions of the LPP continue to expand across the confining magnetic field and drag electrons along with them, the parallel electron current system induced in the background plasma becomes highly asymmetric and evolves rapidly from a pseudo-coaxial system to two antiparallel current sheets. However, the net current through any cross-section has been measured to be zero within experimental error. This can be understood by the formed charge distribution in the perpendicularly expanding LPP due to the Lorentz force pushing ions to one side and electrons toward the other. While expanding in the perpendicular direction, the LPP simultaneously expands along the magnetic field at an average velocity $v_\parallel \simeq 18 c_s \simeq 0.18 v_A$.

The varying current system described above can be expected to radiate KAWs. The measured radiation KAW magnitude increases with v_{T_e}/v_A. For this dependence, a reasonable explanation is because the escaping and return currents having higher magnitudes for larger values of v_{T_e}/v_A. This indicates that it is possible to increase the radiated KAW amplitude simply by increasing the background plasma $\beta \propto v_{T_e}^2/v_A^2$. Further analysis shows that the relative magnetic fluctuation $\delta B_w/B_0$ can be fitted approximately by

$$\frac{\delta B_w}{B_0} = c_{0b} \left(\frac{v_{T_e}}{v_A} \right)^{c_{1b}} \tag{2.25}$$

within the range $0.2 < v_{T_e}/v_A < 2$, where the fitting parameters $c_{0b} = (3.6 \pm 0.2) \times 10^{-3}$ and $c_{1b} = 0.32 \pm 0.06$.

Generation by Alfvén maser

It is well known that laser or maser action requires a combination of a resonator cavity containing a nonequilibrium medium which is able to amplify selectively the wave over a narrow frequency interval and a partial reflector which allows the amplified wave to be repetitively reflected by the resonator cavity back and forth through the medium as well as the wave beam to be transmitted into the outside region. Because of the spatial variation of the ionospheric plasma parameters, the region between the lower E region of the ionosphere and the topside ionosphere can form a natural Alfvén resonator. The topside ionosphere acts as a partial reflector to allow amplified AWs to propagate into the magnetosphere. Possible nonequilibrium sources to excite the ionospheric Alfvén resonator include the natural magnetospheric convection, electron beams, lightning discharges, and powerful high-frequency radio waves, and so on.

In the upgraded LAPD, an equivalent Alfvén resonator cavity has been realized by applying a nonuniform magnetic field to a plasma region bounded between a cathode and a semitransparent mesh anode (Maggs and Morales 2003; Maggs et al. 2005). The semitransparent mesh anode acts as a partial reflector and the nonequilibrium source is the plasma discharge current. When the current exceeds a threshold, the selective amplification results in a highly coherent with $\delta\omega/\omega \simeq 5 \times 10^{-3}$, large amplitude wave to be excited spontaneously from ambient noise and to propagate out of the resonator cavity, through the semitransparent mesh anode, into an adjacent plasma column, where the magnetic field is uniform. The Alfvén maser is observed to evolve in time from an initial $m = 0$ to $m = 1$ mode structure in the transition to the late steady state. The frequency of the maser mode is continuously tunable by changing the strength of the confinement field, which is determined by the cathode-anode separation and the dispersion relation (Maggs et al. 2005).

Carter et al. (2006) experimentally investigated nonlinear interaction between two KAWs, which were produced by the Alfvén resonator cavity of the upgraded LAPD. They found that the nonlinear interaction led to the presence of a pseudomode at the beat frequency, which represents a driven oscillation and does not correspond to a linear normal mode but effectively scatters the KAWs and forms a series of sidebands. A discharge current ($\sim$5 kA) between the cathode and the semitransparent anode (0.55 m separation) generates a beam of primary electrons (typical energy 50 eV) in the Alfvén resonator cavity, which enters the main LAPD chamber and then ionizes and heats the helium gas there to form the helium plasma column, in which the current due to the injected primary electrons can be balanced by a return current due to back-drifting population of thermal electrons (Maggs and Morales 2003). In the main plasma column, typical parameters are $n_e \sim 1 \times 10^{12}$ cm^{-3}, $T_e \sim 6$ eV, $T_i \sim 1$ eV, $B < 2$ kG, and hence $\beta \gtrsim m_e/m_i$, implying that the KAWs in the experiment have short-wavelength dispersion due to finite $k_\perp \rho_s$ (here $\rho_s \sim 0.5 - 1.5$ cm) and finite-frequency dispersion due to finite ω/ω_{ci}.

In the steady state phase of the discharge, the power spectrum of the magnetic fluctuation $\delta B_\perp$ shows that there are two strong modes (identified as $m = 0$ and $m = 1$), which are surrounded by a number of sidebands. Simultaneously, the saturation current I_s fluctuations are also observed at the sideband separation frequency and harmonics (see Fig. 2 in Carter et al. 2006). They thought that when the two large amplitude KAWs (e.g., $m = 1$ and $m = 0$) emitted from the cavity with differing frequency simultaneously propagate in the main plasma column at slightly different phase velocities, they pass through one another and beat together, driving a fluctuation at the beat frequency. The driven fluctuation then scatters the incident KAWs, leading to the generation of a series of sidebands due to the nonlinear beat-wave interaction between the two incident KAWs. The steady-state power spectrum also showed that the beat-wave interaction is very strong, with the amplitude of the driven pseudomodes at the beat frequency and harmonic frequencies comparable to that of the interacting KAWs.

Auerbach et al. (2010) studied the nonlinear beat-wave interaction between two independent KAWs with different frequencies propagating along a narrow, field-aligned density depletion embedded in the center of the LAPD plasma column. As demonstrated in the previous LAPD experiments by Maggs and Morales (1996, 1997), the density gradient associated with such a density striation will drive the drift-KAW instability. To investigate the interaction between the beating KAWs and the drift-KAW instability driven by the density depletion, the beat frequency was set near the instability frequency (6 kHz) in the experiments. The experimental results show that the strongest beat-driven response appears at the beat frequency close to this instability frequency, with peaks centered at 5.5 and 8.25 kHz. Moreover, the measured beat-wave response has a peak amplitude, $\delta I_s/I_s \sim 15\%$, significantly stronger than a beat-wave amplitude of $\delta I_s/I_s \sim 5\%$ measured in the absence of the density depletion. The most striking feature of this interaction, however, is that when the beat KAW has sufficient amplitude, the original unstable drift-KAW is suppressed, leaving only the beat-KAW response, generally at lower amplitude. This provides a novel approach to depressing instabilities driven by density (or temperature) gradients in magnetized laboratory plasmas (Auerbach et al. 2010).

Generation by strong microwave pulses

The physics of the interaction between plasmas and high power waves at the Langmuir frequency is of importance in many areas of space and plasma physics. A number of laboratory research has been done on the interaction of microwaves in a density gradient when $\omega = \omega_{pe}$ in unmagnetized plasmas (Stenzel et al. 1974; Kim et al. 1974; Wong and Stenzel 1978). In the case of magnetized plasmas, which are capable of supporting KAWs, the laboratory experimental investigation of the interaction between a high power microwave pulse at the electron plasma frequency and a magnetized plasma was carried out first by Van Compernolle et al. (2005, 2006) on the upgraded LAPD. In their experiment, the incident wave vector of the microwaves is perpendicular to the background magnetic field, and parallel to the density gradient. The plasma-microwave interaction first leads to the production of field-aligned suprathermal electrons and then KAWs in the electron inertial length scale are excited by these field-aligned suprathermal electrons.

The ambient plasma is formed by a pulsed discharge (with current $I_d \simeq 3.5$ kA, lasting for 8 – 10 ms and pulsed rate at 1 Hz) between cathode and mesh anode apart in 52 cm. After the electron beam formed by the discharge enters the main chamber of the LAPD they collisionally ionize the helium gas there and form the main plasma column of 18 m in long and 50 cm in diameter, which is more than a hundred ion gyroradii at 1.0 kG. The background magnetic field is regulated in a range from 1.0 to 2.25 kG. The experiments were done in the afterglow (~0.5 ms after the discharge termination), so that the electron and ion populations are both cold ($T_e = T_i \lesssim 0.5$ eV). The microwave source generates 0.5 (or 2.5) μs pulses with a power of 70 kW at a frequency of $\omega_{mw} = 9$ GHz, corresponding to a critical plasma density $n_c = 10^{12}$ cm^{-3} determined by $\omega_{pe}(n) = \omega_{mw} = 9$ GHz. This critical layer is typically 10 to 15 cm from the center axis of the LAPD plasma column and the peak plasma density is 1.52×10^{12} cm^{-3}, so that the center of the plasma column is optically thick for the incoming microwaves. The launched microwaves propagate radially, across the background magnetic field, into the radial density gradient of the plasma column. The polarization of the microwaves can be chosen as $\mathbf{E} = E_z\hat{\mathbf{z}}$ ($\parallel \mathbf{B}_0$) for O mode or as $\mathbf{E} = E_y\hat{\mathbf{y}}$ ($\perp \mathbf{B}_0$) for X mode.

The measurements of the localized electric field strength, which is caused by the incident microwaves, showed that the amplitude associated with the O mode peaks at densities just below the O mode cutoff layer where $\omega_{pe}(n) \lesssim \omega_{mw} = 9$ GHz, as expected from the theoretical dispersion relation. For the X mode, however, the associated electric field intensity peaks on the overdense side of the upper hybrid layer where $\sqrt{\omega_{pe}^2(n) + \omega_{ce}^2(n)} \gtrsim \omega_{mw}$, implying that tunneling to the upper-hybrid resonance is achieved. The measurement also indicates the microwaves in the X mode can propagate up to the L cutoff layer where $\sqrt{\omega_{pe}^2(n) + \omega_{ce}^2(n)/4} - \omega_{ce}/2 \lesssim \omega_{mw}$, which is consistent with the theoretical dispersion relation of the X mode.

Signatures of suprathermal electrons can be found in the data of the ion saturation current collected by the Mach probe, which was biased negatively, with a bias typically larger than 50 V. Evidence of suprathermal electrons was also obtained from the field-aligned current, which can be calculated based on the magnetic field data. The main difference between the two measurements is that the magnetic field measured the current due to all electrons, not just the current due to the suprathermal electrons. The two measurements about the suprathermal electrons have reasonable consistence. In particular, the measured results clearly display that both the microwaves and the suprathermal electrons have their peaks at the same locations with respect to the cutoff layer, implying that the suprathermal electrons are accelerated field-aligned in the vicinity of these strong electric fields and have an average drift velocity $v_{ez} \gtrsim v_A$. Further to compare the measurements of the suprathermal electrons accelerated by the O mode and by the X mode, it can be found that they both are linearly proportional to the microwave input power, but there is a threshold value for the microwave power, below which no suprathermal electrons are observed.

Moreover, the threshold is higher for X mode polarized waves by a factor of 2 than for O mode polarized waves. This possibly is associated with the different propagation characteristics for the O mode and X mode. However, the origin of this threshold is still an open question.

The experimental measurements further showed that the suprathermal electrons were followed by a series of magnetic field oscillations recorded by a 3-axis magnetic loop probe. These magnetic perturbations can be identified as arising from KAWs of the inertial regime. For different magnetic fields and different microwave powers, the frequency of the KAWs always remains $\omega \lesssim \omega_{ci}$ and their generation can be explained by Cherenkov radiation of KAWs by the suprathermal electrons (Van Compernolle et al. 2008).

2.6 Laboratory Experiments Versus Space Observations

Since discovered in 1970's, KAWs, as dispersive and dissipative AWs in the kinetic scales of plasma particles, have been believed extensively to play important roles in the auroral electron acceleration, the anomalous transport of the magnetopause particles, the dynamical dissipation and evolution of the solar wind AW turbulence spectrum, the inhomogeneous heating of the solar corona plasma, and the formation of fine magnetic-plasma structures in space and astrophysical plasmas. Professor Hannes Alfvén had believed that theories of cosmic phenomena must also agree with known results from laboratory experiments on Earth, because the same laws of Nature must apply everywhere (Fälthammar 1995). Indeed, it appears that the phenomena observed in laboratory experiments show striking similarities to what has been observed in space (Gekelman 1999). For example, comparing the in situ space explorations and the laboratory experimental measurements of KAWs, it can be found that there are extensive similarities between them, including the two dimensional vortex structures of KAWs in the space exploration (Chmyrev et al. 1988; Volwerk et al. 1996; Wu et al. 1996a, 1997) and in the laboratory experiment (Burke et al. 2000a, b), the plasma striations associated with KAWs in the laboratory experiment (Maggs and Morales 1996, 1997) and in the space exploration (Stasiewicz et al. 1997, 2001; Bellan and Stasiewicz 1998; Sharma and Singh 2009; Kumar et al. 2011), and the broadband KAW turbulence in the laboratory experiment (Burke et al. 2000a, b) and in the space exploration (Stasiewicz 2000a; Stasiewicz et al. 2001, 2004; Chaston et al. 2007b; Pace et al. 2008a, b; Salem et al. 2012).

The greatest advantage of laboratory experiments is that the experimental conditions and parameters are repeatable and controllable. In addition, laboratory devices can be reconfigured to perform different experiments and probes can be removed and repaired. Therefore, when an interesting phenomenon has been discovered by a satellite, but its basic physics is not well understood, the laboratory can provide us an ideal place further to study it. In laboratory plasmas, however, probes are easily larger than both electron gyroradii and the Debye length, so that they can perturb the plasma. On the other hand, although satellite is usually larger than these fundamental lengths, but onboard detectors are generally smaller than the fundamental lengths. Thus, the satellite imposes a minimum perturbation on its plasma environment. Moreover, there must be some phenomena that naturally occur in space plasmas, but can not be produced in the laboratory plasma device, even in a device large as LAPD, because laboratory devices can not provide an enough large size or enough long time to produce those phenomena naturally to occur. Therefore, the joint complementarity between the space exploration and laboratory experiment is necessary for us comprehensively to understand the physics of KAWs and their applications in space plasmas.

As Gekelman (1999) pointed out: "In the past few years the quantum jump in data collection on the Freja and FAST missions have led to the reevaluation of the importance of these waves in the highly structured plasma that was probed." Measurements by satellites in space plasmas show that KAW, especially KAW turbulence is ubiquitous in various space plasma environments from the lower-altitude magnetosphere with a plasma $\beta < Q$ (Stasiewicz et al. 2000b; Chaston et al. 2007b) to the magnetopause with a plasma $\beta > Q$ (Stasiewicz et al. 2001), the magnetosheath (Stasiewicz et al. 2004), and the solar wind (Salem et al. 2012) with a plasma $\beta \sim 1$. The observed KAW turbulent spectra are most often associated with broadband ELF emissions and extend from much lower the local ion gyrofrequency f_{ci}, up to the lower hybrid frequency f_{LH}. Suprathermal electron bursts with energies reaching a few hundred eV and transverse ion heating to similar energies often are observed to be associated with the KAW turbulence and are situated in larger scale cavity regions depleted in plasma Stasiewicz et al. (2000a).

The rest chapters of this book all devote to the observations and applications of KAWs in space and astrophysical plasmas, including the terrestrial auroral and magnetosheath plasmas and the solar wind and coronal plasmas, as well as remoter extrasolar plasmas. The plasma β parameter, the kinetic to magnetic pressure ratio, can vary remarkably from much less than the electron to ion mass ratio $Q \ll 1$ to larger than one in these various astrophysical plasma environments above. In consequence, KAWs play significantly different roles in these various plasma environments because of the sensitive dependence of their physics on the plasma β parameter, moreover, the space explorations of KAWs from the various space plasma environments provide natural laboratories for us to study comprehensively the physics of KAWs, as you will find in the rest chapters of this book.

References

Alfvén, H. (1942). Existence of electromagnetic-hydrodynamic waves. *Nature*, *150*, 405–406.

Auerbach, D. W., Carter, T. A., Vincena, S., & Popovich, P. (2010). Control of gradient-driven instabilities using shear Alfvén beat waves. *Physical Review Letters*, *105*, 135005.

Bellan, P. M., & Stasiewicz, K. (1998). Fine-scale cavitation of ionospheric plasma caused by inertial Alfvén wave ponderomoive force. *Physical Review Letters*, *80*, 3523–3526.

Bhatnagar, P. L., Gross, E. P., & Krook, M. (1954). A model for collision processes in gases. I. Small amplitude processes in charged and neutral one-component systems. *Physical Review*, *94*, 511–525.

Boehm, M. H., Carlson, C. W., McFadden, J. P., et al. (1990). High resolution sounding rocket observations of large amplitude Alfvén waves. *Journal of Geophysical Research*, *95*, 12157–12171.

Borg, G. G., Brennan, M. H., Cross, R. C., et al. (1985). Guided propagation of Alfvén waves in a toroidal plasma. *Plasma Physics and Controlled Fusion*, *27*, 1125–1149.

Burke, A. T., Maggs, J. E., & Morales, G. J. (2000a). Spontaneous fluctuations of a temperature filament in a magnetized plasma. *Physical Review Letters*, *84*, 1451–1454.

Burke, A. T., Maggs, J. E., & Morales, G. J. (2000b). Experimental study of fluctuations excited by a narrow temperature filament in a magnetized plasma. *Physics of Plasmas*, *7*, 1397–1407.

Cao, J. B., Ma, Y. D., Parks, G., et al. (2006). Joint observations by cluster satellites of bursty bulk flows in the magnetotail. *Journal of Geophysical Research*, *111*, 2741–2760.

Carter, T. A., Brugman, B., Pribyl, P., & Lybarger, W. (2006). Laboratory observation of a nonlinear interaction between shear Alfvén waves. *Physical Review Letters*, *96*, 155001.

Chaston, C. C., Hull, A. J., Bonnell, J. W., et al. (2007b). Large parallel electric fields, currents, and density cavities in dispersive Alfvén waves above the aurora. *Journal of Geophysical Research*, *112*, A05215.

Chen, L., & Hasegawa, A. (1974a). Plasma heating by spatial resonance of Alfvén wave. *Physics of Fluids*, *17*, 1399–1403.

Chen, L., & Hasegawa, A. (1974b). A theory of long-period magnetic pulsations, 1. Steady state excitation of filed line resonance. *Journal of Geophysical Research*, *79*, 1024–1032.

Chen, L., & Hasegawa, A. (1974c). A theory of long-period magnetic pulsations, 2. Impulse excitation of surface eigenmode. *Journal of Geophysical Research*, *79*, 1033–1037.

Chmyrev, V. M., Bilichenko, S. V., Pokhotelov, O. A., et al. (1988). Alfvén vortices and related phenomena in the ionosphere and the magnetosphere. *Physica Scripta*, *38*, 841–854.

Cross, R. C. (1983). Experimental observations of localized Alfvén and ion acoustic waves in a plasma. *Plasma Physics*, *25*, 1377–1387.

Cross, R. C. (1988). *An introduction to Alfvén Waves*. Bristol, England: Adam Hilger.

Demirkhanov, R. A., Kirov, A. G., Lozovskii, S. N., et al. (1977). Plasma heating in a toroidal system by a helical quadrupole RF field with $\omega < \omega_{Bi}$. *Plasma Physics Controlled Nuclear Fusion Research*, *3*, 31–37.

Erlandson, R. E., Zanetti, L. J., Acuna, M. H., et al. (1994). Freja observations of electromagnetic ion cyclotron waves and tranverse oxygen ion acceleration on auroral field lines. *Geophysical Research Letters*, *21*, 1855–1858.

Fälthammar, C. G. (1995). In memoriam: Hannes Alfvén. *Astrophysics and Space Science*, *234*, 173–175.

Gekelman, W. (1999). Review of laboratory experiments on Alfvén waves and their relationship to space observations. *Journal of Geophysical Research*, *104*, 14417–14435.

Gekelman, W., Leneman, D., Maggs, J. E., & Vincena, S. (1994). Experimental observation of Alfvén cones. *Physics of Plasmas*, *1*, 3775–3783.

Gekelman, W., Pfister, H., Lucky, Z., et al. (1991). Design, construction, and properties of the large plasma research device - The LAPD at UCLA. *Review of Scientific Instruments*, *62*, 2875–2883.

Gekelman, W., Van Zeeland, M., Vincena, S., & Priby, P. (2003). Laboratory experiments on Alfvén waves caused by rapidly expanding plasmas and their relationship to space phenomena. *Journal of Geophysical Research*, *108*, 1281–1291.

Golovato, S. N., Shohet, J. L., & Tataronis, J. A. (1976). Alfvén wave heating in the Proto-Cleo stellarator. *Physical Review Letters*, *37*, 1272–1274.

Gross, E. P. (1951). Plasma oscillations in a static magnetic field. *Physical Review*, *82*, 232–242.

Hasegawa, A. (1976). Particle acceleration by MHD surface wave and formation of aurora. *Journal of Geophysical Research*, *81*, 5083–5090.

Hasegawa, A., & Chen, L. (1974). Plasma heating by Alfvén wave phase mixing. *Physical Review Letters, 32*, 454–456.

Hasegawa, A., & Chen, L. (1975). Kinetic process of plasma heating due to Alfvén wave excitation. *Physical Review Letters, 35*, 370–373.

Hasegawa, A., & Chen, L. (1976a). Kinetic process of plasma heating by resonant mode conversion of Alfvén wave. *Physical Fluids, 19*, 1924–1934.

Hasegawa, A., & Uberoi, C. (1982). *The Alfvén Waves*. Tech: Inf. Center, US Dept. of Energy, Oak Ridge.

Keller, R., & Pochelon, A. (1978). Alfvén wave heating of a theta pinch. *Nuclear Fusion, 18*, 1051–1057.

Kim, H. C., Stenzel, R. L., & Wong, A. Y. (1974). Development of cavitons and trapping of RF field. *Physical Review Letters, 33*, 886–889.

Kletzing, C. A., Bounds, S. R., Martin-Hiner, J., et al. (2003a). Measurements of the shear Alfvén wave dispersion for finite perpendicular wave number. *Physical Review Letters, 90*, 035004.

Kletzing, C. A., Thuecks, D. J., Skiff, F., et al. (2010). Measurements of inertial limit Alfvén wave dispersion for finite perpendicular wave number. *Physical Review Letters, 104*, 095001.

Koch, R. A., & Horton, W. (1975). Effects of electron angle scattering in plasma waves. *Physical Fluids, 18*, 861–865.

Kumar, S., Sharma, R. P., & Singh, H. D. (2011). Cavitation by nonlinear interaction between inertial Alfvén waves and magnetosonic waves in low beta plasmas. *Solar Physics, 270*, 523–535.

Leneman, D., Gekelman, W., & Maggs, J. E. (1999). Laboratory observations of shear Alfvén waves launched from a small source. *Physical Review Letters, 82*, 2673–2676.

Leneman, D., Gekelman, W., & Maggs, J. E. (2000). Shear Alfvén wave radiation from a source with small transverse scale length. *Physical Plasmas, 7*, 3934–3946.

Leneman, D., Gekelman, W., & Maggs, J. E. (2006). The plasma source of the Large Plasma Device at University of California, Los Angeles. *Review of Scientific Instruments, 77*, 015108.

Louarn, P., Wahlund, J. E., Chust, T., et al. (1994). Observations of kinetic Alfvén waves by the Freja spacecraft. *Geophysical Research Letters, 21*, 1847–1850.

Lundin, R., Eliasson, L., Herendel, G., et al. (1994a). Large-scale auroral plasma density cavities observed by Freja. *Geophysical Research Letters, 21*, 1903–1906.

Maggs, J. E., & Morales, G. J. (1996). Magnetic fluctuations associated field-aligned striations. *Geophysical Research Letters, 23*, 633–636.

Maggs, J. E., & Morales, G. J. (1997). Fluctuations associated a filamentary density depletion. *Physical Plasmas, 4*, 290–299.

Maggs, J. E., & Morales, G. J. (2003). Laboratory realization of an Alfvén wave maser. *Physical Review Letters, 91*, 035004.

Maggs, J. E., Morales, G. J., & Carter, T. A. (2005). An Alfvén wave maser in the laboratory. *Physical Plasmas, 12*, 013103.

Mitchell, C., Maggs, J. E., & Gekelman, W. (2002). Field line resonances in a cylindrical plasma. *Physical Plasmas, 9*, 2009–2018.

Mitchell, C., Vincena, S., Maggs, J. E., & Gekelman, W. (2001). Laboratory observation of Alfvén resonance. *Geophysical Research Letters, 28*, 923–926.

Morales, G. J., & Maggs, J. E. (1997). Structure of kinetic Alfvén waves with small transverse scale length. *Physical Plasmas, 4*, 4118–4125.

Morales, G. J., Loritsch, R. S., & Maggs, J. E. (1994). Structure of Alfvén waves at the skin-depth scale. *Physical Plasmas, 1*, 3765–3774.

Obiki, T., Mutoh, T., Adachi, S., et al. (1977). Alfvén-wave heating experiment in the Heliotron-D. *Physical Review Letters, 39*, 812–815.

Pace, D. C., Shi, M., Maggs, J. E., et al. (2008a). Exponential frequency spectrum in magnetized plasmas. *Physical Review Letters, 101*, 085001.

Pace, D. C., Shi, M., Maggs, J. E., et al. (2008b). Exponential frequency spectrum and Lorentzian pulses in magnetized plasmas. *Physical Plasmas, 15*, 122304.

Salem, C. S., Howes, G. G., Sundkvist, D., et al. (2012). Identification of kinetic Alfvén wave turbulence in the solar wind. *The Astrophysical Journal Letters, 745*, L9.

Sharma, R. P., & Singh, H. D. (2009). Density cavities associated with inertial Alfvén waves in the auroral plasma. *Journal of Geophysical Research, 114*, A03109.

Shi, M., Pace, D. C., Morales, G. J., et al. (2009). Structures generated in a temperature filament due to drift-wave convection. *Physical Plasmas, 16*, 062306.

Stasiewicz, K., Bellan, P., Chaston, C., et al. (2000a). Small scale Alfvénic structure in the aurora, Space Sci. *Review, 92*, 423–533.

Stasiewicz, K., Gustafsson, G., Marklund, G., et al. (1997). Cavity resonators and Alfvén resonance cones observed on Freja. *Journal of Geophysical Research, 102*, 2565–2575.

Stasiewicz, K., Khotyaintsev, Y., & Grzesiak, M. (2004). Dispersive Alfvén Waves observed by Cluster at the magnetopause. *Physica Scripta, T107*, 171–179.

Stasiewicz, K., Khotyaintsev, Y., Berthomier, M., & Wahlund, J. E. (2000b). Identification of widespread turbulence of dispersive Alfvén waves. *Geophysical Research Letters, 27*, 173–176.

Stasiewicz, K., Seyler, C. E., Mozer, F. S., et al. (2001). Magnetic bubbles and kinetic Alfvén waves in the high-latitude magnetopause boundary. *Journal of Geophysical Research, 106*, A29503–29514.

Stenzel, R. L., Wong, A. Y., & Kim, H. C. (1974). Conversion of electromagnetic waves to electrostatic waves in inhomogeneous plasmas. *Physical Review Letters, 32*, 654–657.

Thuecks, D. J., Kletzing, C. A., Skiff, F., et al. (2009). Tests of collision operators using laboratory experiments of shear Alfvén wave dispersion and damping. *Physical Plasmas, 16*, 052110.

Van Compernolle, B., Gekelman, W., & Pribyl, P. (2006). Generation of suprathermal electrons and Alfvén waves by a high power pulse at the electron plasma frequency. *Physical Plasmas, 13*, 092112.

Van Compernolle, B., Gekelman, W., Pribyl, P., & Carter, T. A. (2005). Generation of Alfvén waves by high power pulse at the electron plasma frequency. *Geophysical Research Letters, 32*, L08101.

Van Compernolle, B., Morales, G. J., & Gekelman, W. (2008). Cherenkov radiation of shear Alfvén waves. *Physical Plasmas, 15*, 082101.

Van Zeeland, M., & Gekelman, W. (2004). Laser-plasma diamagnetism in the presence of an ambient magnetized plasma. *Physical Plasmas, 11*, 320–323.

Van Zeeland, M., Gekelman, W., Vincena, S., & Dimonte, G. (2001). Production of Alfvén waves by a rapidly expanding dense plasma. *Physical Review Letters, 87*, 105001.

Van Zeeland, M., Gekelman, W., Vincena, S., & Maggs, J. (2003). Currents and shear Alfvén wave radiation generated by an exploding laser-produced plasma: Perpendicular incidence. *Physical Plasmas, 10*, 1243–1252.

Vincena, S., Gekelman, W., & Maggs, J. E. (2004). Shear Alfvén wave perpendicular propagation from the kinetic to the inertial regime. *Physical Review Letters, 93*, 105003.

Volwerk, M., Louarn, P., Chust, T., et al. (1996). Solitary kinetic Alfvén waves-A study of the Poynting flux. *Journal of Geophysical Research, 101*, 13335–13343.

Weisen, H., Appert, K., Borg, G. G., et al. (1989). Mode conversion to the kinetic Alfvén wave in low-frequency heating experiments in the TCA tokamak. *Physical Review Letters, 63*, 2476–2479.

Wong, A. Y., & Stenzel, R. L. (1978). Ion acceleration in strong electromagnetic interactions with plasmas. *Physical Review Letters, 34*, 727–730.

Wu, D. J. (2012). *Kinetic Alfvén Wave: Theory*. Experiment and Application: Science Press, Beijing.

Wu, D. J., & Chao, J. K. (2004b). Recent progress in nonlinear kinetic Alfvén waves. *Nonlinear Processes in Geophysics, 11*, 631–645.

Wu, D. J., Huang, G. L., & Wang, D. Y. (1996a). Dipole density solitons and solitary dipole vortices in an inhomogeneous space plasma. *Physical Review Letters, 77*, 4346–4349.

Wu, D. J., Wang, D. Y., & Huang, G. L. (1997). Two dimensional solitary kinetic Alfvén waves and dipole vortex structures. *Physical Plasmas, 4*, 611–617.

Chapter 3
KAWs in Magnetosphere-Ionosphere Coupling

3.1 Solar Wind-Magnetosphere-Ionosphere Interaction

Besides the sun shines on the earth there is an invisible direct interaction between the sun and the earth, that is the impact of the solar wind on the earth. The conception of intermittent plasma streams consisting of electrons and ions from the sun was introduced by Chapman and Ferraro (1930) to explain the cause of geomagnetic storms, and later the existence of the solar wind continuously blowing away from the sun was proposed successively by Biermann (1948, 1954, 1957) and Alfvén (1957) in their studies on comet tails to explain the deviation of the comet tail direction from the solar-comet radial direction (i.e., the solar radiation pressure direction) found by Hoffmeister in 1943. The theoretical model of the solar wind was presented by Parker (1958) and subsequently was confirmed by in situ measurement of the Explorer 10 satellite (Heppner et al. 1962; Bridge et al. 1962; Bonetti et al. 1963). Now it is generally known that the solar wind is a supersonic plasma stream with a very high Mach number of $M \sim 10$, which consists of electrons, protons, and ionized helium and minor other heavy ions as well as interplanetary magnetic fields originating from the sun and flows continuously at a high velocity ~300–800 km/s from the solar outermost atmosphere, that is, the solar corona.

The earth's magnetosphere, which is propped up by the earth's intrinsic dipolar magnetic field with a roughly north-south orientation, forms a huge umbrella to protect our living environment from the direct impinging of the solar wind plasma flow and other cosmic energetic charged particles. In fact, because of the interaction of the high velocity magnetized plasma flow of the solar wind with the geomagnetic field, the magnetosphere is compressed to $\sim 10R_E$ on the dayside and to $\sim 20R_E$ on the flank. While on the nightside the magnetosphere is stretched out into a long magnetotail extending beyond $\sim 10^3 R_E$. In result, the upper boundary of the magnetosphere, called the magnetopause, is shaped by the solar wind into a windsock, inside which a large magnetic cavity is carved out by the geomagnetic field in the solar wind, called the magnetosphere first named by Gold (1959). Since entering the space age,

D.-J. Wu and L. Chen, *Kinetic Alfvén Waves in Laboratory, Space, and Astrophysical Plasmas*, Atmosphere, Earth, Ocean & Space,
https://doi.org/10.1007/978-981-13-7989-5_3

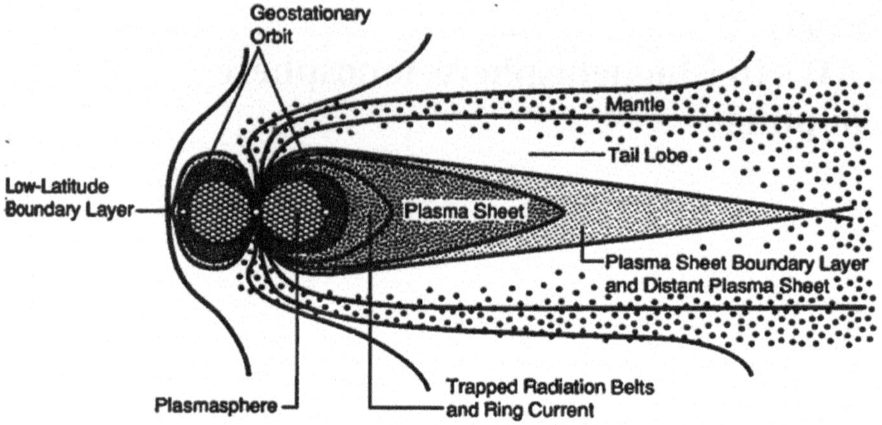

Fig. 3.1 A sketch of the magnetospheric structure as viewed in the noon-midnight meridian plane (from Wolf 1995)

a large number of in situ observations by satellites show that the magnetosphere is a fascinating mixture repository of complex plasma components and electromagnetic waves, including various complex plasma regions, such as the plasmasphere, the van Allen radiation belt, low latitude boundary layer (LLBL), high latitude boundary layer (HLBL, also called plasma mantle), central plasma sheet (CPS, also called plasma sheet), plasma sheet boundary layer (PSBL), plasma tail lobes, and so on. Figure 3.1 shows a sketch of the magnetospheric structure, in which some major plasma regions are briefly introduced below.

The plasmasphere, ranging from $\sim$1000 km above the ground to a geocentric distance $\sim 6R_E$ (or i.e., $\sim 5R_E$ above the ground), is one of the major regions of the magnetosphere, in which plasmas have typical density of $n \sim 10^3$ cm^{-3} and temperature of $T \sim 1$ eV. These dense and cold plasmas are believed to come from so-called the "polar wind" originating from the upper-ionosphere, where the plasma has a temperature close to the gravitational binding energy (Banks and Holzer 1969). The plasma density of the plasmashphere has a sharp decrease at a boundary layer at altitudes of $\sim$3–5R_E, called the "plasmapause". Besides these dense cold plasmas from the ionosphere, however, a large number of energetic particles, as tenuous and hot components, also coexist in approximately this region, that is, so-called the van Allen radiation belts of mainly consisting of energetic particles trapped in the magnetic field of the magnetosphere. These energetic particles have energies typically ranging from keV to MeV and there is a peak at around $\sim$100 keV in the energy spectrum of the ions. Below 10 keV, O^+ is the dominant ion, but H^+ begins to dominate above 50 keV. In particular, these energetic ions contribute the majority of the ring current, in which about half of the ring current is contributed by ions below 85 keV and other half by ions above 85 keV. In addition, energetic electrons inside the radiation belts contribute relatively little to the ring current as well as the total energy of trapped particles. Moreover, unlike the case of energetic ions, the radial distribution of the flux of energetic electrons above 1 MeV has one obvious "slot" centered about $\sim 2.2R_E$, that is, a minimum in fluxes, and the inner and outer parts of this slot are called the "inner radiation belt" and "outer radiation belt", respectively.

On the other hand, the magnetotail on the nightside consists of three major regions. One is the north and south tail lobes between the plasma mantle (i.e., the HLBL) and the PSBL, in which there are open magnetic fields and low-density cold plasma originating from the ionosphere (density below 0.1 cm^{-3}, energy below 1 keV). The CPS is located on the outer side of the nightside plasmasphere and consists of hot plasmas (of keV energies) with densities typically $\sim$0.1–1 cm^{-3}. In the CPS the magnetic fields are closed mostly, although sometimes there are some closed loops of magnetic flux, called "plasmoids", that do not connect to the earth or solar wind. Observations show that the CPS ion population is a mixture of ionospheric and solar wind particles (being mostly of solar-wind origin in quiet times and mostly ionospheric in active times) and has the temperature about several times of the electron temperature and flow velocities much less than the ion thermal velocity. The PSBL is a transition region between the almost empty tail lobes and the hot CPS, in which ions with densities of the order of 0.1 cm^{-3} typically exhibit flow parallel or antiparallel to the local magnetic field at velocities of hundreds of km/s, higher than their thermal velocity. Thus, counterstreaming ion beams of propagating earthward and tailward along the local magnetic field can be observed frequently in the PSBL. In particular, the counterstreams tend to be unstable to various plasma waves. Therefore, various plasma waves and wave-particle interactions are very essential and ubiquitous kinetic processes in collisionless magnetospheric plasmas.

The lower boundary of the magnetosphere is the ionosphere at an altitude range $\sim$100 to $\sim$1000 km above the ground. The existence of the ionosphere, as an electrically conducting layer of the upper atmosphere, was proposed early by Stewart in 1882 to be responsible for the solar-modulated variations of the geomagnetic field that was postulated by Gauss. Later in 1902 Kennely and Heaviside suggested that this conducting layer, initially known as the Kennely-Heaviside layer, is able to reflect radio waves and hence may explain the Marconi radio experiment on transatlantic radio transmissions between England and Canada in 1901. The existence of the ionosphere, however, was verified first by Appleton and Barnett in the United Kingdom in 1924 and by Breit and Tuve in America in 1925 via the reflecting experiment of radio signals, later named the "ionosphere" by Watson-Watt and Appleton in 1926 (see Chian and Kamide 2007). Subsequently, Chapman (1931a, b) developed a theory of upper atmospheric ionization to explain the formation of the ionosphere and its structuring into layers that were found by Appleton, in which, the ionosphere, as a by-product of the sun-earth interaction, is produced mainly via the photoionization by the solar radiation in the ultraviolet and extreme ultraviolet bands. Therefore, the ionosphere consists of dense, collisional, and partially ionized plasmas and acts as a transition layer from the fully ionized magnetospheric plasma to the neutral atmosphere on the ground. In particular, the ionosphere can absorb the major energy of high-energy photons from the solar radiation as well as of energetic particles from cosmic rays, solar energetic particles, and magnetospheric hot particles. Of course, these energetic particles also may lead to the ionosphere ionization via the impact ionization. The ionization in the ionosphere increases with the altitude and the plasma density reaches its peak $n_e \sim 10^5$ to 10^6 cm^{-3} at the altitude of ~250 km.

The electrodynamics of the solar wind-magnetosphere-ionosphere interaction can be described by the so-called global current system, which consists of main six types of large scale currents. One is the magnetopause boundary current flowing on the magnetopause, also called the "Chapman-Ferraro current", which consists of diamagnetic currents via the density and temperature across the boundary layer and was proposed first by Chapman and Ferraro (1930, 1931). The second is the ring current encircling the earth inside the magnetosphere, which is produced mainly by the magnetic gradient and curvature drift motions of energetic ions trapped in the radiation belt of the magnetosphere. The third is the tail current, which flows on the outer boundary of the magnetotail and around the tail lobes. The fourth is the neutral sheet current flowing in the transition region between the north and south tail lobes with nearly antiparallel magnetic fields. The fifth is the ionospheric currents consisting of the Pedersen current along the ionospheric induced electric fields but perpendicular to the magnetic fields and the Hall current perpendicular to both the electric and magnetic fields. The final is the field-aligned current flowing along the earth's magnetic field lines and connecting the magnetosphere to ionosphere, also called the "Birkeland current" or the "auroral current", which was found first by Birkeland (1908) when he analyzed his extensive data on the geomagnetic perturbations associated with auroras and concluded that large electric currents flowed along the local magnetic fields during aurora. A sketch of the global current system shown in Fig. 3.2, in which the ionospheric currents have not been displayed because their sizes are too small to appear. These currents interconnect to establish a global current system and their magnetic fields superpose the earth's intrinsic dipolar field to configure the dynamical magnetosphere.

When impinging on the dayside magnetosphere, a fraction of the solar wind energy flux enters the magnetosphere via magnetic reconnection or viscous-like processes (Akasofu 1981). The solar wind power penetrating the magnetosphere is $\sim 10^{10}$ to 10^{11} W during a quiet solar wind and $\sim 10^{12}$ to 10^{13} W during a strongly disturbed solar wind, which may be caused by an interplanetary coronal mass ejection originating from solar eruptive activity. Although the solar wind power input into the magnetosphere is much less than the radiation power received at earth from the solar shining, known as the solar constant, $\simeq 1.73 \times 10^{17}$ W, it can play an important and crucial role in the dynamics of the magnetosphere.

The solar wind energy entering the magnetosphere, via the global current system, is firstly transferred and distributed, intermediately stored, and eventually released and dissipated in somewhere in the system. For example, some of the energy entering the dayside magnetosphere is directly transferred to the high-latitude ionosphere via field-aligned currents and is dissipated by the ionospheric Joule heating. The remainder is transferred to the magnetotail and stored into the magnetotail current system. Then, one part of the energy stored in the magnetotail is transferred to the ring current where it is dissipated mainly via the charge exchange with neutrals into the atmosphere, another part is transferred to the auroral ionosphere via the field-aligned currents where it is dissipated into auroral radiations and plasma heating, and the

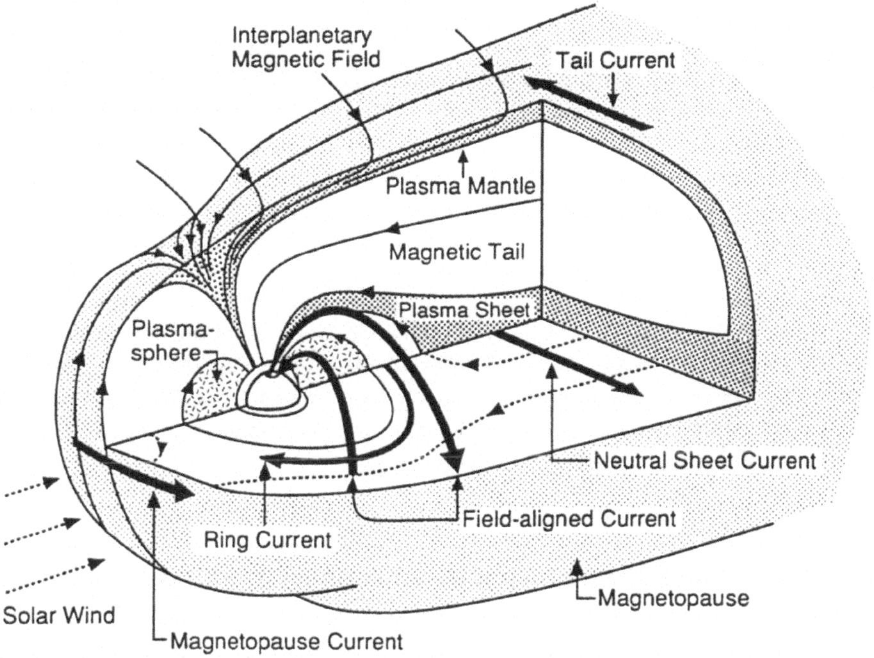

Fig. 3.2 A sketch of the global current system (from Russell 1995)

rest is not dissipated inside the magnetosphere but is released into the downstream solar wind via the down-tail plasmoids in the magnetotail.

When the energy transferring rate from the solar wind into the magnetosphere is low, the magnetosphere responds to these energy processes in a quasi-steady manner, that is, the geomagnetic quiet state. When the energy input rate is relatively high, however, the magnetosphere reacts in a non stationary or explosive manner, such as geomagnetic storms and substorms. Accompanying the electrodynamical processes of the global current system during these storms and substorms, various plasma waves and their wave-particle interactions must appear ubiquitously in the magnetosphere and ionosphere, in particular in the current channels of the global current system, such as the magnetopause, magnetotail, boundary layers, and field-aligned current channels. These current-channel areas have the characteristic mesoscales that typically are much smaller than the macroscales of the global large-scale structures but much larger than the microscales of the particle kinetics. These mesoscale structures can supply plentiful free energy that drives various microscale processes, such as the excitation of various plasma waves and corresponding wave-particle interactions. These microscale waves and wave-particle interaction play essentially and crucially important roles in the energy transport and dissipation of the global current system as well as the energization and diffusion of magnetospheric and ionospheric particles.

In order to comprehensively understand the dynamics of the global current system and its response to the solar wind fluctuations, in principle, it is necessary to consider the solar wind-magnetosphere-ionosphere coupling system as a whole. Because of the complexity and difficulty of the total system, however, current researches basically focus on a few key links of the coupling system, such as the solar wind-magnetosphere coupling in the magnetopause and the magnetotail and the magnetosphere-ionosphere coupling in the polar magnetosphere and auroral plasmas. Therefore, they have been the main objects of satellite exploration since the space age. In particular, a large number of in situ observations by satellites in space plasmas show that KAWs can play a crucially and essentially important role in these key regions of the coupling system. In this chapter we focus on the observation and application of KAWs in the auroral plasma dynamics, especially their application in the auroral electron acceleration, and the next Chap. 4 will discuss KAWs in the solar wind-magnetosphere coupling, with particular emphasis on their application in the transport and heating of plasma particles.

3.2 Aurora and Magnetosphere-Ionosphere Coupling

The aurora, a beautiful and mysterious phenomenon in the Earth's atmosphere, consists of lights in the sky that are emitted by atoms and molecules in the upper atmosphere at the altitude of $\sim$100 km, where neutral gases are ionized and excited by collisions with energetic electrons, called auroral energetic electrons, precipitating from the magnetosphere into the atmosphere. It is most commonly seen in the latitude of $\sim 20°$ from the geomagnetic poles and in the form of auroral arcs, luminous curtains stretching over hundreds of km along the east-west direction, but very much thinner in the north-south extent with a few of km. The microphysics of acceleration of these auroral energetic electrons has long been an open problem. Although it is possible that energetic electrons from the sun are guided directly into the polar region by the geomagnetic field, these energetic electrons can not be responsible for producing auroras because they have neither sufficient energy to produce the magnificent optical displays nor do they strike the atmosphere in the proper locations.

The generation mechanism of auroras, especially the acceleration mechanism of the auroral energetic electrons has been studied over several decades by many ground-based observations as well as by sounding rocket and satellite in situ measurements. These observations have revealed that the auroral phenomena are associated with energetic electrons of 1–10 keV that have relations with enhancements of the upward field-aligned currents during substorms (McIlwain 1960; Mozer et al. 1980). The tendency that there are higher fluxes of the energetic electrons at smaller pitch angles implies that their acceleration is dominated by the field-aligned acceleration processes (Hoffman and Evans 1968). The energy distribution of these auroral energetic electrons is characterized by a precipitous decrease above 10 keV, without a rapid increase towards energies below 4 keV (Bryant 1990). Figure 3.3 presents a typical example of energy distributions of observed auroral energetic electrons

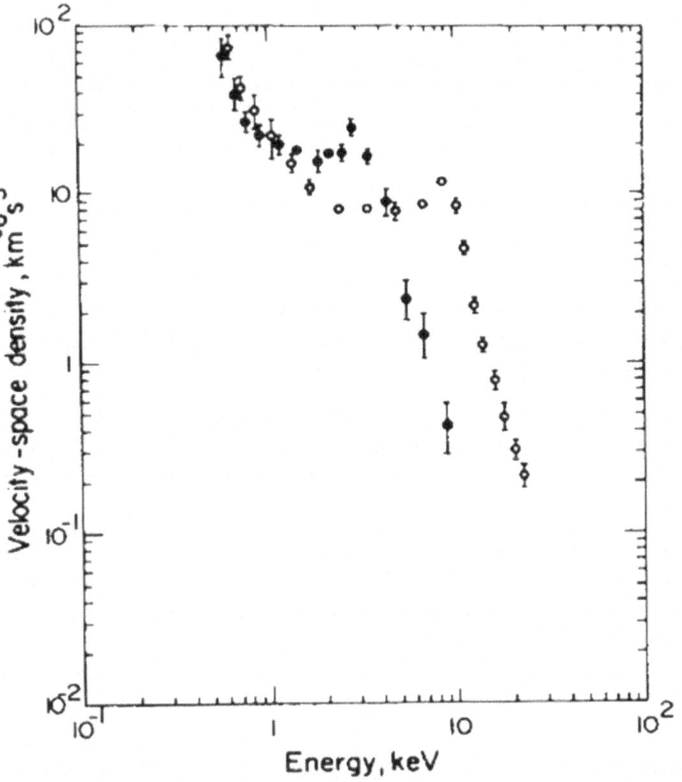

Fig. 3.3 Typical electron energy distributions which are responsible for discrete aurora. The measurements were obtained at the centre (open circles) and edge (solid circles) of an auroral arc with a Skylark rocket, SL 1422, launched from Andoya Norway on 21 Nov. 1976 (from Bryant 1981)

(Bryant 1981), where the open and solid circles give the electron energy distributions measured at the centre and edge of the observed auroral arc. From Fig. 3.3 the auroral energetic electrons of the edge have a similar distribution feature to that of the center but lower energies than that of the center.

In particular, the comparison between the energy distributions of field-aligned electrons simultaneously measured by the DE-1 and 2 satellites in the higher magnetospheric altitude ($\sim 2R_E$) and the lower ionospheric altitude ($\sim 0.1R_E$) clearly demonstrates that the acceleration of auroral energetic electrons occurs in the region of altitudes of a few to ten thousands of km (or i.e., ~ 0.5–$2R_E$), called the auroral acceleration region (Reiff et al. 1988). This indicates that the auroral electron acceleration strongly depends on the environmental condition of the ambient plasma and occurs only in some specific locations in the polar magnetosphere with not only specific latitudes corresponding to the auroral zone but also specific altitudes corresponding to the auroral acceleration region, as well as at specific times corresponding to the duration of substorms.

In fact, besides beautiful and visible auroras these auroral energetic electrons have a mysterious product, that is, so-called auroral kilometric radiation (AKR, Gurnett 1974). AKR is produced by the electron cyclotron maser emission of the auroral energetic electrons at the local electron cyclotron frequency and/or its harmonic frequency (Wu and Lee 1979; Lee and Wu 1980). However, AKR had no been found until space observations onboard satellites became available in 1960s because its frequencies are below the cutoff frequency (i.e., ω_{pe}) of the ionosphere and hence can not enter into the atmosphere and arrive on the ground.

Based on the analysis results of a large number of in situ measurements of the auroral plasma in various altitudes from the low ionosphere ($\sim 10^2$ km) to the high magnetosphere ($\sim 10^5$ km), the auroral energetic electrons have the following main observed properties:

Field-aligned acceleration: The auroral energetic electrons are field-aligned accelerated by parallel electric fields and hence lead to intense field-aligned currents;

Characteristic energy: The auroral energetic electrons have typical energies of several keV and range from 1 to 10 keV;

High-energy cutoff: The auroral energetic electrons have energy spectra characterized by downward extending to the thermal distribution of the background electrons and upward precipitously decreasing above 10 keV;

Acceleration region: The auroral acceleration region is located in the range of altitudes a few to ten thousands of km (or i.e., $\sim$0.5–$2R_E$), which also is the source region of AKR and has very dilute plasma density so that $\omega_{pe} \ll \omega_{ce}$;

Discrete auroral arc: Discrete auroral arc produced by the auroral energetic electrons of precipitating to the ionosphere has a typical width $\sim$ a few km, and the electrons in the auroral arc edges have a similar energy spectrum with that in the auroral arc center but lower energies.

The magnetosphere and ionosphere are strongly coupled by the geomagnetic field and the field-aligned current along the field lines in the Earth's polar region, where the auroral plasma with variable parameters in the magnetosphere and ionosphere directly is vertically penetrated by the geomagnetic field. In particular, the rapidly precipitating auroral energetic electrons must lead to sharply increasing of the field-aligned current and a series of corresponding electrodynamical responses of the polar magnetospheric plasma, also called the auroral plasma, and the visible aurorae is only their response in the ionosphere. Therefore, the auroral plasma dynamics is the base to understand the substorm and auroral phenomena. From the low ionosphere ($\sim 10^2$ km) to the high magnetosphere ($\sim 10^5$ km), however, the magnetosphere-ionosphere coupling system contains a rich variety of plasma populations with the density varying from 10^6 cm^{-3} to less than 10^{-2} cm^{-3}, the temperature from lower than 1 eV to higher than 100 keV, and the magnetic field from 0.5 G to 50 nT with the altitude increasing. It is evident that the altitude distribution feature of the plasma parameters can play an important role in the magnetosphere-ionosphere coupling. Although vast measurements for these polar plasmas have been carried out by ground-based observations and sounding rocket and satellite in situ explorations, it still is very difficult to describe perfectly variations of these plasma parameters with altitudes.

Based on in situ measured data, the plasma density can be calculated by the ion density or by the electron density. In principle, the ion and electron densities should be same because of the charge neutrality condition. The ion components in the auroral plasma consist mainly of oxygen (O^+) and hydrogen (H^+). The oxygen ions originate from the ionosphere and behave at lower altitudes according to the barometric exponent law (Thompson and Lysak 1996), that is,

$$n_O = n_{IO} \exp\left(-h/h_O\right), \tag{3.1}$$

where n_{IO} is the oxygen density in the ionosphere, h is altitude in units of R_E, and h_O is the oxygen scale height (usually a few % of R_E). The hydrogen ions (i.e., protons) originate in the upper ionosphere and magnetosphere and their behavior follows an exospheric power law (Thompson and Lysak 1996),

$$n_H = n_{IH} \left(1 + h\right)^{-m}, \tag{3.2}$$

where n_{IH} is the hydrogen density in the ionosphere and the index m is typically between 1 and 3.

According to the charge neutrality condition, the electron density profile can be modeled by the combination of Eqs. (3.1) and (3.2). Combining the altitudinal density profile measured by the S3-3 satellite in the auroral acceleration region (Mozer et al. 1979) with an exponentially decreasing ionospheric component, Lysak and Hudson (1987) suggested a function to model the distribution of the ambient electron density along the auroral field lines as follows:

$$n(r) = n_0 \exp\left(-\frac{r - r_0}{h_0}\right) + \frac{17\ \text{cm}^{-3}}{(r-1)^{1.5}}, \tag{3.3}$$

where $r = 1 + h$ is the geocentric distance in units of R_E. In this expression, the first term represents the ionospheric density component which can be fitted by density measurements made during a rocket flight and the second term comes from the fitting density determined by the measurements recorded on the S3-3 satellite in the evening auroral zone (Lysak and Hudson 1979). A further comparing with a composite of electron density measurements made by Viking, DE 1, ISIS 1 and Allouette II, Kletzing and Torbert (1994) showed that the modeled profile and the in situ measurements can agree quite well when $n_0 = 6 \times 10^4\ \text{cm}^{-3}$, $r_0 = 1.05$, and $h_0 = 0.06$ in the altitude through $h = 0$ to 4 (R_E) above the ionosphere.

Based on the above analyses of the ion and electron densities, following Kletzing and Torbert (1994), Wu and Chao (2004) takes the ambient plasma density of the auroral plasma, n, as

$$n(h) = 1.38 \times 10^5 \exp\left(-\frac{h}{0.06}\right) + 17h^{-m}\ (\text{cm}^{-3}), \tag{3.4}$$

where m is a fitting parameter used to fit the magnetospheric plasma density and the charge neutrality condition of $n_e = n_i = n$ has been assumed for the ambient plasma density.

For the distribution of the ambient auroral plasma temperature with the altitude, it is generally believed that the magnetospheric temperature is remarkably higher than the ionospheric temperature and that the plasma temperature can change from below 1 eV in the ionosphere up to a few hundreds of eV in the magnetosphere at altitudes of $h \sim 3$–$4R_E$, where the plasma is dominated by the plasma sheet with temperature of ~300–600 eV. Based on measurements of temperature up to the highest altitudes measured by S3-3, Kletzing et al. (1998, 2003) showed that the temperature commonly is below 10 eV at altitudes below 10,000 km ($h \sim 1.5R_E$) and the majority of the measurements gives a temperature lower than 5 eV at altitudes below $1R_E$. Therefore, they inferred that the temperature may begin to increase slightly with the altitude at 7000 km due to the effect of a decline in the cold ionospheric density relative to the hotter CPS component (Kletzing et al. 1998). Taking the magnetospheric temperature as ~300 eV, the temperature profile described by Kletzing et al. (1998) can be modeled by the following formula (Wu and Chao 2004)

$$T_e = 100\,[1 + \tanh\,(h - 2.5)]^{3/2}\ (\text{eV}). \tag{3.5}$$

Unlike the cases of the ambient density and temperature, the ambient magnetic field in the auroral plasma is relatively simple and clear and can be modeled well, at least within a few Earth's radii, by the variation of the Earth's dipolar field as follows:

$$B_0 = \frac{0.6}{(1 + h)^3}\ (\text{G}). \tag{3.6}$$

Figure 3.4 plots the distributions of the ambient plasma parameters with the altitude (h) above the ionosphere, where panels, from the top down, are the electron density n_e (a) in Eq. (3.4) with the fitting parameter $m = 1.5$, temperature T_e (b) in Eq. (3.5), the magnetic field B_0 (c) in Eq. (3.6), the Alfvén velocity v_A (d), and the electron thermal to Alfvén velocity ratio $\alpha_e = v_{T_e}/v_A$ (e), respectively. In the calculation of the Alfvén velocity, the ion mass density of the auroral plasma, ρ_m, is obtained by the sum of oxygen ($16n_{O^+}m_p$) and hydrogen ($n_{H^+}m_p$) ions, that is,

$$\rho_m = 2.208 \times 10^6 e^{-h/0.06} + 17h^{-m}\ (m_p/\text{cm}^3), \tag{3.7}$$

where m_p is the proton mass.

Figure 3.4d, e show clearly that the auroral plasma has a high Alfvén velocity of $v_A > 10^4$ km/s and hence a low electron thermal to Alfvén velocity ratio of $\alpha_e < 1$, especially in the auroral acceleration region, which is located at the altitude ~0.5–$2R_E$, one has $\alpha_e \ll 1$ and $v_A \sim 0.1c$. One of key and difficult problems in understanding the physics of the auroral electron acceleration is how to produce and maintain a large parallel electric field that can accelerate field-aligned electrons to keV-order energies (Fälthammar 2004). In general, the very large parallel

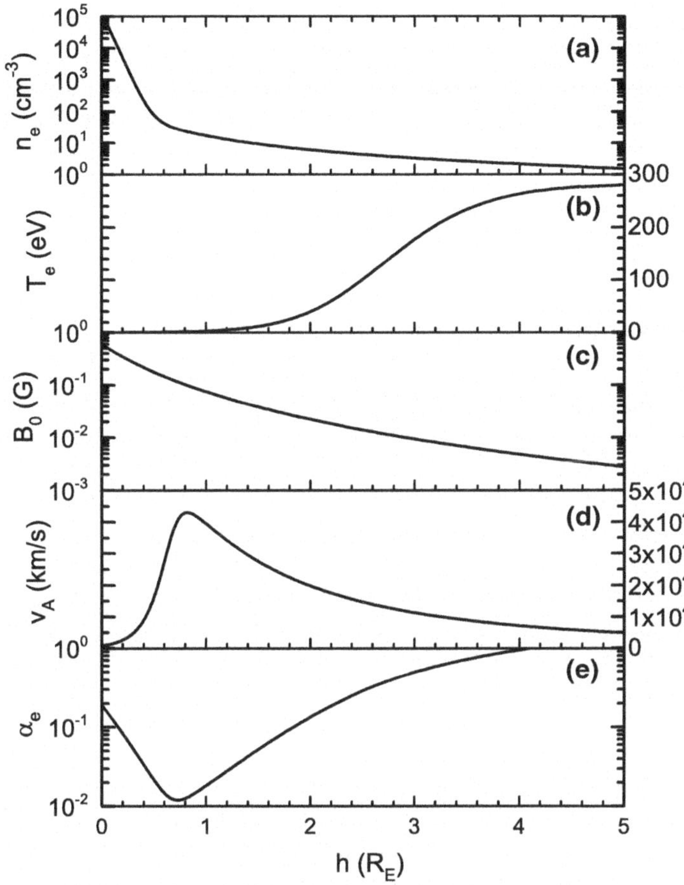

Fig. 3.4 The ambient plasma parameter distributions with the altitude above the ionosphere $h\,(R_E)$: **a** The electron density n_e in cm^{-3}; **b** the electron temperature T_e in eV; **c** the Earth's magnetic field B_0 in G; **d** the Alfvén velocity v_A in km/s; and **e** the electron thermal to Alfvén velocity ratio $\alpha_e \equiv v_{T_e}/v_A$

conductivity can immediately short-circuit any parallel electric fields in a collisionless plasma such as in the magnetosphere. Many "non-ideal" mechanisms, e.g., the "anomalous resistivity" concept (Papadopoulos 1977), the "weak double layer" model (Temerin et al. 1982; Boström et al. 1988; Mölkki et al. 1993), and the direct acceleration by lower-hybrid waves (Bingham et al. 1984) had been invoked as possible explanations.

Wu and Chao (2004) proposed that solitary KAWs (SKAWs) can play an important role in the auroral energetic electrons, in which effective parallel electric fields can be maintained by the inertial motion of electrons. Since 1990s, solitary structures of strong electric field fluctuations have been commonly observed by in situ measurements of the polar-orbit satellites, such as the Freja, FAST and Polar satellites, in the polar ionospheric and magnetospheric plasmas and are identified well as SKAWs (Louarn et al. 1994; Wu et al. 1996a, b, 1997; Chaston et al. 1999). Moreover, the satellite observations also show that the strong KAWs, SKAWs or turbulent KAWs, in the auroral plasma are often associated with the field-aligned acceleration of electrons as well as the cross-field heating of ions (Stasiewicz et al. 2000a). Following, we first introduce the observation and identification of SKAWs in the auroral plasma in Sect. 3.3 and then discuss their theoretical models in Sect. 3.4.

3.3 Observation and Identification of SKAWs

In 1980s, possible applications of KAWs in the formation of discrete auroral arcs had been discussed by some authors (Hasegawa 1976; Goertz and Boswell 1979; Goertz 1981, 1984; Lysak and Dum 1983). However, there were little high resolution in situ measurements that can directly and clearly identify out the fine structures associated with small-scale KAWs. An important example is the identification of electromagnetic fluctuations with a two-dimensional vortex structure (Chmyrev et al. 1988). By the data from the Intercosmos-Bulgaria-1300 satellite (ICB-1300), which measurements of electric and magnetic fields have higher temporal resolution than the previous satellite measurements, Chmyrev et al. (1988) analyzed some strong magnetic and electric field fluctuations on short temporal scales, which occurred on both the polar and equatorial edge of the auroral oval. The measured magnetic fluctuations with $\delta B_{\parallel} \ll \delta B_{\perp}$ and the electric to magnetic fluctuation ratio of $\delta E_{\perp}/\delta B_{\perp} \sim v_A$ indicate these fluctuations to be incompressible Alfvénic modes. In particular, they found, by use of hodograms of the electric field that the perturbed fields have two-dimensional vortex structures and identified them as drift-KAWs. Moreover, the electron flux within the "vortex" region exceeds the background by 2 orders of magnitude, implying that the electrons are trapped within the Alfvénic vortex and travel together with it, from the generation region in the magnetosphere to the observation region in the ionosphere. The authors proposed that the measured drift-KAWs can be directly associated with the structures of auroral arcs. Similar vortex structures associated with KAWs were measured more clearly and refinedly in the laboratory experiments of the LAPD (Burke et al. 2000a, b). The more refined and evident in situ measurement and identification of fine structures associated with KAWs in space plasmas were achieved in following space explorations by the polar-orbit satellites, Freja and FAST, which had the higher temporal resolution to resolve smaller scale structures.

The Freja satellite, a joint Swedish and German scientific satellite, was launched on October 6, 1992 in Jiuquan, China, into a polar orbit traversed the auroral oval almost tangentially in the East-West direction, with 63° inclination and an apogee of 1750 km and perigee of 600 km (see Lundin et al. 1994 for more details). In the auroral plasma environment explored by Freja, the electron density and temperature have a typical value $n_e \sim 10^3$ cm^{-3} and $T_e \sim 1$ eV, respectively, and the geomagnetic field $B_0 \simeq 0.2$ G at the Freja altitude. This indicates the explored low-β plasma

($\beta \sim 10^{-6} \ll Q$) with the electron inertial length $\lambda_e \sim 100$ m and the ion-acoustic gyroradius $\rho_s \sim 10$ m, which are the characteristic scales of KAWs. KAWs have frequencies below the ion gyrofrequency $\omega_{ci}/2\pi \sim$ a few 100 Hz for hydrogen ions or $\sim$ a few 10 Hz for oxygen ions in the Freja exploring environment. The Freja satellite is designed for fine-structure plasma measurements with a high temporal/spatial resolution of the auroral plasma processes. Its experiments allowed three-dimensional measurements of direct-current (at 128 Hz sampling rate) and alternating-current (at 32 kHz sampling rate) magnetic fields with a set of fluxgate and search coil magnetometers. Similarly, two dimensional direct-current (at 768 Hz sampling rate) and alternating-current (at 32 kHz sampling rate) electric fields are measured by a set of probe pairs 7.6–21 m apart from each other. These measurements provided, together with the sampling rate of 32–64 ms resolution electron distribution functions, excellent means of observing the microphysics of KAWs in the auroral plasma environment explored by the Freja satellite.

By use of the data from Freja, Louarn et al. (1994) found that the low-frequency electromagnetic fluctuations with frequencies ~1–20 Hz lower than the local ion gyrofrequency, consist of two distinct kinds. One is the widespread distributed electromagnetic turbulence (typically tens mV/m in the electric fluctuation and tens nT in the magnetic fluctuation) associated with a lower level of the density fluctuation ($|dn/n| \lesssim 10\%$). The other is the solitary strong electric spikes lasting tens ms, which typically have a stronger electric fluctuation ($\delta E \sim$ hundreds mV/m) by one order of magnitude and a similar magnetic fluctuation ($\delta B \sim$ tens nT), and are associated with a strong density fluctuation ($|dn/n| \sim$ tens percent). Two typical examples of them are shown in Fig. 3.5, where panels, from the top down, are electric field, magnetic field, and density fluctuations, respectively. The durations of these two structures both are well below 0.1 s but above 0.05 s, implying their fluctuation frequencies ~10–20 Hz. The event on the left panel has the electric field amplitude ~0.6 V/m, magnetic amplitude ~40 nT, and relative density amplitude ~80%. Other one on the right panel is relative weaker and has the electric field amplitude ~0.2 V/m, magnetic field amplitude ~15 nT, and relative density amplitude ~30%.

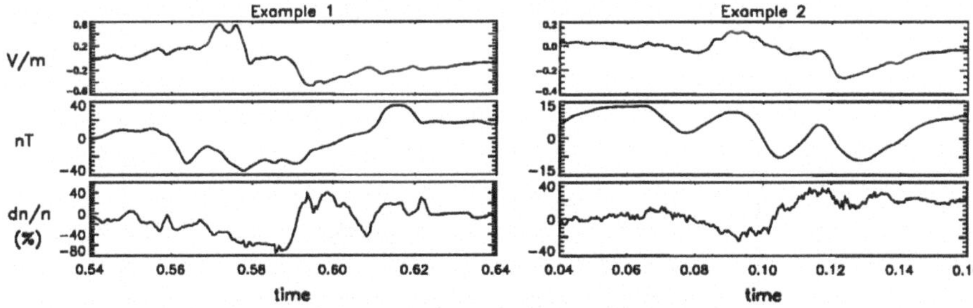

Fig. 3.5 Two examples of SKAWs observed by the Freja satellite during passing the auroral zone at the altitude ~1700 km above the ionosphere on 09/03/1993 (from Louarn et al. 1994)

Louarn et al. (1994) analyzed in detail the electromagnetic features and accompanying density fluctuations of these strong electric spikes and found that their electric to magnetic fluctuation ratio $\delta E/\delta B \sim 5 \times 10^6$ m/s $\sim v_A$ (the local Alfvén velocity), which is the typical characteristic of Alfvénic fluctuations, that is, nonlinear AWs. Their accompanying deep density cavities (i.e., density dips), however, imply that their structures must have been significantly modified by some kinetic effects. Moreover, their short duration of tens ms corresponds a cross-field size of hundreds m, i.e., a few times the local electron inertial length, which is the typically perpendicular wavelength of KAWs in the inertial regime of $\alpha_e < 1$. These clear characteristics of inertial-regime KAWs lead to the conclusion that these solitary strong electric spikes are the nonlinear structures of inertial-regime solitary KAWs (SKAWs, Louarn et al. 1994).

Wu et al. (1996b) further found that these strong electric spikes are accompanied not only by density dips but also often by density humps. According to the theory of one-dimensional SKAWs, inertial-regime SKAWs with $\alpha_e < 1$ are accompanied by only density dip soliton (Shukla et al. 1982), while kinetic-regime SKAWs with $\alpha_e > 1$ are accompanied by only density hump soliton (Hasegawa and Mima 1976). In order to explain the Freja observation of SKAWs accompanied by hump as well as dip solitons, the initial conjecture by Wu et al. (1996b) is that the Freja satellite was traversing the plasma transition region between kinetic and inertial regimes. A more serious conflict between the observation and theory, however, was found soon after by Wu et al. (1996a, 1997). They analyzed some strong density fluctuations ($|dn/n| > 10\%$) accompanying SKAWs observed by the Freja satellite and found that these SKAWs can be accompanied by not only single dip or hump structures, but also composite structures of dip and hump, called the "dipole density soliton".

Two typical examples of the dipole density soliton are shown in Fig. 3.6, their dipole density amplitudes are the dip -50% and hump $+55\%$ for the left one and the dip -40% and hump $+55\%$ for the right one. Table 3.1 lists 21 dipole density solitons accompanying SKAWs (including the two examples at 0935:25.80 and 0935:56.65 shown in Fig. 3.4) presented by Wu et al. (1996a, 1997), which were observed by the Freja satellite when it crossed the auroral plasma at the altitude about 1700 km on days 3 March 1993, 9 March 1993, and 7 March 1994. In Table 3.1, the "time" column lists the universal time (UT) of events, the "dn/n" column presents their

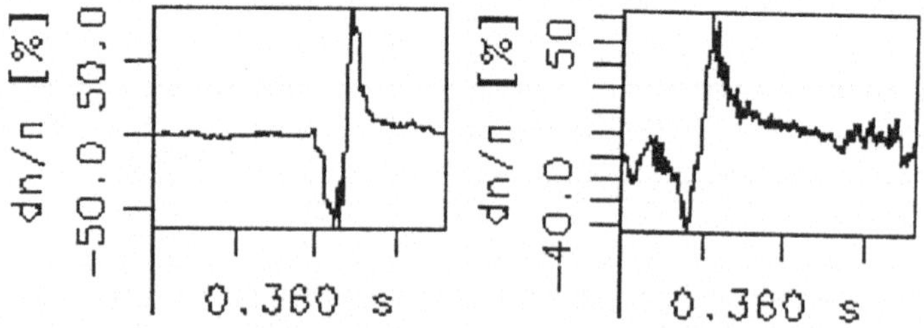

Fig. 3.6 Two examples of dipole density soliton observed by the Freja satellite (from Wu 2010)

Table 3.1 Density solitons associated with the Freja SKAW events (reprinted the table with permission from Wu et al., Phys. Rev. Lett. 77, 4346, 1996a, copyright 1996 by the American Physical Society)

Time (UT)	$dn/n(\%)$	Δt(ms)	Solitons	Time (UT)	$dn/n(\%)$	Δt(ms)	Solitons
0935:16.80	−30(+40)	80	Dipole	0935:53.70	−70	50	Dip
0935:25.80	−50(+70)	70	Dipole	1702:06.95	−30	40	Dip
0935:36.65	−40(+45)	90	Dipole	1703:30.90	−50	50	Dip
0935:54.80	−15(+20)	80	Dipole	1703:33.00	−50	80	Dip
0935:56.85	−40(+15)	80	Dipole	1703:34.85	−30	50	Dip
1703:18.85	−35(+20)	90	Dipole	2039:10.70	−20	60	Dip
1703:52.85	−55(+50)	90	Dipole	1659:03.10	+40	70	Hump
1704:50.85	−20(+15)	70	Dipole	1703:08.80	+60	60	Hump
1704:52.85	−40(+25)	90	Dipole	1704:51.10	+15	40	Hump
0344:59.15	−75(+40)	60	Dipole	2034:03.00	+20	50	Hump
0348:05.60	−25(+35)	60	Dipole				

density amplitudes with the sign "−" or "+" denoting the density "dip" or "hump", respectively, the "Δt" column gives the durations of the SKAWs, and the "solitons" column is the types of density solitons.

From Table 3.1, it can be found that the dipole density solitons are more frequently present than single dip or hump solitons. The number of dipole solitons (11) is about one-half of the total number of SKAWs, and the number of single dip solitons (6) or single hump solitons (4) is about one-fourth of the total number of SKAWs. Moreover, the dipole density solitons have, in average, a larger spatial scale than single dip or hump solitons by a factor of 1.6. The dipole solitons have a typical duration ~80 ms, corresponding the spatial size scale ~560 m and the typical duration of the single dip or hump solitons ~50 ms, corresponding the spatial size scale ~350 m, estimated by the Freja satellite velocity ~7 km/s.

Since the three types of density solitons (dip, hump, and dipole) all are associated with similar local electromagnetic fluctuations with the KAW characteristics, that is, associated with SKAWs, and also all take place in the same plasma environment, that is, the auroral plasma at the altitude ~1700 km, they should be uniformly described by the same physical model. Wu et al. (1996a, 1997) generalized the theory of SKAWs from the one-dimensional case to the two-dimensional case and found that the two-dimensional SKAW model with a dipole vortex structure can uniformly explain well not only the presence of the three kinds of density solitons (i.e., dip, hump, and dipole solitons) but also the rotation character of electric fields associated with the observed SKAWs (Chmyrev et al. 1988; Volwerk et al. 1996).

Volwerk et al. (1996) further analyzed the electromagnetic character of these SKAWs in the Freja observations and their carrying Poynting flux. They found that these strong electromagnetic spikes ($\delta E \sim$ hundreds mV/m and $\delta B \sim$ tens nT) have mainly a rotational character in the plane perpendicular to the geomagnetic field and a tiny magnetic compressional component not exceeding 4% in general. Therefore, they came to the conclusion that these SKAWs are two-dimensional structures associated with small-scale ($\sim$ hundreds m) tubular currents. In addition, they also found that the Poynting flux carried by these SKAWs along the geomagnetic field is typically in the order of $\sim$0.05–0.1 mW/m^2. In both intensity and direction the Poynting flux of these SKAWs is comparable to the typical value required by the auroral acceleration electrons. In particular, their analysis shows that these SKAWs are situated at the edge of large-scale shear regions in the current, where the plasma density appears obviously the inhomogeneity, as observed in the LAPD experiments (Burke et al. 2000a, b).

Other polar-orbit satellite launched soon later, the FAST (Fast Auroral SnapshoT) satellite, also presents similar observation results of SKAWs. The FAST satellite was launched on August 21, 1996 into an 83° inclination polar orbit with 350 km perigee and 4180 km apogee (Carlson et al. 1998) and was designed for studying the microphysics of fine-scale Alfvénic fluctuations observed in the auroral oval. It provides not only three-dimensional electromagnetic field measurements but also multiple baseline electric field measurements thereby removing the temporal/spatial ambiguity inherent in single point measurements and providing direct evidence of the observed spatial structures of KAWs with a perpendicular drift smaller than the satellite velocity. KAWs in the electron inertial length size are observed by the FAST satellite from perigee at 350 km up to apogee at 4180 km. Their observed properties further confirm the observations made by Freja. However, the FAST's high inclination orbit provides a different perspective with a generally north-south cut (rather than east-west) through the flux tubes on which these waves exist. Furthermore, the eccentricity of the orbit allows an altitude profile for these waves to be established extending from just above the ionosphere to the base of the primary auroral electron acceleration potential.

Figure 3.7 shows the transverse electric to magnetic field ratios and the perpendicular sizes of over 100 SKAWs observed by the FAST satellite in similar altitudes (1500–2500 km), where the panel (a) plots the transverse electric to magnetic field ratio versus the local Alfvén velocity and the panel (b) gives the distribution of the perpendicular size of SKAWs in units of the electron inertial length (Chaston et al. 1999). The results show that these SKAWs are characterized typically by: (1) strong electric spikes ($\delta E \sim$ hundreds of mV/m) and magnetic pulses ($\delta B \sim$ tens of nT), which yield the electric-magnetic amplitude ratio, $\delta E/\delta B$, of a few times the local Alfvén velocity, as expected by the theory of the inertial-regime SKAWs (see Sect. 3.4); (2) short durations of tens ms implying a frequency much lower than the local ion gyrofrequency $\sim$ a few hundreds Hz and a transverse scale $\sim$ hundreds of m, which averages to several times the local electron inertial length and is the typical scale of the inertial-regime SKAWs (see Sect. 3.4); (3) accompanying strong density perturbations $\sim$ tens percent. These results are completely similar with that from the Freja observations of SKAWs and confirm again the conclusion that the observed strong electromagnetic spikes are inertial-regime SKAWs.

In fact, the strong electromagnetic spikes (hundreds mV/m electric field fluctuations and tens nT magnetic field fluctuations), accompanying strong density fluctu-

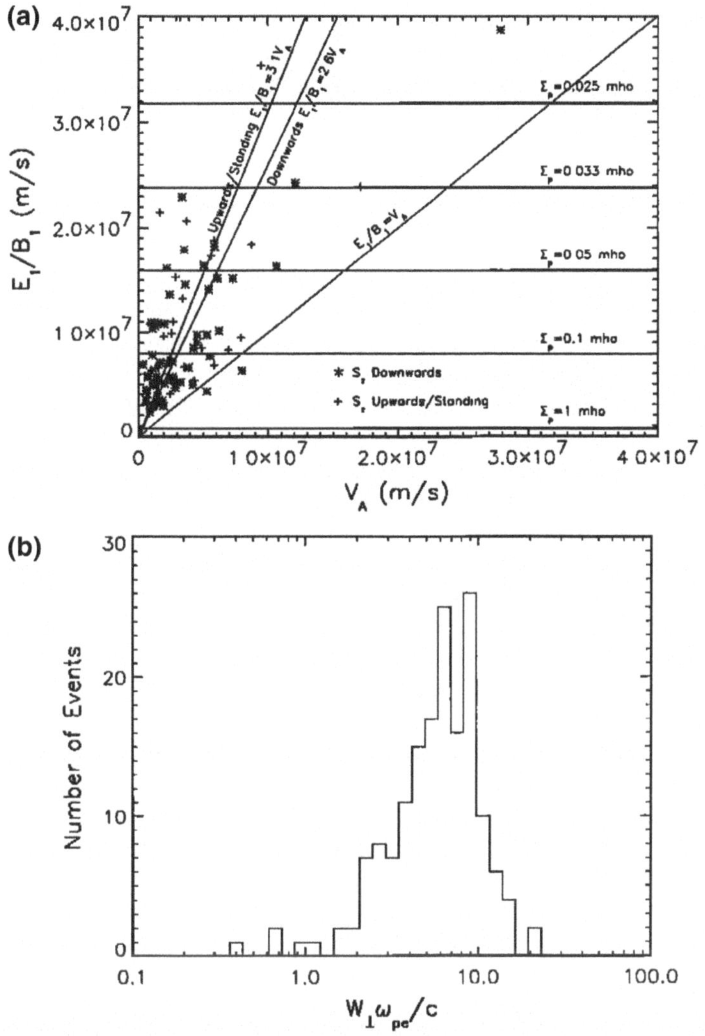

Fig. 3.7 Properties of SKAWs observed by the FAST satellite: **a** the distribution of the transverse electric to magnetic field ratio versus the local Alfvén velocity; **b** the distribution of the perpendicular size of SKAWs in units of the electron inertial length (from Chaston et al. 1999)

ations (tens percent) are very common features in the observations of the Freja and FAST satellites when they cross the auroral plasma at the altitudes ~1000–4000 km (Louarn et al. 1994; Wahlund et al. 1994a; Volwerk et al. 1996; Wu et al. 1996a, b, 1997; Huang et al. 1997; Stasiewicz et al. 1997, 1998, 2000b, 2001; Bellan and Stasiewicz 1998; Chaston et al. 1999). Using novel plasma instruments and a high data rate, the two satellites provided a "first" high resolution plasma diagnostics that permit studies of meso- and micro-size scale phenomena in the 100-meter range for plasma particles and in the 10-meter range for electric and magnetic fields. This provides adequate and reliable experimental data for the detailed studies in their physical property and the establishment of their theoretical models.

3.4 Theoretical Models of SKAWs in Auroral Plasmas

It is evident that the strong electromagnetic spikes, which accompany with strong density fluctuations and present commonly in the observations of the polar-orbit satellites, such as Freja and FAST, during passing through the auroral plasma, are solitary structures of nonlinear KAWs. A variety of nonlinear effects can occur frequently in a plasma because of strongly electromagnetic coupling interactions among charged particles. One of the most typical nonlinear effects is the steepening effect of dispersive waves, in which the wave front steepens and the amplitude enlarges when a fast wave chases and then overtakes a slow wave in a dispersive wavelet. Meanwhile, the dispersive effect in the wavelet also causes the waves to diffuse because of their different phase velocities. When the dispersive effect just balances the steepening effect, a localized and stationary coherent structure can be formed from the wavelet, which can stably propagate without changing its profile, called solitary wave, or soliton (Zabusky and Kruskal 1965).

In the case of KAWs, the existence of exact travelling wave solutions of SKAWs for the kinetic regime of $1 < \alpha_e^2 < Q^{-1}$ (or i.e., $Q < \beta/2 < 1$) was first demonstrated by Hasegawa and Mima (1976) and the corresponding kinetic-regime SKAWs are shown to be accompanied by a hump density soliton. While for the case of the inertial regime of $\alpha_e^2 < 1$ (or i.e., $\beta/2 < Q$) Shukla et al. (1982) also argued the existence of exact solution of SKAWs, but the corresponding inertial-regime SKAWs are accompanied by a dip density soliton. The first exact analytical solution of inertial-regime SKAWs with an arbitrary amplitude was obtained by Wu et al. (1995). In the small amplitude limit, preserving the lowest order nonlinearity, it leads to the standard KdV soliton, as expected.

Wu et al. (1996b) further extended the existence of SKAWs to the more general parametric regime, including the kinetic-inertial transition regime of $\alpha_e^2 \sim 1$. In particular, they presented the universal criterion for the existence of SKAWs (Wu et al. 1996b). In the low frequency approximation of $\omega \ll \omega_{ci}$, the density fluctuation of SKAWs, which correspond to nonlinear wavelets propagating at the direction of $\mathbf{k} = (k_x, 0, k_z)$, can be described by localized solutions (i.e., solitary wave solutions) of the following Sagdeev equation (Wu et al. 1996b):

$$\frac{1}{2}\left(\frac{dn}{d\eta}\right)^2 + K(n; M_z, k_x) = 0 \tag{3.8}$$

with the localized boundary conditions

$$n = 1; \quad v_{ez} = 0; \text{ and } d/d\eta = 0 \text{ for } \eta \to \pm\infty \tag{3.9}$$

in the traveling-wave (or i.e., co-moving) coordinates

$$\eta = k_x x + k_z z - \omega t, \tag{3.10}$$

where $K(n; M_z, k_x)$ is the so-called Sagdeev potential as follows:

$$K(n; M_z, k_x) = -\frac{(1+Q)\, M_z^2 n^4}{k_x^2 \left(M_z^2 - \alpha_e^2 n^2\right)^2} \times \left[\Phi_1 + \frac{\alpha_e^2}{M_z^2}\Phi_2 + \frac{1+\tau}{1+Q}\frac{Q\alpha_e^2}{M_z^2}\left(\Phi_3 + \frac{\alpha_e^2}{M_z^2}\Phi_4\right)\right] \tag{3.11}$$

with

$$\begin{aligned}
\Phi_1 &= \frac{(n-1)^2}{2}\left(M_z^2 - \frac{n+2}{3n}\right), \\
\Phi_2 &= n\left(1 - n + n\ln n\right) - M_z^2 n^2 \left(n - 1 - \ln n\right), \\
\Phi_3 &= \frac{(n-1)^2}{2} - M_z^2 n \left(1 - n + n\ln n\right), \\
\Phi_4 &= M_z^2 n^2 \frac{(n-1)^2}{2} - n^2 \left(n - 1 - \ln n\right),
\end{aligned} \tag{3.12}$$

and $M_z \equiv M/k_z$ is the parallel phase speed in units of v_A. In these expressions, space x, y, z, time t, and density n are normalized by λ_e, $\lambda_e/v_A = \sqrt{Q}/\omega_{ci}$, and n_0, respectively, M is the phase speed in units of v_A, n_0 is the ambient plasma density.

In the expression (3.11) for the Sagdeev potential, the third term (associated with Φ_3 and Φ_4 and to be proportional to $Q\alpha_e^2 = \beta_e/2$) represents the so-called finite-β effect, which can be attributed to the coupling of the ion-acoustic wave with KAWs and is neglectable in low-β plasmas (Yu and Shukla 1978; Kalita and Kalita 1986; Wu and Wang 1996). While the first two terms, associated with Φ_1 and Φ_2, can be attributed to the inertial force and the kinetic pressure of electrons, respectively, and reduce to the inertial-regime case for $\alpha_e < 1$ (Shukla et al. 1982) and the kinetic-regime case for $\alpha_e > 1$ (Hasegawa and Mima 1976).

It is obvious that the necessary condition for a real solution of the Sagdeev equation (3.8) is the Sagdeev potential of Eq. (3.11) less than or equal to zero (i.e., $K(n; M_z, k_x) \le 0$). Thus, a variable range of the density n of the SKAW is determined by two adjacent roots of the Sagdeev potential $K(n; M_z, k_x) = 0$. One is $n = 1$, implying the ambient plasma density and other one should be the nearest by $n = 1$, assumed to be $n_m = n_m(M_z, \alpha_e)$, implying the amplitude of the density variation. Therefore, the so-called nonlinear dispersion relation that describes the relationship between the phase velocity M_z and the amplitude n_m can be obtained from the condition $K(n; M_z, k_x) = 0$. For the case of low-β plasmas, neglecting the finite-β effect, the nonlinear dispersion relation for SKAWs can be given as follows:

$$M_z^2 = \left[\frac{n_m + 2}{6n_m} + \alpha_e^2 n_m^2 \frac{n_m - 1 - \ln n_m}{(n_m - 1)^2}\right] \pm \sqrt{\left[\frac{n_m + 2}{6n_m} + \alpha_e^2 n_m^2 \frac{n_m - 1 - \ln n_m}{(n_m - 1)^2}\right]^2 - 2\alpha_e^2 n_m \frac{1 - n_m + n_m \ln n_m}{(n_m - 1)^2}}. \tag{3.13}$$

In the approximations of $\alpha_e^2 \ll 1$ and $\alpha_e^2 \gg 1$, this leads to the nonlinear dispersion relations

$$M_z^2 = \frac{n_m + 2}{3n_m} \tag{3.14}$$

for inertial-regime SKAWs (Shukla et al. 1982) and

$$M_z^2 = \frac{1 - n_m + n_m \ln n_m}{n_m - 1 - \ln n_m} \frac{1}{n_m} \tag{3.15}$$

for kinetic-regime SKAWs (Hasegawa and Mima 1976), respectively.

It is especially worth noting the analogy between the Sagdeev equation (3.8) and the motion equation of a classical "particle" with the unit mass in the Sagdeev potential well where $K(n; M_z, k_x) \leq 0$ between $n = 1$ and $n = n_m$, and η and n represent the "motion time" and the "spatial position" of the classical "particle", respectively. A soliton solution of Eq. (3.8) indicates the reciprocating motion of the "particle" constrained in the potential well between two points $n = 1$ and $n = n_m$, but with a period of infinity, that is, the "particle" moves only once back and forth between $n = 1$ and $n = n_m$, where n_m is the maximum (or minimum) of the density n in the soliton and determined by the nonlinear dispersion relation $K(n; M_z, k_x)|_{n=n_m\neq 1} = 0$. Based on this analogy, the necessary and sufficient conditions for the existence of the exact solutions for SKAWs can be stated as follows (Wu et al. 1996b):

1. $K(n) \leq 0$ for $n \in [1, n_m]$;
2. $K(n) = 0$ at $n = 1$ and $n = n_m$;
3. $d_n K(n) = 0$ at $n = 1$ and $(n_m - 1) d_n K(n) > 0$ at $n = n_m$.

The condition (1) ensures that the Sagdeev equation (3.8) has a real solution and means that the "particle" moves in a constraining potential well between $n = 1$ and $n = n_m$. The condition (2) means that the "motion velocity" ($d_\eta n = \pm\sqrt{-2K(n)}$) of the "particle" reaches zero at both the two boundaries of the potential well at $n = 1$ and $n = n_m$, otherwise the "particle" does not move back and forth, but moves outside the potential well, implying a shock-like solution instead of a soliton. The condition (3) means that the "particle" is reflected back at the boundary of $n = n_m$ (i.e., the "acceleration" of the "particle" $d_\eta^2 n = -d_n K(n) \neq 0$ and with a direction from the point $n = n_m$ to the point $n = 1$), but not reflected at the other boundary of $n = 1$ (i.e., the "acceleration" $d_\eta^2 n = 0$ at $n = 1$), otherwise the "particle" reciprocates with a finite period, that is, a periodically oscillational solution instead of a soliton solution. The combination of these three conditions gives the general criterion for the existence of the exact solutions for SKAWs.

In the small amplitude limit of $|N_m| \equiv |n_m - 1| \ll 1$, the nonlinear dispersion relation (3.13) reduces to $\delta M_z \equiv M_z - 1 = -N_m/3$, or i.e., $N_m = -3\delta M_z$. In particular, the Sagdeev equation (3.8) reduces to the standard soliton equation, that is, the KdV equation. The above condition (3) of the existent criterion for SKAWs leads to the inequality:

$$\delta M_z \left(1 - \alpha_e^2\right) > 0 \Rightarrow N_m \left(1 - \alpha_e^2\right) < 0, \tag{3.16}$$

implying that kinetic-regime SKAWs with $\alpha_e > 1$ are sub-Alfvénic ($\delta M_z < 0$) and accompanied by a hump ($N_m > 0$) soliton and inertial-regime SKAWs with $\alpha_e < 1$ are super-Alfvénic ($\delta M_z > 0$) and accompanied by a dip ($N_m < 0$) soliton.

From Fig. 3.4e, in the auroral plasma region between 0.2 and $1.8R_E$, which covers well the auroral acceleration region, we have $\alpha_e < 0.1$, implying in which the inertial-regime approximation can be satisfied well. In the inertial-regime approximation of $\alpha_e \ll 1$, an exactly analytical solution of the Sagdeev equation (3.8) can be obtained as follows:

$$N = N_m \frac{1 - \tanh^2 Y}{1 + N_m \tanh^2 Y}, \tag{3.17}$$

where

$$Y \equiv \frac{|\eta|}{k_x D} + \frac{N_m}{1 + N_m} \sqrt{\frac{1 - N/N_m}{1 + N}}, \tag{3.18}$$

and $N \equiv n - 1$ is the relative density fluctuation, $D \equiv -\sqrt{(6 + 2N_m)/N_m}$, and the boundary condition $N = N_m$ at $\eta = 0$ has been used without loss of generality. This solution describes a dip density soliton with an amplitude N_m, characteristic width $\Delta\eta \sim$ a few of $k_x D$, and a symmetrical center at $\eta = 0$.

The perturbed electric and magnetic fields (E_x, E_z, and B_y) of the corresponding inertial-regime SKAW can be given by Wu et al. (1995):

$$E_z = \pm \frac{D}{3} \left(\frac{-N_m}{1 + N_m}\right)^2 \left(1 - \tanh^2 Y\right) \tanh Y \cot\theta, \tag{3.19}$$

$$E_x = \pm D \frac{-N_m}{(1 + N_m)^2} \left(1 + \frac{N_m}{3} \tanh^2 Y\right) \tanh Y, \tag{3.20}$$

$$B_y = \pm D \frac{-N_m}{1 + N_m} \sqrt{\frac{1 + N_m/3}{1 + N_m}} \tanh Y, \tag{3.21}$$

where θ is the propagation angle of the SKAW, that is, $\cot\theta = k_z/k_x$, and the electric and magnetic fields are normalized by $\sqrt{Q} v_A B_0$ and $\sqrt{Q} B_0$, respectively.

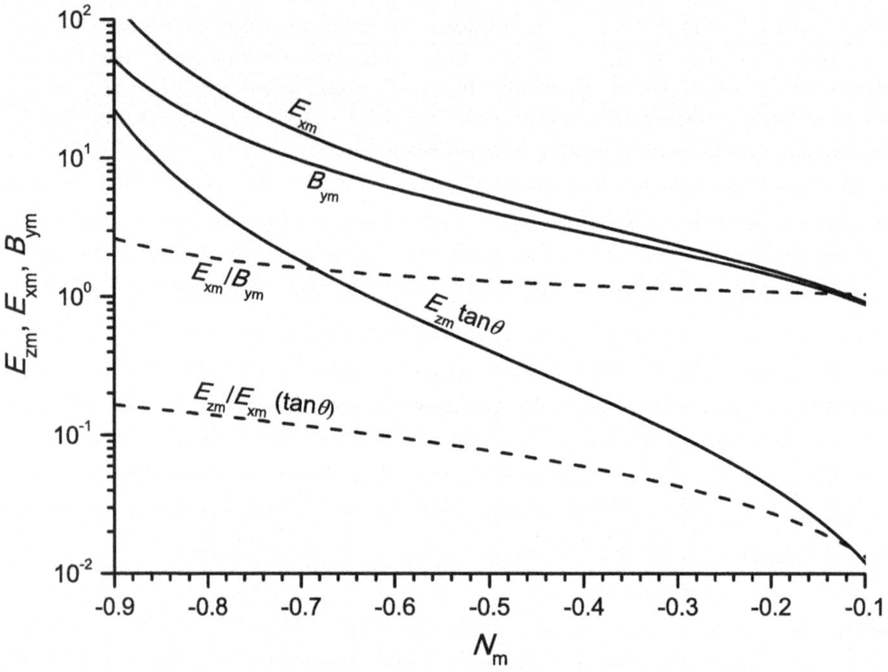

Fig. 3.8 Amplitudes of perturbed electric and magnetic fields of SKAWs versus the density amplitude N_m (from Wu 2012)

Figure 3.8 plots the amplitudes of perturbed electric and magnetic fields of SKAW versus the relative density amplitude N_m, where solid lines are the amplitudes of the parallel and transverse electric fields E_{zm} (amplified by $\tan\theta$) and E_{xm} (in units of $\sqrt{Q}v_A B_0$) and the transverse magnetic field B_{ym} (in units of $\sqrt{Q}B_0$), and two dashed lines are the ratios of the transverse electric to magnetic fields E_{xm}/B_{ym} (in units of the Alfvén velocity v_A) and the parallel to transverse electric fields E_{zm}/E_{xm} (amplified by $\tan\theta$). From Fig. 3.8, in a wide range of the density amplitude (i.e., $0.1 < |N_m| < 0.9$) the transverse electric and magnetic fields of SKAWs have amplitudes, E_{xm} and B_{ym}, several times of $\sqrt{Q}v_A B_0$ and $\sqrt{Q}B_0$, respectively. However, the parallel electric field E_{zm} has a wider varying range from lower than 0.1 when $|N_m| \sim 0.1$ to higher than 10 of $\sqrt{Q}v_A B_0$ when $|N_m| \sim 0.9$. Also Fig. 3.8 shows that the transverse electric to magnetic field ratio, E_{xm}/B_{ym}, is slightly higher than the Alfvén velocity v_A (i.e., $\sim$1–2 v_A for $|N_m|$ between 0.1 and 0.9), as shown by the in situ observations by the FAST satellite (see Fig. 3.7). In addition, the parallel to transverse electric field ratio, E_{zm}/E_{xm} by a factor of $\tan\theta$, is about $\sim$0.1.

In the auroral plasma environment explored by the Freja and FAST satellites at the altitude $\sim 0.3R_E$, one has $B_0 \simeq 0.2$ G, $v_A \simeq 1 \times 10^7$ m/s, and $Q \simeq 3.4 \times 10^{-5}$ (the ion component is dominated by O^+). This leads to $\sqrt{Q}v_A B_0 \sim 1$ V/m and $\sqrt{Q}B_0 \sim 100$ nT. The comparison with the observations by the Freja and FAST satellites shows that the exactly analytical solution of the one-dimensional inertial-regime SKAWs can qualitatively describe the main physical properties of the strong electromagnetic spikes accompanied by strong density fluctuations presented in these satellites in situ observations. However, as shown both by the satellite observations in the space plasma (Chmyrev et al. 1988; Volwerk et al. 1996; Wu et al. 1996a, 1997) and the experimental measurements in the LAPD (Burke et al. 2000a, b), these strong electromagnetic spikes, in fact, are two-dimensional vortex structures, which are accompanied by dipole density solitons (Wu et al. 1996a, 1997).

The satellite observations in the space plasma (Chmyrev et al. 1988; Volwerk et al. 1996) and the experimental measurements in the LAPD (Burke et al. 2000a, b) both show that these two-dimensional vortex structures often present in an inhomogeneous ambient plasma. Taking account of the weak inhomogeneity of the ambient plasma in density, Wu et al. (1996a, 1997) proposed a two-dimensional SKAW model that has a dipole-vortex electromagnetic structure and is accompanied by a dipole density soliton. Its electromagnetic structures can be described by two scalar potentials, ϕ and ψ, as following:

$$\phi(r,\theta) = ur_0B_0\left[1 + a(l_0, l_1)\right]\frac{K_1(l_0R)}{K_1(l_0)}\cos\theta, \text{ for } R > 1$$
$$\phi(r,\theta) = ur_0B_0\left[R + a(l_0, l_1)\frac{J_1(l_1R)}{J_1(l_1)}\right]\cos\theta, \text{ for } R < 1, \tag{3.22}$$

and

$$\psi(r,\theta) = r_0B_0\gamma\left[1 + a(l_0, l_1)\frac{u}{\gamma v_A}\right]\frac{K_1(l_0R)}{K_1(l_0)}\cos\theta, \text{ for } R > 1$$
$$\psi(r,\theta) = r_0B_0\gamma\left[R + a(l_0, l_1)\frac{u}{\gamma v_A}\frac{J_1(l_1R)}{J_1(l_1)}\right]\cos\theta, \text{ for } R < 1, \tag{3.23}$$

where the polar ordinates (r, θ) is determined by

$$x = r\cos\theta \text{ and } \eta = r\sin\theta, \tag{3.24}$$

$\eta \equiv k_y y + k_z z - \omega t$ is the traveling-wave frame, the x-axis directs the inhomogeneity of the ambient plasma density, $R \equiv r/r_0$, r_0 is the characteristic radius for the localized two-dimensional vortex structure, $u = \omega/k_y$ is the characteristic perpendicular propagation velocity, and $\gamma = k_z/k_y$ represents the propagating direction. In addition, in above the expressions (x, y), z, t have been normalized by ρ_s, λ_i, and ω_{ci}^{-1}, respectively, the parameter $a(l_0, l_1)$ is determined by

$$a(l_0, l_1) = -\frac{l_0^2}{\left(2 + l_0^2 + l_1^2\right)J_1(l_1)}, \tag{3.25}$$

l_0 and l_1 are related by:

$$\frac{K_0(l_0)}{l_0K_1(l_0)}+\frac{2}{l_0^2}=\frac{l_1}{2+l_1^2}\frac{J_0(l_1)}{J_1(l_1)}-\frac{2}{2+l_1^2}, \tag{3.26}$$

and $J_{1(0)}$ and $K_{1(0)}$ are the first (zero) order Bessel function and the first (zero) order modified second kind Bessel function, respectively. Finally, the electric and magnetic fields of the vortex structure can be given by the two scalar potentials as follows:

$$\mathbf{E}_\perp=-\nabla_\perp\phi,\;\; E_\parallel=\hat{\mathbf{b}}\cdot\mathbf{E}=-\frac{d\phi}{dz}-\frac{\partial\psi}{\partial t},$$
$$\mathbf{B}_\perp=\nabla\times\left(\psi\hat{\mathbf{z}}\right)=\nabla_\perp\psi\times\hat{\mathbf{z}}. \tag{3.27}$$

In particular, the relative perturbed density $N\equiv n-1$ can be obtained as follows (Wu et al. 1996a, 1997):

$$N=\left[R-(1+a(l_0,l_1))\frac{K_1(l_0R)}{K_1(l_0)}\right]\frac{r_0\cos\theta}{L},\quad \text{for } R>1$$
$$N=-a(l_0,l_1)\frac{J_1(l_1R)}{J_1(l_1)}\frac{r_0\cos\theta}{L},\quad \text{for } R<1, \tag{3.28}$$

where, assuming the weak inhomogeneity of the ambient plasma with the characteristic scale $L\gg r_0$ in the x direction, the boundary condition $N=x/L$ for $r\rightarrow\infty$ has been used, the density continuity condition at $r=r_0$ leads to the perpendicular propagation velocity u as follows:

$$u=\frac{2}{l_1^2}\frac{r_0^2}{\rho_sL}v_s=\frac{2}{l_1^2}\frac{\omega_{ci}r_0^2}{L}. \tag{3.29}$$

Figure 3.9 shows the local density distribution inside the dipole vortex structure with the typical parameters $l_1=4$, $l_0=1.6$, $a(l_0,l_1)=1.9$, and the density amplitude $N_m\sim 50\%$, where (a) is the whole picture of its two dimensional structure, (b) its contours, and (c) observed three kinds of density solitons when the satellite crosses the dipole vortex at the three different positions and directions denoted by S_1, S_2, and S_3 in (b), which correspond to the dip (S_1), hump (S_2), and dipole (S_3) density solitons, respectively. For the ambient plasma of the Freja satellite environment, the typical parameters may be taken as $T_e\sim 10$ eV, $B_0\sim 0.3$ G, $L\sim 10$ km, $n_0\sim 10^3$ cm^{-3}, and the average mass number of ions (mainly by the oxygen ions and the hydrogen ions) $\bar{\mu}\sim 7$, hence $v_s\sim 10$ km/s, $\rho_s\sim 30$ m. The typical scale of SKAWs may be taken as $r_0\sim 10\rho_s\sim 300$ m. From Fig. 3.7, it can be found that these density solitons have a spatial scale length of $\sim$1–2r_0, which is comparable with observations of the Freja satellite.

The perturbed electric and magnetic fields inside the dipole vortex structure can be obtained by Eq. (3.27) as follows:

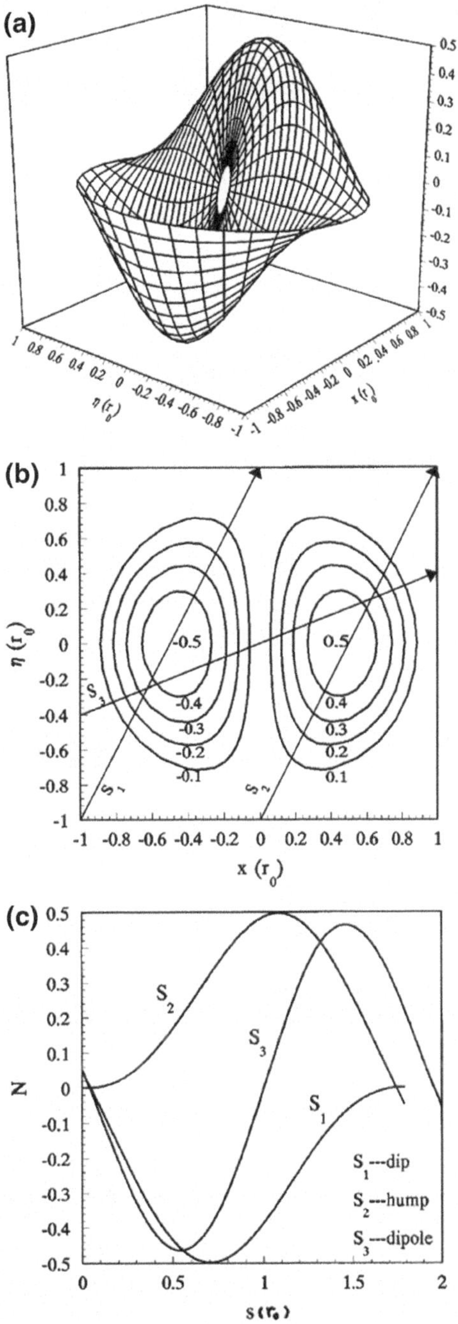

Fig. 3.9 **a** The two-dimensional density distribution in a dipole vortex structure; **b** The contours of the density distribution and three possible cases of the satellite crossing the dipole vortex structure at loci S_1, S_2 and S_3; **c** The dip, hump, and dipole density solitons observed by the satellite, respectively, at loci S_1, S_2, and S_3 (reprinted from Wu et al., Phys. Plasmas, 4, 611–617, 1997, with the permission of AIP Publishing)

$$E_y = -uB_0 \frac{a(l_0, l_1)}{2} \left(\frac{l_1 J_0(l_1 R)}{J_1(l_1)} - \frac{J_1(l_1 R)}{R J_1(l_1)} \right) \sin 2\theta,$$
$$E_x = -uB_0 \left[1 + a(l_0, l_1) \left(\frac{l_1 J_0(l_1 R)}{J_1(l_1)} \cos^2\theta - \frac{J_1(l_1 R)}{R J_1(l_1)} \cos 2\theta \right) \right],$$
$$E_z = \left(\gamma - \frac{u}{v_A} \right) E_y,$$
$$B_x = -\frac{E_y}{v_A}, \quad B_y = \frac{E'_x}{v_A} - \gamma B_0. \tag{3.30}$$

In particular, the peak values of the perturbed electric fields can be obtained from Eq. (3.30) as follows:

$$E'_{xm} \equiv E_{xm} + uB_0 = 2E_{ym} = 2uB_0 \frac{a(l_0, l_1) l_1}{4J_1(l_1)} = 2uB_0 \frac{l_1 N_m}{4J_{1m}} \frac{L}{r_0}. \tag{3.31}$$

From Eq. (3.30), one has the field-aligned component of perturbed electric fields, E_z, much less than their cross-field components, E_x and E_y, that is, $|E_\parallel / E_\perp| \ll 1$, which is consistent with that in one-dimensional SKAWs. Unlike to those of one-dimensional SKAWs, however, the perturbed electric fields of two-dimensional SKAWs are no longer plane polarization. Figure 3.10 shows the "rotation" curves of the perturbed electric fields in the plane perpendicular to the magnetic field. It can be found that the "rotation" direction of perturbed electric fields of dip density solitons is opposite to that of hump density solitons, while perturbed electric fields of dipole density solitons twice undergo opposite "rotation."

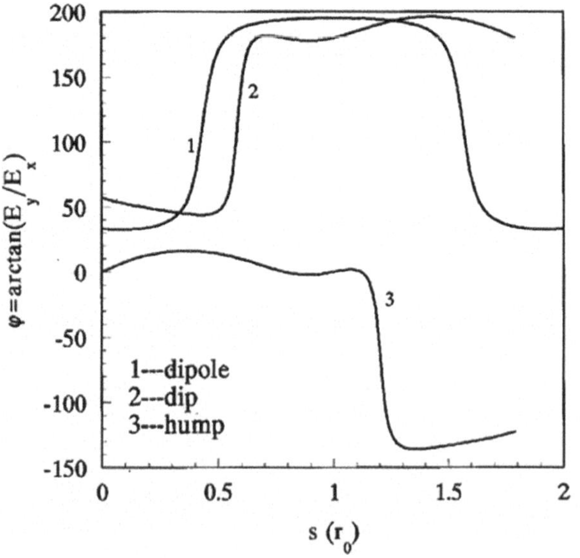

Fig. 3.10 The "rotation" curves of perturbed electric fields associated with dip, hump, and dipole density solitons (reprinted from Wu et al., Phys. Plasmas, 4, 611–617, 1997, with the permission of AIP Publishing)

In comparison to one-dimensional SKAWs, the most important unique feature of two-dimensional SKAWs is that they have vortex structures. In particular, the two-dimensional SKAWs with dipole vortex structures can permit the presence of not only single dip or hump density solitons but also dipole density solitons, while the latter can not present in one-dimensional SKAWs. Second, the polarization of their perturbed electromagnetic fields is no longer plane polarization, which is obviously distinct from that of one-dimensional SKAWs, and has a "rotation" feature, as shown by analyses of the Freja observations (Volwerk et al. 1996). These results indicate that the model of two-dimensional SKAWs with dipole vortex structures can provide us more self-consistent physical explanation for SKAW phenomena in space plasmas, although the one-dimensional SKAW model can well explain their main properties observed by satellites in space plasmas.

3.5 Auroral Electron Acceleration by SKAWs

In a collisionless plasma environment such as the auroral plasma, especially in the auroral acceleration region, in general, the very large parallel conductivity can immediately short-circuit any parallel electric fields. Therefore, one of key difficulties in understanding the physics of the auroral electron acceleration is how to produce and maintain a large parallel electric field that can accelerate field-aligned electrons to keV-order energies (Fälthammar 2004). Now that a field-aligned electric field can develop within KAWs observed commonly in the auroral plasma, they have been proposed by many authors as one of possible acceleration mechanisms for producing auroral electrons (Chmyrev et al. 1988; Goertz 1981, 1984; Goertz and Boswell 1979; Hasegawa 1976; Hasegawa and Mima 1978; Hui and Seyler 1992; Kivelson and Southwood 1986; Kletzing 1994; Lee et al. 1994; Lysak and Carlson 1981; Lysak and Dum 1983; Thompson and Lysak 1996). However, the detailed physical mechanism has been an open problem. Wu (2003a, b) and Wu and Chao (2003, 2004) proposed that SKAWs can play an important role in the auroral electron acceleration, in which effective parallel electric fields can be maintained by the inertial motion of electrons along the magnetic field.

In fact, some further data analyses have clearly revealed that SKAWs observed in the auroral plasma are often associated with enhanced broadband electrostatic fluctuations, which can be identified well as ion-acoustic turbulence. The accompanied ion-acoustic turbulence will possibly cause the dissipation of the SKAWs and lead to their dynamical evolution and deformation. For example, based on the analysis of a large number of events associated with SKAWs observed by Freja, Wahlund et al. (1994a) found that these associated SKAWs could be classified into three different observational phases and proposed that they are possibly responsible for three different stages in the dynamical evolution of SKAWs. The first stage is the "ordinary" SKAW, the second stage is the SKAW accompanied with enhanced broadband electrostatic fluctuations in the ion-acoustic wave mode, and in the third stage, the SKAW has been transformed into an electrostatic-like structure.

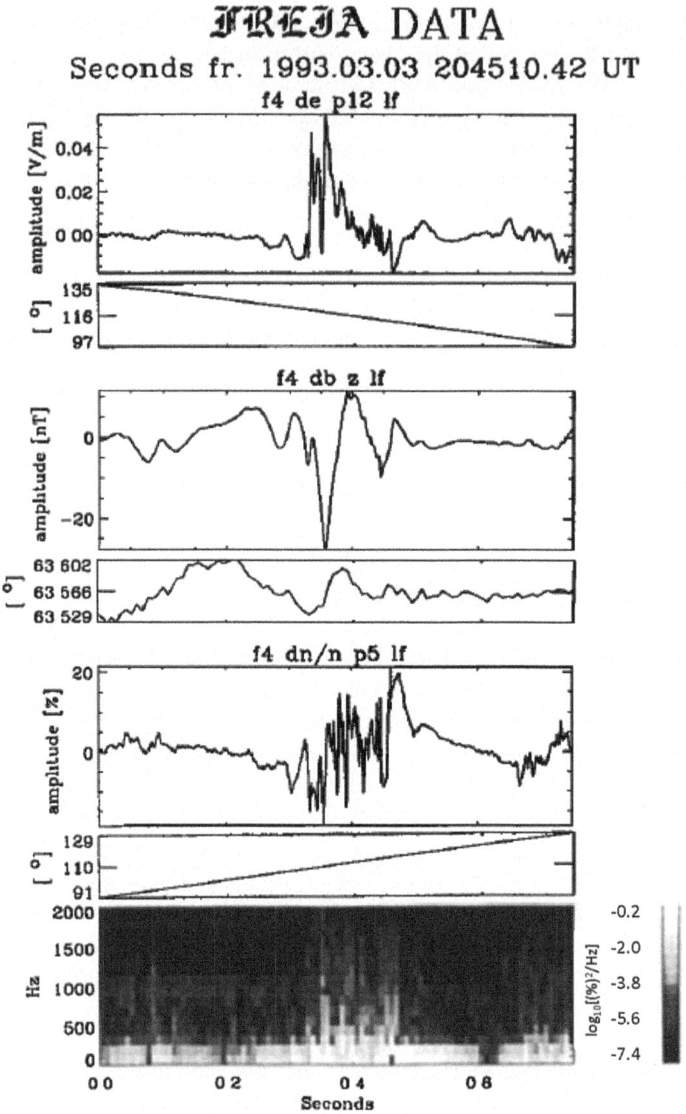

Fig. 3.11 An example of a SKAW filled by enhanced broadband electrostatic turbulence in the ion-acoustic mode (from Wahlund et al. 1994a)

Figure 3.11 shows a clear example of the second stage SKAW, in which a well defined SKAW with $\delta E \sim 60$ mV/m, $\delta B \sim 15$ nT, and $dn/n \sim 30\%$ is detected between 0.3 and 0.5 s, but in contrast to the first stage ordinary SKAW, it also contains enhanced broadband electrostatic and density fluctuations, which can be identified well as enhanced ion-acoustic turbulence (Wahlund et al. 1994a, b). Based on these observations, Wu (2003a, b) proposed that the enhanced ion-acoustic turbulence of accompanying the second stage SKAW can be excited by the electrostatic instability of the electrons trapped inside the SKAW, which are accelerated by the parallel oscillational electric field of the SKAW to an v_A-order field-aligned velocity, especially in the inertial-regime SKAW case, these trapped electrons can have a field-aligned velocity larger than the thermal velocity of the ambient electrons. Then the enhanced ion-acoustic turbulence causes effective dissipation of the SKAW through the wave-particle interaction and leads to the SKAW further evolving into an electrostatic-like structure, that is, the third stage SKAW.

In order to take account of the effect of the enhanced ion-acoustic turbulence on the dynamical evolution of SKAWs, Wu (2003a, b) invoked an anomalous collisional damping term of the electrons in their field-aligned momentum equation, which is caused by the wave-particle interaction due to the enhanced ion-acoustic turbulence and may be described by the following "effective collisional frequency", ν_e, Hasegawa (1975):

$$\nu_e = \frac{W_{ia}}{n_0 T_e}\omega_{pe}, \tag{3.32}$$

where W_{ia} is the energy density of the enhanced ion-acoustic turbulence. Including this "anomalous dissipation" effect due to the enhanced ion-acoustic turbulence, the Sagdeev equation of governing the SKAW dynamics can be written as follows (Wu 2003a, b):

$$\frac{d}{d(-\eta)}\left[\frac{1}{2}\left(\frac{d}{d\eta}\frac{2}{n^2}\right)^2 + \frac{K(n)}{n^6}\right] = -\gamma\left(\frac{d}{d\eta}\frac{2}{n^2}\right)^2, \tag{3.33}$$

where the Sagdeev potential for the inertial-regime SKAWs

$$K(n; M_z, k_x) = -\frac{n^3(n-1)^2}{3M_z^2 k_x^2}\left(\frac{n}{n_m} - 1\right) \tag{3.34}$$

with the density amplitude

$$n_m = \frac{2}{3M_z^2 - 1} < 1 \tag{3.35}$$

and the "damping coefficient" of describing the anomalous dissipation effect

$$\gamma \equiv \frac{\sqrt{Q}}{M_z k_z}\frac{\nu_e}{\omega_{ci}}. \tag{3.36}$$

It is clear that when neglecting the anomalous dissipation effect ($\gamma = 0$) the Sagdeev equation for the inertial-regime SKAWs can be recovered from the integral of Eq. (3.33).

In analogy to the Sagdeev equation (3.8), Eq. (3.33) also describes the motion of a "classical particle" in the "potential well" $K(n)/n^6$, with "time" $t = -\eta$, "space"

$x = 2/n^2$, hence "velocity" $dx/dt = -d\left(2/n^2\right)/d\eta$. The "classical particle" can conserve its energy when $\gamma = 0$. Only if $\gamma > 0$ (independent of specific values of γ), the "damping" $-\gamma\,(dx/dt)^2 = -\gamma\left(d\left(2/n^2\right)/d\eta\right)^2$ will make the "particle" lose gradually its "energy" and ultimately stay at the bottom of the "potential well" $K(n)/n^6$ when $t = -\eta \to +\infty$. The "position" $n = n_d$ of the bottom can be obtained by the condition that the potential energy $K(n)/n^6$ reaches its minimum as follows:

$$n_d = \frac{3n_m}{2+n_m} = \frac{1}{M_z^2} > n_m. \tag{3.37}$$

This result indicates that the density structure of the dissipated SKAW (DSKAW), described by Eq. (3.33) has the asymptotic values $n \to n_d$ and $n \to 1$ when $\eta \to -\infty$ and $\eta \to \infty$, respectively. In other words, the density profile of DSKAWs has a shock-like structure with a density jump

$$\Delta n \equiv 1 - n_d = 1 - \frac{1}{M_z^2} = \frac{2|N_m|}{3+N_m} < |N_m|, \tag{3.38}$$

where $N_m \equiv n_m - 1$.

Figure 3.12 illustrates density behaviors of DSKAWs with the damping coefficient $\gamma = 1.00, 0.50, 0.10, 0.05, 0.02$, and 0.01 from the top down, where the parameters $k_x/k = \sin 89°$ for the quasi-perpendicular propagating angle $\theta = 89°$, and $M_z = 1.29$ for the density amplitude $n_m \simeq 0.5$ have been used. For the sake of comparison, dashed lines in Fig. 3.12 show the density soliton solutions of SKAWs with the same parameters except for $\gamma = 0$.

From Fig. 3.12, it can be found that the density of DSKAWs behaves itself like an ordinary "shock" with a density jump $\Delta n \simeq 0.4$ for a strong dissipation case of $\gamma \sim 1$. For a weakly dissipative regime of $\gamma \ll 1$, however, the density waveform appears to be a train of oscillatory waves with amplitude decreasing in the downstream of $\eta < 0$ and converges upon an ultimate downstream density $n_d \simeq 0.6$ when $\eta \to -\infty$ and, as a result, a local shock-like structure with the same density jump ($\Delta n = 1 - n_d \simeq 0.4$) is formed. It is worth noticing that the ultimate density n_d and hence the density jump Δn are independent of the damping coefficient γ. In fact, the magnitude of γ affects only the oscillating strength and the converging speed in the manner that the lower γ leads to the stronger oscillation and the slower convergence.

At the ultimate downstream state of DSKAWs, the field-aligned "escape" velocity of the electrons trapped inside DSKAWs is given by Wu (2003a, b)

$$v_{ed} \equiv v_{ez}|_{n=n_d} = \left(1 - M_z^2\right) M_z v_A = -\frac{\Delta n}{(1-\Delta n)^{3/2}} v_A, \tag{3.39}$$

which implies the electrons accelerated inside the DSKAW can escape from the downstream at the velocity v_{ed}, in contrast to the case of SKAWs, where the accelerated electrons are, at all, trapped inside the SKAW. These escaping electrons are

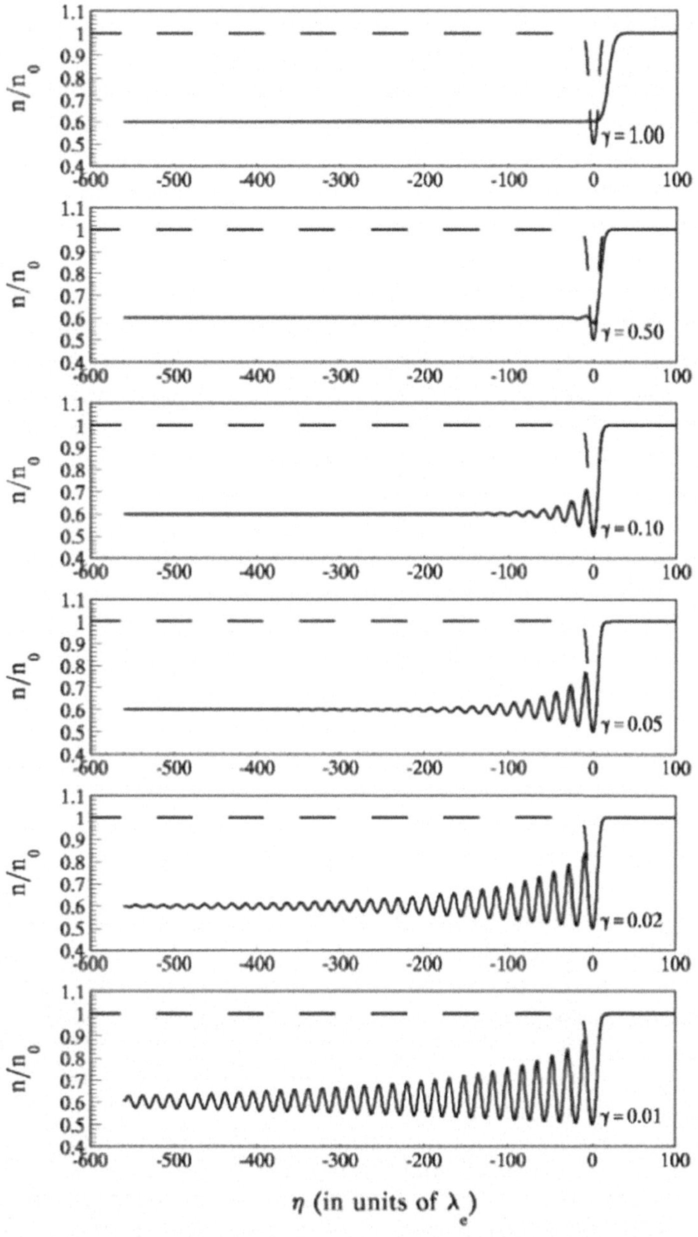

Fig. 3.12 Density distributions of DSKAWs (normalized to n_0) with the parameters $\theta = 89°$, $M_z = 1.29$, and $\gamma = 1, 0.5, 0.1, 0.05, 0.02$, and 0.01, from the top down. Dashed lines represent the density of SKAWs with the same parameters as DSKAWs except for $\gamma = 0$ (reprinted from Wu, Phys. Plasmas, 10, 1364–1370, 2003b, with the permission of AIP Publishing)

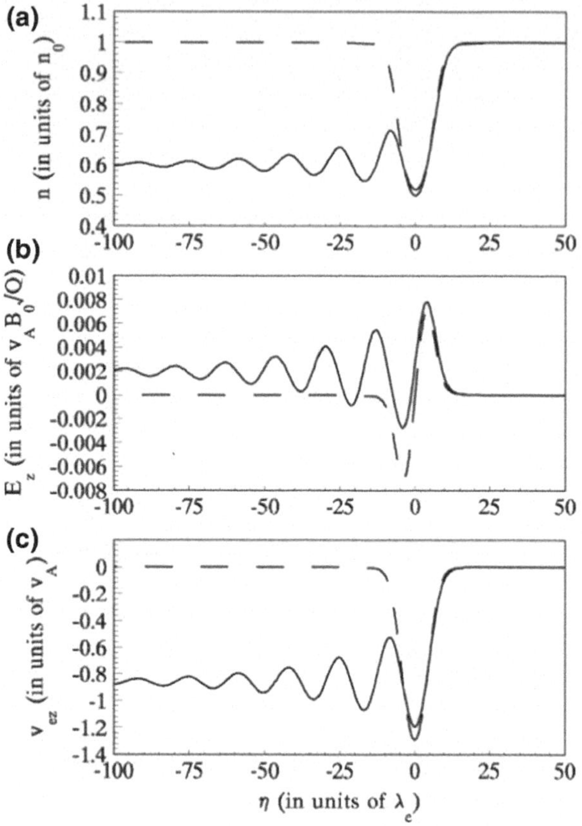

Fig. 3.13 Sketch of the shock-like structure of the DSKAW for the parameters $\theta = 89°$, $M_z = 1.29$, and $\gamma = 0.1$: **a** the density n (normalized to n_0); **b** the field-aligned electric field E_z (normalized to $\sqrt{Q} v_A B_0$); and **c** the electron velocity v_{ez} (normalized to v_A). Dashed lines represent the corresponding solutions of the SKAW with the same parameters except for $\gamma = 0$ (reprinted the figure with permission from Wu, Phys. Rev. E 67, 027402, 2003a, copyright 2003 by the American Physical Society)

accelerated by the field-aligned electric field E_z of the DSKAW, which nonsymmetrically distributes about the center $\eta = 0$.

Figure 3.13 shows the distributions of the density n (a), the field-aligned electric field E_z (b), and the electron velocity v_{ez} (c) in DSKAW with parameters $\theta = 89°$, $M_z = 1.29$, and $\gamma = 0.1$. For the sake of comparison, the solutions of the corresponding SKAW with the same parameters are presented by the dashed lines in Fig. 3.13. From Fig. 3.13b, it can be found that the electric field E_z varies no longer symmetrically about the center $\eta = 0$, and has a nonsymmetrical structure. In particular, the electric field E_z, although oscillating, has a preferential direction of $E_z > 0$ at the downstream, and leads to the electrons to be accelerated towards the downstream, then to get over the collisional resistance at the downstream, and ultimately to escape from the downstream at the velocity v_{ed}, as shown in Fig. 3.13c.

From Eq. (3.39), the escaping velocity of the electrons accelerated by DSKAWs, and hence their escaping energies increase with the increase of the density jump Δn of the DSKAWs and have typically the order of v_A, as expected. In fact, the escaping velocity v_{ed} has magnitudes of ~0.1–10 v_A for a wider Δn through 0.1–0.8. From Eq. (3.39), the energy of the escaping electrons can be given by

$$\varepsilon_{ed} \equiv \frac{1}{2} m_e v_{ed}^2 = \left(M_z^2 - 1\right)^2 M_z^2 \varepsilon_{eA} = \frac{\Delta n^2}{(1 - \Delta n)^3} \varepsilon_{eA}, \quad (3.40)$$

where $\varepsilon_{eA} \equiv m_e v_A^2/2$ is the energy of the electron moving at the Alfvén velocity, called the "electron Alfvén energy", which represents the characteristic energy of the escaping electrons accelerated by a typical DSKAW and is determined by the local Alfvén velocity of the ambient plasma. This indicates that the energy of the electrons accelerated by DSKAWs evidently depends on the local Alfvén velocity (i.e., directly proportional to the electron Alfvén energy ε_{eA}) and ranges ~$0.1–10\varepsilon_{eA}$ for the DSKAWs with density jumps $\Delta n \sim$ 22–65%.

Figure 3.14a presents the variation of the electron Alfvén energy ε_{eA} (in keV), based on the altitude model of the ambient auroral plasma presented in Fig. 3.4. From Fig. 3.14a it can be found that the electron Alfvén energy ε_{eA} has higher values ~1–5 keV in the observed acceleration region of $h \sim 0.5–2R_E$, which is the typical characteristic energies of the auroral energetic electrons as shown by observed (see Fig. 3.3), and reaches the maximal value ≃5.3 keV at $h \simeq 0.82R_E$. It is evident that the higher local Alfvén velocity, and hence the higher electron Alfvén energy, leads to the higher escaping energy of the accelerated electrons in the DSKAW acceleration mechanism. However, in the observations of SKAWs the strong DSKAWs with $\Delta n > 50\%$ usually are rather rare, which leads to the escaping energy of the accelerated $\varepsilon_{ed} \simeq 2\varepsilon_{eA}$, that is, $\varepsilon_{ed} \sim 10.6$ keV for the maximal $\varepsilon_{eA} \simeq 5.3$ keV in the auroral acceleration region. This implies that the highest energy of the auroral energetic electrons accelerated by DSKAWs is ~10.6 keV, and above this energy the auroral energetic electron flux will precipitously decrease because of the absence of extremely strong DSKAWs with density jump $\Delta n > 50\%$ (Bryant 1981). The result presented in Fig. 3.14a indicates that, in the observed auroral acceleration region, the observed DSKAWs accompanied by moderately strong density jumps can effectively accelerate the electrons to the typical keV-order energies of the auroral energetic electrons.

On the other hand, the enhanced electrostatic ion-acoustic turbulence plays an important role in the dynamical evolution of SKAWs towards DSKAWs (Wahlund et al. 1994a). In the case of inertial-regime SKAWs, the enhanced electrostatic turbulence can be effectively excited by the electrostatic instability of the electrons that are trapped in the potential well of the SKAWs and have the typical v_A-order field-aligned velocity evidently higher than the local thermal speed of the ambient plasma (Wu 2003a, b; Wu and Chao 2003, 2004). In particular, the lower parameter α_e (hence the higher velocity ratio v_A/v_{T_e}) leads to the more efficient excitation of the enhanced electrostatic turbulence and the more effectively dynamical evolution of SKAWs towards DSKAWs. Figure 3.4e clearly shows that there is the lower parameter $\alpha_e \ll 1$ in the observed auroral acceleration region, implying that the plasma environment in the observed acceleration region is more favorable of the formation

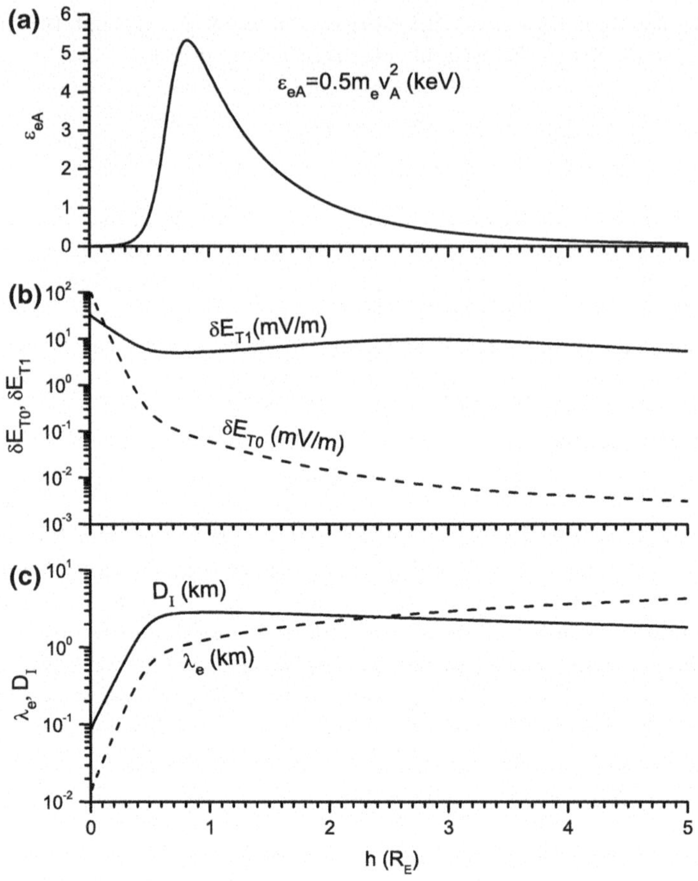

Fig. 3.14 The altitude distribution of the characteristic parameters of the DSKAW acceleration mechanism: the electron Alfvén energy ε_{eA} in keV (**a**), the electrostatic ion-acoustic turbulence δE_{T1} (dashed line) and δE_{T0} (dashed line) in mV/m (**b**), and field-aligned projection of the characteristic width of DSKAWs on the ionosphere D_I (solid line) and the electron inertial length λ_e (dashed line) (**c**)

of DSKAWs. In order to produce the sufficient "anomalous dissipation" to cause SKAWs efficiently evolving into DSKAWs, the strength of the enhanced electrostatic ion-acoustic turbulence, δE_T, is required to be considerably higher than its thermal noise level, δE_{T0}, on which the effective collision frequency in Eq. (3.32) is equal to the classic collision frequency, that is, $\nu_e = \nu_c$ (Hasegawa 1975):

$$\delta E_{T0} \simeq 0.94 \times 10^{-2} T_e^{-1/4} n_0^{3/4} \text{ mV/m}, \quad (3.41)$$

where T_e is in units eV and n_0 is in cm^{-3}. For example, in order to produce the damping coefficient of $\gamma = 0.1$, from Eq. (3.36) the requisite strength of the ion-acoustic turbulence δE_{T1} can be estimated by (Wu and Chao 2003, 2004)

$$\delta E_{T1} \simeq 5.53 n_0^{1/4} T_e^{1/2} B_0^{1/2} \text{ mV/m}. \tag{3.42}$$

Fig. 3.14b shows the variations of δE_{T0} (dashed line) and δE_{T1} (solid line) with the altitude h above the ionosphere. The former (i.e, the thermal noise level) corresponds to the turbulent wave strength that is able to produce an effective collision frequency $\nu_e = \nu_c$ (the classic collision frequency), and the latter does the strength that is able to produce an effective collisional frequency ν_e to result in a damping coefficient $\gamma = 0.1$. From Fig. 3.14b, one can find that $\delta E_{T1} < \delta E_{T0}$ in the ionosphere below $h \simeq 0.13 R_E$, implying that the classic collision dominates the collisional damping and can lead to the damping coefficient $\gamma > 0.1$ in the denser ionosphere below $h \simeq 1.3 R_E$. While in the tenuous magnetosphere above $h \simeq 0.13 R_E$ the dissipation damping is dominated by the effective collision due to the ion-acoustic turbulence. In particular, for the auroral acceleration region of $h \sim 0.5$–$2 R_E$, Fig. 3.14b shows that the damping coefficient of $\gamma = 0.1$ needs the enhanced ion-acoustic turbulence strength $\delta E_{T1} \sim 5$–10 mV/m, which can be well satisfied by the observations of turbulent electric fields in the auroral plasma (Vaivads et al. 1998; Wahlund et al. 1994a, b, 1998).

The width of discrete auroral arcs produced by DSKAWs acceleration mechanism can be estimated from the field-aligned projection of the DSKAW width on the ionosphere because the accelerated auroral electrons precipitate field-aligned into the ionosphere. Both theory and observation (Chaston et al. 1999; Wu 2003b) show DSKAWs at the altitude h have a typical width $D \sim 2\pi\lambda_e(h)$, where $\lambda_e(h) = c/\omega_{pe}(h)$ is the local electron inertial length at the altitude h. The field-aligned projection of the DSKAW with the width $D(h)$ on the ionosphere $D_I(h)$ can be obtained by:

$$D_I(h) = \frac{2\pi\lambda_e(h)}{(1+h)^{3/2}}, \tag{3.43}$$

where the dipole field model of the Earth's magnetic field and the magnetic flux conservation along the polar field lines have been used. Figure 3.14c displays the altitude variations of the electron inertial length $\lambda_e(h)$ (dashed line) and the width of discrete auroral arcs produced by the DSKAW acceleration mechanism $D_I(h)$ (solid line). From Fig. 3.14c, it can be found that the auroral arcs produced by the electrons accelerated by DSKAWs from the auroral acceleration region of $h \sim 0.5$–$2 R_E$ can be expected to have a typical width ~1–3 km in the ionosphere, which is comparable to the characteristic width of observed auroral arcs.

Based on the DSKAW acceleration mechanism of the auroral energetic electrons described above, it is possible that the main observed properties of the auroral energetic electrons, which had been mentioned in the beginning of this chapter, may be explained reasonably and uniformly as follows:

Field-aligned acceleration: In the DSKAW acceleration mechanism, it is the parallel electric field of the DSKAW that accelerates field-aligned the electrons to escape from the downstream of the shock-like structure of the DSKAW and to precipitate along the geomagnetic field on the ionosphere;

Characteristic energy: The characteristic energy of the electrons accelerated by the DSKAW acceleration mechanism is typically in the order of the electron Alfvén energy ε_{eA}, which reaches highest values in the auroral acceleration region of $h \sim 0.5$–$2R_E$ and has the typical energies of the auroral energetic electrons, that is, several keVs for the moderately strong DSKAWs observed most frequently in the auroral plasma;

High-energy cutoff: For the moderately strong DSKAWs with the density jump $\Delta n \sim 50\%$, the highest energy of the electrons accelerated by the DSKAW is ~ 10.6 keV in the auroral acceleration region, it is probably because of the absence of extremely strong DSKAWs with the density jump $\Delta n > 50\%$ that the auroral energetic electron flux precipitously decreases above 10 keV;

Acceleration region: The auroral acceleration region is located in the altitude of $h \sim 0.5$–$2R_E$ because both the Alfvén velocity and the ratio of the Alfvén velocity to the electron thermal speed reach highest there. The former leads to the DSKAW acceleration mechanism having a higher acceleration efficiency, that is, being able to accelerate the electrons to higher energies directly proportional to the square of the Alfvén velocity (i.e., the electron Alfvén energy), and the latter is more favorable of the excitation of the enhanced electrostatic ion-acoustic turbulence and hence of the formation of DSKAWs;

Discrete auroral arc: An individual DSKAW produces a discrete auroral arc, which width is the projection of the width of the corresponding DSKAW on the ionosphere along the geomagnetic field lines, that is, the observed width of typical auroral arcs $\sim$ a few km, as shown in Fig. 3.12c. A large number of observations show that special acceleration source regions of auroral energetic electrons usually come from some cavities with a lower density and hence with a higher Alfvén velocity in the center than that in the edge, and in result the auroral electrons accelerated by DSKAWs in the auroral arc center have higher energies than that in the auroral arc edge.

The auroral energetic acceleration mechanism is an essentially and crucially important problem, but not the whole problem in the auroral and substorm dynamics. Above the discussions show that the DSKAW acceleration mechanism can well explain reasonably and uniformly the main observed properties of the auroral energetic electrons. Based on this explanation we try to propose the following scenario for the auroral phenomena. Strong AWs or KAWs are produced by some generators, for instance, shear flows or pressure gradients in the CPS and PSBL (Borovsky 1993), which propagate into the auroral zone along the geomagnetic field lines and evolve into nonlinear structures, such as SKAWs. When entering into the auroral acceleration region of $h \sim 0.5$–$2R_E$ where is a lower density cavity, a higher ratio $v_A/v_{T_e} > 1$ leads to a higher field-aligned velocity of the electrons trapped inside the SKAWs, $v_{ez} \sim v_A > v_{T_e}$. These fast electrons can effectively excite electrostatic instabilities and lead to the formation of the enhanced electrostatic turbulence, such as the Langmuir wave or ion-acoustic wave turbulence. In consequence, the SKAWs dynamically evolve into the DSKAWs with shock-like structures due to the anomalous dissipation caused by the wave-particle interaction between the turbulent waves and electrons. These DSKAWs are trapped in the auroral acceleration region due to the reflection by the ionosphere or by the Alfvén velocity gradient in the upper magnetosphere (Vogt and Haerendel 1998) and can effectively accelerate field-aligned electrons to escape from the downstream of the shock-like structures. Therefore, when propagating downwards or upwards these DSKAWs can accelerate electrons upwards or downwards and produce upgoing or downgoing energetic electron flows (i.e., field-aligned currents downwards or upwards, Boehm et al. 1995). The fast electrons travelling in the low-density cavity (i.e., the acceleration source region) produce AKR phenomena due to the electron-cyclotron maser instability, and when downgoing and impacting the dense ionosphere or the upper atmosphere they lead to visible auroras.

References

Akasofu, S.-I. (1981). Energy coupling between the solar wind and the magnetosphere. *Space Science Reviews*, *28*, 121–190.

Alfvén, H. (1957). On the theory of comet tails. *Tellus*, *1*, 92–96.

Banks, P. M., & Holzer, T. E. (1969). High-latitude plasma transport: The polar wind. *Journal of Geophysical Research*, *74*, 6317–6332.

Bellan, P. M., & Stasiewicz, K. (1998). Fine-scale cavitation of ionospheric plasma caused by inertial Alfvén wave ponderomoive force. *Physical Review Letters*, *80*, 3523–3526.

Biermann, L. (1948). Über die Ursache der chromosphärischen Turblenz und des UV-Exzesses der Sonnenstrahlung. *Zeitschrift fur Astrophysik*, *25*, 161–169.

Biermann, L. (1951). Kometenschweife und solare Korpuskular Strahlung. *Zeitschrift fur Astrophysik*, *29*, 274–286.

Biermann, L. (1957). Solar corpuscular radiation and the interplanetary gas. *Observatory*, *77*, 109–110.

Bingham, R., Bryant, D. A., & Hall, D. S. (1984). A wave model for the aurora. *Geophysical Research Letters*, *11*, 327–330.

Birkeland, K. R. (1908). *The norwegian aurora polaris expedition 1902–1903: On the cause of magnetic storms and the origin of terrestrial magnetism*. London: Longmans Green & Co.

Boehm, M. H., Clemmons, J., Wahlund, J. E., et al. (1995). Observations of an upward-directed electron beam with the perpendicular temperature of the cold ionosphere. *Geophysical Research Letters*, *22*, 2103–2106.

Bonetti, A., Bridge, H. S., Lazarus, A. J., Lyon, E. F., Rossi, B., & Scherb, F. (1963). Explorer 10 plasma measurements. *Journal of Geophysical Research*, *68*, 4017–4063.

Borovsky, J. E. (1993). Auroral arc thicknesses as predicted by various theories. *Journal of Geophysical Research: Space Physics*, *98*, 6101–6138.

Boström, R., Gustafsson, G., Holback, B., et al. (1988). Characteristics of solitary waves and weak double layers in the magnetospheric plasma. *Physical Review Letters*, *61*, 82–85.

Bridge, H. S., Dilworth, C., Lazarus, A. J., Lyon, E. F., Rossi, B. & Scherb, F. (1962). Direct observations of the interplanetary plasma. *Journal of the Physical Society of Japan*, *17*(Suppl. A-II), 553.

Bryant, D. A. (1981). Rocket studies of particle structure associated with auroral arcs. In Akasofu & Kan (Eds.), *AGU geophysical monographs 25: Physics of auroral arc formation* (p. 103).

Bryant, D. A. (1990). Two theories of auroral electron acceleration. In Buti (Ed.), *Solar and planetary plasma physics* (pp. 58–91). London: World Scientific.

Burke, A. T., Maggs, J. E., & Morales, G. J. (2000a). Spontaneous fluctuations of a temperature filament in a magnetized plasma. *Physical Review Letters*, *84*, 1451–1454.
Burke, A. T., Maggs, J. E., & Morales, G. J. (2000b). Experimental study of fluctuations excited by a narrow temperature filament in a magnetized plasma. *Physics of Plasmas*, *7*, 1397–1407.
Carlson, C. W., Pfaff, R. E., & Watzin, J. G. (1998). The fast auroral SnapshoT (FAST) mission. *Geophysical Research Letters*, *25*, 2013–2106.
Chapman, S. (1931a). The absorption and dissociative or ionizing effect of monochromatic radiation in an atmosphere on a rotating earth. *Proceedings of the Physical Society*, *43*, 26–45.
Chapman, S. (1931b). The absorption and dissociative or ionizing effect of monochromatic radiation in an atmosphere on a rotating earth part II. Grazing incidence. *Proceedings of the Physical Society*, *43*, 483–501.
Chapman, S., & Ferraro, V. C. A. (1930). A new theory of magnetic storms. *Nature*, *126*, 129–130.
Chapman, S., & Ferraro, V. C. A. (1931). A new theory of magnetic storms. *Terrestrial Magnetism and Atmospheric Electricity*, *36*, 77.
Chaston, C. C., Carlson, C. W., Peria, W. J., et al. (1999). FAST observations of inertial Alfvén waves in the dayside aurora. *Geophysical Research Letters*, *26*, 647–650.
Chian, C. L., & Kamide, Y. (2007). An overview of the solar-terrestrial environment. In Y. Kamide & A. Chian (Eds.), *Handbook of the solar-terrestrial environment* (pp. 1–23). Berlin: Springer.
Chmyrev, V. M., Bilichenko, S. V., Pokhotelov, O. A., et al. (1988). Alfvén vortices and related phenomena in the ionosphere and the magnetosphere. *Physica Scripta*, *38*, 841–854.
Fälthammar, C. G. (2004). Magnetic-field aligned electric fields in collisionless space plasmas-a brief review. *Geofisica Internacional*, *43*, 225–239.
Goertz, C. K. (1981). Discrete breakup arcs and kinetic Alfvén waves. In Akasofu & Kan (Eds.), *AGU geophysical monographs 25: Physics of auroral arc formation* (p. 451).
Goertz, C. K. (1984). Kinetic Alfvén waves on auroral field lines. *Planetary and Space Science*, *32*, 1387–1392.
Goertz, C. K., & Boswell, R. W. (1979). Magnetosphere-ionosphere coupling. *Journal of Geophysical Research*, *84*, 7239–7246.
Gold, T. (1959). Motions in the magnetosphere of the earth. *Journal of Geophysical Research*, *64*, 1219–1224.
Gurnett, D. A. (1974). The earth as a radio source-Terrestrial kilometric radiation. *Journal of Geophysical Research*, *79*, 4227–4238.
Hasegawa, A. (1975). *Plasma instabilities and nonlinear effects* (p. 93). New York: Springer.
Hasegawa, A. (1976). Particle acceleration by MHD surface wave and formation of aurora. *Journal of Geophysical Research*, *81*, 5083–5090.
Hasegawa, A., & Mima, K. (1976). Exact solitary Alfvén wave. *Physical Review Letters*, *37*, 690–693.
Hasegawa, A., & Mima, K. (1978). Anomolous transport produce by kinetic Alfvén wave turbulence. *Journal of Geophysical Research*, *83*, 1117–1123.
Heppner, J. P., Ness, N. F., Skillman T. L. & Scearce, C. S. (1962). Magnetic field measurements with the Explorer 10 satellite. *Journal of the Physical Society of Japan*, *17*(Suppl. A-II), 546.
Hoffman, R. A., & Evans, D. S. (1968). Field-aligned electron bursts at high latitude observed by OGO-4. *Journal of Geophysical Research*, *73*, 6201–6214.
Hoffmeister, C. (1943). Physikalische Untersuchungen an Kometen. I. Die Beziehungen des primären Schweifstrahls zum Radiusvektor. *Zeitschrift fur Astrophysik*, *22*, 265–285.
Huang, G. L., Wang, D. Y., Wu, D. J., et al. (1997). The eigenmode of solitary kinetic Alfvén waves by Freja satellite. *Journal of Geophysical Research*, *102*, 7217–7224.
Hui, C. H., & Seyler, C. E. (1992). Electron acceleration by Alfvén waves in the magnetosphere. *Journal of Geophysical Research*, *97*, 3953–3963.
Kalita, M. K., & Kalita, B. C. (1986). Finite-amplitude solitary Alfvén waves in a low-beta plasma. *Journal of Plasma Physics*, *35*, 267–272.
Kivelson, M. G., & Southwood, D. J. (1986). Coupling of global magnetospheric MHD eigenmodes to field line resonance. *Journal of Geophysical Research*, *91*, 4345–4351.

Kletzing, C. A. (1994). Electron acceleration by kinetic Alfvén waves. *Journal of Geophysical Research, 99*, 11095–11103.
Kletzing, C. A., & Torbert, R. B. (1994). Electron time dispersion. *Journal of Geophysical Research, 99*, 2159–2172.
Kletzing, C. A., Mozer, F. S., & Torbert, R. B. (1998). Electron temperature and density at high latitude. *Journal of Geophysical Research, 103*, 14837–14845.
Kletzing, C. A., Scudder, J. D., Dors, E. E., & Curto, C. (2003). Auroral source region: Plasma properties of the high-latitude plasma sheet. *Journal of Geophysical Research, 108*, 1360–1375.
Lee, L. C., & Wu, C. S. (1980). Amplification of radiation near cyclotron frequency due to electron population inversion. *The Physics of Fluids, 23*, 1348.
Lee, L. C., Johnson, J. R., & Ma, Z. W. (1994). Kinetic Alfvén waves as a source of plasma transport at the dayside magnetopause. *Journal of Geophysical Research, 99*, 17405–17411.
Louarn, P., Wahlund, J. E., Chust, T., et al. (1994). Observations of kinetic Alfvén waves by the Freja spacecraft. *Geophysical Research Letters, 21*, 1847–1850.
Lundin, R., Haerendel, G., & Grahn, S. (1994). The Freja science mission. *Space Science Reviews, 70*, 405–419.
Lysak, R. L., & Carlson, C. W. (1981). Effect of microscopic turbulence on magnetosphere-ionosphere coupling. *Geophysical Research Letters, 8*, 269–272.
Lysak, R. L., & Dum, C. T. (1983). Dynamics of magnetosphere-ionosphere coupling including turbulent transport. *Journal of Geophysical Research, 88*, 365–380.
Lysak, R. L., & Hudson, M. K. (1979). Coherent anomalous resistivity in the region of electrostatic shocks. *Geophysical Research Letters, 6*, 661–663.
Lysak, R. L., & Hudson, M. K. (1987). Effect of double layers on magnetosphere-ionosphere coupling. *Laser Particle Beams, 5*, 351–366.
Mölkki, A., Eriksson, A. I., Dovner, P. O., et al. (1993). A statistical survey of auroral solitary waves and weak double layers-1. Occurrence and net voltage. *Journal of Geophysical Research, 98*, 15521–15530.
Mcllwain, C. E. (1960). Direct measurements of particles producing visible auroras. *Journal of Geophysical Research, 65*, 2727–2747.
Mozer, F. S., Cattell, C. A., Hudson, M. K., et al. (1980). Satellite measurements and theories of low altitude auroral particle acceleration. *Space Science Reviews, 27*, 155–213.
Mozer, F. S., Cattell, C. A., Temerin, M., et al. (1979). The dc and ac electric field, plasma density, plasma temperature, and field-aligned current experiments on the S3–3 satellite. *Journal of Geophysical Research, 84*, 5875–5884.
Papadopoulos, K. (1977). A review of anomalous resistivity for the ionosphere. *Reviews of Geophysics, 15*, 113–127.
Parker, E. N. (1958). Interaction of the solar wind with the geomagnetic field. *The Physics of Fluids, 1*, 171–187.
Reiff, P. H., Collin, H. L., Craven, J. D., et al. (1988). Determination of auroral electrostatic potentials using high-and low-altitude particle distributions. *Journal of Geophysical Research, 93*, 7441–7565.
Russell, C. T. (1995). A brief history of solar-terrestrial physics. In M. G. Kivelson & C. T. Russell (Eds.), *Introduction to space physics* (pp. 1–26). New York: Cambridge University Press.
Shukla, P. K., Rahman, H. D., & Sharma, R. P. (1982). Alfvén soliton in a low-beta plasma. *Journal of Plasma Physics, 28*, 125–131.
Stasiewicz, K., Bellan, P., Chaston, C., et al. (2000a). Small scale Alfvénic structure in the aurora. *Space Science Reviews, 92*, 423–533.
Stasiewicz, K., Gustafsson, G., Marklund, G., et al. (1997). Cavity resonators and Alfvén resonance cones observed on Freja. *Journal of Geophysical Research, 102*, 2565–2575.
Stasiewicz, K., Holmgren, G., & Zanetti, L. (1998). Density depletions and current singularities observed by Freja. *Journal of Geophysical Research, 103*, 4251–4260.
Stasiewicz, K., Khotyaintsev, Y., Berthomier, M., & Wahlund, J. E. (2000b). Identification of widespread turbulence of dispersive Alfvén waves. *Geophysical Research Letters, 27*, 173–176.

Stasiewicz, K., Seyler, C. E., Mozer, F. S., et al. (2001). Magnetic bubbles and kinetic Alfvén waves in the high-latitude magnetopause boundary. *Journal of Geophysical Research, 106*, A29503–29514.

Stewart, B. (1882). On the connexion between the state of the sun's surface and the horizontal intensity of the Earth's magnetism. *Proceedings of the Royal Society of London, 34*, 406–409.

Temerin, M., Cerny, K., Lotko, W., & Mozer, F. S. (1982). Observations of double layers and solitary waves in the auroral plasma. *Physical Review Letters, 48*, 175–1179.

Thompson, B. J., & Lysak, R. L. (1996). Electron acceleration by inertial Alfvén waves. *Journal of Geophysical Research, 101*, 5359–5369.

Vaivads, A., Rönnmark, K., Oscarsson, T., & André, M. (1998). Heating of beam ions by ion acoustic waves. *Annales Geophysicae, 16*, 403–412.

Vogt, J., & Haerendel, G. (1998). Reflection and transmission of Alfvén waves at the auroral acceleration region. *Geophysical Research Letters, 25*, 277–280.

Volwerk, M., Louarn, P., Chust, T., et al. (1996). Solitary kinetic Alfvén waves-a study of the Poynting flux. *Journal of Geophysical Research, 101*, 13335–13343.

Wahlund, J. E., Eriksson, A. I., Holback, B., et al. (1998). Broadband ELF plasma emission during auroral energiztion 1. Slow ion acoustic waves. *Journal of Geophysical Research, 103*, 4343–4375.

Wahlund, J. E., Louarn, P., Chust, T., et al. (1994a). On ion-acoustic turbulence and the nonlinear evolution of kinetic Alfvén waves in aurora. *Geophysical Research Letters, 21*, 1831–1834.

Wahlund, J. E., Louarn, P., Chust, T., et al. (1994b). Observations of ion acoustic fluctuations in the auroral topside ionosphere by the Frejia S/C. *Geophysical Research Letters, 21*, 1835–1838.

Wolf, R. A. (1995). Magnetospheric configuration. In M. G. Kivelson & C. T. Russell (Eds.), *Introduction to space physics* (pp. 288–329). New York: Cambridge University Press.

Wu, C. S., & Lee, L. C. (1979). A theory of the terrestrial kilometric radiation. *The Astrophysical Journal, 230*, 621–626.

Wu, D. J. (2003a). Model of nonlinear kinetic Alfvén waves with dissipation and acceleration of energetic electrons. *Physical Review E, 67*, 027402.

Wu, D. J. (2003b). Dissipative solitary kinetic Alfvén waves and electron acceleration. *Physics of Plasmas, 10*, 1364–1370.

Wu, D. J. (2010). Kinetic Alfvén waves and their applications in solar and space plasmas. *Progress in Physics, 30*, 101–172.

Wu, D. J. (2012). *Kinetic Alfvén wave: Theory, experiment and application*. Beijing: Science Press.

Wu, D. J., & Chao, J. K. (2003). Auroral electron acceleration by dissipative solitary kinetic Alfvén waves. *Physics of Plasmas, 10*, 3787–3789.

Wu, D. J., & Chao, J. K. (2004). Model of auroral electron acceleration by dissipative solitary kinetic Alfvén wave. *Journal of Geophysical Research, 109*, A06211.

Wu, D. J., & Wang, D. Y. (1996). Solitary kinetic Alfvén waves on the ion-acoustic velocity branch in a low-β plasma. *Physics of Plasmas, 3*, 4304–4306.

Wu, D. J., Huang, G. L., & Wang, D. Y. (1996a). Dipole density solitons and solitary dipole vortices in an inhomogeneous space plasma. *Physical Review Letters, 77*, 4346–4349.

Wu, D. J., Huang, G. L., Wang, D. Y., & Fälthammar, C. G. (1996b). Solitary kinetic Alfvén waves in the two-fluid model. *Physics of Plasmas, 3*, 2879–2884.

Wu, D. J., Wang, D. Y., & Fälthammar, C. G. (1995). An analytical solution of finite-amplitude solitary kinetic Alfvén waves. *Physics of Plasmas, 2*, 4476–4481.

Wu, D. J., Wang, D. Y., & Huang, G. L. (1997). Two dimensional solitary kinetic Alfvén waves and dipole vortex structures. *Physics of Plasmas, 4*, 611–617.

Yu, M. Y., & Shukla, P. K. (1978). Finite amplitude solitary Alfvén waves. *The Physics of Fluids, 21*, 1457–1458.

Zabusky, N. J., & Kruskal, M. D. (1965). Interaction of "solitons" in a collisionless plasma and the recurrence of initial states. *Physical Review Letters, 15*, 240–243.

Chapter 4
KAWs in Solar Wind-Magnetosphere Coupling

4.1 Solar Wind-Magnetosphere Interacting Boundary Layers

Unlike the magnetosphere-ionosphere coupling is concentrated mainly in the polar magnetosphere, especially in the auroral plasma and is performed via field-aligned currents, the solar wind-magnetosphere coupling can occur in all outer boundaries of the magnetosphere from the magnetopause on the day side to the magnetotail on the night side. On the day side, in fact, before the supersonic magnetized plasma flow of the solar wind directly impacts on the magnetosphere, a shock can be first formed in front of the magnetosphere due to the blunt obstacle of the magnetosphere, called the bow shock, and the magnetopause is the interface between the shocked solar wind and the magnetosphere. The transition layer from the bow shock to the magnetopause is called the magnetosheath, which has a typical width $\sim$2–3 R_E. In the magnetosheath the shocked solar wind has been reduced to subsonic magnetized plasma flows, which is deflected and rounds the magnetosphere and flows on the upper boundary of the magnetosphere, that is, the magnetopause. The upper magnetosphere adjoining the magnetopause is the LLBL, which has a typical scale wider than the magnetopause but narrower than the magnetosheath and can contain a mixture of magnetosheath and magnetosphere plasmas, implying that there are the transport processes of the magnetosheath plasma crossing the magnetopause and entering the magnetosphere.

In the simplest approximation, the location and shape of the magnetopause can be estimated by the balance between the magnetic pressure of the magnetosphere and the plasma pressure of the magnetosheath, determined in turn by the solar wind dynamic pressure, that is,

$$\rho_{sw} v_{sw}^2 = B_{ms}^2/2\mu_o, \tag{4.1}$$

where the subscripts "sw" and "ms" denote the solar wind and magnetosphere, respectively, ρ_{sw} and v_{sw} are the mass density and flow velocity of the solar

D.-J. Wu and L. Chen, *Kinetic Alfvén Waves in Laboratory, Space, and Astrophysical Plasmas*, Atmosphere, Earth, Ocean & Space,
https://doi.org/10.1007/978-981-13-7989-5_4

wind, respectively, and B_{ms} is the magnetic field of the magnetosphere. Since the magnetopause separates the two regions of distinctly different magnetic fields, the geomagnetic field in the magnetosphere and the shocked solar wind magnetic field (i.e., interplanetary magnetic field) in the magnetosheath, it is inevitable that the magnetic field strength and direction change substantially across this interface. This implies that the magnetopause carries a substantial electric current, that is, the Chapman-Ferraro current. Therefore, the magnetopause also can be treated as a thin current sheet, which separates two different but relatively uniform magnetized plasma regimes, the magnetosphere and the magnetosheath (i.e., the shocked solar wind), and through which neither magnetic fluxes nor plasma flows crosses. Therefore, under such ideal balance condition there is no coupling of energy and momentum across the magnetopause, and the Lorentz forces of the currents flowing on the magnetopause may prevent the magnetosheath plasma flows (i.e., the shocked solar wind flows) from crossing the magnetopause, that is, the $\mathbf{j} \times \mathbf{B}$ force must direct the outward normal of the magnetopause and deflect the shocked solar wind to flow round the magnetosphere, implying that the magnetopause current has opposite directions on the day side and on the flank of the magnetopause, that is, the duskward current on the day side and the dawnward current on the flank.

In the above fluid viewpoint of the balance condition, the magnetopause is treated as a infinite thin plane and the magnetic field and plasma flow both change discontinuously across this plane. The same balance condition also can be derived from the particle viewpoint, but not a self-consistent way (see Kivelson and Russell 1995). In a self-consistent particle viewpoint, when the electrons and ions of the solar wind flow penetrate into the magnetosphere from the magnetosheath across the magnetopause, the local charge separation and the corresponding electric field can be caused by the different gyroradius of the electrons and ions in the magnetospheric field, which modifies the motions of the electrons and ions to preserve the charge neutrality. This results in the characteristic width scale of the electron inertial length (i.e., $\lambda_e \sim$ several km) for the balance boundary layer. However, in situ measurements by satellites show that the magnetopause has a characteristic width of several hundreds km, which is in the order of several ion gyroradii. It is possible that the magnetosheath magnetic field and the magnetospheric plasma, which have been ignored, also may have important effects on the structure of the magnetopause.

In fact, from the bow shock, magnetosheath, magnetopause, to the LLBL, there are a series of transition regions and boundary layers formed by the solar wind-magnetosphere interaction, which are characterized by strong gradients in the characteristic plasma quantities, such as the plasma density, temperature, flow velocity, and the magnetic field. These strong gradient structures with the mesoscale features contain much free energy that can be released through various ways such as magnetic reconnection and plasma wave instabilities and drive various microscale processes such as so-called "flux transfer event" (FTE, Russell and Elphic 1978) and turbulent wave-particle interaction in the kinetic scales of particles, which in turn can be responsible for the transport of mass, momentum, and energy across the magnetopause.

The magnetic reconnection on the magnetopause was proposed first by Dungey (1961) to explain the transport of plasma flow and magnetic flux across the magnetopause and the solar wind-magnetosphere coupling. When the interplanetary magnetic field is directed predominantly southward, the magnetic field driven by the solar wind flow against the front of the magnetosphere will be approximately antiparallel to the geomagnetic field on the other side of the magnetopause, that is, a X-line magnetic configuration is formed on the local plane of the magnetopause and the magnetic reconnection between the interplanetary magnetic field and geomagnetic field occurs there. The reconnected geomagnetic field on the dayside, which was closed field with both ends attached to the earth near the north and south poles on the ionosphere, is broken into two open fields, of which one end is still attached to the earth near the north or south pole and the other end is stretched tailward by the solar wind flow into the interplanetary space and integrates into and becomes a part of the magnetotail. The motion of these broken fields will cause an electric field directed from dawn toward dusk, which drives flow from noon toward midnight on the ionosphere, as observed.

It is obvious that the magnetic reconnection on the magnetopause leads to the decrease of the closed magnetic flux of the geomagnetic field. Therefore, it is necessary that some open flux can be returned into the closed flux of the magnetosphere to compensate the loss of the closed flux. Such magnetic-flux compensation can be achieved via the magnetic reconnection at another X-line magnetic configuration in the magnetotail distant CPS, where two open magnetic fields from the north and south tail lobes, respectively, reconnect to form a newly closed geomagnetic field in the earthward of the X-line and a purely interplanetary magnetic field in the tailward of the X-line. It is that the plasma accelerated by the magnetotail reconnection forms the PSBL, which is characterized mainly by sustained field-aligned ion and electron flows, directed both earthward (accelerated by the magnetotail reconnection) and tailward (reflected by the geomagnetic mirror). The observed ion beams have energies ranging from a few of keV to tens of keV and consist of a mix of species of H^+, He^+, He^{2+}, and O^+, and the electron beams have energies typically lower by a factor of 2 or 3.

In principle, these two reconnection rates at the magnetopause and magnetotail must be equal in the time-averages sense, although they probably rarely are on an instantaneous basis. Comparing these two cases, however, one can find some major differences between them. First, on the aspect of the inflow regions the plasma inflows from the north and south lobes on the two sides of the CPS in the magnetotail have approximately identical conditions, but in the case of the magnetopause the magnetosheath plasma inflow in general is much denser and cooler than that from the magnetosphere (or i.e., the LLBL) and they have entirely different physical environments. Second, on the outflows also there is a major difference between these two cases. Unlike the magnetopause magnetic reconnection has two approximately symmetric outflows, in the magnetotail case the tailward outflow and flux tube of the X-line are unattached to the earth and are pulled away to rejoin the solar wind by magnetic tension but the earthward outflow and flux tube of the X-line are restrained by the slower moving dense plasma and the stronger closed geomagnetic field of the CPS. This causes a bifurcation of the earthward outflow, in which although the new closed flux tube itself moves slowly, the plasma outflow accelerated by the magnetotail reconnection moves rapidly along the new closed flux tube toward the earth in both directions, that is, the north and south polar regions, and forms the earthward field-aligned ion and electron flows observed in the PSBL.

Although the magnetic reconnection has been widely investigated in both theory and experiment, there are still many unsolved problems in physical mechanisms. According to different driven mechanisms, the magnetic reconnection processes can be classified into two types, that is, “driven reconnection” and “spontaneous reconnection”. The former refers to the magnetic reconnection that is driven by conditions external to the reconnection current sheet such as inflows, and hence the reconnection rate is determined by the external parameters such as the inflow velocity. In order to match with the corresponding reconnection rate the current sheet is required to have a proper finite resistivity during the reconnection occurs, but the plasmas in these current sheets or boundary layers are nearly collisionless. This indicates it is necessary that other processes provide so-called “anomalous resistivity” during the reconnection. However, it has been an open problem that what is the source of the anomalous resistivity and how it changes accordingly when the external condition varies. Similarly, in the case of so-called the spontaneous reconnection that is driven by the resistive tearing instability of the current sheet oneself, a finite anomalous resistivity also is a necessary prerequisite for triggering the reconnection because the resistive tearing mode instability is a MHD instability caused by a finite resistivity (Furth et al. 1963). In some cases the scattering of particles by plasma waves or turbulence can lead to an anomalous resistivity. Moreover, such wave-particle interaction oneself also can directly contribute to the plasma transport process, called “anomalous transport”.

In fact, besides the magnetic reconnection, another important transport mechanism at the magnetopause and magnetotail is the anomalous viscous by the wave-particle interaction. In particular, the complex transition regions and boundary layers formed by the interaction between the solar wind and magnetosphere inherently are not only nonsteady but also strongly nonuniform because of the nature of strongly turbulent plasma flow of the solar wind. The strongly inhomogeneous structures in the boundary layers contain a large number of free energy that can drive various plasma waves and wave-particle interaction processes. These waves and wave-particle interactions in turn may be responsible for the anomalous plasma transport and diffusion across the boundary layers and achieve the effective coupling between the solar wind and magnetosphere.

KAWs are the most extensively studied waves applicable for the anomalous plasma transport in the magnetosphere because not only the wave-particle interaction of KAWs can provide effectively a matching transport efficiency but also KAW turbulence is observed ubiquitously in the magnetosphere as well as the solar wind. Observations of KAW turbulence in the magnetosphere are introduced in the next section (see the next chapter for KAW turbulence in the solar wind). The Sect. 4.3 presents the diffusion equation of particles due to KAW turbulence, and then based on the diffusion equation Sects. 4.4 and 4.5 discuss the anomalous transport of particles by KAWs in the magnetopause and the particle energization by KAWs in the magnetotail, respectively.

4.2 Observations of KAW Turbulence in Magnetosphere

The last chapter presented in detail observations of nonlinear solitary structures associated with KAWs as strong electric spikes in the observation data of the Freja and FAST satellites in the auroral plasma, which are identified well as SKAWs (see Sect. 3.3). In practice, KAWs exist in the nature mostly in a turbulent form because they can quickly spread their spectrum at a rate $\omega_{ci}\delta B_\perp/B_0$ by a decay to an ion acoustic wave or by nonlinear ion Landau damping (Hasegawa and Chen 1976b; Hasegawa and Mima 1978). Observations of KAW turbulence in various magnetospheric environments with different plasma parameters confirm further not only their ubiquity as pointed out by Hasegawa and Mima (1978) but also the sensitive dependence of their physical property on the local plasma parameters as seen in the Chap. 1.

It has been believed commonly that widespread electromagnetic fluctuations below the ion gyrofrequency in auroral plasmas are associated with AW turbulence because the observed electric to magnetic fluctuation ratio is about the local Alfvén velocity, that is, $\delta E_\perp/\delta B_\perp \sim v_A$. Although a characteristic linear dispersion relation is the best "identification card" of wave modes, which describes the propagation property of the wave (see Sect. 1.2), it is not easy to identify the dispersion relation of these low-frequency electromagnetic fluctuations associated with KAWs because KAWs can develop quickly into a nonlinear broadband continuous spectrum, in which the linear dispersion relation has been distorted badly (Hasegawa and Chen 1976b). On the other hand, the Doppler frequency due to the satellite moving at a typical velocity $\sim$10 km/s has usually an order same with or even higher than the local ion gyrofrequency and hence is much larger than the characteristic frequency of the Alfvénic mode. Therefore, broadband KAW turbulence is identified mainly through its electromagnetic polarization property, which remarkably depends on the local plasma parametric conditions (see Sect. 1.3).

The auroral plasma environment explored by the Freja and FAST when their passing through the auroral oval at the altitudes of $h \sim 1500$–2500 km (i.e., $h \sim 0.25$–$0.4\,R_E$) is typical inertial parametric regime of KAWs with the parameter $\alpha_e \ll 0.1$ (see the panel (e) of Fig. 3.2). In general, the ion component of the polar plasma is dominated by O^+ ions in altitudes lower than $\sim$3000 km (i.e., $h < 0.5\,R_E$) and by H^+ ions in altitudes higher than $\sim$3000 km (i.e., $h > 0.5\,R_E$). In the data of low-frequency electromagnetic fluctuations measured by the Freja and FAST satellites, besides strong electric spikes which have been identified well as SKAWs (see Sect. 3.3), strong broadband electromagnetic fluctuations more often present. These broadband electromagnetic fluctuations have feature electric fields lower than that of the strong electric spikes by one order of magnitude, that is, tens of mV/m. Moreover, transversely heating ions are most often associated with these broadband fluctuations (Norqvist et al. 1996; Knudsen and Wahlund 1998; Wahlund et al. 1998). Stasiewicz et al. (2000b) and Chaston et al. (2007b) analyzed two typical events observed by the Freja and FAST satellites, respectively.

Stasiewicz et al. (2000b) analyzed a typical example of broadband electromagnetic fluctuations observed by the Freja satellite on March 07, 1994 when it passed through the dawnside auroral oval at the altitude of 1700 km. A strong broadband electromagnetic fluctuation event presented in the duration 09 : 46 : 00–09 : 46 : 55, which has a wide frequency range ~1–500 Hz in the satellite frame. Stasiewicz et al. (2000b) argued that the observed apparent frequency can be attributed to the Doppler shifted spatial structures, that is, $\omega = \omega' + \mathbf{k} \cdot \mathbf{v}_s \simeq \mathbf{k} \cdot \mathbf{v}_s$, where the true frequency of the wave $\omega' \ll \omega$ (the apparent frequency), $\mathbf{k}$ is the wave number of the observed spatial structures, and $\mathbf{v}_s$ is the satellite velocity. For the typical Freja plasma environment at the altitude ~1700 km, one has the geomagnetic field $B_0 \sim 0.25$ G, the ion gyrofrequency is $f_{O^+} = \omega_{O^+}/2\pi \sim 24$ Hz for the O^+ dominant plasma or $f_{H^+} \sim 380$ Hz for the H^+ dominant plasma, the electron density $n_e \sim 2000$ cm^{-3}, and the Alfvén velocity $v_A \sim 3000$ km/s. The bulk electron temperature $T_e \sim 1$ eV, but the bulk ion temperature $T_i > T_e$ because of the bulk ion heating by the strong broadband electromagnetic turbulence. Thus, one has the electron inertial length $\lambda_e \sim 100$ m, the ion gyroradius $\rho_{O^+} \sim 20$ m ($< \lambda_e$) for the O^+ dominant case or $\rho_{H^+} \sim 5$ m ($\ll \lambda_e$) for the H^+ dominant case. This indicates that the electric to magnetic fluctuation ratio $\delta E/\delta B$ can be given by Eq. (1.36) for the KAW turbulence, that is,

$$\frac{\delta E}{\delta B} = v_A \sqrt{\left[1 + \left(\frac{2\pi\lambda_e}{v_s}\right)^2 f^2\right]\left[1 + \left(\frac{2\pi\rho_{O^+}}{v_s}\right)^2 f^2\right]}, \tag{4.2}$$

where $k_\perp = 2\pi/\lambda_\perp \simeq 2\pi f/v_s$ has been used. For KAWs with the frequency $f' \ll f$, the measured Doppler shifted frequency range $f \sim$ 1–500 Hz implies that the perpendicular wavelength of the observed broadband KAW turbulence ranges $\lambda_\perp \simeq v_s/f \sim$ 14–7000 m for the Freja velocity $v_s \simeq 7$ km/s.

Using snapshot wave measurements from two perpendicular electric and magnetic antennas, Stasiewicz et al. (2000b) extend the frequency coverage and construct the time-averaged ratio $\delta E/\delta B$ for the frequency range 0–16 kHz. Figure 4.1 shows this ratio plotted as a function of the perpendicular wavelength $\lambda_\perp \simeq v_s/f$, where the fitted line marked by "*" represents the theoretical prediction given by Eq. (4.2), which is valid for an inertial-regime KAW mode. From Fig. 4.1, it can be found that in the range of spatial scales $\lambda_\perp \sim$ 30–7000 m, that is, the frequency domain $f \sim$ 1–200 Hz, the measurement is very well consistent with the theoretical prediction of the inertial-regime KAW polarization relation of Eq. (4.2). Some small departures are possibly caused due to a finite true frequency $f' < f$, which can modify the polarization relation of Eq. (4.2). On the other hand, in the small-scale range below 30 m the measured spatial-scale spectrum has a plateau lasting to ~7 m and remarkably decreases in lower scales. This could probably be related to the stochastic dissipation in the electric-drift scale $\rho_d = \delta E_\perp/\omega_{ci} B_0$ (Stasiewicz et al. 2000b). In addition, the electric field measured by a 21 m double probe antenna onboard the Freja satellite also may be strongly attenuated in the smaller spatial scales.

In the measurements by the FAST satellite, similar KAW turbulence also is often observed when crossing the auroral plasma. For example, using the FAST data mea-

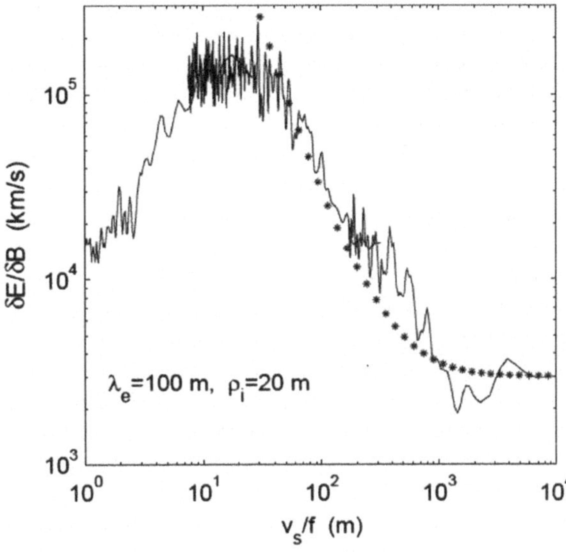

Fig. 4.1 The electric to magnetic fluctuation ratio $\delta E/\delta B$ for a 15 s time interval (09:46:35–09:46:50 on 7 March, 1994 by Freja). The line marked by "$*$" represents the theoretical prediction given by Eq. (4.2) for the inertial-regime KAW mode at $\lambda_\perp \sim 30$–$10{,}000$ m. Other parameters $v_A = 3000$ km/s, $\lambda_e = 100$ m, and $\rho_{O^+} = 20$ m have been used (from Stasiewicz et al. 2000b)

sured on 22 June, 1997, Chaston et al. (2007b) analyzed the spectral distribution of the electric to magnetic fluctuation ratio E_x/B_y for a low-frequency broadband electromagnetic fluctuation event, which was recorded by the FAST satellite when passing through the dayside auroral oval at the almost same altitude with the Freja satellite ($h \sim 1600$ km) and lasted an interval of 0.1 s from 13:55:20.51 to 13:55:20.61 UT. Figure 4.2a and b show the power spectrum E_x^2 and the spectral distribution of the ratio E_x/B_y in $k_x = 2\pi f/v_s$, respectively, where f is the apparent frequency measured in the FAST satellite frame and v_s the moving velocity of the FAST satellite.

As pointed out by the analysis of Chaston et al. (2007b), the variation in phase with satellite rotation indicates that the measured spatial spectrum k is approximately perpendicular to $\mathbf{B}_0$, that is, $k \simeq k_x$. In consequence, they concluded a very similar result with that of Stasiewicz et al. (2000b), that is, that the observed electric to magnetic fluctuation ratio, $\delta E/\delta B$, for the measured frequency $f < 100$ Hz, could be fitted very well by the theory of the inertial-regime KAW if it is assumed that the measured frequency f is mainly due to the satellite Doppler shifted spatial spectrum (i.e., $k_x \simeq 2\pi f/v_s$), as shown by Fig. 4.2. This very good agreement between the observed result and the theoretical expectation provides further evidence for that the nature of the low-frequency broadband electromagnetic fluctuations is the inertial-regime KAW turbulence.

The kinetic parametric regime of KAWs with $\alpha_e > 1$, in general, presents in the upper magnetosphere at altitudes $h > 4\,R_E$ (see the bottom panel of Fig. 3.2). Therefore, it can be expected that the kinetic-regime KAW turbulence may be observed by high-orbit satellites. The Polar satellite also is another polar-orbit satellite but moving at higher orbits than both of the Freja and FAST satellites, which was launched

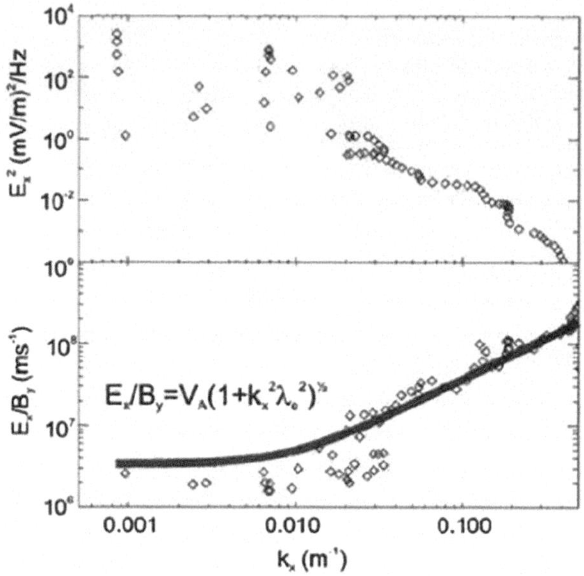

Fig. 4.2 Spectral properties averaged over the interval from 13:55:20.51 to 13:55:20.61 UT on 22 June, 1997, when low-frequency Alfvénic fluctuations were observed. **a** Observed power spectrum of E_x^2. **b** Observed spectral distribution of E_x/B_y and that predicted by the inertial-regime KAW model (bold line) based on Eq. (4.2), where the measured local plasma parameters $n_e \simeq 2000\,\text{cm}^{-3}$, $\lambda_e \simeq 120$ m, $v_A \simeq 3500$ km/s, and $k_x = 2\pi f/v_s$ have been used (from Chaston et al. 2007b)

on February 24, 1996, into a highly elliptical orbit, with apogee at 9 R_E and perigee at 1.8 R_E. Its orbit inclination is 86° and the orbit period about 18 h. Besides general instruments to measure in-situ the fluxes of charged particles, electric and magnetic fields, and electromagnetic waves, there are three instruments onboard the Polar satellite to image the aurora in various wavelengths when the satellite is near apogee, high over the northern polar region (see Harten and Clark (1995) for more details).

In high-β (i.e., $\alpha_e \gg 1$) plasma environments such as the solar wind, magnetosheath, and magnetopause, strong depressions of the magnetic field with $\delta B/B_0 \sim$ 90% often are observed to appear as irregular trains of magnetic dips or an isolated magnetic dip, called "magnetic holes" or "magnetic bubbles" (Kaufmann et al. 1970; Turner et al. 1977; Lühr and Klöcker 1987; Treumann et al. 1990). These magnetic structures usually are believed to result from the mirror instability, which may develop in high-β plasmas with a temperature anisotropy of $T_\perp/T_\parallel > 1 + \beta_\perp^{-1}$ (Fazakerley and Southwood 1994; Winterhalter et al. 1994; Lucek et al. 1999). In the magnetosheath such ion anisotropy can be produced either by the bow shock compression (Crooker and Siscoe 1977) or by field-aligned draping close to the magnetopause (Denton et al. 1994). Magnetic holes, however, also present often in mirror mode stable environments (Winterhalter et al. 1994). Stasiewicz et al. (2001) presented an example of magnetic bubbles observed by the Polar satellite when crossing the magnetopause current layer and entering the magnetosheath during periods of high solar wind pressure.

Figure 4.3 shows the magnetometer data during 14 : 00–15 : 00 UT on April 11, 1997, from which it can be found that the Polar satellite was in the high-latitude compressed lobe untill 14:28 UT, when it encountered three magnetic bubble layers

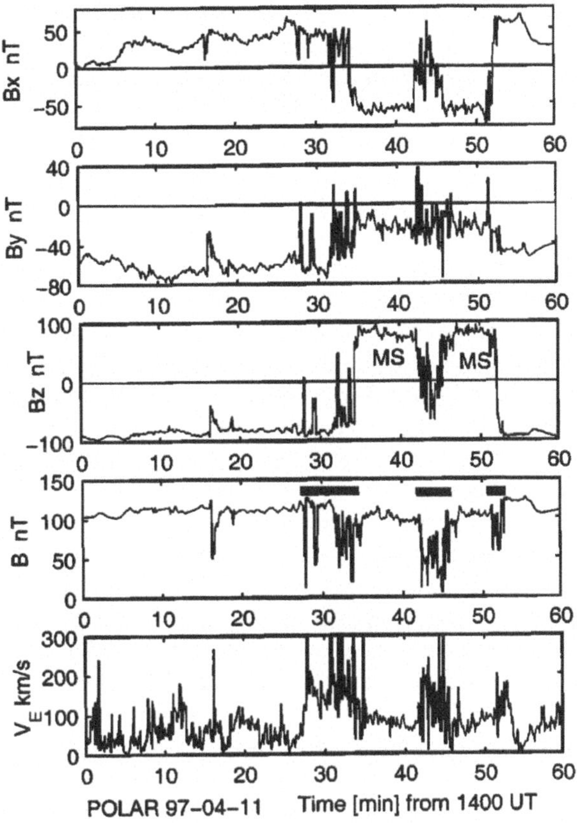

Fig. 4.3 Magnetopause crossings by Polar on April 11, 1997: three components of magnetic field (B_x, B_y, B_z) in the GSM (geocentric solar magnetospheric) coordinates, the strength B, and the convection velocity v_E (from Stasiewicz et al. 2001)

charactered by strong magnetic dips. The magnetic bubble layers are located adjacent to magnetopause current sheets characterized by reversals of the z-component of the magnetic field. The measured minimum magnetic field during the bubble layers is 1.4 nT, implying that the bubble layers are very strong depressions up to 98% of the ambient magnetic field with $B_0 \sim 100$ nT. However, the further analysis of the plasma measurements showed that in all regions explored during 13 : 30–15 : 30 UT, including the magnetosheath, magnetopause, boundary bubble layers, and magnetosphere, the temperature anisotropic parameter $\left(T_\perp/T_\parallel\right)/\left(1+\beta_\perp^{-1}\right)$ is well below one (Stasiewicz et al. 2001). This implies that the observed plasmas all are mirror mode stable in all explored regions (Southwood and Kivelson 1991).

Based on the fact that the bubbles are closely related to the strong current sheets nearby the magnetopause and there is a high correlation between the electric and magnetic fluctuations in the bubble layers, Stasiewicz et al. (2001) proposed that the bubble layers can be produced by tearing mode reconnection processes associated the strong current sheets and the low-frequency broadband electromagnetic fluctuations in the bubble layers probably are spatial turbulence of KAWs by Doppler shifted to higher frequency range (~0–30 Hz) due to convective plasma flows v_E. These KAWs can be excited by the Hall instability created by mesoscale gradients in the magnetic field and plasma pressure and their spatial spectrum may extend from several ion gyroradii (∼500 km) down to the electron inertial length (∼5 km). For example, the electric and magnetic fluctuation components, E_y and B_z, in the first bobble layer display well correlation, and the ratio $\delta E/\delta B \simeq 200$ km/s (for the amplitudes $\delta E \simeq 30$ mV/m and $\delta B \simeq 150$ nT) is well close to the local Alfvén velocity (v_A). These typical characteristics both indicate the Alfvénic nature of the electromagnetic fluctuations observed inside the bubble layer.

However, in comparison with the plasma environments observed by the Freja and FAST satellites at altitudes $h < 1\ R_E$, there are two considerably different aspects in the environment explored by the Polar satellite nearby the magnetopause. One is that the magnetopause plasma has the parameter $\alpha_e \gg 1$, that is, $\rho_i > \rho_s \gg \lambda_e$, in which the dispersion of KAWs is caused mainly by the ion gyroradius effect of finite $k_\perp^2 \rho_i^2$, instead of the electron inertial length effect of finite $k_\perp^2 \lambda_e^2$. The other difference is that the convective velocity of plasma flows v_E at Freja altitudes is much smaller than the satellite moving velocity v_s and hence the Doppler shift of the spatial turbulence of KAWs is dominated by the satellite velocity v_s. While at the Polar apogee nearby the magnetopause, the plasma convective flows $v_E \sim 100$–200 km/s is much larger than the satellite velocity $v_s \sim 3$ km/s. Thus, the Doppler shift of KAWs is contributed mainly by the convective velocity v_E, instead of the satellite velocity v_s. Therefore, the polarization spectrum of KAWs in the Polar plasma environments can be fitted approximately by Stasiewicz et al. (2001):

$$\frac{\delta E}{\delta B} = v_A \sqrt{1 + \left(\frac{2\pi \rho_i}{v_E \cos\theta}\right)^2 f^2}, \tag{4.3}$$

where $f = \mathbf{k} \cdot \mathbf{v}_E/2$ is the Doppler frequency in the satellite frame, θ is the angle between the wave vector $\mathbf{k}$ and the convective velocity $\mathbf{v}_E$. In addition, the in situ measurements show the local ion temperature $T_i \simeq 200$ eV $\gg T_e \simeq 40$ eV (the local electron temperature), and hence the ion acoustic gyroradius ρ_s ($< \rho_i$) has been neglected in the above fitted expression.

Figure 4.4 presents the average power spectrum of the electric field fluctuations (upper panel) and the spectral distribution of the ratio $\delta E/\delta B$ (lower panel), which were measured in the frequency of 0–10^4 Hz when the Polar passed through the magnetic bobble layers during 14:30–14:53 on April 11, 1997. The frequency spectrum is covered by three instruments, that is, the dc magnetic field, low-, and high-frequency waveform receiver are sampled at the rate 8.3, 100, and 50000 per second, respectively. The mismatch of the plots of the low- and high-frequency waveform receivers at the transition frequency of ∼50 Hz possibly is caused by the different filter characteristics of the wave instruments and the different locations of their snapshots. Anyway, the power spectrum and the waveforms show that the low-frequency broadband turbulence has a significant power drop above ∼30 Hz, which is close to the lower hybrid frequency $f_{LH} \simeq \sqrt{f_{ci} f_{ce}} \sim 32$ Hz.

The asterisk curve in the lower panel of Fig. 4.4 represents the theoretical prediction by the polarization relation of the kinetic-regime KAWs, which is calcu-

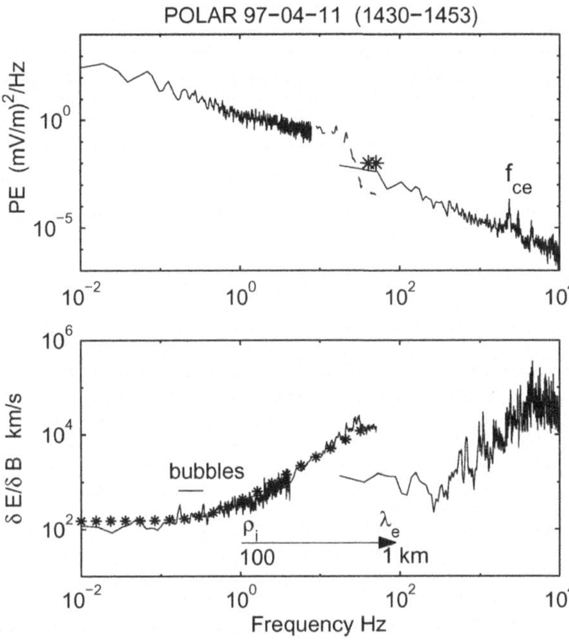

Fig. 4.4 Power spectrum of the electric field fluctuations and the ratio $\delta E/\delta B$ observed by Polar during 14:30–14:53 on 1997-04-11 when crossing the magnetic bubble layers, where the asterisk plot shows the theoretical prediction by Eq. (4.3) with fitted parameters $v_A = 150\,\text{km/s}$ and $\rho_i/v_E = 0.4$ s (from Stasiewicz et al. 2001)

lated by Eq. (4.3) with the average fitted parameters $v_A = 150\,\text{km/s}$, $\rho_i/v_E = 0.4$ s ($\rho_i = 60\,\text{km}$ and $v_E = 150\,\text{km/s}$), and $\cos\theta = 1$. From Fig. 4.4, a good agreement between the theoretical prediction and the measurements by Polar provides indirect evidence that the physical nature of the low-frequency broadband electromagnetic fluctuations is the spatial turbulence of kinetic-regime KAWs by Doppler shifted to the frequency range (~0–30 Hz) due to the local convective plasma flows $v_E = 150\,\text{km/s}$. For the corresponding spatial scale $\lambda \sim v_E/f$, one has ~100 km (i.e., the order of the local ion gyroradius) at the lower frequency $f \sim 1$ Hz and ~ 1 km (i.e., the order of the local electron inertial length) at the higher frequency $f \sim$ 100 Hz. This indicates that the scale spectrum of the KAW spatial turbulence extends from ρ_i down to λ_e. For the bubble structures, in particular, the typical observed frequency ~0.2 Hz implies their spatial scales ~500 km, that is, the typical scale of SKAWs. While above 30 Hz the spatial scales approach the electron inertial length and dissipative processes related to the electron energization possibly lead to the dropout of the electric field power seen in the upper panel of Fig. 4.4. This is different from the case observed by Freja, in which the inertial-regime KAW is dissipated mainly due to the stochastic ion acceleration because the spatial scale spectrum of the inertial-regime KAW turbulence extends from the electron inertial length $\lambda_e \sim 300$ m down to the ion gyroradius $\rho_i \sim 20$ m (Stasiewicz et al. 2000c)

Low-frequency KAW turbulence extending from below the ion gyrofrequency f_{ci} to above f_{ci}, even up to the lower hybrid frequency f_{LH} is rather common in satellite observations of space plasmas (Stasiewicz et al. 2000a). For instance, in the magnetosheath and the solar wind, observations by the Cluster satellites also show clear evidence for broadband KAW turbulence. The Cluster consists of four spacecraft (referred to as C1, C2, C3, and C4) and was launched by ESA (European Space Agency) in 2000 (Escoubet et al. 2001). The main goal of the Cluster is to study small-scale plasma structures in three dimensions in the key plasma regions, such as the magnetotail, polar cusps, auroral zones, magnetopause, bow shock, and solar wind. The separation distances between the satellites have varied between 100 km and 20,000 km, to address the relevant spatial scales. The most important advantage of Cluster is to allow a three-dimensional mapping of space and to distinguish between temporal and spatial structures. The Cluster has an elliptical polar orbit with a perigee altitude $\sim 3\, R_E$ and an apogee altitude $\sim 19\, R_E$. The satellites are spin-stabilized with a spin period of about 4 s, and have an orbital period of 57 h.

The Cluster experienced a 7-hours journey of passing through in the magnetopause boundary during 09:00–16:00 UT on December 31, 2000, and the separations of the four satellites were between 400–1000 km. The satellite C3, which located at the GSE (geocentric solar ecliptic) coordinate position of (3, 11, 9) R_E at 09:00 UT and traveled to (6, 15, 7) R_E at 16:00 UT, observed over 120 boundary crossings between the magnetosphere (plasma density $\sim$0.1 cm^{-3}) and the magnetosheath (plasma density $\sim$10 cm^{-3}) during this journey. The satellites move at a velocity of 3 km/s, which is much less than the convective velocity of plasma flows nearby the boundary $\sim$50–300 km/s. Therefore, the observed multiple magnetopause crossings should be attributed to the rapidly oscillational motions of the magnetopause boundary, which can be caused by not only the strong perturbations of the interplanetary magnetic field or solar wind plasma flows but also the surface waves on the boundary driven by Kelvin-Helmholtz instability with a period of a few minutes.

Figure 4.5 shows a typical example of the magnetopause crossings by C3 on December 31, 2000, in which the satellite electric potential (top panel shown with

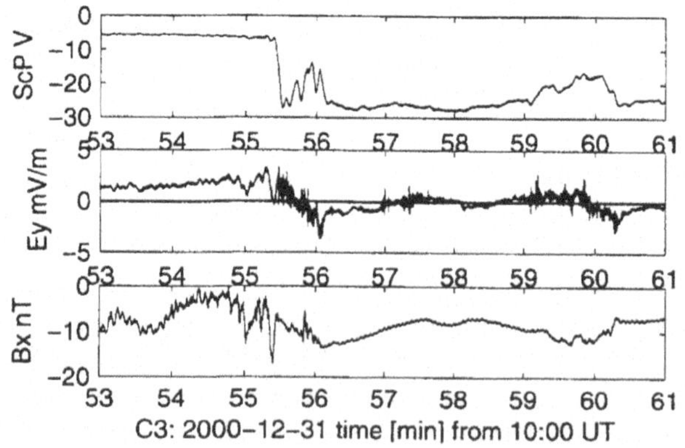

Fig. 4.5 An example of magnetopause crossings by C3 in the duration between 10:53:00 to 11:01:00 UT on December 31, 2000 and the crossing occurs at $\sim$10:55:30 UT, where top panel presents the satellite potential (to represent the plasma density), and middle and bottom panels plot the electric (E_y) and magnetic (B_x) field components, respectively (from Stasiewicz et al. 2004)

a reversed sign) represents an approximately logarithmic measure of the plasma density and the typical potential 5 (25) V on the magnetosheath (magnetosphere) side corresponds to a plasma density of 10 (0.1) cm^{-3}. By use of measurements of the four satellites, the velocity of the boundary motion can be estimated to be $\sim$150 km/s, and the thickness of the boundary layer characterized by a strong density gradient can be estimated to be $\sim$1000 km by the crossing time of $\sim$10 s. In particular, some evidently enhanced broadband electromagnetic fluctuations can be observed during the boundary crossings, although the fluctuations are quite regular on both the magnetosheath and magnetospheric sides. Moreover, these enhanced broadband fluctuations are clearly co-located within the strong density gradients. In general, at the magnetopause boundary layer, the typical kinetic scales of plasma particles are the ion gyroradius $\rho_i \sim 100$ km $\sim$ 2–3 times the ion acoustic gyroradius ρ_s (because $T_i \gg T_e$), the ion inertial length $\lambda_i \sim \rho_i \sim 100$ km (because $\beta_i \sim 1$), and the electron inertial length $\lambda_e \sim 2$ km. The typical convective velocity of plasma flows $v_E \sim 100$ km/s in the magnetopause boundary layer, much larger than the satellite velocity of $v_s \sim 3$ km/s. Thus, low-frequency plasma waves with frequencies $f' < f_{ci} \sim 0.3$ Hz and perpendicular scales $\lambda_\perp \sim \rho_i \sim \lambda_i$ or $\sim \lambda_e$ in the boundary layer would be observed as Doppler shifted spatial structures at an apparent frequency 1 or 50 Hz, respectively, both above the local ion gyrofrequency f_{ci}.

Stasiewicz et al. (2004) analyzed low-frequency broadband turbulence in the frequency range between $f \sim 0.03$ and 10 Hz, extending from below the proton gyrofrequency ($f_{ci} \simeq 0.3$ Hz) to the low hybrid frequency $f_{LH} \simeq 13$ Hz), which was observed by the Cluster in the boundary layer when crossing the magnetopause into the magnetosheath in the time interval from 10:55:00 to 10:56:00 UT. Figure 4.6 displays the spectral distribution of the ratio $\delta E/\delta B$ in the boundary layer, where the measured density and ambient magnetic field are $n \simeq 1\,\mathrm{cm}^{-3}$ and $B_0 \simeq 18$ G, respectively. From Fig. 4.6, the observed polarization spectrum can be well fitted by Eq. (4.3)

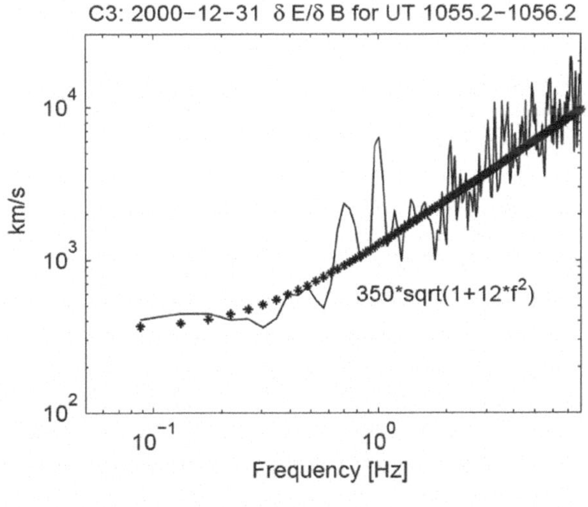

Fig. 4.6 The spectral distribution of the ratio $\delta E/\delta B$ in the magnetopause boundary layer. The asterisk plot shows equation (4.3) computed for $v_A = 350$ km/s, $\rho_i = 80$ km, $v_E = 150$ km/s, and $\cos\theta = 1$ (from Stasiewicz et al. 2004)

with the fitted parameters $v_A = 350\,\text{km/s}$, $\rho_i = 80\,\text{km}$, $v_E = 150$ km/s, and $\cos\theta = 1$ in the rather wide frequency range from $f \sim 0.1 f_{ci}$ up to $f \sim f_{LH} \sim 40 f_{ci}$. The measurements by the Cluster demonstrated again that the physical nature of the enhanced low-frequency broadband electromagnetic fluctuations is the KAW turbulence.

4.3 Diffusion Equation of Particles Due to KAW Turbulence

The magnetopause separates the two regions of distinctly different plasma environments, the shocked solar wind and the magnetosphere. In the steady state the magnetopause is a pressure balance interface between the two regions and prevents the shocked solar wind plasma flows from crossing the magnetopause and entering the magnetosphere. However, the dynamical coupling relationship between solar activities and geomagnetic storms or substorms and a large number of in situ satellite measurements both show that there are the transport processes crossing the magnetopause, which lead to the shocked solar wind plasma flows in the magnetosheath directly entering the magnetosphere (Hones et al. 1972; Eastman and Hones 1979). At the dayside magnetopause the main mechanisms for the transport processes can be classed into the two kinds, one is magnetic reconnection due to current sheet instability on the magnetopause (Dungey 1961) and the other one is viscous or viscous-like processes due to wave-particle interaction (Axford and Hines 1961). The most important evidence for the reconnection process is observations of FTEs (Russell and Elphic 1978), in which the plasma flows along the reconnected magnetic fields and transports directly into the magnetosphere.

However, the conditions favorable for reconnection are not always present. For instance, when the interplanetary magnetic field is northward the reconnection process possibly is diminished significantly. Therefore, the fact that the plasma transport crossing the magnetopause has always existed indicates that there must be other mechanisms for the plasma transport process. The viscous-like processes actually are the anomalous diffusion of plasma particles in the presence of the enhanced electromagnetic wave turbulence, in which the wave-particle interaction causes the particles to random walk across the boundary layer. In fact, the LLBL extending from the dayside magnetopause to the magnetotail is formed through the anomalous diffusion processes, where the plasma density, temperature, and velocity are observed to be intermediate between magnetosheath and magnetosphere values (Eastman and Hones 1979). The diffusion coefficient (or kinematic viscosity) required for maintaining the boundary layer can be estimated as $D \sim 10^9$ m^2/s (Sonnerup 1980).

In an inhomogeneous plasma environment such as the magnetopause layer the stochastic Coulomb collisions between particles indeed may lead to diffusion and transport of plasma particles across the magnetic field. In the case of crossing the magnetopause, the diffusion coefficient caused by the Coulomb collision is really too low to explain the formation of the diffusion boundary layer. While, the fluctuation fields of turbulent waves have stochastically varying phases and hence may produce the stochastic forces acting on particles. In an inhomogeneous plasma, similar to the Coulomb collision, the stochastic forces by turbulence waves also can lead to the diffusion of particles both in velocity and coordinate spaces, called "anomalous diffusion". Various mode waves have been suggested to be responsible for the anomalous diffusion in the boundary layer, such as ion acoustic wave (Galeev and Sagdeev 1984), lower hybrid wave driven by density gradients (Huba et al. 1977), modified two-stream instability driven by currents (Papadopoulos 1979), electron cyclotron drift wave (Haerendel 1978), and whistler waves driven by current gradients (Drake et al. 1994). Based on the amplitudes of the observed electromagnetic fluctuations, however, the saturation diffusion coefficients associated with these high-frequency waves with frequencies above the ion gyrofrequency all almost are significantly below the required diffusion coefficient (Labelle and Treumann 1988).

On the other hand, low-frequency KAWs with frequencies below the ion gyrofrequency have been found throughout the magnetosphere. In particular, the enhanced low-frequency broadband electromagnetic fluctuations, which are observed frequently nearby the magnetopause boundary layer, have been identified well as KAW turbulence. Unlike MHD AWs, KAWs can have wave-particle resonant interaction with thermal or nonthermal particles through the Landau resonance of $\omega = k_\parallel v_\parallel$ because of having nonzero parallel electric fields, $\delta E_\parallel \neq 0$, caused by the short perpendicular wavelength comparable to the ion gyroradius (i.e., $\lambda_\perp \sim \rho_i$). This implies that KAWs also can contribute significantly to anomalous transport processes through the Landau resonance of $\omega = k_\parallel v_\parallel$. Moreover, because of their large parallel wavelength much longer than the kinetic scales of particles, the plasma volume affected by KAWs is much larger than that affected by those high-frequency waves, such as electrostatic waves or cyclotron waves. In consequence, KAW turbulence can produce a higher diffusion coefficient than the high-frequency waves. In addition, KAWs may be excited effectively by the resonant mode conversion of MHD surface waves (Hasegawa 1976) or by the drift instability (Mikhailovskii 1967) in inhomogeneous plasmas, and hence they can present ubiquitously in various boundary layers with strong gradients in plasma density or magnetic field, as shown by observations. Meanwhile, an enhanced broadband turbulence of KAWs can be very easy formed once excited because KAW can spread its spectrum very quickly at a rate $\omega_{ci}\delta B_\perp/B_0$ by decay to an ion acoustic wave or by nonlinear ion Landau damping (Hasegawa and Chen 1976b).

Since KAWs have these advantages above, it is very reasonable that KAW turbulence is considered to be responsible for most of the anomalous transport processes. The quasi-linear theory has been commonly used in deriving anomalous diffusion coefficients caused by turbulent wave-particle interactions (Sagdeev and Galeev 1969). In consideration of the ignorable smallness of the electron gyroradius, the diffusion coefficient of electrons caused by KAW turbulence can be derived from the drift kinetic equation of the reduced distribution function of electrons, $f_e(\mathbf{r}, v_z, t)$, as follows (Hasegawa and Mima 1978):

$$(\partial_t + v_z\partial_z) f_e + \nabla_\perp \cdot (\mathbf{v}_d f_e) - \frac{e}{m_e}\left[E_z + (\mathbf{v}_d \times \mathbf{B}_\perp)\cdot\hat{z}\right]\partial_{v_z} f_e = 0, \quad (4.4)$$

where ∂ denotes the partial derivative with respect to the subscripts, $\mathbf{E}$ and $\mathbf{B}$ are the wave electric and magnetic fields with subscripts z and $\perp$ indicating components parallel and perpendicular to the ambient magnetic field $\mathbf{B}_0 = B_0\hat{z}$, respectively, $\hat{z}$ is the unit vector along the z-axis, and $\mathbf{v}_d$ is the drift velocity perpendicular to $\mathbf{B}_0$ and given by

$$\mathbf{v}_d = \frac{\mathbf{E}_\perp \times \hat{z}}{B_0} + \frac{\mathbf{B}_\perp}{B_0} v_z \tag{4.5}$$

for electrons.

Assuming that the ambient plasma density has a gradient in the x direction, for the turbulent fields $\mathbf{E_k}$ and $\mathbf{B_k}$ the quasi-linear equation of the ensemble average distribution $\langle f_e \rangle$ is

$$\partial_t \langle f_e \rangle + \left\langle \frac{E_y + v_z B_x}{B_0} \partial_x f_e^{(1)} \right\rangle - \frac{e}{m_e} \left\langle E_z \partial_{v_z} f_e^{(1)} \right\rangle = 0, \tag{4.6}$$

where the angle brackets denote the ensemble average,

$$f_e^{(1)} = \mathrm{Re} \sum_{\mathbf{k}} f_{e\mathbf{k}}^{(1)}(x) \exp\left[-i\left(\omega t - \mathbf{k}\cdot\mathbf{r}\right)\right] \tag{4.7}$$

is the linear fluctuation distribution function, and "Re" represents the real part. The Fourier amplitude can be given by

$$f_{e\mathbf{k}}^{(1)}(x) = i\mathrm{P}\left[\frac{E_{z\mathbf{k}}/B_0}{k_z v_z - \omega}\left(\frac{k_y v_z}{\omega}\partial_x - \omega_{ce}\partial_{v_z}\right) - \frac{E_{y\mathbf{k}}}{\omega B_0}\partial_x\right]\langle f_e \rangle +$$
$$+\pi\delta\left(k_z v_z - \omega\right)\left(\frac{eE_{z\mathbf{k}}}{m_e}\partial_{v_z} - \frac{k_z v_z E_{z\mathbf{k}}}{\omega B_0}\partial_x\right)\langle f_e \rangle, \tag{4.8}$$

where "P" represents the principal value and "δ" denotes the Dirac delta function from the imaginary part of the Plemelj formula, $\mathrm{Im}(\omega - k_z v_z)^{-1} = -\pi\delta(\omega - k_z v_z)$, which represents the contribution of the wave-particle resonance interaction (i.e., the Landau damping).

The quasi-linear diffusion coefficients in velocity and coordinate spaces are contributed by the first and second terms following the Dirac delta function in Eq. (4.8), respectively. Taking the second term and substituting into Eq. (4.6), one has

$$\partial_t \langle f_e \rangle = \partial_x \left(\sum_{\mathbf{k}} \pi\delta\left(k_z v_z - \omega\right) \frac{k_y^2 v_z^2 \| E_{z\mathbf{k}} \|^2}{2\omega^2 B_0^2} \partial_x \langle f_e \rangle \right), \tag{4.9}$$

where the cross terms of ∂_x and ∂_{v_z} have vanished because the coefficients are equal zero in the $\mathbf{k}$ summation. Assuming the Gaussian distribution for $\langle f_e \rangle$, that is,

$$\langle f_e \rangle = \frac{n_e(x,t)}{\sqrt{2\pi} v_{T_e}} \exp\left(-\frac{v_z^2}{2v_{T_e}^2}\right), \tag{4.10}$$

the diffusion equation for the electron density $n_e(x,t)$ can be obtained by integrating Eq. (4.9) over v_z as follows:

$$\partial_t n_e = \partial_x \left(D_{ex} \partial_x n_e\right) \tag{4.11}$$

with the diffusion coefficient

$$D_{ex} = \sqrt{\frac{\pi}{8}} \sum_{\mathbf{k}} \frac{k_y^2 \| E_{z\mathbf{k}} \|^2}{k_z^2 B_0^2} \frac{1}{|k_z| v_{T_e}} \tag{4.12}$$

for $v_{T_e} \gg v_A$, that is, $\alpha_e \gg 1$.

The dispersion relation of KAWs can be written as (Hasegawa, 1976)

$$\begin{aligned} \frac{\omega^2}{k_z^2 v_A^2} &= \frac{\chi_i}{1 - I_0(\chi_i)\exp(-\chi_i)} + (1 - i\delta_e)\chi_s \\ &\simeq 1 + \frac{3}{4}\chi_i + (1 - i\delta_e)\chi_s, \end{aligned} \tag{4.13}$$

where $\chi_s = \chi_i/\tau = k_\perp^2 \rho_s^2$, the dissipation factor δ_e due to the Landau damping is

$$\delta_e = \sqrt{\frac{\pi}{2}} \frac{\omega}{|k_z| v_{T_e}} \sim \alpha_e^{-1}, \tag{4.14}$$

and the approximation in Eq. (4.13) is taken for $\chi_i \ll 1$. On the other hand, from the kinetic polarization relation of KAWs (Hasegawa 1976), one has

$$\frac{k_y E_{z\mathbf{k}}}{\omega B_{x\mathbf{k}}} = \frac{1 - I_0(\chi_i)\exp(-\chi_i)}{1 - I_0(\chi_i)\exp(-\chi_i) + \tau}. \tag{4.15}$$

By use of Eqs. (4.13) and (4.15), the electron diffusion coefficient of Eq. (4.12) can be expressed in the term of the wave magnetic field $B_{x\mathbf{k}}$ as follows:

$$\begin{aligned} D_{ex} &= \sqrt{\frac{\pi}{8}} \sum_{\mathbf{k}} \frac{\chi_s v_A^2}{|k_z| v_{T_e}} \frac{B_{x\mathbf{k}}^2}{B_0^2} \frac{1 - I_0(\chi_i)\exp(-\chi_i)}{1 - I_0(\chi_i)\exp(-\chi_i) + \tau} \\ &\simeq \sqrt{\frac{\pi}{8}} \sum_{\mathbf{k}} \frac{\chi_s^2}{1+\chi_s} \frac{v_A^2}{|k_z| v_{T_e}} \frac{B_{x\mathbf{k}}^2}{B_0^2} = D_0 \sqrt{\frac{\pi}{8}} \sum_{\mathbf{k}} \frac{2\chi_s^2}{1+\chi_s} \frac{\omega_{ci}}{|k_z| v_{T_e}} \frac{1}{\beta_e} \frac{B_{x\mathbf{k}}^2}{B_0^2} \\ &= D_0 \sum_{\mathbf{k}} \frac{\chi_s^2}{1+\chi_s} \frac{\omega_{ci}}{\omega} \frac{\delta_e}{\beta_e} \frac{B_{x\mathbf{k}}^2}{B_0^2}, \end{aligned} \tag{4.16}$$

where $\beta_e = 2v_s^2/v_A^2$ and $D_0 \equiv \rho_s^2 \omega_{ci}$ is the diffusion coefficient measured in terms of the ion acoustic gyroradius steps per ion cyclotron period. In general, the value under the summation has the unit order and hence $D_{ex} \sim D_0$. It is especially worth noting that $D_0 = 16 D_B$, where $D_B = \rho_s^2 \omega_{ci}/16$ is the well known Bohm diffusion coefficient, which is possibly caused by various plasma turbulence (Bohm 1949). This indicates that the KAW diffusion coefficient D_{ex} is relative to the Bohm diffusion coefficient D_B.

In the low-frequency approximation of $\omega \ll \omega_{ci}$, the Fourier amplitude of the ion linear fluctuation distribution function, $f_{i\mathbf{k}}^{(1)}(x, \mathbf{v})$, can be obtained by the Maxwell-Vlasov equation (Hasegawa 1975):

$$f_{i\mathbf{k}}^{(1)} = -i \sum_n \frac{e^{-in\theta}}{\omega_{\mathbf{k}} - k_z v_z} J_0\left(\frac{k_y v_\perp}{\omega_{ci}}\right) J_n\left(\frac{k_y v_\perp}{\omega_{ci}}\right) \times \left[\left(1 - \frac{k_z v_z}{\omega_{\mathbf{k}}}\right) \frac{E_{y\mathbf{k}}}{B_0} \partial_x + \frac{k_y v_z}{\omega_{\mathbf{k}}} \frac{E_{z\mathbf{k}}}{B_0} \partial_x + \frac{\omega_{ci} E_{z\mathbf{k}}}{B_0} \partial_{v_z}\right] f_i, \tag{4.17}$$

where $f_i = f_i\left(x + v_y/\omega_{ci}, v_\perp^2 + v_z^2\right)$ is the unperturbed ion distribution function and $\theta = \tan^{-1}(v_y/v_x)$. In the local approximation, assuming that the perpendicular wave vector has only the k_y component, one can have

$$\partial_{v_y} f_i = \sin\theta \partial_{v_\perp} f_i + \frac{\cos\theta}{v_\perp} \partial_\theta f_i + \frac{1}{\omega_{ci}} \partial_x f_i. \tag{4.18}$$

The quasi-linear response of f_i can be given by the ensemble average of the quasi-linear wave-particle interaction as follows (Hasegawa 1987):

$$\begin{aligned} \partial_t f_i &= -\omega_{ci} \left\langle \frac{\mathbf{E}_{\mathbf{k}} + \mathbf{v} \times \mathbf{B}_{\mathbf{k}}}{B_0} \cdot \partial_{\mathbf{v}} f_{i\mathbf{k}}^{(1)} \right\rangle \\ &= -\omega_{ci} \left\langle \left[\frac{k_z v_y}{\omega_{\mathbf{k}}} \frac{E_{y\mathbf{k}}}{B_0} + \left(1 - \frac{k_y v_y}{\omega_{\mathbf{k}}}\right) \frac{E_{z\mathbf{k}}}{B_0}\right] \partial_{v_z} f_{i\mathbf{k}}^{(1)} \right. \\ &\quad \left. + \left[\left(1 - \frac{k_z v_z}{\omega_{\mathbf{k}}}\right) \frac{E_{y\mathbf{k}}}{B_0} + \frac{k_y v_z}{\omega_{\mathbf{k}}} \frac{E_{z\mathbf{k}}}{B_0}\right] \partial_{v_y} f_{i\mathbf{k}}^{(1)} \right\rangle \\ &= \frac{\pi \omega_{ci}^2}{2} \sum_{\mathbf{k}} \partial_{iD} \left(\delta(\omega_{\mathbf{k}} - k_z v_z) I_0(\chi_i) e^{-\chi_i} \frac{E_{z\mathbf{k}}^2}{B_0^2} \partial_{iD}\right) f_i, \end{aligned} \tag{4.19}$$

where ∂_{iD} is the ion diffusion operator defined as follows:

$$\partial_{iD} \equiv \partial_{v_z} + \frac{k_y v_z}{\omega_{ci} \omega_{\mathbf{k}}} \partial_x. \tag{4.20}$$

By use of the following diffusion coefficients, D_{iv_z} and D_{ix}, in velocity and coordinate spaces,

$$D_{iv_z} \equiv \frac{\pi\omega_{ci}^2}{2}\sum_{k_z}\delta(k_z v_z - \omega) I_0(\chi_i) e^{-\chi_i}\frac{E_{z\mathbf{k}}^2}{B_0^2}$$

$$D_{ix} \equiv \frac{\pi}{2}\sum_{k_z}\delta(k_z v_z - \omega) I_0(\chi_i) e^{-\chi_i}\frac{k_y^2 v_z^2}{\omega_{\mathbf{k}}^2}\frac{E_{z\mathbf{k}}^2}{B_0^2}, \tag{4.21}$$

The quasi-linear diffusion equation (4.19) of ions can read as

$$\partial_t f_i = \left[\partial_{v_z}\left(D_{iv_z}\partial_{v_z}\right) + \partial_x\left(D_{ix}\partial_x\right)\right] f_i. \tag{4.22}$$

By the way, in the limit of $\chi_i \to 0$ the ion diffusion coefficients have the same expression of Eq. (4.12) for electrons. However, if a local dynamic balance between electrons and ions can be maintained, the ambipolarity results in the diffusion coefficient of the ions identical to that of the electrons, that is, $D_{ix} = D_{ex}$.

It is necessary to point out that the "diffusion" discussed here includes only the contribution from the resonant wave-particle interaction, but does not include the contribution from the nonresonant wave-particle interaction. Therefore, the diffusion coefficient is attributed mainly to the E_z component of fluctuation fields because for low-frequency wave with frequencies below ω_{ci}, in general, the resonant wave-particle interaction can occur only through the field-aligned motion of the particles, that is, the Landau resonance. However, it is also possible that the nonresonant wave-particle interaction contributes significantly to the diffusion processes of particles in coordinate space as well as in velocity space (Wu et al. 1997; Lu et al. 2006).

4.4 Particle Transport by KAWs at Magnetopause

Lee et al. (1994) first proposed KAWs to be responsible for the anomalous transport at the magnetopause, in which the large density and magnetic field gradients in the magnetopause boundary can effectively couple large-scale AWs with small-scale KAWs and enhance the turbulence strength of KAWs via the resonant mode conversion of the MHD AWs. In particular, the parallel electric field of the enhanced KAW turbulence can break down the "frozen-in" condition and decouple the plasma from the magnetic field. In consequence, the anomalous plasma transport across the magnetic field can be caused by the enhanced KAW turbulence at the magnetopause boundary and the corresponding diffusion coefficient can be estimated as order $\sim 10^9\ \mathrm{m}^2/\mathrm{s}$, which is the typical value required by the model of Sonnerup (1980). Meanwhile, they also showed that the parallel electric field of the enhanced KAW turbulence may accelerate electrons along the magnetic field to energies of order ~100 eV, and can provide a reasonable acceleration mechanism for counterstreaming electron beams observed in the magnetopause boundary.

Following closely the work of Lee et al. (1994), Prakash (1995) calculated the anomalous diffusion coefficient due to the enhanced KAW turbulence at the dayside magnetopause based on the quasilinear theory of Hasegawa and Mima (1978). The enhanced KAW turbulence was generated at the magnetopause by the resonant mode conversion of the surface AW originating from the shocked solar wind and has an enhanced amplitude greater than the surface AW, as expected by the theory (Hasegawa 1976; Hasegawa and Chen 1976a). By use of the wave parameters measured by the GEOS-2 satellite, Prakash (1995) estimated the diffusion coefficient to be comparable to the required typical value $\sim 10^9\,\mathrm{m^2/s}$ and inferred the boundary layer thickness $\sim 10^3$ km, as shown by observations.

Subsequently, Johnson and Cheng (1997) further proposed a scenario as shown in Fig. 4.7, in which compressional MHD waves in the magnetosheath propagate to the magnetopause and couple with transverse KAWs at the magnetopause. The large ambient gradients lead to the Alfvén velocity, v_A, monotonically increasing across the magnetopause by about a factor of 10, for instance, from ~400 km/s at the magnetosheath side to ~4000 km/s at the magnetosphere side for a typical parallel wavelength $\lambda_\parallel \sim R_E = 0.1 L_\parallel$ (see Fig. 4.7). This indicates that the magnetopause layer can have an AW continuous spectrum between ~60–600 mHz, which may overlap with the frequency band of magnetosheath compressional MHD waves of maximum wave power. In consequence, when the frequency of the magnetosheath MHD compressional wave matches the magnetopause AW frequency, the compressional wave couples strongly with the surface AW at the resonant surface and then the kinetic effect of the finite ion gyroradius leads to the resonant mode conversion of the surface AW into KAW near the resonant surface. Moreover, the amplitude of the

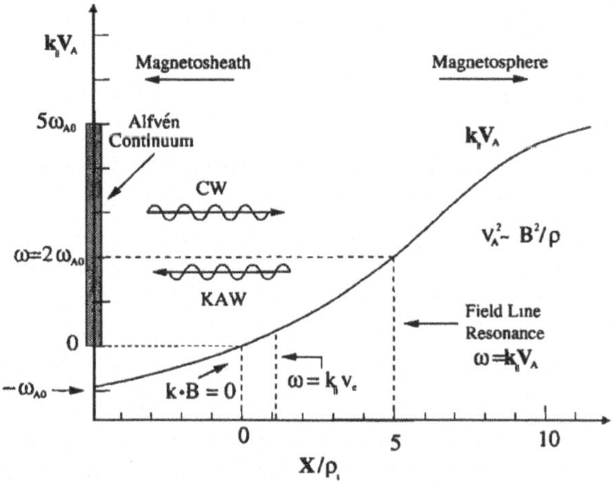

Fig. 4.7 A sketch of the resonant mode conversion of compressional MHD waves into KAWs in the magnetopause. The dayside magnetopause layer has an extension width $L_\parallel \sim 10\,R_E$ and steep thickness $L_\perp \sim 10\rho_i$ ($\rho_i \sim 50$ km). The Alfvén velocity v_A and hence the Alfvén frequency $\omega_A = k_\parallel v_A$ increase by a factor of 10 across the magnetopause. A compressional MHD wave with frequency ω propagates from the magnetosheath to the resonant surface of $\omega_A = \omega$, for example, at $x = 5\rho_i$, and mode converts to KAWs (from Johnson and Cheng 1997)

mode converted KAW in the magnetopause layer can be strongly enhanced over that of the magnetosheath compressional wave by a factor ~10 and the enhanced KAW may quickly spread its spectrum to form the enhanced KAW turbulence according to the theoretical prediction (Hasegawa 1976; Hasegawa and Mima 1978).

By use of the hybrid code model, in which ions are treated as fully kinetic particles but electrons as a massless fluid, Lin et al. (2010; 2012) numerically simulated the resonant mode conversion of a fast mode compressional wave to KAW at the magnetopause based on a tangential discontinuity model of the magnetopause. The ambient density over the magnetopause $n(x)$ is modeled by

$$n(x) = \frac{n_m + n_s}{2} + \frac{n_m - n_s}{2} \tanh \frac{x}{D}, \tag{4.23}$$

where n_m and n_s (=$10n_m$) are the densities in the magnetosphere ($x > 0$) and magnetosheath ($x < 0$), respectively, $D = 7.5\,\lambda_{is}$ (the ion inertial length in the magnetosheath) is the half thickness of the magnetopause layer, and the x axis is normal to the magnetopause. The ambient magnetic field $\mathbf{B}(x)$ is parallel to the magnetopause plane (i.e., the magnetopause is a tangential discontinuity) and the magnetic field strength variation across the magnetopause layer is determined by the total pressure balance. Therefore, the Alfvén velocity can be obtained as follows:

$$v_A^2(x) = \frac{B^2(x)}{\mu_0 m_i n(x)} = v_{As}^2 \frac{1 + \left(1 + \tau^{-1}\right)\left(1 - n(x)/n_s\right)\beta_i^{(s)}}{n(x)/n_s}, \tag{4.24}$$

where v_{As} and $\beta_i^{(s)}$ are the Alfvén velocity and the ion kinetic to magnetic pressure ratio in the magnetosheath, respectively.

For initial sinusoidal waves with frequencies $\omega_0 \sim 0.1$–$0.4\,\omega_{ci}^{(s)}$ (the magnetosheath ion gyrofrequency) and wave numbers $k_0 \sim 0.1$–$0.3\,\lambda_{is}^{-1}$ (or wavelengths $\lambda \sim 20$–$60\,\lambda_{is}$), which satisfy the dispersion relation of the MHD fast mode, the simulation results show that coherent waves with large wave numbers $k_x\rho_i \sim 1$ are excited apparently and enhanced quickly by a factor ~10 in a duration $\sim 30/\omega_{ci}^{(s)}$ at the resonant surface. The further analysis shows that these enhanced waves can satisfy well the dispersion and polarization relations of KAWs (Lin et al. 2010), as expected by the theory of the resonant mode conversion. Meanwhile, the simulated nonlinear evolution of the first stage KAWs clearly demonstrates that a driven KAW nonlinearly decays into a daughter KAW and an ion acoustic wave (Lin et al. 2012), confirming the parametric decay and spectral spreading processes (Hasegawa and Chen 1976b; Hasegawa and Mima 1978). In particular, the three-dimensional simulation by Lin et al. (2012) further shows that the enhanced KAWs can have large azimuthal (k_y) as well as normal wave numbers (k_x), which are generated nonlinearly by parametric decay of the linearly excited primary KAWs. This nonlinearly parametric decay process is accompanied by the simultaneous excitation of zonal flow modes with similarly large k_y (Lin et al. 2012). Thus, the combination of the linear mode conversion and nonlinear parametric decay can lead to more efficient plasma transport processes across the magnetopause.

Another possible generation mechanism for KAWs in the magnetosheath is the nonlinear evolution of the mirror instability proposed by Wu et al. (2001). Mirror waves and their instabilities are one of the most common fluctuations in the magnetosheath due to the interaction of the solar wind and the magnetosphere (Lin et al. 1998). The perturbed velocity $\delta\mathbf{v}$ and magnetic field $\delta\mathbf{B}$ of mirror waves are usually coplanar with the wave vector $\mathbf{k}$ and unperturbed magnetic field $\mathbf{B}_0$. By using hybrid simulations, however, Wu et al. (2001) showed that in the nonlinear evolution stage of the mirror instability driven by the temperature anisotropy (i.e., $T_\perp/T_\parallel > 1 + 1/\beta_\perp$), the extra non-coplanar components in $\delta\mathbf{v}$ and $\delta\mathbf{B}$ associated with the mirror waves can be excited in high wavenumber modes, which may possibly be attributed to the Hall or kinetic effects due to the kinetic-scale coupling with the mirror waves. In particular, their results showed that along with the nonlinear evolution, these high wavenumber perturbations can be decoupled from the mirror waves and form a pair of KAWs propagating in opposite directions (Wu et al. 2001). This indicates these KAWs generated in the dayside magnetosheath can propagate tailward along the magnetic field lines draped around the magnetopause and contribute to the plasma transport into the magnetosphere.

In observation, Chaston et al. (2007c) gave an example of the resonant mode conversion of AWs to KAWs in the magnetopause layer, which was observed by Cluster on January 20, 2001 when traveling sunward on the duskside flank of the magnetosphere. The C4 satellite observed the step-like variations in the magnetic field, implying that the satellite is traversing through a series of anti-sunward propagating surface wave vortices driven by a Kelvin-Helmholtz instability (Hasegawa et al. 2004). While the crossings are caused mainly by the boundary moving at a velocity $v_b \simeq 95$ km/s and the surface wave propagating at a velocity $v_s \simeq 240$–255 km/s. The Alfvén velocity v_A was observed to decrease across the boundary from $\sim$500 km/s at the magnetosphere side to lower than 200 km/s at the magnetosheath side. It is particularly worth noting that when the surface wave velocity v_s matches the local Alfvén velocity v_A at the Alfvén resonant surface within the boundary, great enhancements of electric field fluctuations of containing sharp spikes are observed.

Their analysis shows that these enhanced electric field fluctuations can be identified well to belong to KAWs, which are driven by the resonant mode conversion of the surface AWs or Kelvin-Helmholtz vortices at the resonant surface (Chaston et al. 2007c). Meanwhile, the particle observations by the C3 satellite provide clearly evidence for plasma transport across the magnetopause boundary layer, which identify the mixture of the magnetosheath ions with energies centered on $\sim$0.4 keV and $T_\perp > T_\parallel$ and the magnetospheric ions with energies above 2 keV and containing significant O^+ population. Moreover, the bursty field-aligned electrons also are observed throughout the crossings and the largest energies and fluxes present at the magnetosphere. These results suggest that these enhanced KAWs excited by the resonant mode conversion of the surface AWs within the magnetopause boundary layer also may produce efficient anomalous plasma transport across the magnetopause and field-aligned electron acceleration, which are similar to that observed in magnetic reconnection events near the subsolar (Mozer et al. 2005) and high latitude magnetopause (Chaston et al. 2005), although here there is no the operation of magnetic reconnection.

Besides the resonant mode conversion, magnetic reconnection also can be effectively exciting sources of KAWs. Chaston et al. (2005) showed an example of the generation of KAWs associated with the magnetic reconnection X line, which was observed by the Cluster satellite on March 18, 2002 when passing through the high latitude magnetopause. In this observation, enhanced low-frequency electromagnetic fluctuations were observed in the vicinity of the magnetic reconnection X line and the corresponding reconnection event was identified from the plasma jet reversal and associated diffusion region observed by the four Cluster satellites separated by $\sim$ 100 km (Phan et al. 2003). Chaston et al. (2005) identified these enhanced electromagnetic fluctuations as the drift-KAW mode because the diamagnetic drifts of ions and electrons across the magnetopause density gradient

$$\mathbf{v}_{di(e)} = \rho_{i(e)}^2 \omega_{ci(e)} |\partial_x \ln n(x)| \hat{y} \tag{4.25}$$

can influence significantly the dispersion relation of KAWs propagating at the y direction (Mikhailovskii 1967), as shown in the three-dimensional simulation by Lin et al. (2012). In fact, in a low-frequency wave with frequency $\omega \ll \omega_{ci}$ and perpendicular wave number k_y the perpendicular drift of ions is given by

$$\mathbf{v}_{i\perp} = v_{di}\hat{y} - i\frac{\omega - \omega_i^*}{\omega_{ci}}\frac{c_i(\chi_i)E_y}{B_0}\hat{y} - \frac{I_0(\chi_i)e^{-\chi_i}E_y}{B_0}\hat{x}, \tag{4.26}$$

where the second and third terms are the polarization and electric drifts, respectively, and $\omega_i^* = k_y v_{di}$ is the ion drift frequency. While the field-aligned drift of ions can be given by

$$v_{iz} = i\frac{\omega_{ci}}{\omega}\left(\frac{B_x}{B_0}v_{di} + I_0(\chi_i)e^{-\chi_i}\frac{E_z}{B_0}\right). \tag{4.27}$$

The drift-KAW dispersion and polarization relations can be obtained, respectively, as follows (Chaston et al. 2005):

$$\left(\frac{\omega - \omega_i^*}{\omega} - \frac{1}{M_z^2 c_i(\chi_i)}\right)\left(\frac{M_z^2}{\Omega} - \frac{\beta_e}{2}I_0(\chi_i)e^{-\chi_i}\right) = k_y^2\rho_s^2 \tag{4.28}$$

and

$$\frac{E_y}{B_x} = \frac{\left(1 - \Omega M_z^{-2} I_0(\chi_i)e^{-\chi_i}\beta_e/2\right) M_z v_A}{1 - \Omega\left(M_z^{-2} I_0(\chi_i)e^{-\chi_i}\beta_e/2 - c_i(\chi_i)k_y^2\rho_s^2\right)}, \tag{4.29}$$

where $\Omega \equiv \left(\omega - \omega_i^*\right) / \left(\omega - \omega_e^*\right)$. The wave vector $\mathbf{k}$ can be determined by the multiple measurements of the Cluster satellites and the diamagnetic drift velocities may be estimated from the density measurements, that is, $v_{di} \simeq 42.5$ km/s and $v_{de} \simeq 3$ km/s. The analysis of Chaston et al. (2005) shown that the dispersion and polarization relations of the drift-KAW mode in Eqs. (4.28) and (4.29) can be demonstrated well by the enhanced low-frequency electromagnetic fluctuations observed in the vicinity of the reconnection X line, implying that the magnetic reconnection can be important exciting source of KAWs. In particular, by use of the k-filtering technique (Pincon 1995) they first time demonstrated directly the KAW dispersion relation by the in situ measurement of satellite in space plasmas.

In another observation of the multi-point measurements by the THEMIS satellites the generation of enhanced KAWs also is shown to have relations with magnetic reconnection, in which enhanced low-frequency broadband electromagnetic fluctuations were observed by THEMIS during its traversing near the magnetopause on June 3, 2007 and identified by Chaston et al. (2008) as enhanced turbulent spectra of frequency shifted KAWs. In particular, the most intensity of these enhanced KAWs presents along the reconnected flux tubes in the magnetosheath just outside the magnetopause. Combining the measurements of particles, they showed that these enhanced KAWs are coincident with increases in $T_{e\parallel}$ and $T_{i\perp}$, implying the electrons and ions are heated field-aligned and cross-field, respectively, by the enhanced KAWs. Correspondingly, the energy spectrum of the enhanced KAW turbulence displays distinct power-law scalings, in which the spectral energy density $B_x^2 \propto k_p^{-1.5}$ and $k_p^{-3.0}$ for $k_p\rho_s < 1$ and > 1, respectively, where $k_p = \omega_{sf}/v_p$ is the wave number along the plasma flow at the velocity $\mathbf{v}_p$, and the shifted frequency $\omega_{sf} = \omega + \mathbf{k} \cdot \mathbf{v}_p \simeq \mathbf{k} \cdot \mathbf{v}_p$. These distinct power-law scalings possibly are attributed to the Landau and transit time damping on ions and electrons, respectively. Moreover, they also showed that the strength of the enhanced KAW turbulence can exceed the threshold for stochastic ion scattering or acceleration, and therefore these KAWs can produce an anomalous diffusion coefficient $D_w \sim D_B$ if their wave-normal angles are sufficiently large.

4.5 Formation of Ion Beams by KAWs in Magnetotail

The magnetotail is another important region for the solar wind-magnetosphere coupling, in particular, the PSBL between the CPS and the magnetotail lobe, which is believed originating from magnetic reconnection in the magnetotail neutral current sheet and is characterized mainly by sustained energetic field-aligned ion and electron beams at flow velocities remarkably higher their thermal velocities. Moreover, these beams are observed to propagate both earthward and tailward. In general, the beams propagating earthward are believed to originate in the magnetotail by direct field-aligned acceleration (Decoster and Frank 1979), by current sheet acceleration (Lyons and Speiser 1982), magnetic reconnection (Cowley et al. 1984), or by injection of magnetosheath plasma onto closed tail field lines and convection toward the earth (Whelan and Goertz 1987). The tailward beams are believed to be either the geomagnetic polar mirror reflected earthward beams (Williams 1981) or the upgoing energetic beams originating within the aurora acceleration region (Sharp et al. 1981). In fact, accompanying earth's magnetospheric activities, energetic ion and electron beams can be observed commonly in various magnetospheric boundary layers, such as the LLBL near the magnetopause (Ipavich et al. 1984; Eastman et al. 1985a, b) and the PSBL in the magnetotail (Lui et al. 1978; Eastman et al. 1985a; Baker et al. 1986; Takahashi and Hones 1988; Baumjohann et al. 1990). For example, counterstreaming electron beams are often observed simultaneously with strong low-frequency wave activity in the LLBL (Ogilvie et al. 1984; Takahashi et al. 1991; Song et al. 1993). Lee et al. (1994) proposed that KAWs, which can be generated by the resonant mode conversion of the surface AWs, may be responsible for the acceleration of these electron beams.

As pointed out by Hasegawa (1987), the anomalous diffusion process of plasma particles caused by the enhanced KAW turbulence can have comparable diffusion time scales in the field-aligned velocity space diffusion (i.e., with the diffusion coefficient $D_{iv_\parallel}$) and in the cross-field coordinate space diffusion (i.e., with the diffusion coefficient $D_{i\perp}$) because of the anisotropic dispersion relation of KAWs. In general, the particle diffusion time scales can be estimated by $t_{D\perp} \sim L_\perp^2/D_{i\perp}$ in the coordinate space and $t_{Dv_\parallel} \sim v_{T_i}^2/D_{iv_\parallel}$ in the velocity space, where $L_\perp$ is the characteristic length scale of the ambient cross-field inhomogeneity. By use of the δ-function resonant condition $v_z = \omega_{\mathbf{k}}/k_z$, from Eq. (4.21) one has $D_{iv_z}/D_{ix} \sim \omega_{ci}^2 \langle k_z^2 \rangle / \langle k_y^2 \rangle$ and hence

$$\frac{t_{Dx}}{t_{Dv_z}} \sim \frac{L_x^2}{v_{T_i}^2}\frac{D_{iv_z}}{D_{ix}} \sim \frac{L_x^2}{\rho_i^2}\frac{\langle k_z^2 \rangle}{\langle k_y^2 \rangle}. \tag{4.30}$$

Although $L_x \gg \rho_i$ in general it is possible that $t_{Dx}/t_{Dv_z} \sim 1$ because the short-wavelength kinetic effect of KAWs in the plane direction perpendicular to the ambient magnetic field leads to the strong anisotropy of KAWs with $\langle k_z^2 \rangle \ll \langle k_y^2 \rangle$. This possibly leads to the so-called double diffusion process of plasma particles, in which the particle diffuses simultaneously in the velocity and coordinate spaces.

For the double diffusion process of $t_{Dx} \simeq t_{Dv_z}$, the quasi-linear diffusion equation (4.22) has a unique feature that the simultaneous velocity and coordinate diffusions may lead to the distribution function forming a plateau at the resonant velocity. In the x-v_z phase plane, this plateau appears along the diffusion path determined by the operator $\partial_{iD} = 0$, that is,

$$\partial_x v_z = \frac{\omega_{ci}\omega_{\mathbf{k}}}{k_y v_z} = 0 \Rightarrow \frac{v_z^2}{2} = w_0 + \int \frac{\omega_{ci}\omega_{\mathbf{k}}}{k_y} dx. \tag{4.31}$$

In an inhomogeneous plasma with a density gradient along the x-direction, the local Alfvén velocity $v_A(x)$, the dispersion relation of KAWs $M_z^2(x) = K_A(\chi_i)$, and the initial distribution function $f_i(x + v_y/\omega_{ci}, v_\perp^2 + v_z^2)$ all depend on the x coordinate. Accompanying the ion diffusion process in the coordinate space from the high density region at the plasma center with a low $v_A(x)$ (assuming $x = 0$) to the low density region at the plasma edge with a high $v_A(x)$, the ions simultaneously diffuse in the velocity space along the plateau path in the $x - v_z$ plane, which asymptotic line is along the direction $v_z = v_A(x)$ at the edge where $v_A(x) \gg v_z(0) > v_A(0)$, implying that the double diffusion process caused by enhanced KAW turbulence can lead to the formation of fast ion beams at the plasma edge (Hasegawa 1987).

By use of a two-dimensional quasi-nonlinear code based on the Ritz-Galerkin method and finite elements (Strang and Fix 1973; Segerlind 1976), Moghaddam-Taaheri et al. (1989) numerically simulated and performed the double diffusion process of ions by an enhanced KAW turbulence, which was suggested by Hasegawa (1987). For the case of the ion diffusion across the PSBL from the dense CPS to the tenuous tail lobe, they consider the PSBL as a slab between $x = 0$ (the CPS side) and $x = L_b$ (the lobe side), in which the ambient density $n(x)$ and magnetic field $B_0(x)$ (along the z axis) satisfy the total pressure balance (note: for the sake of consistency, here the coordinate axes x and z are interchanged with those in Moghaddam-Taaheri et al. 1989):

$$\left(1+\tau^{-1}\right) n_i(x)T_i + \frac{B_0^2(x)}{2\mu_0} = \left(1+\tau^{-1}\right) n_i(0)T_i + \frac{B_0^2(0)}{2\mu_0}, \tag{4.32}$$

and assume the plasma β varying from $\beta_i(0) = 1$ at the CPS side to $\beta_i(L_b) \ll 1$ at the lobe side, but with constant ion and electron temperatures. As the plasma density decreases and the magnetic field increases the Alfvén velocity $v_A(x) = B_0(x)/\sqrt{\mu_0 m_i n_i(x)}$ increases linearly from $v_A(0) = \sqrt{2/\beta_i(0)}v_{T_i} = \sqrt{2}v_{T_i}$ at the CPS to $v_A(L_b) = \sqrt{2/\beta_i(L_b)}v_{T_i} = v_A(0)/\sqrt{\beta_i(L_b)}$ at the lobe, that is,

$$v_A(x) = v_A(0) + \frac{v_A(L_b) - v_A(0)}{L_b}x. \tag{4.33}$$

Taking the spectral energy density of the enhanced KAW turbulence as

$$w_{\mathbf{k}} = \frac{\epsilon_0 E_{z0}\delta(\phi-\phi_0)}{4\pi k_\perp \Delta k_\perp \Delta k_z} \exp\left[-\frac{(k_\perp - k_{\perp 0})^2}{2\Delta k_\perp^2} - \frac{(k_z - k_{z0})^2}{2\Delta k_z^2}\right], \tag{4.34}$$

where $k_\perp^2 = k_x^2 + k_y^2, \phi = \tan^{-1}(k_y/k_x), k_{z0} = 2\pi L_B^{-1}, k_{\perp 0} = \rho_i^{-1}(0), \Delta k_z = 0.2k_{z0}$, L_B is the field-aligned characteristic size scale, and the spectral width $\Delta k_\perp$ and the propagation direction in the cross-field plane (i.e., the x-y plane), ϕ_0, are left as the stimulating parameters, the diffusion coefficients of Eq. (4.21) can be rewritten as:

$$\begin{aligned} D_{iv_z} &= \frac{\pi\omega_{ci}^2 c^2}{2}\int d^3\mathbf{k}\delta(k_z v_z - \omega_{\mathbf{k}}) I_0(\chi_i)e^{-\chi_i}\frac{w_{\mathbf{k}}}{B_0^2/2\mu_0} \\ D_{ix} &\equiv \frac{\pi c^2}{2}\int d^3\mathbf{k}\delta(k_z v_z - \omega_{\mathbf{k}}) I_0(\chi_i)e^{-\chi_i}\frac{v_z^2 k_\perp^2 \sin^2\phi}{\omega_{\mathbf{k}}^2}\frac{w_{\mathbf{k}}}{B_0^2/2\mu_0}. \end{aligned} \tag{4.35}$$

Based on observations (Eastman et al. 1985a), other parameters can be taken as $L_b = 2R_E, L_B = 10L_b = 20R_E, B_0 = 20\,\text{nT}, n_i(0) = 1\,\text{cm}^{-3}, n_i(L_b) = 0.05\,\text{cm}^{-3}, T_i = 1$ keV, and hence $\beta_i(0) = 1$ and $\rho_i \simeq 160$ km $\simeq R_E/40$. The initial distribution function is fitted by a Maxwellian distribution with a high-energy tail,

$$f_i(v_z, x, t = 0) = \frac{n_i(x)}{\sqrt{2\pi}} \left[\left(1 - \frac{n_t}{n_i}\right) e^{-v_z^2/2} + \frac{n_t}{\sqrt{6} n_i} e^{-v_z^2/12} \right], \tag{4.36}$$

where $n_t \sim 0.01 n_i$ is the contribution of the high-energy tail. They modeled the two cases with or without the particle loss at $x = L_b$, that is, the magnetic field is open or closed in the lobe.

Their simulation results show that, as expected by Hasegawa (1987), KAWs can diffuse ions simultaneously in velocity space along the magnetic field and in coordinate space across the magnetic field with comparable transport rates when $k_y \sim \rho_i^{-1} \gg k_x \gg k_z$ (i.e., $\phi_0 \simeq \pi/2$) and lead to the formation of beam-like ion distributions with a feature velocity centered at $v_z(x) \sim K_A(\chi_i) v_A(x)$ (where K_A comes from the KAW dispersion relation $M_z^2 = K_A$). Moreover, the wider spectrum of the KAW turbulence, $\Delta k_\perp$, leads to the wider beam. However, the diffusion across the magnetic field is slower when $k_x \sim \rho_i^{-1} \gg k_y \gg k_z$ (i.e., $\phi_0 \simeq 0$). The beam density $n_b(x)$ in the PSBL is approximately constant in time after forming the plateau for the closed field lobe side or after reaching a steady state for the open field lobe side. However, the relative density $n_b(x)/n_i(x)$ ranges from $\sim$0.01 near the boundaries to $\sim$0.1 near the center of the PSBL for the closed field lobe and from $\sim$0.01 near the boundaries to $\sim$0.4 near the center of the PSBL for the open field lobe, similar to the observed ion beam density $n_b \sim 0.1$–$0.8 n_i$ in the PSBL (Eastman et al. 1984). Since the plasma β decreases from $\beta \sim 1$ at the CPS side (i.e., $x = 0$) to $\beta \ll 1$ at the lobe side (i.e., $x = L_b$) in the PSBL, the beam velocity $v_b \sim K_A v_A(x) \sim v_A(x)$ will increase with away from the CPS (i.e., x increases), as shown by observations (Takahashi and Hones 1988). Taking the ion temperature $T_i \sim 1$–2 keV at the CPS side, the simulation result gives the beam energy ranging from $\sim$2 to $\sim$240 keV, well comparable with the observed values of 1 to > 100 keV (Eastman et al. 1984; Williams 1981). In addition, the scenario of the ions diffusing from the PSBL into the lobe also indicates that the beam plasma has the same composition in the lobe as that in the PSBL, in agreement with the result by Huang et al. (1987).

On the other hand, in the case of the open field lobe the diffusion ions arriving at the lobe will leave the PSBL and lost into the distant tail along the open magnetic fields and the distribution function will reach a steady state when the balance between the ion diffusion rate into the PSBL from the CPS and the ion loss rate into the distant tail from the PSBL is satisfied. This indicates that the PSBL can act as an intermediate agent in transferring ions from the CPS to the distant magnetotail if the PSBL has an open field boundary at the lobe side. In particular, this transferring process of the PSBL may avoid the pressure dilemma which results from the loss-free adiabatic convection of particles earthward on closed geomagnetic fields because these loss-free convection particles earthward would result in the excess pressure buildup in the CPS.

Low-frequency electromagnetic fluctuations in the magnetotail, especially in the PSBL have been identified extensively as AWs and KAWs. Based on the observations by the Polar satellite, Wygant et al. (2002) found that enhanced low-frequency ($\omega <$ 1 Hz) electromagnetic fluctuations in the PSBL at altitudes of 4–6 R_E, where the local ion gyroradius $\rho_i \sim 20$ km and the electron inertial length $\lambda_e \sim 10$ km, i.e., implying the kinetic parametric regime of KAWs with $\alpha_e > 1$, have large electric to magnetic amplitude ratio (larger than the local Alfvén velocity) and small cross-field scale sizes $\sim$20–120 km (i.e., $\sim$1–6ρ_i). It is very evident that these observed characteristics can be strong evidence that these enhanced fluctuations are KAWs (Wygant et al. 2002). Based on 5-year observation data by the Geotail satellite in the magnetotail, Takada et al. (2005) investigated the statistical properties of low-frequency (0.01–0.1 Hz) electromagnetic waves and their relations to ion beams in the PSBL. Their results indicate that these low-frequency AWs can be the effective sources of KAWs and have clear correlation with the field-aligned ion beams in the PSBL.

In fact, the PSBL in high latitudes can magnetically project on the ionospheric auroral regions. By use of the simultaneous observations by the Polar satellite in the PSBL at altitude $\sim 7\,R_E$ and the FAST satellite in the auroral region at altitude $\sim$3500 km, Dombeck et al. (2005) analyzed the dynamical evolution of the observed enhanced electromagnetic fluctuations when propagating earthward along the geomagnetic fields from the PSBL to the auroral region during the main phase of a major geomagnetic storm on 22 October 1999. They found that these fluctuations have evident characteristics of the KAW mode mixed with AWs, their Poynting flux has a mean net loss of $\sim$2.1 mW/m^2 in the propagation process, and simultaneously the electron energy flux earthward has a mean net increase of $\sim$1.2 mW/m^2 (Dombeck et al. 2005). This implies that the lager amplitude AWs driven by the magnetotail CPS activity (for instance, magnetic reconnection) propagate earthward and convert into enhanced KAWs in the PSBL, while these enhanced KAWs accelerate electrons to travel earthward. Keiling et al. (2002) compared the Alfvénic Poynting flux originating from the CPS at geocenter distances of 4–7 R_E and the energy flux of magnetically conjugate precipitating electrons at 100 km altitude and found very evident correlation between the magnetotail Alfvénic Poynting flux and the ionospheric electron energy flux, implying that enhanced KAWs propagate throughout the PSBL from the magnetotail to the auroral region.

In consideration of the strong inhomogeneity in the PSBL similar to that in the magnetopause, the resonant mode conversion of enhanced AWs into KAWs is one of possible mechanisms for the generation of KAWs in the PSBL. The enhanced AWs excited by the Kelvin-Helmholtz instability of sheared flows between outside and inside the magnetopause have the characteristic phase velocity $v_{p\|} \simeq v_{sw}/2 \gg v_{mp}$, where v_{sw} and v_{mp} are the plasma flow velocities outside and inside the magnetopause, respectively. These enhanced AWs can propagate along the magnetic field into the tenuous magnetotail lobe and subsequently propagate as the slow compressible mode in the tail lobe because their parallel phase velocity is much less than the local Alfvén velocity in the lobe, that is, $v_{p\|} \ll v_{Ab} \sim v_A(L_b)$. After arriving at the PSBL, where the plasma is inhomogeneous and the local Alfvén velocity $v_A(x)$ varies from $v_{Ab} \gg v_{p\|}$ in the lobe to $v_{Ap} \sim v_A(0) \ll v_{p\|}$ in the CPS, they possibly encounter the Alfvén resonant surface in the PSBL, where the parallel phase velocity of the incident slow wave matches to the local Alfvén velocity, that is, $v_A(x) = v_{ph}$. In the neighborhood of this resonant surface the incident slow wave becomes the resonant surface AW and further converts into KAWs via the kinetic effect of the finite ion gyroradius.

Another possible mechanism for the conversion of AWs to KAWs in the PSBL may occur between the CPS and the PSBL, in which the dawn-dusk convection electric field causes the $\mathbf{E} \times \mathbf{B}$ drift near the equatorial plane and the corresponding earthward plasma flow is braked by the strong dipole-like magnetic field in the CPS. A two-dimensional global hybrid simulation by Guo et al. (2015) showed that this flow braking in the low latitude CPS can lead to the excitation of AWs in the CPS and then the excited AWs propagate along the geomagnetic fields toward the polar region and convert into KAWs in the high latitude magnetotail due to the inhomogeneity of the magnetic field and plasma density along the propagation path.

It is worth paying attention to that both observation and theory show that the resonant mode conversion of AWs to KAWs can be one of important and efficient mechanisms for the generation KAWs in the magnetosphere, especially in the outer boundary layers when enhanced AWs propagate throughout the magnetospheric inhomogeneous plasmas.

Like the magnetopause, the magnetotail CPS is another prime location for magnetic reconnection, which also can be effectively exciting sources of KAWs in the magnetotail. The THEMIS (Time History of Events and Macroscale Interactions during Substorms) mission, launched on February 17, 2007, consists of five micro-satellites (termed "probes": p1, p2, p3, p4, and p5), which line up along the magnetotail (Angelopoulos 2008). Its five probes can track the motion of particles, plasma and waves from one point to another in key regions of the magnetosphere to elucidate plasma processes responsible for substorms, such as the local disruption of the plasma sheet current or its interaction with the rapid plasma influx originating from magnetic reconnection in the magnetotail. Duan et al. (2016) analyzed two similar events of enhanced low-frequency electromagnetic fluctuations during substorm expansion phases observed by the THEMIS satellites when they traversed from the PSBL to the CPS in the magnetotail at the geocenter distance $\sim 10\,R_E$ on February 3 and 7, 2008. In these two events, the local Alfvén velocities were observed both to decrease sharply from $\sim$2000 km/s in the PSBL to $\sim$1000 km/s in the CPS. While their transverse electric to magnetic ratios of the enhanced electromagnetic fluctuations range between 2000 and 6000 km/s and both are evidently larger than the local Alfvén velocity. Meanwhile their Poynting fluxes both are almost along the magnetic field. The observed characteristics of these enhanced fluctuations can be well consistent with their KAW mode nature (Duan et al. 2016).

Based on the physics of KAWs, Dai et al. (2017) further proposed a more in detail theoretical explanation for these enhanced low-frequency electromagnetic fluctuations, in which the excitation of KAWs may be attributed to the strongly inhomogeneity in the local plasma, such as the local ion pressure gradient. By use of the observation by the Cluster satellites when crossing the magnetotail neutral sheet at the geocenter distance $\sim 18\,R_E$ and encountering a magnetic reconnection diffusion region on September 19, 2003, Chaston et al. (2009) compared the explanation by the KAW physics to that by the whistler wave physics and found that the observed enhanced electromagnetic fluctuations can be fitted better by the prediction of the KAW mode, implying that their results also confirmed further the KAW nature of these enhanced low-frequency electromagnetic fluctuations.

References

Angelopoulos, V. (2008). The THEMIS mission. *Space Science Reviews, 141*, 5.
Axford, W. I., & Hines, C. O. (1961). A unifying theory of high-latitude geophysical phenomena and geomagnetic storms. *Canadian Journal of Physics, 39*, 1433.
Baker, D. N., Bame, S. J., Feldman, W. C., et al. (1986). Bidirectional electron anisotropies in the distant tail-ISEE 3 observations of polar rain. *Journal of Geophysical Research, 91*, 5637–5662.
Baumjohann, W., Paschmann, G., & Lühr, H. (1990). Characteristics of high-speed ion flows in the plasma sheet. *Journal of Geophysical Research, 95*, 3801.
Bohm, D. (1949). Note on a theorem of bloch concerning possible causes of superconductivity. *Physical Review, 75*, 502–504.
Chaston, C. C., Hull, A. J., Bonnell, J. W., et al. (2007b). Large parallel electric fields, currents, and density cavities in dispersive Alfvén waves above the aurora. *Journal of Geophysical Research, 112*, A05215.
Chaston, C. C., Johnson, J. R., Wilber, M., et al. (2009). Kinetic Alfvén wave turbulence and transport through a reconnection diffusion region. *Physical Review Letters, 102*, 015001.
Chaston, C. C., Peticolas, L. M., Carlson, C. W., et al. (2005). Energy deposition by Alfvén waves into the dayside auroral oval: Cluster and FAST observations. *Journal of Geophysical Research, 110*, A02211.
Chaston, C. C., Wilber, M., Mozer, F. S., et al. (2007c). Mode conversion and anomalous transport in kelvin-helmholtz vortices and kinetic Alfvén waves at the earth's magnetopause. *Physical Review Letters, 99*, 175004.
Chaston, C., Bonnell, J., Mcfadden, J. P., et al. (2008). Turbulent heating and cross-field transport near the magnetopause from THEMIS. *Geophysical Research Letters, 35*, L17S08.
Cowley, S. W. H., Hynds, R. J., Richardson, I. G., et al. (1984). Energetic ion regions in the deep geomagnetic tail: ISEE-3. *Geophysical Research Letters, 11*, 275–278.
Crooker, N. U., & Siscoe, G. L. (1977). A mechanism for pressure anisotropy and mirror instability in the dayside magnetosheath. *Journal of Geophysical Research, 82*, 185–186.
Dai, L., Wang, C., Zhang, Y., et al. (2017). Kinetic Alfvén wave explanation of the Hall fields in magnetic reconnection. *Geophysical Research Letters, 44*, 634–640.
Decoster, R. J., & Frank, L. A. (1979). Observations pertaining to the dynamics of the plasma sheet. *Journal of Geophysical Research, 84*, 5099–5121.
Denton, R. E., Anderson, B. J., Fuselier, S. A., et al. (1994). Ion anisotropy-driven waves in the Earth's magnetosheath and plasma depletion layer. *Solar System Plasmas in Space and Time, 84*, 111–119.
Dombeck, J., Cattell, C., Wygant, J. R., et al. (2005). Alfvén waves and Poynting flux observed simultaneously by Polar and FAST in the plasma sheet boundary layer. *Journal of Geophysical Research. 110*, A12S90.
Drake, J. F., Kleva, R. G., & Mandt, M. E. (1994). Structure of thin current layers: Implications for magnetic reconnection. *Physical Review Letters, 73*, 1251–1254.
Duan, S. P., Dai, L., Wang, C., et al. (2016). Evidence of kinetic Alfvén eigenmode in the near-Earth magnetotail during substorm expansion phase. *Journal of Geophysical Research, 121*, 4316–4330.
Dungey, J. W. (1961). Interplanetary magnetic field and the auroral zones. *Physical Review Letters, 6*, 47–48.
Eastman, T. E., & Hones, E. W. (1979). Characteristics of the magnetospheric boundary layer and magnetopause layer as observed by IMP 6. *Journal of Geophysical Research, 84*, 2019–2028.
Eastman, T. E., Frank, L. A., & Huang, C. Y. (1985a). The boundary layers as the primary transport regions of the earth's magnetotail. *Journal of Geophysical Research, 90*, 9541–9560.
Eastman, T. E., Frank, L. A., Peterson, W. K., & Lennartsson, W. (1984). The plasma sheet boundary layer. *Journal of Geophysical Research, 89*, 1553–1572.
Eastman, T. E., Popielawska, B., & Frank, L. A. (1985b). Three-dimensional plasma observations near the outer magnetospheric boundary. *Journal of Geophysical Research, 90*, 9519–9539.

Escoubet, C. P., Fehringer, M., & Goldstein, M. (2001). The cluster mission. *Annales Geophysicae, 19*, 1197–1200.
Fazakerley, A. N., & Southwood, D. J. (1994). Mirror instability in the magnetosheath. *Advances in Space Research, 14*, 65–68.
Furth, H. P., Killeen, J., & Rosenbluth, M. N. (1963). Finite-resistivity instabilities of a sheet pinch. *Physics of Fluids, 6*, 459.
Galeev, A. A., & Sagdeev, R. Z. (1984). Current instabilities and anomalous resistivity of plasma. In A. A. Galeev & R. N. Sudan (Eds.), *Basic plasma physics 2* (pp. 271–303). Amsterdan: North-Holland Physics Publishing.
Guo, Z., Hong, M., Lin, Y., et al. (2015). Generation of kinetic Alfvén waves in the high-latitude near-earth magnetotail: A global hybrid simulation. *Physics of Plasmas, 22*, 022117.
Haerendel, G. (1978). Microscopic plasma processes related to reconnection. *Journal of Atmospheric and Terrestrial Physics, 40*, 343–353.
Harten, R., & Clark, K. (1995). The design features of the GGS wind and polar spacecraft. *Space Science Reviews, 71*, 23–40.
Hasegawa, A. (1975). *Plasma instabilities and nonlinear effects* (p. 93). New York: Springer-Verlag.
Hasegawa, A. (1976). Particle acceleration by MHD surface wave and formation of aurora. *Journal of Geophysical Research, 81*, 5083–5090.
Hasegawa, A. (1987). Beam production at plasma boundaries by kinetic Alfvén waves. *Journal of Geophysical Research, 92*, 11221–11223.
Hasegawa, A., & Chen, L. (1976a). Kinetic process of plasma heating by resonant mode conversion of Alfvén wave. *Physics of Fluids, 19*, 1924–1934.
Hasegawa, A., & Chen, L. (1976b). Parametric decay of Kinetic Alfvén Wave and its application to plasma heating. *Physical Review Letters, 36*, 1362–1365.
Hasegawa, A., & Mima, K. (1978). Anomolous transport produce by kinetic Alfvén wave turbulence. *Journal of Geophysical Research, 83*, 1117–1123.
Hasegawa, H., Fujimoto, M., Phan, T. D., et al. (2004). Transport of solar wind into earth's magnetosphere through rolled-up kelvin-helmholtz vortices. *Nature, 430*, 755–758.
Hones, E. W., Akasofu, S. I., Bame, S. J., & Singer, S. (1972). Outflow of plasma from the magnetotail into the magnetosheath. *Journal of Geophysical Research, 77*, 6688–6695.
Huba, J. D., Gladd, N. T., & Papadopoulos, K. (1977). The lower-hybrid-drift instability as a source of anomalous resistivity for magnetic field line reconnection. *Geophysical Research Letters, 4*, 125–126.
Huang, C. Y., Frank, L. A., Peterson, W. K., et al. (1987). Filamentary structures in the magnetotail lobes. *Journal of Geophysical Research, 92*, 2349–2363.
Ipavich, F. M., Gosling, J. T., & Scholer, M. (1984). Correlation between the He/H ratios in upstream particle events and in the solar wind. *Journal of Geophysical Research, 89*, 1501–1507.
Johnson, J. R., & Cheng, C. Z. (1997). Kinetic Alfvén waves and plasma transport at the magnetopause. *Geophysical Research Letters, 24*, 1423–1426.
Kaufmann, R. L., Horng, J. T., & Wolfe, A. (1970). Large-amplitude hydromagnetic waves in the inner magnetosheath. *Journal of Geophysical Research, 75*, 4666–4676.
Keiling, A., Wygant, J. R., Cattell, C., et al. (2002). Correlation of Alfvén wave poynting flux in the plasma sheet at 4–7 R_E with ionospheric electron energy flux. *Journal of Geophysical Research, 107*, 1132–1145.
Kivelson, M. G., & Russell, C. T. (1995). *Introduction to space physics*. Cambridge: Cambridge University Press.
Knudsen, D. J., & Wahlund, J.-E. (1998). Core ion flux bursts within solitary kinetic Alfvén waves. *Journal of Geophysical Research, 103*, 4157–4170.
Lühr, H., & Klöcker, N. (1987). AMPTE-IRM observations of magnetic cavities near the magnetopause. *Geophysical Research Letters, 14*, 186–189.
Labelle, J., & Treumann, R. A. (1988). Plasma waves at the dayside magnetopause. *Space Science Reviews, 47*, 175–202.

Lee, L. C., Johnson, J. R., & Ma, Z. W. (1994). Kinetic Alfvén waves as a source of plasma transport at the dayside magnetopause. *Journal of Geophysical Research, 99*, 17405–17411.

Lin, C. H., Chao, J. K., Lee, L. C., Wu, D. J., et al. (1998). Identification of mirror moves by the phase difference between perturbed magnetic field and plasmas. *Journal of Geophysical Research, 103*, 6621–6631.

Lin, Y., Johnson, J. R., & Wang, X. Y. (2010). Hybrid simulation of mode conversion at the magnetopause. *Journal of Geophysical Research, 115*, A04208.

Lin, Y., Johnson, J. R., & Wang, X. Y. (2012). Three-dimensional mode conversion associated with kinetic Alfvén waves. *Physical Review Letters, 109*, 125003.

Lu, Q. M., Wu, C. S., & Wang, S. (2006). The nearly isotropic velocity distributions of energetic electrons in the solar wind. *The Astrophysical Journal, 638*, 1169–1175.

Lucek, E. A., Dunlop, M. W., Balogh, A., et al. (1999). Identification of magnetosheath mirror modes in equator-s magnetic field data. *Annales Geophysicae, 17*, 1560–1573.

Lui, A. T. Y., Frank, L. A., Ackerson, K. L., et al. (1978). Plasma flows and magnetic field vectors in the plasma sheet during substorms. *Journal of Geophysical Research, 83*, 3849–3858.

Lyons, L. R., & Speiser, T. W. (1982). Evidence for current sheet acceleration in the geomagnetic tail. *Journal of Geophysical Research, 87*, 2276.

Mikhailovskii, A. B. (1967). Oscillations of an inhomogeneous plasma. In Leontovich, (Ed.), *Reviews of plasma physics* (pp. 159–172). New York: Consultants Bureau.

Moghaddam-Taaheri, E., Goertz, C. K., & Smith, R. A. (1989). Ion beam generation at the plasma sheet boundary layer by kinetic Alfvén waves. *Journal of Geophysical Research, 94*, 10047–10060.

Mozer, F. S., Bale, S. D., Mcfadden, J. P., & Torbert, R. B. (2005). New features of electron diffusion regions observed at subsolar magnetic field reconnection sites. *Geophysical Research Letters, 32*, L24102.

Norqvist, P., André, M., Eliasson, L., et al. (1996). Ion cyclotron heating in the dayside magnetosphere. *Journal of Geophysical Research, 101*, 13179–13193.

Ogilvie, K. W., Fitzenreiter, R. J., & Scudder, J. D. (1984). Observations of electron beams in the low-latitude boundary layer. *Journal of Geophysical Research, 89*, 10723–10732.

Papadopoulos, K. (1979). The role of microturbulence on collisionless reconnection. In S. I. Akasofu (Ed.), *Dynamics in the Magnetosphere, 289*. Dordrecht, Holland: D. Reidel Publ. Co.

Phan, T., Frey, H. U., Frey, S., et al. (2003). Simultaneous cluster and IMAGE observations of cusp reconnection and auroral proton spot for northward IMF. *Geophysical Research Letters, 30*, 1509.

Pincon, J. L. (1995). Cluster and the K-Filtering. In *Proceedings of the Cluster Workshops, Data Analysis Tools and Physical Measurements and Mission-Oriented Theory*.

Prakash, M. (1995). Anomalous plasma diffusion due to kinetic Alfvén wave fluctuations at the dayside magnetopause, in cross-scale coupling in space plasmas. *AGU, 249*.

Russell, C. T., & Elphic, R. C. (1978). Initial ISEE magnetometer results: Magnetopause observations. *Space Science Reviews, 22*, 681–715.

Sagdeev, R. Z., & Galeev, A. A. (1969). *Nonlinear plasma theory benjamin*. New York.

Segerlind L. J. (1976). Applied finite element analysis (2nd Ed.). *Journal of Vibration Acoustics Stress and Reliability in Design, 109*(3), 329.

Sharp, R. D., Carr, D. L., Peterson, W. K., & Shelley, E. G. (1981). Ion streams in the magnetotail. *Journal of Geophysical Research, 86*, 4639.

Song, P., Russell, C. T., Fitzenreiter, R. J., et al. (1993). Structure and properties of the subsolar magnetopause for northward interplanetary magnetic field: Multiple-instrument particle observations. *Journal of Geophysical Research, 98*, 11319–11337.

Sonnerup, B. U. O. (1980). Theory of the low-latitude boundary layer. *Journal of Geophysical Research, 85*, 2017–2026.

Southwood, D. J., & Kivelson, M. G. (1991). An approximate description of field-aligned currents in a planetary magnetic field. *Journal of Geophysical Research, 96*, 67–75.

Stasiewicz, K., Bellan, P., Chaston, C., et al. (2000a). Small scale Alfvénic structure in the aurora. *Space Science Reviews, 92*, 423–533.

Stasiewicz, K., Khotyaintsev, Y., & Grzesiak, M. (2004). Dispersive Alfvén waves observed by cluster at the magnetopause. *Physica Scripta*, *T107*, 171–179.

Stasiewicz, K., Khotyaintsev, Y., Berthomier, M., & Wahlund, J. E. (2000b). Identification of widespread turbulence of dispersive Alfvén waves. *Geophysical Research Letters*, *27*, 173–176.

Stasiewicz, K., Lundin, R., & Marklund, G. (2000c). Stochastic ion heating by orbit chaotization on electrostatic waves and nonlinear structures. *Physica Scripta*, *T84*, 60–63.

Stasiewicz, K., Seyler, C. E., Mozer, F. S., et al. (2001). Magnetic bubbles and kinetic Alfvén waves in the high-latitude magnetopause boundary. *Journal of Geophysical Research*, *106*, A29503–29514.

Strang, G., & Fix, G. (1973). A Fourier Analysis of the Finite Element Variational Method. *Constructive Aspects Of Functional Analysis*, 795–840.

Takada, T., Seki, K., Hirahara, M., et al. (2005). Statistical properties of low-frequency waves and ion beams in the plasma sheet boundary layer: Geotail observations. *Journal of Geophysical Research*, *110*, A02204.

Takahashi, K., & Hones, E. W, Jr. (1988). ISEE 1 and 2 observations of ion distributions at the plasma sheet-tail lobe boundary. *Journal of Geophysical Research*, *93*, 8558–8582.

Takahashi, K., Sibeck, D. G., Newell, P. T., & Spence, H. E. (1991). ULF waves in the low-latitude boundary layer and their relationship to magnetospheric pulsations: A multisatellite observation. *Journal of Geophysical Research*, *96*, 9503–9519.

Treumann, R. A., Güdel, M., & Benz, A. O. (1990). Alfvén wave solitons and solar intermediate drift bursts. *Astronomy & Astrophysics*, *236*, 242–249.

Turner, J. M., Burlaga, L. F., Ness, N. F., & Lemaire, J. F. (1977). Magnetic holes in the solar wind. *Journal of Geophysical Research*, *82*, 1921–1924.

Wahlund, J. E., Eriksson, A. I., Holback, B., et al. (1998). Broadband ELF plasma emission during auroral energiztion 1. Slow ion acoustic waves. *Journal of Geophysical Research*, *103*, 4343–4375.

Whelan, T., & Goertz, C. K. (1987). A new model for the ion beams in the plasma sheet boundary layer. *Geophysical Research Letters*, *14*, 68–71.

Williams, D. J. (1981). Energetic ion beams at the edge of the plasma sheet-ISEE 1 observations plus a simple explanatory model. *Journal of Geophysical Research*, *86*, 5507–5518.

Winterhalter, D., Neugebauer, M., Goldstein, B. E., et al. (1994). Ulysses field and plasma observations of magnetic holes in the solar wind and their relation to mirror-mode structures. *Journal of Geophysical Research*, *99*, 23371–23381.

Wu, D. J., Wang, D. Y., & Huang, G. L. (1997). Two dimensional solitary kinetic Alfvén waves and dipole vortex structures. *Physics of Plasmas*, *4*, 611–617.

Wu, B. H., Wang, J. M., & Lee, L. C. (2001). Generation of kinetic Alfvén waves by mirror instability. *Geophysical Research Letters*, *28*, 3051–3054.

Wygant, J. R., Keiling, A., Cattell, C. A., et al. (2002). Evidence for kinetic Alfvén waves and parallel electron energization at 4–6 R_E altitudes in the plasma sheet boundary layer. *Journal of Geophysical Research*, *107*, 1201–1215.

Chapter 5
KAW Turbulence in Solar Wind

5.1 Solar Wind: Natural Laboratory for Plasma Turbulence

Plasma turbulence is a stochastic state of the plasma and its electromagnetic fluctuations in spatial and temporal structures. Unlike the randomness of thermal fluctuations (or i.e., thermal noise) that is caused by the discrete randomness of microscopic particles (or i.e., the hypothesis of molecular chaos), however, the turbulent stochasticity of the plasma turbulence should be attributed to the chaotic nondeterminacy of plasma collective modes when dynamically evolving into the chaos state via their nonlinear coupling (Ruelle and Takens 1971; Gollub and Swinney 1975). Plasmas, consisting of charged particles, are intrinsically different from neutral fluids, consisting of neutral atoms or molecules, and can have a number of various collective eigenmodes due to the interparticle electromagnetic interaction. The nonlinear coupling between these collective modes not only is the most distinctive and essential characteristic of the plasma dynamics but also can make the plasma more easily develop into a turbulent state. Therefore, plasma turbulence is ubiquitous phenomena and may play important and crucial roles in various plasma processes such as plasma diffusion and transport, plasma acceleration and heating, plasma intermittent structures and energy dissipation.

The solar wind continuously expands from the solar outmost atmosphere (i.e., the solar corona) into the interplanetary space in the form of supersonic and super-Alfvénic plasma flows and directly impacts on the earth's magnetosphere. Observations and theory both show that as its high-Mach expands the solar wind has developed into a strongly turbulent state filled by large-amplitude electromagnetic fluctuations from low frequencies much less than the ion gyrofrequency to high frequencies near the electron Langmuir frequency in the interplanetary space. Therefore, the solar wind not only plays an important linking and transport role in the solar-terrestrial coupling system, but also is a natural laboratory for studying plasma turbulence (Tu and Marsch 1995; Bruno and Carbone 2013).

D.-J. Wu and L. Chen, *Kinetic Alfvén Waves in Laboratory, Space, and Astrophysical Plasmas*, Atmosphere, Earth, Ocean & Space,
https://doi.org/10.1007/978-981-13-7989-5_5

The study of ordinary fluid turbulence can be traced back to the experimental investigation by Osborne (1883) more than one century ago, who observed and investigated experimentally the transition from laminar to turbulent flow inside a pipe and found that this transition depends on a single parameter,

$$R_e = \frac{UL}{\nu}, \tag{5.1}$$

now called the Reynolds number, where U is the characteristic flow velocity, L is the characteristic length scale of the velocity field $\mathbf{u}(\mathbf{r}, t)$, and ν is the viscosity coefficient. The Reynolds number R_e, in fact, represents the relative strength between the non-linear convection term, $(\mathbf{u} \cdot \nabla)\mathbf{u}$, and the viscosity term, $\nu\nabla^2\mathbf{u}$, in the Navier-Stokes equation:

$$\partial_t \mathbf{u} + (\mathbf{u} \cdot \nabla)\,\mathbf{u} = -\rho^{-1}\nabla p + \nu\nabla^2\mathbf{u}, \tag{5.2}$$

where the incompressible condition of $\nabla \cdot \mathbf{u} = 0$ has been used for the sake of simplicity and ρ and p are the mass density and the kinetic pressure of the fluid, respectively. A larger Reynolds number represents stronger nonlinearity and the turbulent flow usually occurs in the case of high Reynolds numbers, in which the non-linear term dominates the dynamical behavior of the flow.

For the incompressible MHD case of a conducting fluid in a magnetic field, the Navier-Stokes equations may be reduced to

$$\begin{aligned} \partial_t \mathbf{u} + (\mathbf{u} \cdot \nabla)\,\mathbf{u} &= \left(\mathbf{b}' \cdot \nabla\right)\mathbf{b}' - \rho^{-1}\nabla p_t + \nu\nabla^2\mathbf{u} \\ \partial_t \mathbf{b}' + (\mathbf{u} \cdot \nabla)\,\mathbf{b}' &= \left(\mathbf{b}' \cdot \nabla\right)\mathbf{u} + \nu_m\nabla^2\mathbf{b}', \end{aligned} \tag{5.3}$$

where $p_t \equiv p + B^2/2\mu_0$ is the total pressure, $\mathbf{b}' \equiv \mathbf{B}/\sqrt{\mu_0\rho} = \mathbf{v}_A + \mathbf{b}$ the total magnetic field in the velocity form, $\mathbf{v}_A = \mathbf{B}_0/\sqrt{\mu_0\rho}$ the Alfvén velocity, $\mathbf{B}_0 = B_0\hat{z}$ the uniform ambient magnetic field, $\mathbf{b} \equiv (\mathbf{B} - \mathbf{B}_0)/\sqrt{\mu_0\rho}$ is the fluctuation magnetic field in the velocity form, $\nu_m \equiv 1/\sigma\mu_0$ the magnetic viscosity coefficient, and σ the conductivity. Similar to the Reynolds number R_e, a magnetic Reynolds number that represents the relative strength of the nonlinear convection of the conducting fluid, R_m, can be defined as follows:

$$R_m = \frac{UL}{\nu_m} \sim \frac{v_A L}{\nu_m}, \tag{5.4}$$

In the solar wind, the magnetic Reynolds number can be very large as $R_m > 10^{10}$, implying that the turbulence can be ubiquitous states of solar wind plasma flows.

From the MHD equation (5.3), the plasma flow $\mathbf{u}$ and the magnetic field $\mathbf{b}$ are coupled together each other through the Lorentz force. Introducing the so-called Elsässer variables:

$$\mathbf{z}^{\pm} \equiv \mathbf{u} \pm \mathbf{b}, \tag{5.5}$$

the MHD equation (5.3) can be rewritten as symmetrized form as follows:

$$\left(\partial_t \mp \mathbf{v}_A \cdot \nabla + \mathbf{z}^{\mp} \cdot \nabla\right) \mathbf{z}^{\pm} = -\rho^{-1}\nabla p_t + \nu^{\pm}\nabla^2\mathbf{z}^{\pm} + \nu^{\mp}\nabla^2\mathbf{z}^{\mp}, \tag{5.6}$$

where $2\nu^{\pm} = \nu \pm \nu_m$. The total pressure p_t can be determined by the condition $\nabla \cdot \mathbf{z}^{\pm} = 0$, or given by the solution of the following equation

$$\nabla^2 p_t = -\nabla\nabla \cdot \cdot \mathbf{z}^+\mathbf{z}^-. \tag{5.7}$$

Neglecting the viscous term, the linearization of Eq. (5.6) leads to

$$(\partial_t \mp \mathbf{v}_A \cdot \nabla)\, \mathbf{z}^{\pm} = 0, \tag{5.8}$$

which general solutions with the "traveling wave" form, $\mathbf{z}^{\mp}(\mathbf{r} \mp \mathbf{v}_A t)$, describe Alfvénic fluctuations propagating parallel ("$-$") and antiparallel ("$+$") to the uniform ambient magnetic field $\mathbf{B}_0$, respectively.

Although the experimental investigation of turbulence had started more than one century ago, so far the most important achievement is still the legacy of A. N. Kolmogorov based on some phenomenological considerations and simple dimensional analysis, that is, so-called the Kolmogorov inertial spectrum of turbulence driven by the nonlinear energy cascade. According to the scenario proposed by Kolmogorov (1941), the structurized energy at the injection scale L is transferred by nonlinear interactions, throughout a string of intermediate scales λ, to the small dissipation scale $l_D \ll \lambda \ll L$, and eventually is dissipated at this dissipation scale l_D. According to the first similarity hypothesis by Kolmogorov (1941), the whole dynamical process from the energy injection at the large scale L to the energy dissipation at the small dissipation scale l_D is controlled mainly by the two parameters: (1) the energy transferring rate $\varepsilon_{T\lambda} \sim u_\lambda^2/\tau_{T\lambda} \sim u_\lambda^3/\lambda$ and (2) the energy dissipation rate $\varepsilon_{D\lambda} \sim u_\lambda^2/\tau_{D\lambda} \sim u_\lambda^2\nu/\lambda^2$ (or by their ratio, that is, the local Reynolds number $R_{e\lambda} = \varepsilon_{T\lambda}/\varepsilon_{D\lambda} = u_\lambda\lambda/\nu$), where $\tau_{T\lambda} \sim \lambda/u_\lambda$ and $\tau_{D\lambda} \sim \lambda^2/\nu$ are the characteristic transferring and dissipation times at the scale λ, respectively. The two parameters $\varepsilon_{T\lambda}$ and $\varepsilon_{D\lambda}$ represent the relative strengths of the nonlinear convection and the viscosity, respectively, and in general, both are dependent on the scale λ. In particular, in the large injection scale L one has $\varepsilon_{TL}/\varepsilon_{DL} = UL/\nu = R_e \gg 1$, implying that in the case of large Reynolds numbers, the fluid is unable to dissipate the whole injection energy at the scale L and the excess energy must be transferred to smaller scales $\lambda < L$ until it is dissipated entirely at the dissipation scale $\lambda \sim l_D \ll L$. This is the physical reason for the energy turbulent cascade toward small scales.

For a steady case, the energy injection rate $\varepsilon_{TL} \sim U^3/L$ at the large injection scale L must be balanced by the energy dissipation rate $\varepsilon_{Dl_D} \sim u_{l_D}^2\nu/l_D^2$ at the small dissipation scale l_D, that is,

$$\varepsilon_{TL} \sim \varepsilon_{Dl_D} \Rightarrow \frac{l_D^2}{L^2} \sim \frac{1}{R_e}\frac{u_{l_D}^2}{U^2}. \tag{5.9}$$

In particular, in order to avoid the energy piled up at some scales λ, it is a plausible hypothesis, that is, that second similarity hypothesis by Kolmogorov (1941), that the energy transferring rate $\varepsilon_{T\lambda}$ is independent from the scale λ in the so-called inertial scale region of $L \gg \lambda \gg l_D$ where the viscosity has a negligible smallness (i.e., the infinite Reynolds number approximation), that is, one has

$$\varepsilon_{T\lambda} = u_\lambda^3/\lambda = \text{constant} \Rightarrow u_\lambda \sim \varepsilon_T^{1/3}\lambda^{1/3}, \tag{5.10}$$

where ε_T is the constant energy transferring rate. Thus, from Eq. (5.9) the effective dissipation scale l_D can be estimated as

$$\frac{l_D}{L} \sim R_e^{-1/2}\frac{u_{l_D}}{U} \sim R_e^{-1/2}\left(\frac{l_D}{L}\right)^{1/3} \Rightarrow l_D \sim R_e^{-3/4}L, \tag{5.11}$$

which depends on the Reynolds number R_e as well as the energy injection scale L and is much less than the injection scale L in the case of high Reynolds numbers.

In the case of high Reynolds numbers $R_e \gg 1$ there can be a very wide inertial region of structurized scales λ (i.e., $L \gg \lambda \gg l_D \sim R_e^{-3/4}L$) in which the flow motions u_λ at almost all scales λ from L to l_D are excited by the nonlinear convection interaction through driving the large-scale structurized energy (U^2) to be fragmented into small-scale structurized energy (u_λ^2) at smaller and smaller scales $\lambda \ll L$ until approaching the effective dissipation scale l_D. In this inertial region, based on the requirement of the constant energy transferring rate ε_T, the energy of fragmented structures at the scale λ, u_λ^2, may be estimated by

$$u_\lambda^2 \sim \varepsilon_T^{2/3}\lambda^{2/3} \Rightarrow u_k^2 \sim \varepsilon_T^{2/3}k^{-2/3}, \tag{5.12}$$

where $k \sim 1/\lambda$ represents the wave number at the fragmented scale λ. Introducing the spectral energy density E_k in the wave number k space, one has

$$\delta u_k^2 = E_k\delta k \Rightarrow E_k \sim \partial_k u_k^2 \sim \varepsilon_T^{2/3}k^{-5/3}. \tag{5.13}$$

This inertial-scale spectrum E_k is the famous Kolmogorov spectrum and the fragmented process of the large-scale structurized energy toward the small-scale structurized energy also is called the nonlinear energy cascade process or the turbulent cascade process, which is driven possibly by nonlinear interactions. However, the physical mechanism of the nonlinear interactions that leads to the energy cascade toward small-scale structures is still unclear.

Based on the theory of classical nonlinear dynamical systems originated by Poincaré at the end of the 19th century, the early studies on the laminar-turbulent transition mechanism were modeled by a sequence of Hopf bifurcations. The first transition scenario was proposed by Landau (1944), in which as the Reynolds number increases the transition is attributed to a infinite series of Hopf bifurcations and each subsequent bifurcation adds a new incommensurate frequency to the flow motions, in consequence, the quasi-periodic motions involving the infinite number of degree of freedom resemble the state of a turbulent flow (Hopf 1948). However, this Landau scenario is very difficult to be realized because the incommensurate frequencies cannot exist without coupling between them (Smale 1967).

Ruelle and Takens (1971) described another scenario of the birth to turbulence, in which after a few, usually three, Hopf bifurcations the flow becomes suddenly chaotic state characterized by a very intricate attracting subset, called a strange attractor. The flow in the chaotic state appears a highly irregular and very complicated topological structure, and its trajectories in phase space look to be unpredictable because the extremely strong dependence of their behavior on initial conditions leads to the exponential divergence of neighboring trajectories. This chaotic feature also is called the "butterfly effect" and represents the deterministic chaos. A few years later, the experimental measurements by Gollub and Swinney (1975) further confirmed that the transition to turbulence in a flow between co-rotating cylinders does not agree with the scenario proposed by Landau (1944), but is more consistent with that described by Ruelle and Takens (1971).

In the Ruelle-Takens scenario, however, the formation of wide continuous spectra of turbulence is still an open problem. The complexity and difficulty of the turbulence study both are beyond imagination, as implied by its name "turbulence". Up to now there are few basic principles that have been definitely established, exactly confirmed, and convincingly accepted in this community. As pointed out by Lumley and Yaglom (2001), "even after 100 years, turbulence studies are still their infancy".

In the case of plasma turbulence, the situation becomes more intricate because a lot of eigenmodes of plasma and electromagnetic field fluctuations can present simultaneously in the plasma, which have various and different dispersion relations and polarization states. For instance, multiple fields and species, different kinetic processes perpendicular and parallel to the local magnetic field, the coupling between various eigenmodes, and so on, all inevitably increase greatly the complexity of plasma turbulent cascade processes. In particular, the collisionless characteristic of major space and astrophysical plasmas leads to the classical scenario of the turbulence transition from the inertial to dissipative regimes, where the injected turbulent energy is converted eventually into the thermal energy of particles by collisional dissipation, no longer being valid. In collisionless plasma environments plasma turbulence with variously continuous spectra has been extensively observed (e.g., in the solar wind, Bale et al. 2005; Bruno and Carbone 2005; Sahraoui et al. 2009, 2010; He et al. 2012a; Salem et al. 2012; Podesta 2013; Goldstein et al. 2015). While the kinetic theory obviously is necessary to understand the physics of collisionless plasma turbulence.

In this chapter, our concerns focus on the kinetic scales of solar wind turbulent spectra, in particular, on the spectral transition from large MHD scales to small kinetic scales of particles, where KAWs can play an important and crucial role. After briefly introducing the anisotropic cascade of MHD turbulence from the MHD scales to the particle kinetic scales in Sects. 5.2 and 5.3 describes some basic concepts of the gyrokinetics, which is proper model for KAWs. Recent progresses of studies of KAW turbulence in the solar wind in theory and observation will be discussed in Sects. 5.4 and 5.5, respectively.

5.2 Anisotropic Cascade of AW Turbulence Towards Small Scales

5.2.1 *Iroshnikov-Kraichnan Theory of AW Turbulence*

Iroshnikov (1963) and Kraichnan (1965) earliest realized that the presence of a magnetic field could significantly influence the turbulence of a conducting fluid, in which the fluctuations of flow and magnetic field, $\mathbf{u}_\lambda$ and $\mathbf{b}_\lambda$ at the scale λ can be related each other via the Lorentz force $\mathbf{j} \times \mathbf{b}$ and the related fluctuations propagate in the form of the eigenmode (i.e., AWs) at the Alfvén velocity v_A along the ambient magnetic field $\mathbf{B}_0$. From the reduced MHD equation (5.6), the nonlinear interaction between the fluctuations $\mathbf{z}^\pm$ is included in the term $\mathbf{z}^\mp \cdot \nabla \mathbf{z}^\pm$, which describes the interaction of the head-on "collision" between opposite propagating fluctuations $\mathbf{z}^+$ and $\mathbf{z}^-$. Similar to the case of ordinary fluids, the energy transferring rate can be estimated by $\varepsilon_{T\lambda} \sim u_\lambda^2/\tau_{T\lambda}$. Unlike the case of ordinary fluids, however, the characteristic time of the head-on interaction $\tau_{T\lambda}$ is no longer simple flow turnover time $\sim\lambda/u_\lambda$, rather is lengthened by a factor $\sqrt{N}$ because of the stochastic nature of the collision process in which $N \sim v_A^2/u_\lambda^2$ is the random collision number for a λ-scale wave to lose memory of its initial state (Dobrowolny et al. 1980). In fact, from Eq. (5.6) the amplitude δu_λ during one collision can be estimated by $\delta u_\lambda \sim (\lambda/v_A)\left(u_\lambda^2/\lambda\right) \sim u_\lambda^2/v_A$, and hence the number of random collisions $N \sim \langle u_\lambda^2 \rangle / \langle (\delta u_\lambda)^2 \rangle \sim v_A^2/u_\lambda^2$, where $\langle * \rangle$ represents the random average. In result, the characteristic time $\tau_{T\lambda} \sim \lambda v_A/u_\lambda^2$ and hence the energy transferring rate $\varepsilon_{T\lambda} \sim u_\lambda^4/v_A\lambda$. For the inertial scale region of λ, adopting the Kolmogorov hypothesis of the constant energy transferring rate, that is, $\varepsilon_{T\lambda} = \varepsilon_T$ independent of the scale λ, one has

$$u_\lambda^2 \sim (\varepsilon_T v_A \lambda)^{1/2} \Rightarrow E_k \sim (\varepsilon_T v_A)^{1/2} k^{-3/2}. \tag{5.14}$$

This is so-called the Iroshnikov-Kraichnan spectrum, an isotropic MHD turbulent spectrum, which was viewed as self-evidently correct for 30 years until the mid-1990s, Sridhar and Goldreich (1994) directly challenged the correctness of the Iroshnikov-Kraichnan theory and proposed the resonant 4-wave coupling scenario for weak turbulence of AWs.

5.2.2 *Goldreich-Sridhar Theory of AW Turbulence*

The ideal MHD or reduced MHD equations (5.3) or (5.6) with $\nu = \nu_m = 0$ have an important feature that they can have steady and stable nonlinear traveling wave solutions with arbitrary amplitude and form of $\mathbf{u}(x, y, z \pm v_A t) = \pm \mathbf{b}(x, y, z \pm v_A t)$, which propagate along (for the sign $-$) or opposite to (for the sign $+$) the magnetic field $\mathbf{B}_0 = B_0\hat{z}$ at the Alfvén velocity $v_A = B_0/\sqrt{\mu_0\rho_0}$ (see e.g., Parker 1979).

The only possible nonlinear coupling is the "collision" between oppositely propagating waves. The Iroshnikov-Kraichnan scenario of AW turbulence describes the "isotropically" random walking process of an AW in the "sea" of other AWs with random phases. Each collision of a λ-scale AW with other oppositely propagating AW (the characteristic collision time $\tau_A \sim \lambda/v_A$) causes only a small variation of its amplitude (i.e., $\delta u_\lambda \sim u_\lambda^2/v_A \ll u_\lambda$), until these variations build up to be comparable with its amplitude so that it has entirely lost memory of its initial state after undergoing successive-N random collisions, implying the effective energy transfer or cascade from the λ-scale AW to smaller scale AWs is finished. Sridhar and Goldreich (1994) argued that this process is equivalent to the resonant 3-wave interaction but does not work because the 3-wave coupling coefficients vanish in isotropic case.

In a magnetized plasma, the parallel and perpendicular variations are caused mainly by the second and third terms of Eq. (5.6), respectively, the former represents the propagation of AW with the characteristic Alfvén time

$$\tau_A \sim \frac{\lambda_z}{v_A} \sim \frac{1}{k_z v_A} \sim \frac{1}{\omega_k}, \tag{5.15}$$

and the latter does the nonlinear interaction between opposite propagation AWs with the characteristic nonlinear time

$$\tau_N \sim \frac{\lambda_\perp}{u_{\lambda_\perp}} \sim \frac{1}{k_\perp u_{\lambda_\perp}}, \tag{5.16}$$

where λ_z (k_z) and $\lambda_\perp$ ($k_\perp$) are the parallel and perpendicular scales (wave numbers) of AWs, respectively. Based on the weak turbulence approximation of $\tau_A \ll \tau_N$ (or i.e., $u_{\lambda_z} \sim u_{\lambda_\perp} \ll v_A$) and the resonant 4-wave interaction description, Sridhar and Goldreich (1994) derived the three dimensional inertial-region energy spectrum with form as follows:

$$\sum_\lambda u_\lambda^2 = \int E(k_z, k_\perp)\frac{\mathrm{d}^3\mathbf{k}}{(2\pi)^3} \Rightarrow E(k_z, k_\perp) \sim \varepsilon_T^{1/3} v_A k_\perp^{-10/3}, \tag{5.17}$$

where the energy transferring rate $\varepsilon_T \sim u_\lambda^2/\tau_T$ can depend only on the parallel wave number k_z and the characteristic transferring time $\tau_T \sim \sqrt{N}\tau_N \sim N\tau_A$.

The three dimensional energy spectrum $E(k_z, k_\perp) \propto k_\perp^{-10/3}$ leads to $u_\lambda^2 \propto k_\perp^{-4/3}$ and hence $N \sim \tau_T/\tau_A \propto k_\perp^{-4/3}$ because of the $k_\perp$-scale independence of ε_T. This implies that as the cascade proceeds to high $k_\perp$ (i.e., short perpendicular scale $\lambda_\perp$), N decreases for fixed k_z. In particular, when N decreases to approaching to the order of unity the weak turbulence approximation will be invalid. In fact, when $N \sim 1$ the inertial region range of the turbulence cascade by the resonant 4-wave interaction will shrink to zero and the turbulence is called the strong turbulence.

Goldreich and Sridhar (1995) introduced the "critical balance" condition to character the strong turbulence of AWs as follows:

$$N \sim 1 \Rightarrow \tau_T \sim \tau_N \sim \tau_A \Rightarrow k_\perp u_{k_\perp} \sim k_z v_A, \tag{5.18}$$

where $k_\perp = 2\pi/\lambda_\perp$. For the inertial region of the strong turbulence, the second Kolmogorov hypothesis of the scale independence of the energy transferring rate, that is, $\varepsilon_T \sim u_{k_\perp}^2/\tau_T \sim u_{k_\perp}^2/\tau_N \sim$ constant leads to

$$u_{k_\perp}^3 \sim \varepsilon_T k_\perp^{-1} \Rightarrow E_{k_\perp} \sim \varepsilon_T^{2/3} k_\perp^{-5/3}, \tag{5.19}$$

to return back to the Kolmogorov spectrum again, but an anisotropic version.

Simultaneously, from the critical balance condition of Eq. (5.18) one has

$$k_z v_A \sim k_\perp u_{k_\perp} \Rightarrow k_z \sim v_A^{-1} \varepsilon_T^{1/3} k_\perp^{2/3}, \tag{5.20}$$

implying the anisotropic cascade of the strong AW turbulence towards small scales, now known as the Goldreich-Sridhar theory of AW turbulence, an anisotropic model of AW turbulent cascade. This scale dependent anisotropy also is the most important result of the Goldreich-Sridhar theory, which predicts that as the turbulent cascade proceeds towards smaller and smaller scales the anisotropy of the wave becomes stronger and stronger, that is, the anisotropic ratio $k_\perp/k_z = \lambda_z/\lambda_\perp \sim v_A \varepsilon_T^{-1/3} k_\perp^{1/3}$ increases with $k_\perp \sim 1/\lambda_\perp$.

The result of the strong turbulence theory by Goldreich and Sridhar (1995) may be universal although it is established on the base of the strong turbulence condition, that is, the critical balance condition. In fact, for an even initially isotropic excitation of small fluctuation of $u_L \ll v_A$ with $k_z \sim k_\perp \sim L^{-1}$ and hence $N \gg 1$ at the injection scale L, the initially weak turbulence develops towards and eventually into the strong turbulence with $N \sim 1$, in which the colliding waves are strongly anisotropic AWs with $k_\perp \gg k_z$ (i.e., $\lambda_\perp \ll \lambda_z$) and faster split into small scale waves at the direction perpendicular than at the parallel to the magnetic field (Goldreich and Sridhar 1995). For the intermediate case between the weak and strong turbulence, Goldreich and Sridhar (1997) further investigated the spectral transition from the weak to strong turbulence and found that the interactions of all orders, i.e., the 3-wave and 4-wave interactions, can have the same contribution to the energy cascade transferring rate, although the 3-wave interaction dominates over all higher order interactions during individual collisions (Ng and Bhattacharjee 1996).

Subsequently, a series of numerical simulations further confirmed the scale dependence of the anisotropy predicted by Goldreich and Sridhar (1995). For instance, Cho and Vishniac (2000) performed directly three-dimensional numerical simulations for MHD turbulence by use of a pseudospectral code to solve the incompressible MHD equation and found that the anisotropy almost is scale independence when the components of the wave vector **k**, k_z and $k_\perp$, are calculated straightforwardly at the parallel and perpendicular to the direction of the large-scale magnetic field, respectively.

However, for the wave numbers $\tilde{k}_z$ and $\tilde{k}_\perp$ measured relative to the local magnetic field direction, the results can be consistent with the scaling law of the anisotropy, that is, $\tilde{k}_z \propto \tilde{k}_\perp^{2/3}$, predicted by Goldreich and Sridhar (1995). This indicates that the energy transfer of the AW turbulence towards small scales exactly is a local spectral cascade process whose anisotropy depends on the local magnetic field.

Maron and Goldreich (2001) simulated incompressible MHD turbulence for the strong magnetic field case with $\gamma \equiv v_A^2/u_L^2 \gg 1$. Similar to the result by Cho and Vishniac (2000), the simulation result by Maron and Goldreich (2001) showed also that the anisotropy increases with increasing $k_\perp$ such that excited modes are confined inside a cone bounded by $k_z \propto k_\perp^{2/3}$ as predicted by Goldreich and Sridhar (1995). However, their results also have a notable discrepancy that the one-dimensional energy spectra determined from their simulations display $E_k \propto k_\perp^{-3/2}$ predicted by the Iroshnikov-Kraichnan theory, rather than $E_k \propto k_\perp^{-5/3}$ expected by the Goldreich-Sridhar model. In particular, the results of numerical simulations of decaying and forced MHD turbulence without and with mean magnetic field presented by Müller et al. (2003) showed a gradual transition of the perpendicular energy spectrum from the Goldreich-Sridhar form at the weak field case of $\gamma \ll 1$ to the Iroshnikov-Kraichnan spectrum at the strong field case of $\gamma \gg 1$.

5.2.3 Boldyrev Theory of AW Turbulence

Motivated by these intriguing numerical findings, Boldyrev (2005, 2006) proposed a new phenomenological model for MHD turbulence, which can lead to an external field dependent energy spectrum such that in the limiting cases of a weak and strong external field the new model can reproduce the Goldreich-Sridhar and Iroshnikov-Kraichnan spectra, respectively. Boldyrev (2005, 2006) further analyzed the nonlinear interaction term in the reduced MHD equation (5.6), $\mathbf{z}^{\mp} \cdot \nabla \mathbf{z}^{\pm}$, and proposed that the nonlinear interaction possibly is reduced by a factor $\left(u_{\lambda_\perp}/v_A\right)^\alpha$ due to the dynamically aligned effect between $\mathbf{u}$ and $\mathbf{b}$, where α is some undetermined exponent with the range $0 \le \alpha \le 1$. Moreover, this dynamic alignment effect can be the scale dependent, that is, increases with the scale decreases, so that turbulent structures in small scales become locally anisotropic in the plane perpendicular to the large-scale ambient magnetic field $\mathbf{B}_0$.

Following Boldyrev (2005, 2006), the nonlinear interaction reduced by the dynamic alignment effect in Eq. (5.6) can be estimated by

$$\mathbf{z}^{\mp} \cdot \nabla \mathbf{z}^{\pm} \sim \frac{u_{\lambda_\perp}}{\lambda_\perp} \left(\frac{u_{\lambda_\perp}}{v_A} \right)^\alpha . \tag{5.21}$$

Thus, the characteristic time of the nonlinear interaction τ_N may be given by

$$\tau_{N\lambda_\perp} \sim \frac{\lambda_\perp}{u_{\lambda_\perp}} \left(\frac{v_A}{u_{\lambda_\perp}} \right)^\alpha . \tag{5.22}$$

Assuming the strong turbulence condition in small scales, that is, the energy transferring time $\tau_T \sim \tau_N$, the constant energy transferring rate condition, $\varepsilon_T \sim u_{\lambda_\perp}^2/\tau_T \sim$ constant, leads to the inertial-region turbulent energy spectrum as follows

$$u_{\lambda_\perp}^2 \sim \varepsilon_T v_A^\alpha \lambda_\perp u_{\lambda_\perp}^{-(1+\alpha)} \Rightarrow u_{k_\perp}^2 \propto k_\perp^{-2/(3+\alpha)} \Rightarrow$$
$$E_{k_\perp} \propto du_{k_\perp}^2/dk_\perp \propto k_\perp^{-(5+\alpha)/(3+\alpha)}. \tag{5.23}$$

The anisotropy of the scaling law can be obtained by the critical balance condition or equivalently by the causality principle (Boldyrev 2005), that is,

$$\frac{\lambda_z}{v_A} \sim \tau_{N\lambda_\perp} \Rightarrow k_z \propto k_\perp^{2/(3+\alpha)}. \tag{5.24}$$

The condition $E_{k_\perp} dk_\perp = E_{k_z} dk_z$ gives the parallel energy spectrum

$$E_{k_z} = E_{k_\perp} \frac{dk_\perp}{dk_z} \propto k_z^{-2}, \tag{5.25}$$

independent of α and hence independent of the external field strength.

In comparison with the numerical simulation results (Cho and Vishniac 2000; Maron and Goldreich 2001; Müller et al. 2003), the undetermined exponent α may be associated with the external field strength parameter $\gamma = v_A^2/u_L^2$. In consequence, the Goldreich-Sridhar spectrum is obtained by $\alpha \to 0$ for the weak field limit of $\gamma \ll 1$ and the Iroshnikov-Kraichanan spectrum is given by $\alpha \to 1$ for the strong field limit of $\gamma \gg 1$. It is difficult, however, that the alignment exponent α is determined by the simple dimensional analysis.

In fact, the dynamic alignment effect of $\alpha \neq 0$ leads to the local anisotropy of AW structures in the plane perpendicular the large-scale mean magnetic field $\mathbf{B}_0$ (i.e., the l direction in Fig. 5.1), as shown in Fig. 5.1 (Boldyrev 2006). The perpendicular wave vector $k_\perp \sim 1/\lambda_\perp$ (i.e., the λ direction in Fig. 5.1) is along the maximum gradient direction of AWs in the plane perpendicular to $\mathbf{B}_0$, also which is approximately perpendicular to the AW field $\mathbf{b}$ (i.e., the ξ direction in Fig. 5.1). Meanwhile, the wave field $\mathbf{b}$ causes the distortion of the field line and the corresponding field line displacement along $\mathbf{b}$, $\lambda_\times$, can be estimated by

$$\lambda_\times \sim \frac{b_{\lambda_\perp}}{v_A} \lambda_z \propto \lambda_\perp^{3/(3+\alpha)}, \tag{5.26}$$

where $b_{\lambda_\perp} \sim u_{\lambda_\perp}$ and the scaling relations (5.23) and (5.24) have been used. Usually one has $\lambda_\times \gg \lambda_\perp$, implying that individual small-scale AWs appear anisotropic turbulent "eddy" structures.

In the case of decaying MHD turbulence, magnetic and velocity fluctuations can approach their configuration so that $\mathbf{u} = \mathbf{b}$ or $\mathbf{u} = -\mathbf{b}$ and hence the nonlinear interaction vanish in the reduce MHD equation (5.6), called the dynamic alignment

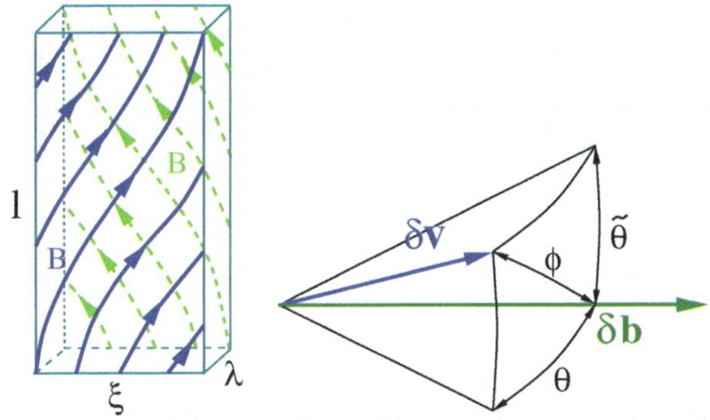

Fig. 5.1 Left: Structures of anisotropic turbulent eddies, in which the large-scale mean magnetic field is in the vertical direction; Right: Sketch of three-dimensional angular alignment relation between velocity and magnetic field fluctuations (reprinted figure with permission from Boldyrev, Phys. Rev. Lett. 96, 115002, 2006, copyright 2006 by the American Physical Society)

effect (Dobrowolny et al. 1980; Grappin et al. 1982; Pouquet et al. 1986). As pointed out by Boldyrev (2006), in the case of driven turbulence the tendency of the dynamic alignment should be preserved although the exact alignment can not be reached because the energy cascade towards mall scales needs to be maintained by non-zero nonlinear interaction. In the steady case, the alignment between the velocity and magnetic field fluctuations should be consistent with maintaining a constant energy transferring rate. This leads to the scaling relation

$$\theta_{\lambda_\perp} \sim \frac{\lambda_\perp}{\lambda_\times} \propto \lambda_\perp^{\alpha/(3+\alpha)}, \tag{5.27}$$

where the scaling relation (5.26) and the alignment approximation $\theta_{\lambda_\perp} \ll 1$ have been used (see the right panel of Fig. 5.1). From the right panel of Fig. 5.1, on the other hand, we have

$$\tilde{\theta}_{\lambda_\perp} \sim \frac{\lambda_\times}{\lambda_z} \propto \lambda_\perp^{1/(3+\alpha)}, \tag{5.28}$$

where the scaling relations (5.24) and (5.26) have been used. Therefore, the misaligned angle between velocity and field fluctuations

$$\phi_{\lambda_\perp} = \sqrt{\theta_{\lambda_\perp}^2 + \tilde{\theta}_{\lambda_\perp}^2} \propto \lambda_\perp^{\alpha/(3+\alpha)} \sqrt{1 + \lambda_\perp^{\alpha}}. \tag{5.29}$$

If we require further the best alignment to be reached, that is, the minimal misaligned angle $\phi_{\lambda_\perp}$, the condition $\theta_{\lambda_\perp} = \tilde{\theta}_{\lambda_\perp}$ should be satisfied, which minimizes the "uncertainty" of the misaligned angle (Boldyrev 2006). This leads to the alignment exponent $\alpha = 1$, which is corresponding to the Iroshnikov-Kraichnan spectrum in the strong field case of $\gamma \gg 1$. Perhaps this is a universal phenomenon in the small-scale AW turbulence spectra. In fact, for the small-scale fluctuations the so-called "local external field" actually is the mean fluctuation in larger scales and the larger-scale fluctuation always is stronger relative to the smaller-scale fluctuation. In consequence, for the small-scale AW turbulence, the local external field always can satisfy the strong field condition of $\gamma > 1$.

In addition, the results of some numerical simulations show that the dissipative structures in MHD turbulence are microcurrent sheets rather than filaments (Biskamp and Müller 2000; Maron and Goldreich 2001; Biskamp 2003). This also is consistent with anisotropic structures of turbulent eddies at small scales due to the dynamic alignment effect in the driven turbulence. These small-scale eddies can be viewed as strongly anisotropic sheets or ribbons stretched along the local mean and fluctuation magnetic field lines described by Boldyrev (2006), rather than filaments along the local mean magnetic field but approximate isotropy in the plane perpendicular to the local mean field expected in the Goldreich-Sridhar theory (Goldreich and Sridhar 1995).

5.2.4 *Observations of Spectral Anisotropy of AW Turbulence*

The solar wind, as supersonic and super-Alfvénic plasma flows outwards from the active solar corona, is ubiquitously in the strong turbulence, in which the majority is dominated by AW turbulence (Tu and Marsch 1995). Also observations of solar wind turbulence further confirmed the theoretical expectation of the anisotropic nature of AW turbulence. For instance, by use of Ulysses satellite observations of the polar fast wind, Horbury et al. (2008) estimated quantitatively the anisotropic power and scaling of magnetic field fluctuation in inertial range MHD turbulence and found for the first time the magnetic power spectra index ~ -2 in the quasi-parallel direction and ~ -1.67 in the quasi-perpendicular direction.

Based on data of high-speed streams measured by the Stereo satellites (A and B) in the ecliptic plane, Podesta (2009) found that overall, the power-law indices resulted from high-speed streams in the ecliptic plane are qualitatively and quantitatively similar to these resulted from the polar fast solar wind by Ulysses observations. However, for high-speed streams in the ecliptic plane the scaling exponent of the perpendicular to parallel power ratio, whose precise value is important for comparisons with turbulence theories, is difficult to determine because there is not sufficient data to obtain reliable measurements at low frequencies where the record length is limited by the relatively short lifetime of high-speed streams in the ecliptic plane. By using the same Ulysses data used by Horbury et al. (2008), Podesta (2009) reproduced the similar results in the inertial range, that is, the spectral index of solar wind turbulence in the inertial range varies continuously from ~ -2 in the quasi-parallel direction to ~ -1.6 in the quasi-perpendicular direction. Meanwhile, he also found the power spectrum has an evident spectral break near $k_\perp \rho_i \sim 1$ when extended the spectrum into the kinetic scales of plasma particles, implying the transition from the inertial to dissipative range, that is, the turbulent cascade transition from AW cascade to KAW cascade process.

Figure 5.2 shows a typical example, in which the upper panel shows clearly the typical characteristic of the variation of the power-law exponent continuously from ~ -2 in the quasi-parallel direction to ~ -1.6 in the quasi-perpendicular direction. While the lower panel presents the perpendicular to parallel power ratio versus the satellite frame frequency ν, which is reversely proportional to the spatial scale, or i.e., directly proportional to the wave number. The scaling law exponent of 0.35 is well consistent with the value $1/3$ expected by the Goldreich-Sridhar theory (or i.e., the weak-field approximation of $\alpha \ll 1$ in the Boldyrev theory). In particular, an evident spectral break occurs between 0.1 and 0.2 Hz near the transition regime from the inertial to dissipative range, implying that the kinetic effects begin to play an

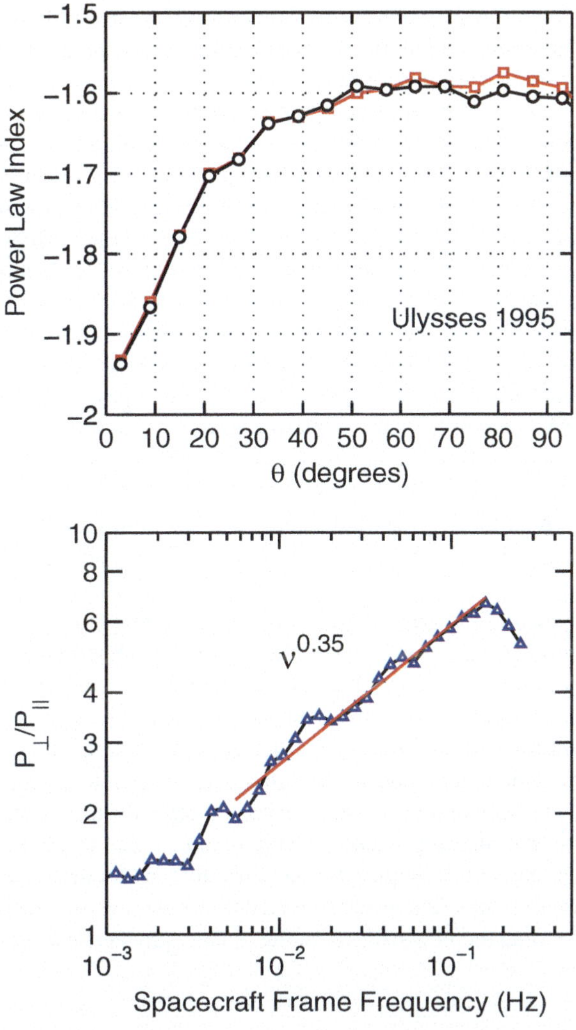

Fig. 5.2 Upper panel: Power-law exponent versus the angle θ between the local mean magnetic field and the mean flow direction (radial direction) for the Ulysses data on 1995 DOY 100 − 130. The red and black curves are obtained by the power averaged with and without directing selection, respectively; Lower panel: The ratio of the perpendicular to parallel power versus the frequency ν in the satellite frame, in which the red line represents the linear least-squares fit with the slope 0.35 ± 0.04 (from Podesta 2009)

important role in the AW turbulent cascade and the AW turbulent cascade starts to turn into the KAW turbulent cascade.

Wicks et al. (2010) further extended the measurements of the power and spectral index anisotropy of solar wind magnetic field fluctuations from scales larger than the outer scale down to the ion gyroscale, covering the entire inertial range. Their results show that the power spectrum above the outer scale of turbulence is approximately isotropic, while at smaller scales the turbulent cascade causes the power spectral anisotropy: close to ~ -2 along and $\sim -5/3$ across the local magnetic field consistent with a critically balanced Alfvénic turbulence. In particular, they found that an enhancement of the parallel power coincident with a decrease in the perpendicular power presents at the smallest scales of the inertial range, that is, close to the ion gyroscale. They conjectured that this is most likely related to energy injection by ion kinetic modes such as the kinetic firehose instability driven by the ion temperature anisotropy and marks the beginning of the kinetic range, that is, the dissipation range of solar wind turbulence.

Introducing the scale-dependent local mean magnetic field, Luo and Wu (2010) directly calculated for the first time the scaling index of the anisotropy of solar wind turbulence based on the scale-dependent local mean field and resulted in $k_\parallel \propto k_\perp^{0.61}$ with the index $\sim$0.61 between 2/3 and 1/2, that is, the weak-field limit of $\alpha \to 0$ and the strong-field limit of $\alpha \to 1$, where α is the "depletion" index of the nonlinear interaction due to the dynamic alignment effect (Boldyrev 2005, 2006). In fact, the above observations and others (see Podesta 2013) all show that the "depletion" of the nonlinear interaction due to the dynamic alignment effect can possibly play an important role in the formation of the observed anisotropic power spectra of solar wind AW turbulence, in which the parallel power spectra have the same index ~ -2, independent from the depletion index α, and the perpendicular power spectral index ranges well between $-5/3$ and $-3/2$, that is, the weak-field limit of $\alpha \to 0$ and the strong-field limit of $\alpha \to 1$ (Boldyrev 2005, 2006).

Moreover, detailed analyses of numerical simulations show that the scale-dependent dynamic alignment also can possibly be responsible for the presence of the intermittency in solar wind AW turbulence, especially in small scales (Beresnyak and Lazarian 2006; Mason et al. 2006; Chandran et al. 2015). In more general unbalanced cases between the energy transferring rate and the nonlinear interaction driven rate, the unbalanced turbulent cascade can possibly lead to the formation of intermittent structures in the unbalanced scales, in which the depletion index of the nonlinear interaction $\alpha < 1$. Besides the dynamic alignment, the unbalance turbulent cascade can be caused by other scale-dependent processes, such as the dissipation and dispersion of plasma waves, which both sensitively depend on the wave scales in small scales approaching to the kinetic scales of plasma particles. These will greatly increase the complexity and difficulty of theoretical studies of AW turbulence in small scales. Therefore, the KAW turbulence, the kinetic-scale AW turbulence, must become the new frontier in the plasma physics and turbulence physics communities.

5.3 Gyrokinetic Description of KAWs

5.3.1 Gyrokinetic Approximations

The theories of incompressible MHD turbulence, from the early isotropic Iroshnikov-Kraichnan scenario (Iroshnikov 1963; Kraichnan 1965), to the later Goldreich-Sridhar theory with two-dimensional anisotropy (Sridhar and Goldreich 1994; Goldreich and Sridhar 1995, 1997), and recently the Boldyrev model with three-dimensional anisotropy (Boldyrev 2005, 2006), all propose an anisotropically turbulent cascade process towards small scales. The anisotropic scaling law, $k_\parallel \propto k_\perp^{2/(3+\alpha)}$ with the parameter $0 < \alpha < 1$, indicates that the energy cascades primarily by developing preferentially small scales perpendicular to the local magnetic field, i.e., with $k_\perp \gg k_z$, as schematically shown in Fig. 5.3. As the AW turbulent cascade towards smaller scales proceeds to approaching to the kinetic scales of particles, such as the ion gyroradius ρ_i or the electron inertial length λ_e, AWs inevitably

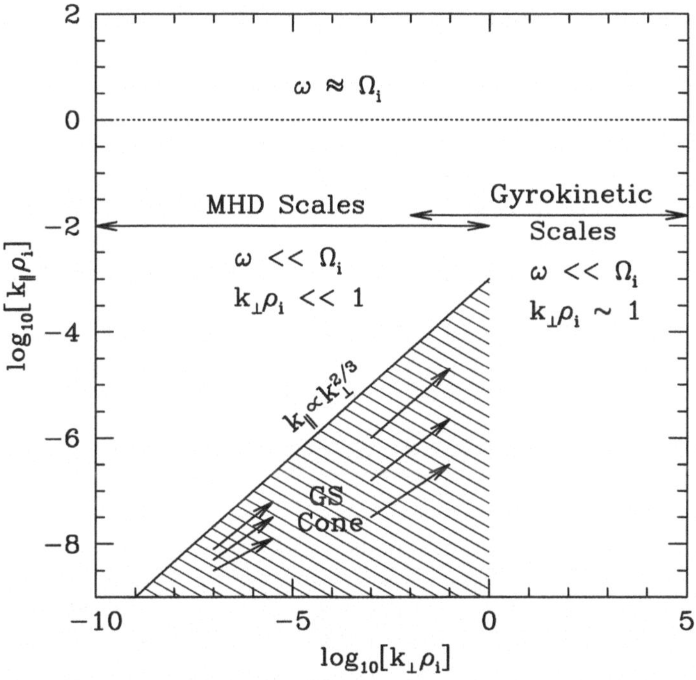

Fig. 5.3 Schematic diagram of the low-frequency, anisotropic AW cascade in wave number space. The horizontal axis is perpendicular wave number; the vertical axis is the parallel wave number, proportional to the frequency. MHD is valid only in the limit $\omega \ll \omega_{ci}$ and $k_\perp \rho_i \ll 1$; gyrokinetic theory remains valid when the perpendicular wave number is of the order of the ion gyroradius, $k_\perp \rho_i \sim 1$. Note that $\omega \to \omega_{ci}$ only when $k_\perp \rho_i \to 1$, so gyrokinetics is applicable for $k_z \ll k_\perp$ (from Howes et al. 2006)

enter the scale range of KAWs where the AW turbulent cascade becomes the KAW turbulent cascade and the MHD theory will be no longer valid.

In the solar wind the mean collision free path of particles $\lambda_c \sim 1$ AU, much larger than the parallel wavelength of AWs, this implies that the solar wind plasma typically is collisionless and the kinetic processes of particles can play an important role in the turbulent cascade of small scale AWs. In particular, due to the intrinsic anisotropy of AW turbulence, low-frequency AWs with frequencies well below the ion gyrofrequency, that is, $\omega \ll \omega_{ci}$ can have short perpendicular wavelengths comparable to the ion gyroradius (i.e., $k_\perp \rho_i \sim 1$), much less than the parallel wavelength of AWs (i.e., $k_\parallel \ll k_\perp$). The motion of individual particles in the small sacle wave fields can be separated well into the fast cyclotron motion around the local magnetic field (with the characteristic time scale $t_f \sim 1/\omega_{ci}$) and the slow drift motion of the gyrocenter (with the characteristic time scale $t_s \sim 1/\omega \gg t_f$). Gyrokinetics is a reduced kinetic theory by averaging over the fast cyclotron motion of charged particles and describes the particle as the charged ring centering at the gyrocenter and moving in the ring-averaged electromagnetic field (Rutherford and Frieman 1968; Frieman and Chen 1982; Howes et al. 2006; Schekochihin et al. 2009). The gyrokinetic approximation retains incompressible AW and the slow magnetosonic wave with the finite Larmor radius effects as KAWs and KSWs (Chen and Wu 2011a, b), the collisionless dissipation due to parallel Landau damping, as well as the Coulomb collisions (if so), but excludes the fast magnetosonic wave and the cyclotron resonance.

Gyrokinetics, a very valuable tool in the study of laboratory plasmas (Chen and Zonca 2016), is particularly suitable for describing the low-frequency, small-scale, and anisotropic AW turbulence, that is, the KAW turbulence, in the solar wind and other astrophysical plasma environments (Howes et al. 2006; Schekochihin et al. 2009). The simplifications of equations for gyrokinetics are mainly based on the low-frequency ($\omega \ll \omega_{ci}$) and small-scale ($\rho_i \ll l_0$) approximations, where l_0 is the scale length of the parallel wavelength $\lambda_\parallel$ much larger than the scale size of the perpendicular wavelength $\lambda_\perp$ (i.e., ρ_i). The former approximation allows to average all quantities over the cyclotron period of particles, and the latter approximation allows to expand all averaged quantities into the series of powers with the small parameter

$$\epsilon = \frac{\rho_i}{l_0} \ll 1. \tag{5.30}$$

Thus, there are three relative scales in the gyrokinetics: the fast gyromotion scale at ω_{ci} and $\rho_i = v_{T_i}/\omega_{ci}$ for the microscopic motions of particles; the intermediate fluctuation scale at ω and l_0 associated with the fluctuations of distribution function and electromagnetic fields in waves; and the slow equilibrium scale at T and L connected to the system macroscopic processes such as heating and inhomogeneity, that is, the outer or injection scale for the turbulence system. For their relative orders one has:

$$\frac{k_\parallel}{k_\perp} \sim \frac{\rho_i}{l_0} \sim O(\epsilon),$$
$$\frac{\omega}{\omega_{ci}} \sim \frac{k_\parallel v_A}{\omega_{ci}} \sim \frac{k_\perp \rho_i}{\sqrt{\beta_i}}\epsilon \sim O(\epsilon), \tag{5.31}$$

where $\beta_i \sim 1$ (or i.e., $v_{T_i} \sim v_A$) has been assumed for the solar wind plasma environment. Figure 5.4 depicts schematically the relative scaling relation in the gyrokinetics. The perpendicular flow velocity $\mathbf{u}_\perp$ is roughly in order of the $\mathbf{E}\times\mathbf{B}/B_0^2$ drift velocity, that is,

$$u_\perp \sim \frac{\mathbf{E}\times\mathbf{B}}{B_0^2} \sim \frac{E}{b}\frac{b}{B_0} \sim O(\epsilon)v_A \sim O(\epsilon)v_{T_i}. \tag{5.32}$$

If using the critical balance and constant energy transferring rate conditions, for the microscopic kinetic scale ρ_i and the macroscopic injected scale L we can have the scaling relation as follows:

$$\epsilon \sim \frac{u_\perp}{v_A} \sim \left(\frac{\rho_i}{L}\right)^{1/3} \Leftrightarrow \frac{l_0}{L} \sim \epsilon^2, \tag{5.33}$$

where the constant energy transferring rate condition of Eq. (5.19) have been used for the injection and kinetic scales, $\lambda \sim L$ and $\lambda \sim \rho_i$. For the case of the solar wind, we have typically $L \sim 10^8$ km and $\rho_i \sim 10^2$ km, and hence $\epsilon \sim (\rho_i/L)^{1/3} \sim 10^{-2}$ and $l_0 \sim 10^4$ km.

Gyrokinetics is most naturally described in gyrocenter coordinates as shown in Fig. 5.4, where the position of a particle $\mathbf{r}$ and velocity $\mathbf{v}$ are given by

$$\mathbf{r} = \mathbf{R}_s - \omega_{cs}^{-1}\mathbf{v}\times\hat{\mathbf{z}}$$
$$\mathbf{v} = v_z\hat{\mathbf{z}} + v_\perp\left(\cos\theta\hat{\mathbf{x}} + \sin\theta\hat{\mathbf{y}}\right), \tag{5.34}$$

where $\mathbf{R}_s$ is the position of the gyrocenter of a s-species particle, v_z and $v_\perp$ are parallel and perpendicular velocities of the particle, and θ is the gyrophase angle. Gyrokinetic averages all quantities over the gyrophase angle θ associated with particles at a fixed gyrocenter $\mathbf{R}_s$ and all quantities associated with electromagnetic fields at a fixed position $\mathbf{r}$, that is,

$$\langle A(\mathbf{r},\mathbf{v},t)\rangle_{\mathbf{R}_s} = \frac{1}{2\pi}\oint A(\mathbf{R}_s - \omega_{cs}^{-1}\mathbf{v}\times\hat{\mathbf{z}},\mathbf{v},t)d\theta,$$
$$\langle A(\mathbf{R}_s,\mathbf{v},t)\rangle_{\mathbf{r}} = \frac{1}{2\pi}\oint A(\mathbf{r} + \omega_{cs}^{-1}\mathbf{v}\times\hat{\mathbf{z}},\mathbf{v},t)d\theta, \tag{5.35}$$

where the θ integration is done keeping v_z and $v_\perp$ constant.

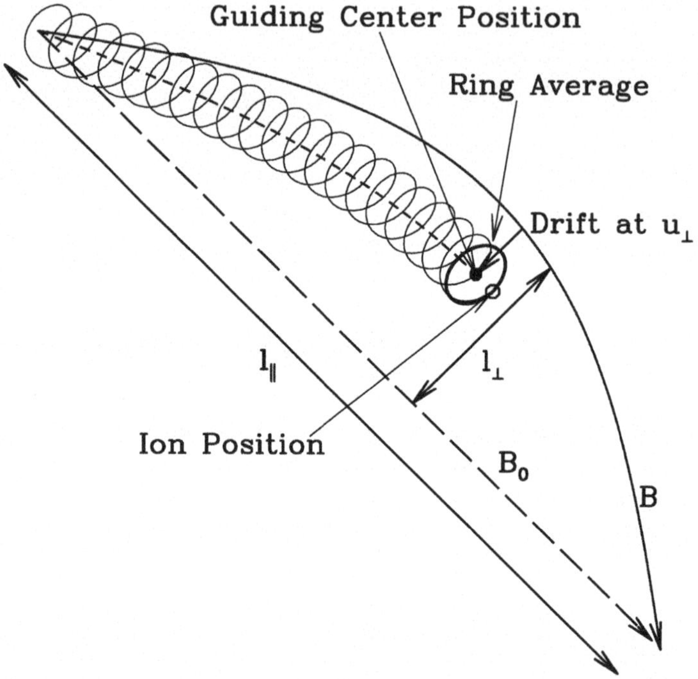

Fig. 5.4 Gyrocenter coordinates and scales in gyrokinetics. The open and filled circles are the particle and its gyrocenter positions, respectively; $l_\perp \sim \lambda_\perp \sim \rho_i$ and _ $\| \sim \lambda_\| \sim l_0$ are the characteristic perpendicular and parallel length scales in gyrokinetics, respectively, and $\mathbf{B}_0$ (dashed line) and $\mathbf{B}$ (solid line) are the ambient and perturbed magnetic field, respectively (from Howes et al. 2006)

5.3.2 Gyrokinetic Equations

In the gyrokinetic approximations, the distribution function of the s-species particles can be expanded into power series of ϵ as follows (Howes et al. 2006):

$$f_s = F_{s0}(v,t)\exp\left(-\frac{q_s\Phi(\mathbf{r},t)}{T_{s0}}\right) + h_s(\mathbf{R}_s, v, v_\perp, t) + \delta f_{s2} + \ldots, \tag{5.36}$$

where $v = \sqrt{v_z^2 + v_\perp^2}$ and the equilibrium distribution function, i.e., the zero-order term, is assumed to be an isotropic Maxwellian distribution:

$$F_{s0}(v,t) = \frac{n_{s0}}{(2\pi)^{3/2}\, v_{T_s}^3}\exp\left(-\frac{v^2}{2v_{T_s}^2}\right). \tag{5.37}$$

The first-order distribution function consists of the Boltzmann term caused by the fluctuation electric potential Φ,

$$\left[\exp\left(-\frac{q_s\Phi(\mathbf{r},t)}{T_{s0}}\right)-1\right]F_{s0}(v,t)\simeq-\frac{q_s\Phi(\mathbf{r},t)}{T_{s0}}F_{s0}(v,t), \tag{5.38}$$

and the "charged ring" distribution $h_s(\mathbf{R}_s, v, v_\perp, t)$.

By gyroaveraging the following kinetic equation,

$$\left\langle\frac{\partial f_s}{\partial t}+\mathbf{v}\cdot\nabla f_s+\frac{q_s}{m_s}\left(\mathbf{E}+\mathbf{v}\times\mathbf{B}\right)\cdot\frac{\partial f_s}{\partial\mathbf{v}}=\left(\frac{\partial f_s}{\partial t}\right)_c\right\rangle_{\mathbf{R}_s}, \tag{5.39}$$

the so-called gyrokinetic equation that governs the ring distribution function $h_s(\mathbf{R}_s, v, v_\perp, t)$ can be derived as follows (Frieman and Chen 1982; Howes et al. 2006; Chen and Zonca 2016):

$$\frac{\partial h_s}{\partial t}+\left\langle\frac{d\mathbf{R}_s}{dt}\right\rangle_{\mathbf{R}_s}\cdot\frac{\partial h_s}{\partial\mathbf{R}_s}=\left\langle\frac{d\mathcal{E}_s}{dt}\right\rangle_{\mathbf{R}_s}\frac{F_{s0}}{T_{s0}}+\left(\frac{\partial h_s}{\partial t}\right)_c, \tag{5.40}$$

where the second term on the right-hand side (i.e., $(\partial h_s/\partial t)_c$) denotes the effect of collisions on the perturbed ring distribution function, called the gyrokinetic collision operator, and the first term,

$$\left\langle\frac{d\mathcal{E}_s}{dt}\right\rangle_{\mathbf{R}_s}\frac{F_{s0}}{T_{s0}}=-\left\langle\frac{\partial f_s}{\partial\mathcal{E}_s}\frac{d\mathcal{E}_s}{dt}\right\rangle_{\mathbf{R}_s}, \tag{5.41}$$

represents the effect of collisionless work done the charged rings by the fluctuation fields, in which $\mathcal{E}_s\equiv m_s v^2/2+q_s\Phi$ is the total energy of the particle and the first adiabatic invariant condition $\langle d\mu_s/dt\rangle_{\mathbf{R}_s}=0$ with $\mu_s=m_s v_\perp^2/2B_0$ has been used.

The drift velocity of the gyrocenter in Eq. (5.40) is given by

$$\left\langle\frac{d\mathbf{R}_s}{dt}\right\rangle_{\mathbf{R}_s}=v_z\hat{\mathbf{z}}-\frac{\partial\langle\Phi\rangle_{\mathbf{R}_s}}{\partial\mathbf{R}_s}\times\frac{\hat{\mathbf{z}}}{B_0}+\frac{\partial\langle\mathbf{v}\cdot\mathbf{A}\rangle_{\mathbf{R}_s}}{\partial\mathbf{R}_s}\times\frac{\hat{\mathbf{z}}}{B_0}, \tag{5.42}$$

where the second and third terms on the right-hand side are the electric and magnetic drifts, respectively, Φ and $\mathbf{A}$ are the scalar and vector electromagnetic potentials with the Coulomb gauge condition of $\nabla\cdot\mathbf{A}=0$ and are related to the perturbed electromagnetic fields by

$$\mathbf{E}=-\nabla\Phi-\frac{\partial\mathbf{A}}{\partial t},$$
$$\delta\mathbf{B}=\nabla\times\mathbf{A}. \tag{5.43}$$

Correspondingly, the fluctuation fields satisfy the Maxwell equations under the gyrokinetic approximations as follows:

$$\begin{aligned}\sum_s q_s \delta n_s &= \sum_s \left(-\frac{q_s^2 n_{s0}}{T_{s0}} \Phi + q_s \int \langle h_s \rangle_{\mathbf{r}} d^3\mathbf{v} \right) = 0, \\ \nabla_\perp^2 A_z &= -\mu_0 \sum_s q_s \int v_z \langle h_s \rangle_{\mathbf{r}} d^3\mathbf{v}, \\ \nabla_\perp \delta B_z &= -\mu_0 \sum_s q_s \int \left\langle (\mathbf{v}_\perp \times \hat{\mathbf{z}}) h_s \right\rangle_{\mathbf{r}} d^3\mathbf{v}, \end{aligned} \tag{5.44}$$

which are, in turn, the quasi-neutrality condition, the parallel and perpendicular components of the Ampere law, respectively.

5.3.3 *Gyrokinetic Dispersion Relation of KAWs*

The coupled Eqs. (5.40), (5.42), and (5.44) establish the base of gyrokinetics. Following Howes et al. (2006), the collisionless linear gyrokinetic dispersion equation can be obtained by the straightforward linearization procedure as follows:

$$\left(\frac{\chi_i}{M_z^2} A - AB + B^2 \right) \left(\frac{2A}{\beta_i} - AD + C^2 \right) = (AE + BC)^2, \tag{5.45}$$

where

$$\begin{aligned} A &= 1 + I_0(\chi_i) e^{-\chi_i} \zeta_i Z(\zeta_i) + \tau_0 \left(1 + I_0(\chi_e) e^{-\chi_e} \zeta_e Z(\zeta_e) \right), \\ B &= 1 - I_0(\chi_i) e^{-\chi_i} + \tau_0 \left(1 - I_0(\chi_e) e^{-\chi_e} \right), \\ C &= c_{i2}(\chi_i) \zeta_i Z(\zeta_i) - c_{e2}(\chi_e) \zeta_e Z(\zeta_e), \\ D &= 2c_{i2}(\chi_i) \zeta_i Z(\zeta_i) + 2\tau_0 c_{e2}(\chi_e) \zeta_e Z(\zeta_e), \\ E &= c_{i2}(\chi_i) - c_{e2}(\chi_e), \end{aligned} \tag{5.46}$$

where $\zeta_s \equiv \omega / \sqrt{2} k_z v_{T_s}$ and $\tau_0 = T_{i0}/T_{e0}$. From the gyrokinetic dispersion equation, the gyrokinetic dispersion relation, M_z^{GK} depends on the three parameters, the perpendicular wave number $k_\perp \rho_i$, the plasma kinetic to magnetic pressure ratio β, and the ion to electron temperature ratio τ_0.

The first and second factors on the left-hand side of the gyrokinetic dispersion equation (5.45) correspond to the KAW and KSW modes, respectively, and the factor on the right-hand side represents the coupling between KAWs and KSWs that is only important at finite ion gyroradius and high β cases. Here we discuss some limiting cases below. In the high-β limit of $\beta_i \gg 1$, by use of the approximation of the plasma dispersion function under the small-argument expansion, $Z(\zeta_s) \simeq i\sqrt{\pi}$, Eq. (5.46) is reduced to

$$\begin{aligned} A &\simeq 1 + \tau_0 + i\sqrt{\pi}\zeta_i \left(I_0(\chi_i)e^{-\chi_i} + \sqrt{Q}\tau_0^{3/2} \right), \\ B &\simeq 1 - I_0(\chi_i)e^{-\chi_i}, \\ C &\simeq i\sqrt{\pi}\zeta_i \left(c_{i2}(\chi_i) - \sqrt{Q\tau_0} \right), \\ D &\simeq i2\sqrt{\pi}\zeta_i \left(c_{i2}(\chi_i) + \sqrt{Q/\tau_0} \right), \\ E &\simeq c_{i2}(\chi_i) - 1. \end{aligned} \tag{5.47}$$

In the limit of $\chi_i \sim 1/\sqrt{\beta_i} \ll 1$, the dispersion equation (5.45) is reduced to

$$\left(B - \frac{\chi_i}{M_z^2} \right) D = E^2 \tag{5.48}$$

with $B \simeq \chi_i$, $D \simeq i2M_z\sqrt{\pi/\beta_i}$, and $E \simeq -3\chi_i/2$. Its solution is

$$M_z = \pm\sqrt{1 - \left(\frac{9}{16}\sqrt{\frac{\beta_i}{\pi}}\chi_i \right)^2} - i\frac{9}{16}\sqrt{\frac{\beta_i}{\pi}}\chi_i. \tag{5.49}$$

In the long wavelength limit $\chi_i \ll 1/\sqrt{\beta_i}$, this leads to the ordinary AW of $M_z = \pm 1$ with weak damping of $\omega_i \ll \omega_r$. When $\chi_i > (16/9)\sqrt{\pi/\beta_i}$, the dispersion relation of Eq. (5.49) leads to a purely damped mode.

On the other hand, for the low-β case of $\beta_i \ll 1$, the gyrokinetic dispersion equation (5.45) is reduced to

$$\left(\frac{\chi_i}{M_z^2}A - AB + B^2 \right) \frac{2A}{\beta_i} = 0. \tag{5.50}$$

The second factor $A = 0$ leads to the ion acoustic wave in the long wavelength limit of $\chi_i \ll 1$, that is,

$$M_z = \pm\sqrt{\frac{\beta_e}{2}} - i\sqrt{\frac{\pi\beta_i}{16}}\tau_0^{-2}e^{-1/2\tau_0}. \tag{5.51}$$

The first factor of Eq. (5.50),

$$\frac{\chi_i}{M_z^2}A - AB + B^2 = 0, \tag{5.52}$$

corresponds to KAWs in low-β plasmas. In the kinetic regime of $Q < \beta_i \ll 1$, expanding the ion and electron dispersion functions in large and small arguments, respectively, one has

$$A \simeq 1 + \tau_0 - I_0(\chi_i)e^{-\chi_i}$$
$$+ iM_z\sqrt{\frac{\pi}{\beta_i}}\left(I_0(\chi_i)e^{-\chi_i - M_z^2/\beta_i} + \sqrt{Q}\tau_0^{3/2}I_0(\chi_e)e^{-\chi_e}\right). \quad (5.53)$$

The resulting dispersion relation may be obtained by the weakly damping approximation of $|\text{Im}M_z| \ll |\text{Re}M_z|$ as follows:

$$\text{Re}M_z = \pm\sqrt{\frac{(1 + \tau_0 - I_0(\chi_i)e^{-\chi_i})\,\chi_i}{\tau_0 I_0(\chi_e)e^{-\chi_e}B}},$$
$$\text{Im}M_z = -\frac{\chi_i e^{\chi_e}}{2I_0(\chi_e)}\sqrt{\frac{\pi}{\beta_i}}\left(\frac{I_0(\chi_i)}{\tau_0^2 I_0(\chi_e)}e^{\chi_e - \chi_i - M_z^2/\beta_i} + \sqrt{\frac{Q}{\tau_0}}\right). \quad (5.54)$$

In the inertial regime of $\beta_i < Q \ll 1$, the expansion of all plasma dispersion functions in large arguments gives approximation

$$A \simeq B - \frac{I_0(\chi_e)e^{-\chi_e}}{M_z^2}\frac{\beta_i}{2Q} + iM_z\sqrt{\frac{\pi}{\beta_i}} \times$$
$$\left(I_0(\chi_i)e^{-\chi_i - M_z^2/\beta_i} + \sqrt{Q}\tau_0^{3/2}I_0(\chi_e)e^{-\chi_e - Q\tau_0 M_z^2/\beta_i}\right). \quad (5.55)$$

This leads to the dispersion relation,

$$\text{Re}M_z = \pm\sqrt{\frac{I_0(\chi_e)e^{-\chi_e}\beta_i\chi_i}{(2Q\chi_i + I_0(\chi_e)e^{-\chi_e}\beta_i)\,B}},$$
$$\text{Im}M_z = -\frac{2Q^2 I_0(\chi_e)e^{-\chi_e}\beta_i\chi_i^3}{(2Q\chi_i + I_0(\chi_e)e^{-\chi_e}\beta_i)^3\,B^2}\sqrt{\frac{\pi}{\beta_i}} \times$$
$$\left(I_0(\chi_i)e^{-\chi_i - M_z^2/\beta_i} + \sqrt{Q}\tau_0^{3/2}I_0(\chi_e)e^{-\chi_e - Q\tau_0 M_z^2/\beta_i}\right). \quad (5.56)$$

For the short-wavelength case of the perpendicular wavelength between the ion and electron gyroradii, $\chi_i \gg 1 \gg \chi_e$, we have $A \simeq 1 + \tau_0\left(1 + i\sqrt{\pi}\zeta_e\right)$, $B \simeq -E \simeq 1$, $C \simeq -i\sqrt{\pi}\zeta_e$, and $D \simeq i2\sqrt{\pi}\zeta_e/\tau_0$. Thus, the solution of the gyrokinetic dispersion equation (5.45) becomes

$$\text{Re}M_z = \pm\frac{\sqrt{2}k_\perp\rho_i}{\sqrt{\beta_i + 2\tau_0/(1 + \tau_0)}},$$
$$\text{Im}M_z = -\frac{\chi_i}{2}\sqrt{\frac{Q\pi}{\tau_0\beta_i}}\left\{1 + \left[\frac{(1 + \tau_0)\,\beta_i}{2\tau_0 + (1 + \tau_0)\,\beta_i}\right]^2\right\}. \quad (5.57)$$

This agrees with the KAW dispersion relation in the short-wavelength limit (Kingsep et al. 1990). In fact, for the short-wavelength case of $k_\perp\rho_i \gg 1$, the low-frequency

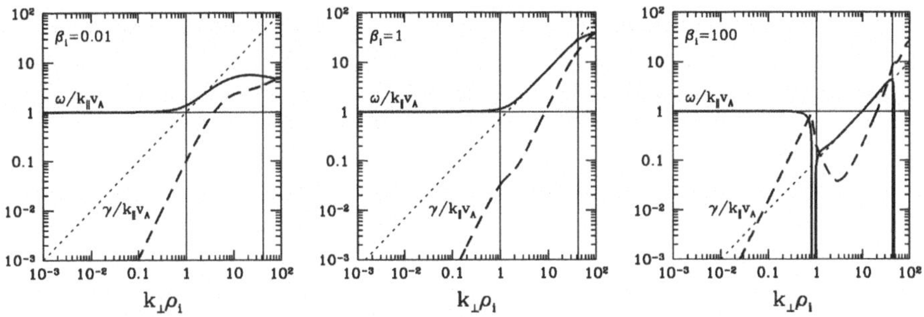

Fig. 5.5 Numerical solutions of the linear gyrokinetic dispersion relation showing the transition from AWs in the inertial range ($k_\perp\rho_i \ll 1$) to KAWs in the ion ($k_\perp\rho_i \sim 1$) and electron ($k_\perp\rho_e \sim 1$) gyrokinetic ranges: Panels, from left to right, show three cases for $\beta_i = 0.01, 1$, and 100, respectively, where bold solid and dashed lines represent the real frequency ω and the damping rate γ, respectively, and dotted lines give the asymptotic KAW solution in the short-wavelength limit case of $k_\perp\rho_i \gg 1$ by Eq. (5.56). All solutions are normalized by the Alfvén frequency $\omega_A \equiv k_z v_A$ (denoted by the horizontal line) and they are functions of $k_\perp$ only in gyrokinetics. Two vertical lines show the ion ($k_\perp\rho_i = 1$) and electron ($k_\perp\rho_e = 1$) gyroradius, respectively. Finally for the sake of simplicity the parameter $\tau_0 = 1$ has been used (from Schekochihin et al. 2009, ©AAS reproduced with permission)

dynamics is dominated by KAWs. When the AW cascade approaches to the wavelength $k_\perp\rho_i \sim 1$, some fraction of the AW energy seeps through to wavelengths smaller than the ion gyroradius and is channeled into a cascade of KAWs. This cascade can extend to smaller wavelengths until approaching the electron gyroradius, $k_\perp\rho_e \sim 1$, at which the KAWs are dissipated by the electron Landau damping.

Figure 5.5 shows the numerical solutions of the linear dispersion relations of the gyrokinetics, which illustrates how the AW becomes a dispersive KAW and the kinetic Landau damping becomes important as the wave scale approaches the ion gyroscale.

5.4 Theory and Simulation of KAWs Turbulence

5.4.1 Basic Properties of KAW Turbulence

When the AW turbulence cascades to small scales comparable to the kinetic scales of plasma particles, such as $k_\perp\rho_i \sim 1$, the AWs become dispersive, that is, KAWs. For the kinetic physics of low-frequency KAW with frequencies $\omega \ll \omega_{ci}$, the gyrokinetics presented in the last section provides a powerful tool and can well describe the low-frequency KAWs in a wide wave-number range from the ion gyroscale ($k_\perp\rho_i \ll 1$) to the electron gyroscale ($k_\perp\rho_e \sim 1$). Based on the gyrokinetics, the last section discusses analytically the basic characteristics of KAWs, including their linear dispersion relations and Landau damping. Besides these small-scale KAWs have strongly anisotropic characteristics, the dispersion of KAWs breaks the original critical balance of the AW turbulence, the polarization of KAWs contains perturbations of plasma density and magnetic field strength, and the collisionless damping of KAWs due to the wave-particle interaction leads to the dissipation of the turbulent wave energy. These factors all can significantly influence the turbulent cascade process and the scaling law of the turbulence spectrum. Moreover, these new characteristics of KAWs all are sensitively dependent on the local plasma parameters. Meanwhile, other modes presented nearly the ion gyroscales, such as whistler, magnetosonic, and ion-Bernstein waves, also could possibly play some role in the dynamical cascading and coupling of turbulent waves. These all inevitably result in the great increase of the complexity and difficulty of KAW turbulence.

In order to simplify further the complexity, another approximation often used in the ion gyroscale range of $\sqrt{Q} \ll k_\perp \rho_i \ll 1/\sqrt{Q}$ is the small-parameter expansion of the electron gyrokinetic equation in powers of $\sqrt{Q}$. This approximation also means

$$k_\perp \rho_e \sim \sqrt{Q} k_\perp \rho_i \ll 1. \tag{5.58}$$

Thus, the gyrokinetics of electrons may be reduced further to the dynamics of a magnetized fluid with the perturbed density δn_e and field-aligned flow velocity u_{ez}, which are given by Schekochihin et al. (2009)

$$\begin{aligned} \frac{\delta n_e}{n_{e0}} &= -\frac{Ze\Phi}{T_{i0}} + \sum_{\mathbf{k}} \frac{e^{i\mathbf{k}\cdot\mathbf{r}}}{n_{i0}} \int J_0(k_\perp \rho_i') h_{i\mathbf{k}} d^3\mathbf{v}, \\ u_{ez} &= \frac{1}{\mu_0 e n_{e0}} \nabla_\perp^2 A_z + \sum_{\mathbf{k}} \frac{e^{i\mathbf{k}\cdot\mathbf{r}}}{n_{i0}} \int v_z J_0(k_\perp \rho_i') h_{i\mathbf{k}} d^3\mathbf{v}, \end{aligned} \tag{5.59}$$

where $Z = q_i/e$ is the ion charge number, $\rho_i' = v_\perp/\omega_{ci}$ and $\mathbf{R}_e \simeq \mathbf{r}$ has been used because of the smallness of the electron gyroradius ρ_e. The perturbed field equations can be written as:

$$\frac{\partial A_z}{\partial t} + \hat{\mathbf{b}} \cdot \nabla \Phi = \hat{\mathbf{b}} \cdot \nabla \left(\frac{T_{e0}}{e} \frac{\delta n_e}{n_{e0}} \right), \tag{5.60}$$

$$\frac{d}{dt} \left(\frac{\delta n_e}{n_{e0}} - \frac{\delta B_z}{B_0} \right) + \hat{\mathbf{b}} \cdot \nabla u_{ez} = -\frac{T_{e0}}{e B_0} \left\{ \frac{\delta n_e}{n_{e0}}, \frac{\delta B_z}{B_0} \right\}, \tag{5.61}$$

and

$$\begin{aligned} \frac{\delta B_z}{B_0} &= \frac{\beta_i}{2} \left(1 + \frac{Z}{\tau_0} \right) \frac{Ze\Phi}{T_{i0}} - \frac{\beta_i}{2} \times \\ &\sum_{\mathbf{k}} \frac{e^{i\mathbf{k}\cdot\mathbf{r}}}{n_{i0}} \int \left(\frac{Z}{\tau_0} J_0(k_\perp \rho_i') + \frac{2 v_\perp^2}{v_{T_i}^2} \frac{J_1(k_\perp \rho_i')}{k_\perp \rho_i'} \right) h_{i\mathbf{k}} d^3\mathbf{v}, \end{aligned} \tag{5.62}$$

where

$$\frac{d}{dt} \equiv \frac{\partial}{\partial t} + \mathbf{u}_E \cdot \nabla = \frac{\partial}{\partial t} + \frac{1}{B_0}\{\Phi, \cdots\},$$
$$\hat{\mathbf{b}} \cdot \nabla \equiv \frac{\partial}{\partial z} + \frac{\delta \mathbf{B}_\perp}{B_0} \cdot \nabla = \frac{\partial}{\partial z} - \frac{1}{B_0}\{A_z, \cdots\}, \tag{5.63}$$

$\mathbf{u}_E = \hat{\mathbf{z}} \times \nabla_\perp \Phi / B_0$, and the Poisson brackets "$\{A, B\}$" are defined by

$$\{A, B\} \equiv \hat{\mathbf{z}} \cdot (\nabla A \times \nabla B) . \tag{5.64}$$

The above Eqs. (5.59)–(5.62), combining the ion gyrokinetic equation (5.40) for $s = i$, give the kinetic description of plasma fluctuations near the ion gyroscale.

For the long-wavelength (or weak dispersion) case of $k_\perp \rho_i \ll 1$, the expansion of the gyrokinetics in $k_\perp \rho_i$ can obtain precisely the similar results with that by the reduced MHD description, in which there is no change in the physical nature until the ion gyrokinetic range of $k_\perp \rho_i \sim 1$. The understanding of the physics of turbulent cascade processes in this transition regime is still very poor. However, on the other side of this transition regime, that is, in the short-wavelength limit of $k_\perp \rho_i \gg 1$, some further simplification is possible because this short-wavelength limit indicates that $k_\perp \rho_i' \gg 1$ and hence the all Bessel functions in Eqs. (5.59) and (5.62) are small. This results in Eqs. (5.59) and (5.62) to be reduced to

$$\frac{\delta n_i}{n_{i0}} = -\frac{Ze\Phi}{T_{i0}} = -\sqrt{\frac{2}{\beta_i}} \frac{\phi}{\rho_i v_A},$$
$$u_{ez} = \frac{1}{\mu_0 e n_{e0}} \nabla_\perp^2 A_z = -\frac{\rho_i \nabla_\perp^2 \psi}{\sqrt{\beta_i/2}},$$
$$\frac{\delta B_z}{B_0} = \frac{\beta_i}{2}\left(1 + \frac{Z}{\tau_0}\right)\frac{Ze\Phi}{T_{i0}} = \sqrt{\frac{\beta_i}{2}}\left(1 + \frac{Z}{\tau_0}\right)\frac{\phi}{\rho_i v_A}, \tag{5.65}$$

where $\phi \equiv \Phi/B_0$ and $\psi \equiv -v_A A_z/B_0$ are the so-called scalar stream and flux functions. The first equation here indicates that the ion response is approximated by the Boltzmann distribution in the electrostatic potential field Φ, the second equation implies that the contribution of the ions on the field-aligned current is ignorable, and the third equation expresses that the magnetic pressure is balanced by the electron and ion kinetic pressures.

Substituting Eq. (5.65) into the field equations (5.60) and (5.61), a closed equation system of the scalar stream and flux functions ϕ and ψ can be obtained as follows:

$$\frac{\partial \psi}{\partial t} = v_A \left(1 + \frac{Z}{\tau_0}\right) \hat{\mathbf{b}} \cdot \nabla \phi,$$
$$\frac{\partial \phi}{\partial t} = -\frac{v_A}{1 + (1 + Z/\tau_0)\,\beta_i/2} \hat{\mathbf{b}} \cdot \nabla \left(\rho_i^2 \nabla_\perp^2 \psi\right). \tag{5.66}$$

This equation system is also referred as the equations of the electron reduced MHD, in which the magnetic field is frozen into the electron flow velocity $\mathbf{v}_e$, while the ions are immobile, that is, $\mathbf{u}_i \sim 0$ (Kingsep et al. 1990). The linear dispersion relation of this electron reduced MHD equation system is given by the dispersion equation (5.57) (instead of τ_0 by τ_0/Z for $Z \neq 1$) and describes KAWs in the short-wavelength limit of $\chi_i \gg 1$. The polarization relations of its two linear eigenmodes can be expressed by

$$\varphi_{\mathbf{k}}^{\pm} = \sqrt{\frac{Z+\tau_0}{\tau_0}\left(1+\frac{Z+\tau_0}{\tau_0}\frac{\beta_i}{2}\right)}\frac{1}{\rho_i}\phi_{\mathbf{k}} \mp k_\perp \psi_{\mathbf{k}}, \tag{5.67}$$

which represent the parallel (i.e., $M_z > 0$ for a single "+") and antiparallel (i.e., $M_z < 0$ for a single "−") propagating KAWs, respectively. In particular, for the electric and magnetic field fluctuations of the KAWs we have

$$\begin{aligned}
\frac{\delta \mathbf{B}_{\mathbf{k}}}{B_0} &= \hat{\mathbf{z}}\sqrt{\frac{1+Z/\tau_0}{2+(1+Z/\tau_0)\beta_i}}\frac{\varphi_{\mathbf{k}}^{+}+\varphi_{\mathbf{k}}^{-}}{2v_A} - i\hat{\mathbf{z}}\times\hat{\mathbf{k}}_\perp\frac{\varphi_{\mathbf{k}}^{+}-\varphi_{\mathbf{k}}^{-}}{2v_A} \\
&= -i\hat{\mathbf{z}}\times\hat{\mathbf{k}}_\perp\frac{k_\perp\psi_{\mathbf{k}}}{v_A} + \hat{\mathbf{z}}\left(1+\frac{Z}{\tau_0}\right)\frac{\phi_{\mathbf{k}}}{\rho_i v_A} \\
\frac{\mathbf{E}_{\perp\mathbf{k}}}{B_0} &= -i\mathbf{k}_\perp\frac{\Phi_{\mathbf{k}}}{B_0} + i\omega_{\mathbf{k}}\frac{\mathbf{A}_{\perp\mathbf{k}}}{B_0} \\
&= \left(-i\mathbf{k}_\perp + \hat{\mathbf{z}}\times\mathbf{k}_\perp\frac{\omega_{\mathbf{k}}}{\omega_{ci}}\frac{\beta_i}{2}\frac{Z+\tau_0}{\tau_0\chi_i}\right)\phi_{\mathbf{k}}.
\end{aligned} \tag{5.68}$$

This indicates that the waves have an elliptically right-hand polarization with an elongated ellipse because $|\omega_{\mathbf{k}}/\omega_{ci}| \ll 1$.

Similarly to the reduced MHD equations, the electron reduced MHD equation system (5.66) with the Elsässer-like variables $\varphi^{\pm}$ in the velocity dimension also allows steady and stable nonlinear traveling wave solutions with arbitrary amplitude, which can be constructed by setting the Poisson brackets to be equal to zero that leads to

$$\begin{aligned}
\{\psi,\ \phi\} = 0 &\Rightarrow \psi = c_1\phi \\
\{\psi,\ \rho_i^2\nabla_\perp^2\psi\} = 0 &\Rightarrow \rho_i^2\nabla_\perp^2\psi = c_2\psi,
\end{aligned} \tag{5.69}$$

where c_1 and c_2 are two undetermined constants. Substituting them into the equation system (5.66), one has

$$c_1^2 = -\frac{1}{c_2}\left(1+\frac{Z}{\tau_0}\right)\left(1+\frac{Z+\tau_0}{\tau_0}\frac{\beta_i}{2}\right), \tag{5.70}$$

implying $c_2 < 0$ for real solutions. In gyrokinetics, in fact, the Poisson bracket nonlinearity vanishes for any monochromatic KAWs in the wave number $\mathbf{k}$ space because the Poisson bracket of two KAWs with wave numbers $\mathbf{k}$ and $\mathbf{k}'$ is directly proportional to $\hat{\mathbf{z}} \cdot (\mathbf{k} \times \mathbf{k}')$. Therefore, any monochromatic linear KAWs can be also an exact nonlinear solution with an arbitrary amplitude and the constant $c_2 = -2k_\perp^2 \rho_i^2$.

However, different from the AW case, the nonlinear interaction of two KAWs can occur not only between counterpropagating KAWs but also between copropagating KAWs because their dispersive nature allows the fast one of the copropagating waves with different wave numbers $k_\perp$ to catch up the slow one and to interact with it.

5.4.2 Power Spectra of KAW Turbulence

Based on the electron reduced MHD equation system (5.66), the scaling law of KAW turbulence can be obtained by combining the constant-flux KAW cascade and the critical balance hypotheses. As the AW turbulence in the inertial range, the critical balance of turbulent cascade is set up by the hypothesis that the parallel correlation length $\lambda_z \sim k_z^{-1}$ is determined by the wave propagating distance in the nonlinear decorrelation time $\tau_{KTk_\perp} \sim k_\perp^{-1} u_E^{-1} \sim k_\perp^{-2} \phi_{k_\perp}^{-1}$, that is, the linear propagating time scale $\tau_L \sim \omega^{-1} \sim \tau_{KTk_\perp}$ (the nonlinear interaction time scale). This leads to

$$\tau_L \sim \tau_{KTk_\perp} \Rightarrow \frac{\sqrt{1+\beta_i}}{\sqrt{2} k_\perp \rho_i k_z v_A} \sim \frac{1}{k_\perp^2 \phi_{k_\perp}}, \tag{5.71}$$

where the scaling relation between the scalar flow and magnetic functions, $\psi_{k_\perp} = c_1 \phi_{k_\perp} \sim \sqrt{1+\beta_i} \phi_{k_\perp} / \rho_i k_\perp$, and $Z = \tau_0 = 1$ have been used. The constant-flux KAW cascade indicates the kinetic energy transferring rate $\varepsilon_{KTk_\perp}$ independent of $k_\perp$, that is,

$$\varepsilon_{KTk_\perp} \sim \frac{k_\perp^2 \psi_{k_\perp}^2}{\tau_{KTk_\perp}} \sim \frac{1+\beta_i}{\rho_i^2} \frac{\phi_{k_\perp}^2}{\tau_{KTk_\perp}} \sim \varepsilon_{KT} = \text{constant}. \tag{5.72}$$

The combination of Eqs. (5.71) and (5.72) leads to the the scaling relations of the scalar potential $\phi_{\mathbf{k}}$ and the anisotropy as follows:

$$\phi_{\mathbf{k}} \sim \frac{\varepsilon_{KT}^{1/3} \rho_i^{2/3}}{(1+\beta_i)^{1/3}} k_\perp^{-2/3};$$
$$k_z \sim \frac{\varepsilon_{KT}^{1/3} (1+\beta_i)^{1/6}}{\rho_i^{1/3} v_A} k_\perp^{1/3}. \tag{5.73}$$

In particular, from the polarization relations in Eq. (5.68) the energy spectral densities of magnetic and electric fields for KAWs can be obtained as follows:

$$E_{Bk_\perp} \sim \frac{d}{dk_\perp}|\phi_{\mathbf{k}}|^2 \sim k_\perp^{-7/3};$$
$$E_{Ek_\perp} \sim \frac{d}{dk_\perp}|k_\perp\phi_{\mathbf{k}}|^2 \sim k_\perp^{-1/3}. \tag{5.74}$$

The results of numerical simulations based on the electron MHD turbulence also showed similar scaling relations (Biskamp et al. 1996, 1999; Cho and Lazarian 2004).

Figure 5.6 illustrates the critical balance for the transition of the KAW turbulent cascade from the low wave number regime of $k_\perp\rho_i \ll 1$ to high wave number regime of $k_\perp\rho_i \gg 1$. The critical balance condition of $\omega = \omega_{\rm nl}$ constrains the turbulent energy flow cascading along the path in the $k_z - k_\perp$ plane, which is given by the solid line in Fig. 5.6. From Fig. 5.6, the KAW turbulent cascading process towards smaller scales has the same anisotropic scaling (i.e., $k_z \propto k_\perp^{2/3}$) with that of the AW turbulence described by the Goldreish-Sridhar theory until the transition point at $k_\perp\rho_i = 1$ is approached. Above this transition point the anisotropy becomes

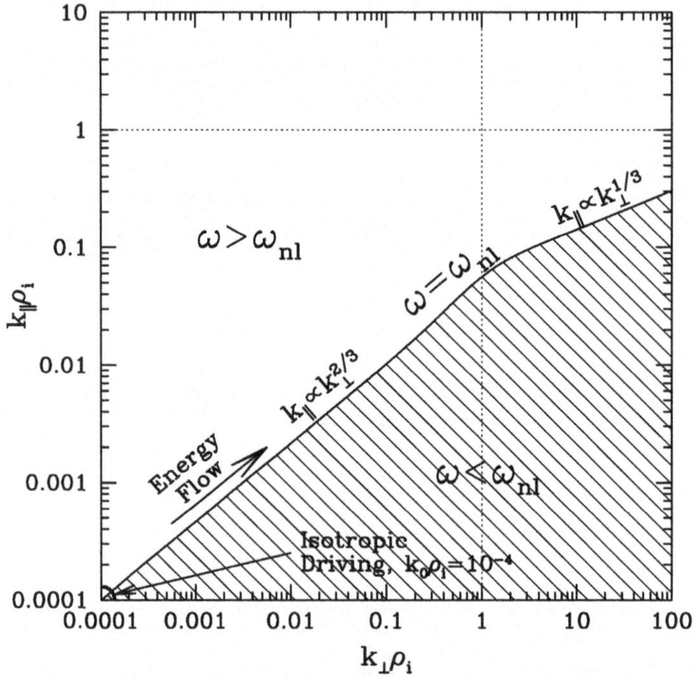

Fig. 5.6 Schematic diagram of the critical balance and anisotropy for the KAW turbulent cascade from low wave numbers of $k_\perp\rho_i \ll 1$ to high wave numbers of $k_\perp\rho_i \gg 1$. The horizontal and vertical axes are the perpendicular ($k_\perp\rho_i$) and parallel ($k_\parallel\rho_i$) wave numbers, respectively. $\omega = M_z k_z v_A$ and $\omega_{\rm nl} = k_\perp v_\perp$ are the linear wave and nonlinear interaction frequencies, respectively. The critical balance is given by $\omega = \omega_{\rm nl}$, and $\omega >$ or $< \omega_{\rm nl}$ correspond to the weak or over-strong turbulence, respectively (from Howes et al. 2008b)

stranger, that is, $k_z \propto k_\perp^{1/3}$, implying that the KAW turbulence faster develops into field-aligned filamentous structures in the kinetic scales of $k_\perp \rho_i > 1$.

5.4.3 AstroGK Simulations of Ion-Scale KAW Turbulence

Based on some fluid codes, the critical balance and its leading to results have been examined and verified extensively in MHD simulations for AW turbulence (Cho and Vishniac 2000; Maron and Goldreich 2001; Müller et al. 2003) and in electron MHD simulations for kinetic turbulence (Biskamp et al. 1996, 1999; Cho and Lazarian 2004). However, these fluid codes can not properly model the effect of wave-particle interactions and the kinetic dissipation in kinetic turbulence. In order to more properly take account of the kinetic effects due to wave-particle interactions in astrophysical plasmas, a kinetic simulation, called the AstroGK code, has been developed based on the gyrokinetic theory (Howes et al. 2006; Schekochihin et al. 2009; Numata et al. 2010). One of benchmarks to check the applicability of numerical simulations is to reproduce analytical results of the linear theory. The applicability of the AstroGK code has been checked extensively to agree well with the linear gyrokinetic theory in the kinetic scales of plasma particles. For example, Fig. 5.7 displays the normalized real frequency ($\omega/k_\parallel v_A$) and damping rate ($\gamma/k_\parallel v_A$)of KAWs produced by the AstroGK (squares) as the function of the normalized perpendicular wave number $k_\perp \rho_i$ and the corresponding analytic results from the linear collisionless gyrokinetic theory (lines) in the typical kinetic scale range from $k_\perp \rho_i = 0.1$ to $k_\perp \rho_i = 10$ (Howes et al. 2008a).

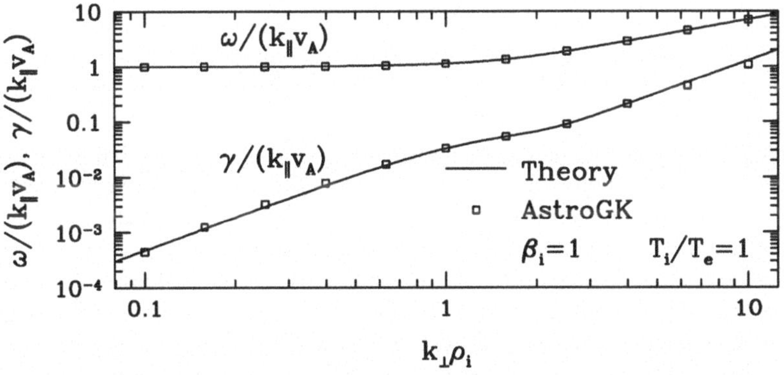

Fig. 5.7 Normalized frequencies $\omega/k_\parallel v_A$ and damping rates $\gamma/k_\parallel v_A$ versus normalized perpendicular wave number $k_\perp \rho_i$ for a plasma with $\beta_i = 1$ and $T_i = T_e$. The AstroGK code (squares) correctly reproduces the analytic results from the linear collisionless gyrokinetic dispersion relation (reprinted the figure with permission from Howes et al., Phys. Rev. Lett. 100, 065004, 2008a, copyright 2008 by the American Physical Society)

From Fig. 5.7, one can find almost exact agreement between the numerical simulation by the AstroGK code and the analytical results from the linear gyrokinetic theory, in which the ordinary AW dispersion relation ($\omega = \pm k_\parallel v_A$) with little damping ($\gamma \ll k_\parallel v_A$) is obtained again for the MHD limit of $k_\perp \rho_i \ll 1$ and for the high wave-number case of $k_\perp \rho_i \gg 1$ the KAW dispersion relation $\omega = \pm k_\parallel v_A k_\perp \rho_i / \sqrt{\beta_i/2 + \tau/(1+\tau)}$ (see Eq. 5.57) with stronger damping can be rediscovered. TenBarge and Howes (2012) further investigated the applicability of the critical balance to kinetic turbulence and found that the results of kinetic simulations can well agree with the theoretical predictions. This implies that the critical balance also is satisfied well in kinetic turbulence and a more strongly anisotropic turbulent cascade may extend into the kinetic scales.

Howes et al. (2008a, 2011) used the AstroGK code more in detail to solve numerically the gyrokinetic and Maxwell coupled equations (5.40), (5.42), and (5.44), in which the equilibrium distributions for electrons (F_{e0}) and protons (F_{i0}) both are assumed are uniform Maxwellian distributions, the real ion-electron mass ratio $m_i/m_e = Q^{-1} = 1836$ is used. The spatial coordinates (x, y) perpendicular to the mean magnetic field ($\mathbf{B}_0$) are dealt with pseudospectrally and the development along the parallel coordinate z is calculated by an upwind finite-difference scheme. Collisional effects are included by a conservative linearized collision operator of consisting of energy diffusion and pitch-angle scattering (Abel et al. 2008; Barnes et al. 2009). For the kinetic simulation of collisionless kinetic turbulence in astrophysical plasmas one of main difficulties is how to model the energy injection at the largest driving scales and how to remove the energy at the smallest dissipating scales because the kinetic scales are much less than the physical driving scale and physical processes of astrophysical plasmas in the kinetic scales have naturally collisionless characteristic.

For a kinetic system to reach a steady state in kinetic simulations, the driving injection power in large scales into the system must be dissipated into heat in small scales. In a collisionless or weak-collision kinetic system, the distributions possibly develop some non-physical small-scale structures near the grid resolution scale of the numerical simulation, which leads to falsely and strongly numerical gradients and disturbs significantly the system evolution, even no reachable steady state. Thus, numerical simulations of kinetic turbulence must include proper collisions as well as sufficient grid resolutions to guarantee the correct relationship between small-scale structures in velocity and position space. In general, it is very difficult that a physical collision operator simultaneously satisfies the grid resolution requirements for ions and electrons. Therefore, a artificially enhanced hypercollisionality (analogous to the hyperviscosity in fluid simulations) often is involved to terminate the cascade of kinetic turbulence at small scales close to the grid scale.

Using the AstroGK code and employing a hypercollisionality with the form of a pitch-angle scattering operator dependent on the collision rate $\nu_h\,(k_\perp/k_{\perp d})^8$, where $k_{\perp d}$ is the grid wave number (i.e., the maximum wave number in the kinetic system), Howes et al. (2008a) presented the first fully electromagnetic gyrokinetic simula-

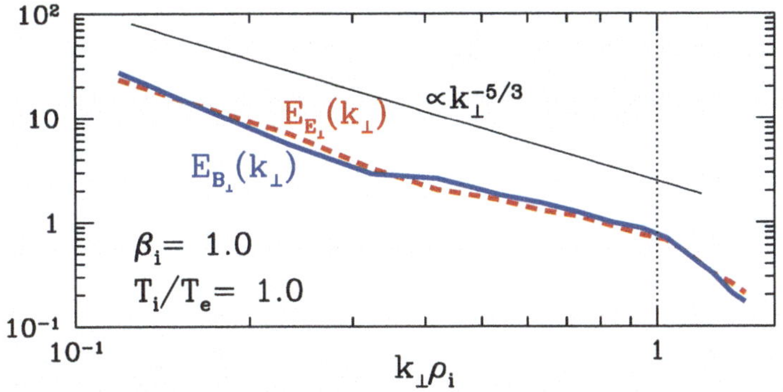

Fig. 5.8 Magnetic (solid line) and electric (dashed line) energy spectra in the MHD regime ($k_\perp \rho_i < 1$) given by the AstroGK simulation. The box size is $L_\perp = 2\pi 10\rho_i$. Electron hypercollisionality is dominant for $k_\perp \rho_i > 1$ denoted by dotted line (reprinted the figure with permission from Howes et al., Phys. Rev. Lett. 100, 065004, 2008a, copyright 2008 by the American Physical Society)

tions of magnetized turbulence in a homogeneous weak-collision plasma near the ion gyroscale (ρ_i). Figure 5.8 shows the normalized perpendicular magnetic (solid line) and electric (dashed line) energy spectra in the inertial range of $k_\perp \rho_i < 1$, where the plasma parameters $\beta_i = T_i/T_e = 1$ have been used. As expected by the Goldreich-Sridhar theory for critically balanced reduced AW turbulence in the inertial regime, these two spectra are nearly coincident and both show a scaling consistent with $\propto k_\perp^{-5/3}$. Moreover, this also, for the first time, demonstrated a MHD turbulence spectrum by a kinetic simulation and further confirmed the applicability of the AstroGK simulation.

From Fig. 5.8, a spectral break at the transition from AW to KAW turbulence can be clearly found at $k_\perp \rho_i \sim 1$. Figure 5.9 shows the perpendicular magnetic (bold solid line) and electric (bold dashed line) and parallel magnetic (bold dash-dotted line) energy spectra obtained by the AstroGK simulation around this transition (Howes et al. 2008a), where the corresponding energy spectra given by a turbulent cascade model based on the assumptions of the local nonlinear energy transfer, the critical balance between linear propagation and nonlinear interaction times, and applicability of linear dissipation rates are displayed in thin lines for comparison (Howes et al. 2008a). From Fig. 5.9, the energy spectra of kinetic turbulence given by the AstroGK simulation exhibited clearly that the evidently spectral break occurs at $k_\perp \rho_i \simeq 2$ and above this breaking point the magnetic and electric energy spectra become steepening ($\sim k_\perp^{-7/3}$) and flattening ($\sim k_\perp^{-1/3}$), respectively. Moreover, the turbulent energy spectra at scales below and above this transition are consistent with the predictions for critically balanced AW (Goldreich and Sridhar 1995) and KAW (Schekochihin et al. 2009) turbulent cascades (see Eq. 5.74), respectively.

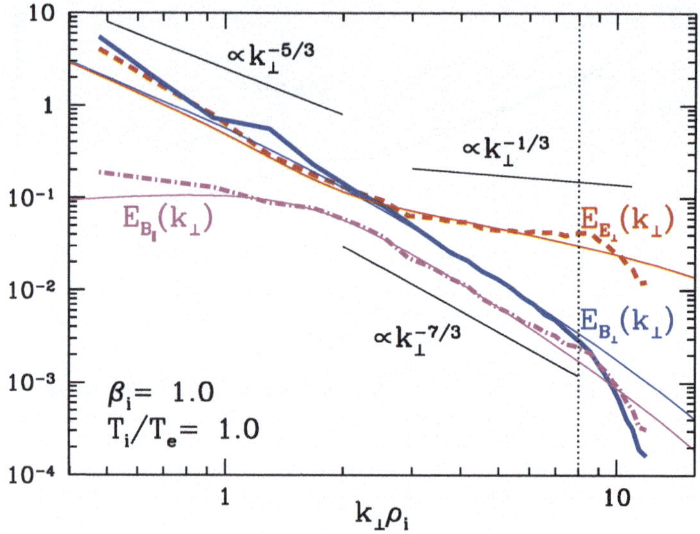

Fig. 5.9 Bold lines: normalized energy spectra for $\delta B_\perp$ (solid line), $\delta E_\perp$ (dashed line), and $\delta B_\parallel$ (dash-dotted line); Thin lines: solution of the turbulent cascade model (Howes et al. 2008b). Dimensions are $(N_x; N_y; N_z; N_\varepsilon; N_\xi; N_s) = (64; 64; 128; 8; 64; 2)$, requiring 5×10^9 computational mesh points, with box size $L_\perp = 5\pi\rho_i$. Electron hypercollisionality is dominant for $k_\perp\rho_i > 8$ denoted by dotted line (reprinted the figure with permission from Howes et al., Phys. Rev. Lett. 100, 065004, 2008a, copyright 2008 by the American Physical Society)

5.4.4 *AstroGK Simulations of Electron-Scale KAW Turbulence*

For the importance of the hypercollisionality involved in Howes et al. (2008a), the collision frequency ν_h has a marginal value for the ions, but a larger ν_h value is required for the electrons, implying that the hypercollisionality can not well model the electron heating in the electron gyroscale ρ_e. In order to extend the AstroGK simulation to the electron gyroscale range, Howes et al. (2011) focused the simulation concerning domain on the kinetic scales between the ion ($k_\perp\rho_i = 1$) and electron gyroradius ($k_\perp\rho_e = 1$), in which the nonlinear transferring of the turbulent energy at scales larger than the largest scales is modeled by six driving modes of parallel currents, $j_{z\mathbf{k}}$, with frequencies $\omega_0 = 1.14\omega_{A0}$ and wave vectors $(k_x\rho_i, k_y\rho_i, k_zL_z/2\pi) = (1, 0, \pm1)$, $(0, 1, \pm1)$, and $(-1, 0, \pm1)$, where $\omega_{A0} \equiv k_{z0}v_A$. The driving amplitudes are determined by the critical balance condition at the largest scales and the energy is injected only at $k_\perp\rho_i = 1$ so that the amplitudes at all higher wave number $k_\perp\rho_i > 1$ can be attributed to the nonlinear turbulent cascade process. The simulation domain is anisotropic and has the sizes $L_\perp = 2\pi\rho_i \simeq 42.8\,(2\pi\rho_e) \ll L_z$ and the plasma parameters $\beta_i = 1$ and $\tau_0 = T_{i0}/T_{e0} = 1$ are used. In addition, a recursive expansion procedure is used to reach a statistically steady state at acceptable numerical calculation cost.

In particular, in order to prevent the non-physical small-scale velocity structures in the velocity space created numerically by wave-particle interactions due to exceeding the velocity space resolution from disturbing the physics of the kinetic damping (such as the Landau damping), they used the collision frequencies of $\nu_i = 0.04\omega_{A0}$ for ions and $\nu_e = 0.5\omega_{A0}$ for electrons, instead of the artificially enhanced hyper-collisionality, to erase these small-scale velocity structures. Thus, all dissipations come from resolved collisionless damping mechanisms and hence the steady state energy spectra at all scales, including the dissipative scales, are obtained by resolved physical processes. Therefore, the results may be more reliably compared directly to observational data.

Figure 5.10 presents the steady state energy spectra of the kinetic simulations (thick lines) for the perpendicular magnetic ($E_{B_\perp}(k_\perp)$, black solid), electric ($E_{E_\perp}(k_\perp)$, green dashed), and parallel magnetic ($E_{B_\parallel}(k_\perp)$, purple dot-dashed) field fluctuations in the kinetic scale range from the ion gyroscale extending to the electron gyroscale. From Fig. 5.10 the most salient feature of the magnetic and electric energy spectra is that they appear excellent power-law spectra over the entire scale range from the ion

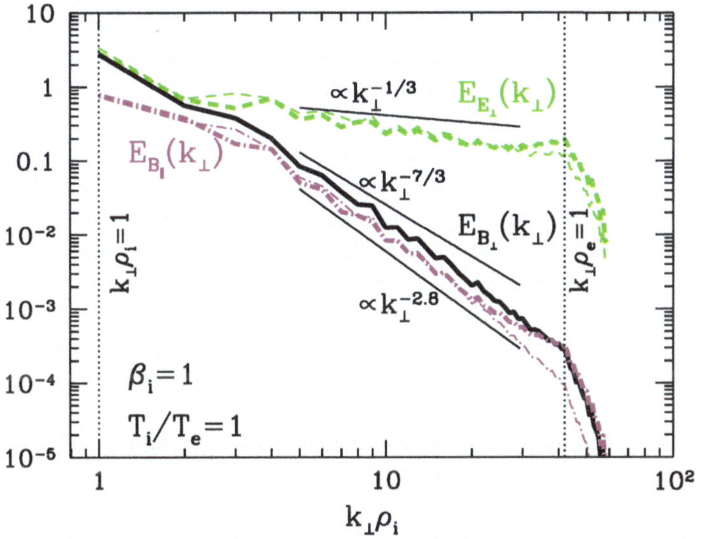

Fig. 5.10 The black thick solid, green thick dashed, and purple thick dot-dashed lines are the energy spectra of the kinetic turbulence for the perpendicular magnetic ($E_{B_\perp}(k_\perp)$), electric ($E_{E_\perp}(k_\perp)$), and parallel magnetic field fluctuations ($E_{B_\parallel}(k_\perp)$), which are given by the kinetic simulations. The green thin dashed and purple thin dot-dashed lines represent the perpendicular electric and parallel magnetic energy spectra predicted theoretically from the simulated perpendicular magnetic energy spectrum based on the polarization relations of the linear colissionless KAWs, which are excellent agreement with the simulation results. The two vertical thin dotted lines denote the positions of the ion and electron gyroradius (i.e., $k_\perp\rho_i$ and $k_\perp\rho_e$), respectively (reprinted the figure with permission from Howes et al., Phys. Rev. Lett. 107, 035004, 2011, copyright 2011 by the American Physical Society)

gyroscale at $k_\perp \rho_i = 1$ to the electron gyroscale at $k_\perp \rho_e = 1$. In general, the role of dissipations (due to collisions or collisionless kinetic damping) is removing the turbulent energy into the heating of plasma particles and a sufficiently strong dissipation at the ion gyroscale may lead to the turbulent cascade process terminating nearly the ion gyroscale. This results in that the turbulent energy spectra exhibit an exponential falloff at the ion gyroscale and can not reach the electron gyroscale (Podesta et al. 2010). However, observations of turbulent energy spectra at the kinetic scales in the solar wind (Bale et al. 2005; Alexandrova et al. 2009; Kiyani 2009; Sahraoui et al. 2009, 2010; Chen et al. 2010; Goldstein et al. 2015) show a continuation of the kinetic turbulent cascade until to the electron gyroscale that is well consistent with the AstroGK simulation results of the KAW turbulence presented in Fig. 5.10.

On the other hand, Fig. 5.10 also shows that the perpendicular electric (green thin dashed line) and parallel magnetic (purple thin dot-dashed line) energy spectra calculated directly by the perpendicular magnetic energy spectrum of the kinetic simulations based on the polarization relations of the linear colissionless KAWs both are excellent agreement with the corresponding simulation results. This indicates that the nature of the KAW turbulence is well consistent with the properties of linear KAWs although these turbulent KAWs are strongly nonlinear waves.

The effect of the Landau damping seems merely to steepen the energy spectra slightly, for instance, the magnetic energy spectrum is closer to $\propto k_\perp^{-2.8}$ rather than $\propto k_\perp^{-7/3}$ predicted by the theory of KAW turbulence without dissipation and the electric energy spectrum also is slightly steeper than the theoretically predicted spectrum $\propto k_\perp^{-1/3}$. One possible explanation for the turbulence energy spectra to extend to the electron gyroscale is that the nonlinear cascade process of the KAW turbulence can sufficiently fast transfer the energy into smaller scale waves before it is exhausted by the linear and nonlinear Landau damping if only these damping rates are lower than the linear wave frequency and hence slower than the nonlinear energy transferring rate. Observations of turbulent energy spectra at kinetic scales in the solar wind (Kiyani 2009; Alexandrova et al. 2009; Chen et al. 2010; Sahraoui et al. 2010) also show very well agreement with the slightly steepened energy spectra and further confirmed the effect of the Landau damping on the kinetic turbulent energy spectra.

Figure 5.11 shows the relative importance of various damping (or heating) processes, where panel (a) compares the nonlinear damping rate $\gamma_{nl} \sim Q_e/E_{B_\perp}$ (dashed line) to the nonlinear energy transferring frequency $\omega_{nl} \sim k_\perp v_\perp (\delta B_\perp/B_0) \sqrt{1 + k_\perp^2 \rho_i^2/2}$ (dotted line), panel (b) is the ion (solid line) and electron (dashed line) collisional heating rates normalized to the generalized energy (including both KAW and ion entropy cascades) transferring rate E, panel (c) presents the relative damping rates of the linear collisionless Landau damping for ions (solid line) and electrons (dashed line) in the gyrokinetic model, and two vertical dotted lines denote the positions of $k_\perp \rho_i = 1$ and $k_\perp \rho_e = 1$ (i.e., $k_\perp \rho_i = 42.8$), respectively. From Fig. 5.11a, the nonlinear energy transferring rate ω_{nl} does dominate the nonlinear damping rate γ_{nl} above the spectral breaking point at $k_\perp \rho_i \sim 1$ until $k_\perp \rho_i \sim 25$, as expected above. Comparing Fig. 5.11b and c, it can be found that there is an evident shift between the peak of the ion collisional heating rate (at the higher wave number $k_\perp \rho_i \sim 20$) and the

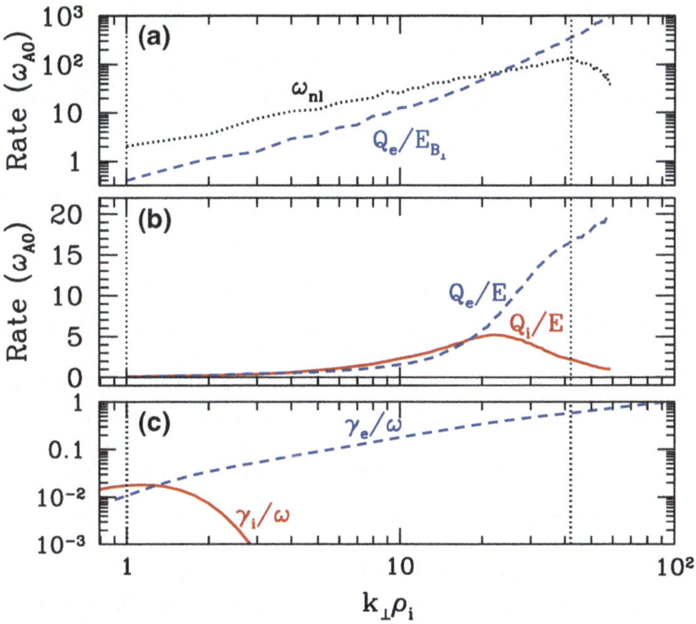

Fig. 5.11 **a** Nonlinear damping rate $Q_e/E_{B_\perp}$ and nonlinear energy transferring frequency ω_{nl}; **b** Ion (solid) and electron (dashed) collisional heating rate; **c** The linear ion (solid) and electron (dashed) Landau damping rates (reprinted the figure with permission from Howes et al., Phys. Rev. Lett. 107, 035004, 2011, copyright 2011 by the American Physical Society)

peak of the linear ion Landau damping rate (at the lower wave number $k_\perp \rho_i \sim 1$). This shift of the peak of ion collisional heating to higher wave number can be attributed to the effect of the ion entropy cascade because the ion entropy cascade can transfer energy to sub-ion gyroscales. Thus, this simulation results also provide the direct evidence of the ion entropy cascade in kinetic turbulence simulation.

In a gyrokinetic simulation covering wider scales from the tail of the MHD range to the electron gyroradius scale, Told et al. (2015) analyzed more in detail nonlinear energy transfer and dissipation in the transition from AW to KAW turbulence and further confirmed the multiscale nature of the dissipation range of the KAW turbulence. Their simulation results show that for typical solar wind parameters at 1 AU, about 30% of the nonlinear energy may be transferred by the nonlinear turbulent cascade process from the tail of the MHD inertial range to the electron gyroradius scale. Their results also indicate that the collisional dissipation could occur across the entire kinetic region and about 70% of the total dissipation is contributed from electron collisions, which exhibit a broad peak around $k_\perp \rho_i \sim 1$–5. On the other hand, the ion free energy can be cascaded to smaller scales and then is dissipated close to the electron gyroradius scale (around $k_\perp \rho_i \sim 25$). This phenomenon is possibly associated with the ion entropy cascade and the fact that $\nu_i \ll \nu_e$ (Tatsuno et al. 2009; Schekochihin et al. 2009; Howes et al. 2011).

5.4.5 Kinetic PIC Simulations of KAW Turbulence

In a three-dimensional particle-in-cell (PIC) simulation of plasma turbulence, resembling the plasma conditions found at kinetic scales of the solar wind, but using a reduced ion to electron mass ratio $m_i/m_e = 64$, Grošelj et al. (2018) also investigated the power spectra of electric and magnetic fluctuations and their ratios with focus on the kinetic region between $k_\perp \rho_i = 1$ and $k_\perp \rho_e = 1$. The simulation domain is a triply periodic box of size $L_\perp^2 \times L_z$ elongated along the mean magnetic field $\mathbf{B}_0 = B_0 \hat{\mathbf{z}}$, where $L_\perp = 16.97\lambda_i$, $L_z = 42.43\lambda_i$, and λ_i is the ion inertial length as usual. The initial fluctuations consist of counterpropagating AWs with different phases and wave numbers $(k_{\perp 0}, 0, \pm k_{z0})$, $(0, k_{\perp 0}, \pm k_{z0})$, and $(2k_{\perp 0}, 0, \pm k_{z0})$, where $k_{\perp 0} = 2\pi/L_\perp$ and $k_{z0} = 2\pi/L_z$. The initial turbulence amplitude is chosen such as to satisfy critical balance. Ions and electrons have an initial Maxwellian velocity distribution with equal isotropic temperature T_0 and uniform density n_0, and hence one has $v_{T_e} = 8v_{T_i} = 0.25c$, $\beta_i = 0.5$, $v_A = 2v_{T_i}$, and $\omega_{pe}/\omega_{ce} = 2.0$. Thus, the dimensionless plasma parameters and the physical setup resemble the plasma conditions inferred from solar wind measurements.

They performed the simulation using the PIC code OSIRIS (Fonseca et al. 2002, 2008) with the spatial resolution $(N_x, N_y, N_z) = (768, 768, 1536)$ and employing, on average, 74 particles per cell per species. Figure 5.12 shows the quasi-steady state power spectra of normalized magnetic (red solid line), electric (green dashed line), and density (blue dotted line) fluctuations in the kinetic region between $k_\perp \rho_i = 1$ and $k_\perp \rho_e = 1$, where the ion (electron) inertial length is denoted by $d_{i(e)}$, instead of λ_i as usual, and λ_D is the Debye length. Although well-defined power-law spectra can

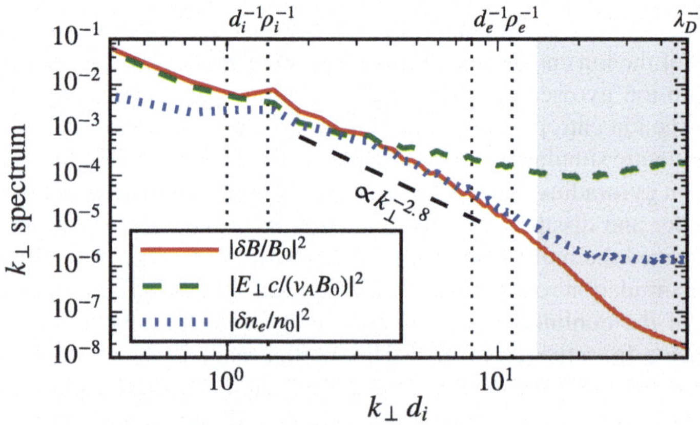

Fig. 5.12 One-dimensional $k_\perp$ spectra of magnetic, perpendicular electric, and density fluctuations at time $t_1 = 0.71 t_A$, where $t_A = L_z/v_A$. The slope of -2.8 is shown for reference. Gray shading is used to indicate the range of scales dominated by particle noise (reprinted the figure with permission from Grošelj et al., Phys. Rev. Lett. 120, 105101, 2018, copyright 2018 by the American Physical Society)

not be established in the reduced kinetic region because of the limitations due to the reduced ion to electron mass ratio, the tendency of the energy spectral distribution still is relatively good agreement with a number of observations (Bale et al. 2005; Alexandrova et al. 2009; Sahraoui et al. 2010; Chen et al. 2013) as well as with the gyrokinetic simulations with the realistic ion to electron mass ratio (Howes et al. 2008a, 2011; TenBarge et al. 2013; Told et al. 2015). From Fig. 5.12, one can find that the electric field spectrum flattens in the kinetic range and separates from the magnetic energy, whereas the density spectrum converges toward a near equipartition with the magnetic spectrum in appropriately normalized units, which has features consistent with solar wind observations and the gyrokinetic simulations.

In particular, the near equipartition among density and magnetic fluctuations in the subion range is a key property of KAWs, differentiating from the weakly compressible whistler waves (i.e., $(|\delta n_e|/n_0)^2 \ll (|\delta B|/B_0)^2$; Gary and Smith 2009; Boldyrev et al. 2013; Chen et al. 2013). In the asymptotic limit of $\rho_i^{-1} \ll k_\perp \ll \rho_e^{-1}$ and $k_\parallel \ll k_\perp$, the analytical approximation of KAWs leads to Howes et al. (2006), Schekochihin et al. (2009), Gary and Smith (2009), Boldyrev et al. (2013):

$$
\begin{aligned}
\frac{(|\delta n_e|/n_0)^2}{(|\delta B|/B_0)^2} &\sim \beta_i + 2\beta_i^2,\\
\frac{(|\delta n_e|/n_0)^2}{\left(|\delta B_\parallel|/B_0\right)^2} &\sim \frac{1}{\beta_i^2},\\
\frac{|\delta B_\parallel|^2}{|\delta B|^2} &\sim \frac{\beta_i}{1+2\beta_i},\\
\frac{|E_\perp^2|}{|\delta B_\perp|^2 v_A^2} &\sim \frac{k_\perp^2\rho_i^2}{4+4\beta_i},
\end{aligned}
\tag{5.75}
$$

where the assumption $Z_i = T_i/T_e = 1$ has been used. Thus, for $\beta_i = 0.5$ the first expression leads to $(|\delta n_e|/n_0)^2 \sim (|\delta B|/B_0)^2$, as shown in Fig. 5.12. To further demonstrate the KAW nature of the measured fluctuations, Fig. 5.13 presents the turbulent spectral ratios in comparison with the above analytical predictions and all spectral ratios exhibit very good agreement between the PIC simulations and the theoretical predictions for KAWs.

Theoretical analyses (Howes et al. 2006; Schekochihin et al. 2009) and numerical simulations (Howes et al. 2008a, 2011; Grošelj et al. 2018) based on gyrokinetics and full kinetics showed that the KAW turbulence has two distinctive features. The one is that the linear physical characteristics in dispersion and polarization of KAWs still continue to exist in nonlinear turbulent KAWs, including their anisotropy of quasi-perpendicular propagation with $k_\perp \gg k_\parallel$ and their electric and magnetic polarized senses, such as $\delta E_\perp/\delta B_\perp$ and $\delta B_\parallel/\delta B_\perp$. The other one is that there are two evident transitions in the energy spectra of the KAW turbulence, which are located near at the wave numbers $k_\perp\rho_i = 1$ and $k_\perp\rho_e = 1$ and caused by the wave-number dependence of the linear Landau damping and the KAW dispersion. In physics, the Landau damping due to the wave-particle interactions for ions (near $k_\perp\rho_i = 1$) and electrons (near

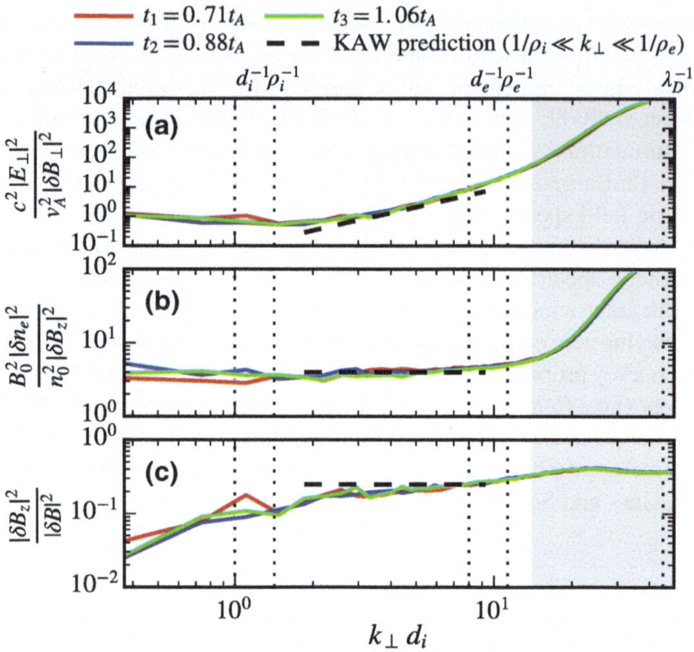

Fig. 5.13 The spectral ratios obtained from the PIC simulation: Solid lines correspond to three different times $t_1 = 0.71t_A$, $t_2 = 0.88t_A$, and $t_3 = 1.06t_A$ and their good coincidence indicates a quasi-steady state. Dashed lines show the analytical predictions by Eq. (5.75) for KAWs (reprinted the figure with permission from Grošelj et al., Phys. Rev. Lett. 120, 105101, 2018, copyright 2018 by the American Physical Society)

$k_\perp \rho_e = 1$) leads to the KAW turbulent energy transferring directly into the resonant ions and electrons, respectively. On the other hand, the wave-number dependence of the KAW dispersion influences the transferring processes of the KAW turbulent energy from large to small kinetic scales. In particular, the KAW turbulence has a steeper energy spectrum $\propto k_\perp^{-7/3}$ (or $\propto k_\perp^{-8/3}$) in the kinetic region between $k_\perp \rho_i = 1$ and $k_\perp \rho_e = 1$, called the "kinetic inertial region", evidently different from the Kolmogorov spectrum $\propto k_\perp^{-5/3}$ in the MHD inertial range of $k_\perp \rho_i \ll 1$.

5.4.6 Fluid-Like Simulations of KAW Turbulence

Different from the gyrokinetic model of kinetic turbulence, which is valid only for the low-frequency regime of $\omega \ll \omega_{ci}$, Boldyrev et al. (2013) analyzed in detail linear modes of electromagnetic fluctuations nearly the ion gyrofrequency with special emphasis on the role and physical properties of KAWs and whistler waves. They found that KAWs exist in the low-frequency regime of $\omega \ll k_\perp v_{T_i}$ and whistler waves occupy a different frequency regime of $k_\perp v_{T_i} \ll \omega \ll k_z v_{T_e}$. The corresponding kinetic damping rates can be given by

$$\frac{\gamma}{\omega_0} = -\sqrt{\frac{\pi}{2}} \frac{1 + (1 + \beta)^2}{(2 + \beta)^{3/2} \beta^{1/2}} k_\perp \rho_e \tag{5.76}$$

for KAWs and

$$\frac{\gamma}{\omega_0} = -\sqrt{\frac{\pi}{2}} \frac{k_\perp}{k} k_\perp \rho_e - \frac{2\sqrt{\pi}|k_z|\lambda_i}{\beta_i^{3/2}} e^{-k_z^2 \lambda_i^2/\beta_i} \tag{5.77}$$

for whistler waves, where $\beta = \beta_i + \beta_e$.

By use of fluid-like numerical simulations of strong KAW turbulence at subion gyroscales, based on the electron MHD equation including a driving force and small dissipation term, they also obtained similar steepened energy spectra $\propto k_\perp^{-8/3}$ for the magnetic and plasma density fluctuations of the KAW turbulence (Boldyrev and Perez 2012; Boldyrev et al. 2013). They proposed, however, that the nonlinear dynamics of KAWs has an important tendency to concentrate magnetic and density fluctuations in two-dimensional current sheet-like structures and this tendency leads to strong spatio-temporal intermittency in the field distributions. In particular, they thought that the energy spectrum steepening of the kinetic turbulence should be attributed mainly to the generation of the strong intermittency in the kinetic scales rather than the kinetic damping and the latter contributes only a small part to the spectrum steepening.

5.4.7 *Effects of Dispersion on KAW Turbulent Spectra*

Based on a phenomenalogical model of the KAW intermittency at kinetic scales, Zhao et al. (2016) further investigated the effect of the intermittency on the turbulent energy spectra of the KAW turbulence below and above the ion gyrofrequency. The generation of intermittent structures in turbulence indicates that the turbulent fluctuations occupy only a fraction of the phase-space volume (Matthaeus et al. 2015) and the occupying probability P_l is of scaling dependent in the form of $P(l) \propto l^{3-D}$, where D is the fractal dimension of the intermittent structures and is a fraction in general cases (Frisch 1995). For instance, one has $D = 0$, 1, and 2 and hence $P(k_\perp) \propto k_\perp^{-3}$, $k_\perp^{-2}$, and $k_\perp^{-1}$ for the ball-like, tube-like, and sheet-like structures, where $k_\perp \sim 2\pi/l$. As KAWs are elongated fluctuations along the local mean magnetic fields (i.e., $k_\perp \gg k_\parallel$), the isotropic ball-like structures can hardly develop. Thus a phenomenological scaling law can be given by $P(k_\perp) \propto k_\perp^{-\kappa}$ with a scaling index between 1 and 2 (i.e., $1 \leq \kappa = 3 - D \leq 2$) for the filling probability (or the occupying probability) of the intermittent structures of KAWs in the kinetic scales. In consequence, the steady-state energy spectra can be obtained by the combination of the scaling law of the filling probability dependent of the intermittent structures and the critical balance condition dependent of the dispersion relation, and the resulting energy spectral index ranges between 7/3 and 3 (Zhao et al. 2016).

As noticed by Zhao et al. (2013), it is very evident that the linear dispersion relation can significantly influence the turbulent energy spectra in the kinetic scales because the critical balance condition and hence the scaling anisotropy both sensitively depend on the dispersion relation. In fact, from the dispersion relation of Eq. (1.20) for KAWs, a general scaling law for the anisotropy of the KAW turbulence can be obtained as follows (Zhao et al. 2013)

$$k_z = \frac{k_{z0}}{k_{\perp 0}^{2/3}} \frac{\left(1+\lambda_e^2 k_\perp^2\right)^{5/6}}{\left(1+\rho_{is}^2 k_\perp^2\right)^{1/6}} k_\perp^{2/3}, \tag{5.78}$$

where k_{z0} and $k_{\perp 0}$ are the driving parallel and perpendicular wave numbers. In the low-$k_\perp$ limit of both $\lambda_e k_\perp$ and $\rho_{is} k_\perp \ll 1$, the Goldreich-Sridhar anisotropic scaling law of $k_z \propto k_\perp^{2/3}$ is recovered. For the kinetic regime KAWs with $\rho_{is}^2 k_\perp^2 \gg 1 \gg \lambda_e^2 k_\perp^2$, the anisotropic scaling law of the KAW turbulence reduces to $k_z \propto k_\perp^{1/3}$ as shown in Eq. (5.73). While for the inertial regime KAWs with $\rho_{is}^2 k_\perp^2 \ll 1 \ll \lambda_e^2 k_\perp^2$ and the transition regime KAWs with $\rho_{is}^2 k_\perp^2 \sim \lambda_e^2 k_\perp^2 \gg 1$, the anisotropic scaling law of the KAW turbulence can read as $k_z \propto k_\perp^{7/3}$ and $k_z \propto k_\perp^2$, respectively, implying that field-aligned small-scale structures can develop faster than cross-field small-scale structures.

Correspondingly, the general steady-state energy spectra of the KAW turbulence with the general dispersion relation of Eq. (1.20) can be derived as follows (Zhao et al. 2013):

$$E_{B_\perp}(k_\perp) \propto \left(1+\lambda_e^2 k_\perp^2\right)^{-1/3} \left(1+\rho_{is}^2 k_\perp^2\right)^{-1/3} k_\perp^{-5/3} \tag{5.79}$$

for the perpendicular magnetic fluctuations,

$$E_{E_\perp}(k_\perp) \propto \left(1+\lambda_e^2 k_\perp^2\right)^{2/3} \left(1+\rho_{is}^2 k_\perp^2\right)^{-4/3} \left(1+\rho_i^2 k_\perp^2\right)^2 k_\perp^{-5/3} \tag{5.80}$$

for the perpendicular electric fluctuations, and

$$E_{E_z}(k_\perp) \propto \left[\alpha_e^2 - \left(1+\rho_i^2 k_\perp^2\right)\right]^2 \left(1+\lambda_e^2 k_\perp^2\right)^{1/3} \left(1+\rho_{is}^2 k_\perp^2\right)^{-5/3} k_\perp^{5/3} \tag{5.81}$$

for the parallel electric fluctuations.

In consideration of the dispersive effect, Voitenko et al. (2011) studied further the turbulent spectra and spectral pattern in the transition regime from the weakly to strongly dispersive KAWs based on the three-wave coupling dynamics (Voitenko 1998a, b). In the weakly dispersive regime of $k_\perp \rho_i < 1$, the nonlinear coupling rates are given by Voitenko (Voitenko 1998a, b)

$$\gamma_{\mathbf{k}}^{NL} \simeq 0.4\omega_{ci} \frac{v_A}{v_{T_i}} \frac{\delta B_{\mathbf{k}}}{B_0} k_\perp^3 \rho_i^3 \tag{5.82}$$

for the interaction of co-propagating KAWs and

$$\gamma_{\mathbf{k}}^{NL} \simeq 0.3\omega_{ci}\frac{v_A}{v_{T_i}}\frac{\delta B_{\mathbf{k}}}{B_0}k_\perp^2\rho_i^2 \tag{5.83}$$

for the interaction of counter-propagating KAWs, which is larger than the coupling rate of co-propagating KAWs for weakly dispersive KAWs with $k_\perp\rho_i < 1$. On the other hand, in the strongly dispersive regime of $k_\perp\rho_i > 1$, the nonlinear coupling rates are (Voitenko 1998a, b)

$$\gamma_{\mathbf{k}}^{NL} \simeq 0.3\omega_{ci}\frac{v_A}{v_{T_i}}\frac{\delta B_{\mathbf{k}}}{B_0}k_\perp^2\rho_i^2 \tag{5.84}$$

for the interaction of co-propagating KAWs and

$$\gamma_{\mathbf{k}}^{NL} \simeq 0.2\omega_{ci}\frac{v_A}{v_{T_i}}\frac{\delta B_{\mathbf{k}}}{B_0}k_\perp^2\rho_i^2 \tag{5.85}$$

for the interaction of counter-propagating KAWs, which is close to the coupling rate of co-propagating KAWs. It is obvious that the nonlinear coupling rates of strongly dispersive KAWs are larger than that of weakly dispersive KAWs.

For sufficiently small perturbed amplitude and sufficiently high linear frequency, the turbulent cascade process can not reach the critical balance and the so-called weakly turbulent cascade process of $\gamma_{\mathbf{k}}^{NL} < \omega_{\mathbf{k}}$ can occur, in which the fluctuations have enough time to set up linear dispersion and polarization relations and conserve their dispersive energy, and hence the energy transferring rate among the fluctuations is relatively slow. Since the nonlinear coupling rate $\gamma_{\mathbf{k}}^{NL}$ is directly proportional to the relative amplitude of the fluctuations $\delta B_{\mathbf{k}}/B_0$, the strongly turbulent cascade process with the critical balance can realize for sufficiently large amplitude or sufficiently low linear frequency. For the weakly turbulent cascade, the turbulent energy spectrum can be estimated from the conservation law of the dispersive energy and the nonlinear coupling rate.

Motivated by the observed multi-kink pattern of kinetic turbulent spectra in the solar wind (Chen et al. 2010; Sahraoui et al. 2010), Voitenko et al. (2011) proposed that the entire energy spectrum from the nondispersive AW turbulence to the dispersive KAW turbulence can have two breaking points due to the variation of the dispersive property of KAWs. The first breaking point occurs at the transition region from the nondispersive AW turbulence to the weakly dispersive KAW turbulence and the second one presents at the transition region between the weakly dispersive KAW turbulence and the strongly dispersive KAW turbulence. Figure 5.14 depicts schematically the pattern of the double-kink energy spectrum. Taking account for possible effects of other dynamical phenomena on the turbulent cascade process, such as the dynamic alignment, the intermittent structure, the kinetic damping, and the turbulent intensity (i.e., the ratio of the nonlinear coupling rate to the linear frequency, $\gamma_{\mathbf{k}}^{NL}/\omega_{\mathbf{k}}$), the spectral indices of the weakly and strongly dispersive KAW

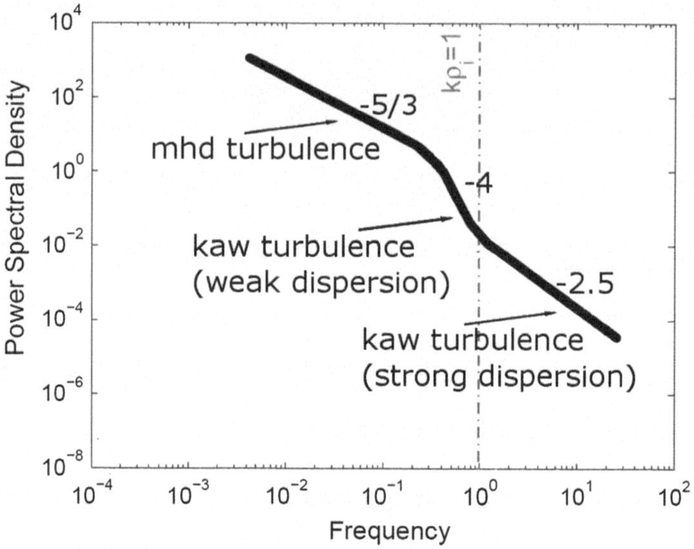

Fig. 5.14 A schematic depiction of the double-kink pattern turbulent spectrum from the nondispersive AW turbulence to weakly and strongly dispersive KAW turbulence. The spectral indices −4 and −2.5 of the weakly and strongly dispersive KAW turbulent spectra should be replaced by −3 and −7/3 for the strong turbulence (from Voitenko et al. 2011)

turbulence can range from −4 to −3 and from −2.5 to −7/3, respectively, dependent on the turbulent intensity.

5.5 Observational Identifications of KAW Turbulence

5.5.1 Initial Identifications of KAW Dispersion and Polarization

The solar wind, a supersonic and super-Alfvénic plasma flow originating from the solar high-temperature corona, develops a strong turbulence during its expanding through the whole heliosphere. The solar wind turbulence spreads a very wide range of scales from MHD scales larger than the size of the Sun to the kinetic scales of plasma particles, such as the ion gyroradius and the electron gyroradius. The dynamics of the solar wind turbulence in the MHD scale range is dominated by incompressible AWs (Coleman 1968; Tu and Marsch 1995) and well described by the Goldreich-Sridhar theory (Goldreich and Sridhar 1995, 1997), which predicts that the AW turbulence drives an energy cascade from large to small scales preferentially in the direction perpendicular to the ambient magnetic field, as shown by simulations (Cho and Vishniac 2000; Maron and Goldreich 2001; Müller et al. 2003), so that the turbulence becomes progressively more anisotropic and the turbulent energy spectrum is concentrated in the wave-vector region of $k_\perp \gg k_\parallel$ as the energy cascade proceeds to higher wave numbers. Moreover, the observed anisotropy of solar wind turbulence in the MHD scales also exhibits an evident wave-number dependence consistent with the theoretical predictions (Horbury et al. 2008; Podesta 2009; Podesta et al. 2010; Wicks et al. 2010; Luo and Wu 2010; Luo et al. 2011; Forman et al. 2011). In consequence, when the energy cascade reaches the kinetic scales it is inevitable that the small-scale kinetic turbulence with $k_\perp \gg k_\parallel$ transfers naturally from the AW turbulence into the KAW turbulence. This implies that the KAW turbulence predominates the solar wind turbulence in the kinetic scales (Howes et al. 2008b; Boldyrev and Perez 2012; Podesta 2013).

In the kinetic scales, however, the dynamics of the KAW turbulence can become much more complicated than that of the MHD AW turbulence because intricate various factors, such as the linear dispersion and polarization relations (Zhao et al. 2013), the kinetic dissipation due to the Landau damping (Howes et al. 2008b, 2010), the dynamical alignment or the intermittency (Boldyrev 2005, 2006; Boldyrev and Perez 2012), and the coupling with other mode waves (Gary and Smith 2009; Mithaiwala et al. 2012; Boldyrev et al. 2013; López 2017; Cerri et al. 2017), interweave each other and influence the dynamics of the KAW turbulence in the kinetic scales. For instance, recent results based on the gyrokinetic simulations (Howes et al. 2011; Told et al. 2015) and the nonlinear wave-wave coupling (Voitenko et al. 2011) show that the turbulent energy spectrum from the MHD to kinetic scales and from the ion to electron gyroradius exhibits the multi-breaking feature. In this section, we present recent observations of the solar wind turbulence in the kinetic scales and discuss their physical nature of KAWs.

Theoretical analyses and numerical simulations in Sect. 5.4 showed that the KAW turbulence has two distinctive features. The one is that the linear physical characteristics in dispersion and polarization of KAWs still continue to have in nonlinear turbulent KAWs, including their anisotropy of quasi-perpendicular propagation with $k_\perp \gg k_\parallel$ and their electric and magnetic polarized senses, such as $\delta E_\perp/\delta B_\perp$ and $\delta B_\parallel/\delta B_\perp$. The other one is that there are two evident transitions in the energy spectra of the KAW turbulence, which are located near at the wave numbers $k_\perp \rho_i = 1$ and $k_\perp \rho_e = 1$ and caused by the wave-number dependence of the linear Landau damping and the KAW dispersion. In physics, the Landau damping due to the wave-particle interactions for ions and electrons leads to the KAW turbulent energy transferring directly into the resonant ions and electrons, respectively, and the dispersion influences the transferring of the KAW turbulent energy from large to small kinetic scales. In particular, the KAW turbulence has a steeper energy spectrum $\propto k_\perp^{-7/3}$ (or $\propto k_\perp^{-8/3}$) in the kinetic inertial region between $k_\perp \rho_i = 1$ and $k_\perp \rho_e = 1$, evidently different from the Kolmogorov spectrum $\propto k_\perp^{-5/3}$ in the MHD inertial range of $k_\perp \rho_i \ll 1$.

The first attempt of measuring the dispersion and polarization of solar turbulent fluctuations in the kinetic scales was made by Bale et al. (2005). Based on the data measured in situ by the four Cluster satellites during the interval 00:07:00–03:21:51 UT (~195 min) on 19 February 2002, when the Cluster was near the apogee ($\sim 19R_E$) of its inclined orbit and spent several hours in the ambient solar wind, Bale et al. (2005) calculated the power, the phase velocity and the polarization of the electric and magnetic field fluctuations as well as their spectral density over the MHD inertial and kinetic dissipative wave number ranges. In order to strengthen the reliability of the calculation results they computed the power spectral density using both the Morlet wavelet and fast Fourier transform (FFT) schemes (Torrence and Compo 1998) and restricted their interpretation to the region where the two results agree.

Figure 5.15 shows their calculation results, where the panel (a) presents the power spectral densities of electric (green) and magnetic (black) field fluctuations computed using the wavelet (upper) and fast Fourier transform (lower) schemes. From the panel (a) of Fig. 5.15, the turbulent energy spectra of electric and magnetic fluctuations

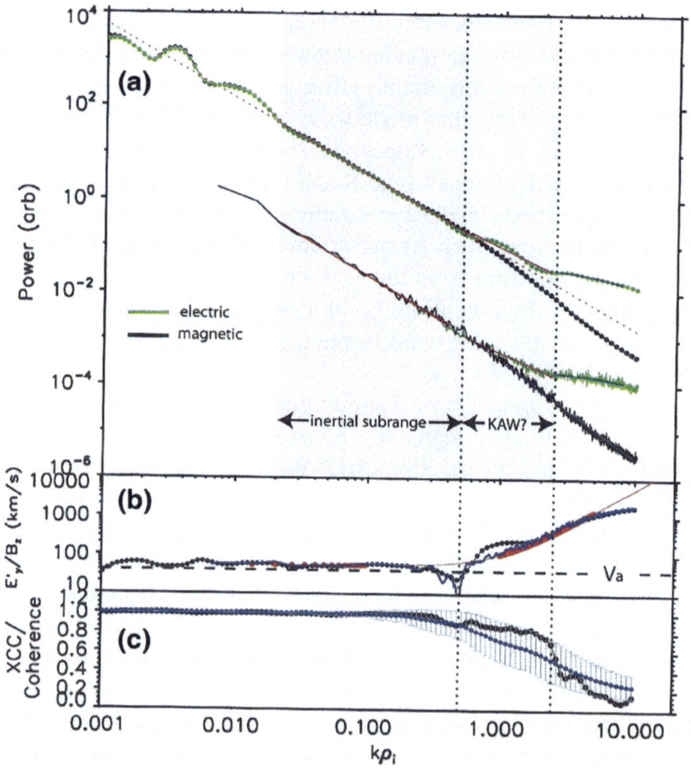

Fig. 5.15 Panel **a** shows the wavelet (upper) and FFT (lower) power spectra of E_y (green) and B_z (black) versus the normalized wave number $k\rho_i$; Panel **b** shows the ratio of the electric to magnetic spectra in the plasma frame, where the average Alfvén speed ($v_A \simeq 40$ km/s) is shown by a horizontal line and the red line is a fitted curve by the KAW linear polarization relation; Panel **c** shows both the cross coherence of E_y with B_z by blue dots with error bars and the correlation between the electric and magnetic power by black dots (reprinted the figure with permission from Bale et al., Phys. Rev. Lett. 94, 215002, 2005, copyright 2005 by the American Physical Society)

agree strikingly each other in the inertial subrange between $k\rho_i \simeq 0.015$ and 0.45 and both show power law behavior with indices of -1.7, which is consistent with the Kolmogorov spectrum of $\propto k^{-5/3}$, as expected by the Goldreich-Sridhar AW turbulence theory, where the wave number is determined by the so-called Taylor hypothesis (Taylor 1938), that is, the Doppler drift, $k = \omega/v_{sw}$, and v_{sw} is the local solar wind velocity much larger than the average Alfvén velocity v_A. At $k\rho_i \simeq 0.45$ the electric and magnetic energy spectra both show evident break and then they start to separate obviously due to the kinetic dispersion effects of KAWs, in which the electric spectrum becomes a flattened power law spectrum with an index of -1.26 and the magnetic spectrum becomes steeper with an index of -2.12.

The panel (b) in Fig. 5.15 shows the ratio of the electric to magnetic fluctuations in the plasma frame, $\delta E'_y/\delta B_z$ (the subscripts y and z denote the y and z components in the GSE (geocentric solar ecliptic) coordinate system, where the black dots and blue line are computed from the wavelet and FFT spectrum, respectively, and well agree. The horizontal line denotes the average Alfvén velocity $v_A \simeq 40$ km/s and the red line is a fitted curve by the KAW polarization relation in the panel (b). It can be found that the wavelet (the black dots) and FFT (the blue line) ratios of electric to magnetic fluctuations nearly coincide and are well consistent with the average Alfvén velocity $v_A \simeq 40$ km/s in the inertial subrange $0.015 < k\rho_i < 0.45$ and in the higher wave number range of $0.5 < k\rho_i < 5$ the FFT ratio also can be well with the fitted curve by the KAW polarization relation. The panel (c) shows both the cross coherence of the electric with magnetic fluctuations (blue dots with error bars) and the correlation between their power spectra (black dots). It is clear that the fluctuations are strongly correlated through the inertial range (with coefficient $\simeq 1$), remain well correlated in the kinetic range between the two breaking points $0.45 < k\rho_i < 2.5$, and begin to lose correlation quickly above the second breaking point at $k\rho_i \simeq 2.5$.

In this analysis, they also compared the observed dispersion to that the prediction of the whistler mode and found that the prediction of the whistler mode is much shallower than the observed dispersion at the kinetic region of $k\rho_i \sim 1$. This strongly suggested the physical nature of KAWs for the observed turbulent fluctuations and indicated that as the turbulent cascade proceeds towards smaller and smaller scales, the AW turbulence in the inertial subrange of $k\rho_i \ll 1$ progressively transfers into the KAW turbulence at $k\rho_i \sim 1$, becoming more electrostatic and eventually damping on the thermal plasma at higher wave numbers of $k\rho_i \gg 1$. In spite of the similar evidence for the KAW nature of solar wind turbulence at the kinetic scales further was confirmed later in other observations (Sahraoui et al. 2009; Kiyani 2009; Alexandrova et al. 2009), there has been a strong debate about the actual nature of the kinetic turbulence in the solar wind, that is, whether or not it is KAW (Bale et al. 2005; Sahraoui et al. 2009) or whistler wave (Gary and Smith 2009; Podesta et al. 2010) turbulence.

5.5.2 Refined Identifications of KAW Dispersion and Spectra

Sahraoui et al. (2010) considered that the difficulty in addressing this problem unambiguously stems from the lack of direct measurements of three-dimensional dispersion relations at the kinetic scales. Indeed, nearly all previous research has used additional assumptions, such as the Taylor frozen-in-flow approximation (i.e., the Taylor hypothesis, Taylor 1938), to infer the spatial properties of the turbulence from the measured temporal ones. The Taylor approximation is valid only if all fluctuation phase velocities are smaller than the solar wind flow velocity v_{sw}, and more importantly, it provides only the wave number spectrum in the direction parallel to the flow $\mathbf{v}_{sw}$. The absence of information on the two other directions prevents a full estimation of the dispersion relations, which compromises the chance of identifying unambiguously the nature of kinetic turbulence.

Using the k-filtering technique (Pincon 1995) on the Cluster multi-satellite data on 10 January 2004 from 06:05 to 06:55 UT, Sahraoui et al. (2010) measured the wave number $\mathbf{k}$ of solar wind fluctuations in the kinetic scales and analyzed the three-dimensional dispersion relation as well as the magnetic energy spectra. This time interval was selected because it has simultaneously several advantages suitable for the measurements of the three-dimensional wave number $\mathbf{k}$ of turbulent fluctuations at the kinetic scales by using the k-filtering technique: (i) The four satellites were located in the solar wind without connection to electron foreshock to avoid as much as possible sharp gradients in the magnetic field components; (ii) Burst mode data (with a sampling 450 Hz) were available, which allows examination of high frequency turbulence; (iii) The magnetic fluctuations had high amplitudes relative to the sensitivity floor of the Cluster search-coil magnetometer; (iv) The Cluster satellites formed a regular tetrahedron configuration, which is a necessary condition for appropriate k-spectra determination; (v) The small separation among the Cluster satellites (by $\sim$ 200 km) is appropriate for exploring subproton scales.

They selected four time periods ($06:06-06:10$; $06:15-06:25$; $06:27-06:41$; $06:50-06:55$, see Table 1 of Sahraoui et al. (2010) for the average plasma parameters during them) and measured the full wave vectors $\mathbf{k}$ for each satellite-frame frequency f_{sc} between 0.04 and 2 Hz by applying the k-filtering technique. Figure 5.16 presents the angles between the measured wave vectors $\mathbf{k}$ and the mean magnetic field $\mathbf{B}_0$ (θ_{kB}) and the solar wind flow $\mathbf{v}_{sw}$ (θ_{kv}). For the selected four time intervals, $06:06-06:10$, $06:15-06:25$, $06:27-06:41$, and $06:50-06:55$, they obtained the propagation angles, on average, $\langle\theta_{kB}\rangle = 86° \pm 6°, 91° \pm 6°, 87° \pm 4°$, and $90° \pm 7°$, respectively. This clearly demonstrated the strongly anisotropic ($k_\perp \gg k_\parallel$) and the quasi-perpendicular propagating ($\theta_{kB} \simeq 90°$) characteristics of the observed turbulent fluctuations at the kinetic scales, which are well consistent with the physical nature of KAWs.

On the other hand, the results in Fig. 5.16 also show that the wave vectors have moderate and relative spread alignment angles, on average, $\langle\theta_{kv}\rangle = 37° \pm 09°, 31° \pm 08°, 14° \pm 10°$, and $37° \pm 11°$ for the four intervals, respectively. Besides the third interval (denoted in red) shows a quasi-alignment (i.e., $\langle\theta_{kv}\rangle = 14° \pm 11°$ with the

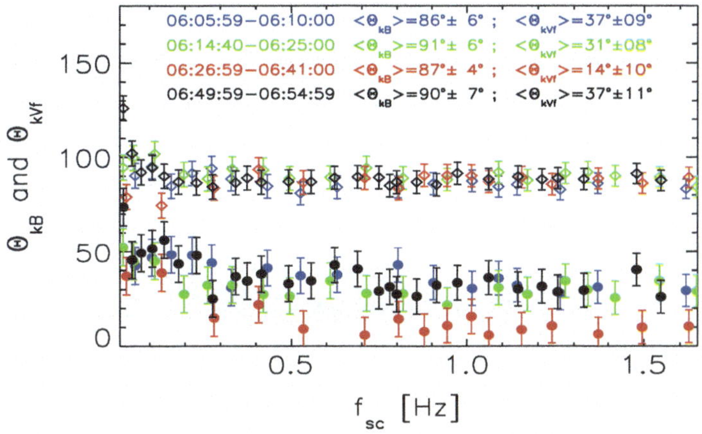

Fig. 5.16 Angles θ_{kB} (diamonds) and θ_{kv} (dots) with related error bars as estimated by using the k-filtering technique (reprinted the figure with permission from Sahraoui et al., Phys. Rev. Lett. 105, 131101, 2010, copyright 2010 by the American Physical Society)

solar wind flow, the finite alignment angles $\langle\theta_{kv}\rangle \sim 40^\circ$ of the wave vectors **k** with respect to the solar wind flow $\mathbf{v}_{sw}$ during the other three intervals indicate that the observed turbulent fluctuations are not frozen in the solar wind flows and propagate at the wave vectors of departure from the flows by a finite angle. This breaks down the Taylor frozen-in-flow condition and hence could lead to significant distortions in the k-spectra if they were calculated by using the Taylor approximation.

On the other hand, using the measured wave vector **k**, the wave frequency in the plasma frame can be given by the satellite-frame frequency and the Doppler frequency shift as follows:

$$\omega = 2\pi f_{sc} - \mathbf{k}\cdot\mathbf{v}_{sw} = 2\pi f_{sc} - kv_{sw}cos\theta_{kv}. \tag{5.86}$$

Figure 5.17 displays the measured dispersion relations compared to the theoretical predictions based on the Vlasov-Maxwellian equations for KAWs (the blue lines) and fast magnetosonic (i.e., whistler) waves (the red lines) at three quasi-perpendicular propagation angles $\theta_{kB} = 85^\circ$, 87°, and 89°. It is very clear and evident from the displayed results in Fig. 5.17 that the measured turbulent fluctuations cascade and propagate following the dispersion relations of the KAW mode but far from those of the fast magnetosonic (or i.e., whistler) mode in the typical kinetic scale range from 0.04 to 2 $k_\perp\rho_i$, covering both the transition and the Kolmogorov inertial regions. This directly demonstrates based on the measured dispersion relation that the observed turbulent fluctuations in the kinetic scales have the physical nature of KAWs, rather than that of whistler waves.

In addition, also Fig. 5.17 shows clearly that the measured wave frequencies remarkably depart from both the proton cyclotron and electron Landau resonant frequencies, but are close to the proton Landau resonant frequency. This indicates

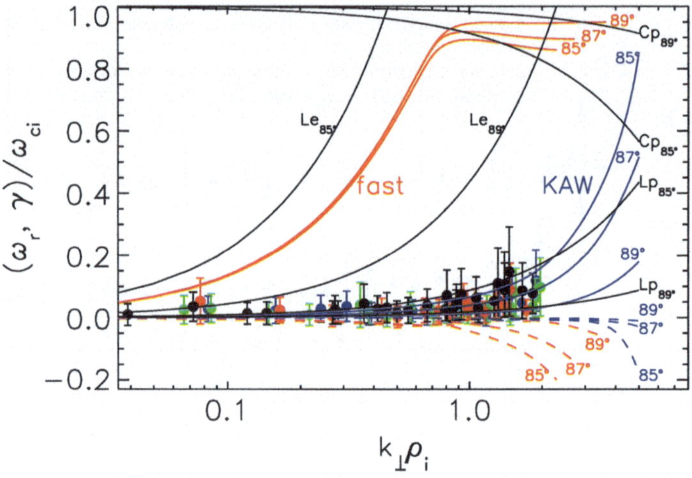

Fig. 5.17 Measured dispersion relations (dots), with estimated error bars, compared to linear solutions of the Maxwell-Vlasov equations for KAWs (the blue lines) and fast magnetosonic (i.e., whistler) waves (the red lines) at three propagation angles $\theta_{kB} = 85°$, $87°$, and $89°$, where the corresponding dashed lines are their damping rates. The black curves ($L_{p(e)}$) are the proton (electron) Landau resonant frequency $\omega = k_\parallel v_{T_{i(e)}}$, and the curves C_p are the proton cyclotron resonant frequency $\omega = \omega_{ci} - k_\parallel v_{T_i}$ (reprinted the figure with permission from Sahraoui et al., Phys. Rev. Lett. 105, 131101, 2010, copyright 2010 by the American Physical Society)

that the proton Landau damping dominates over the electron Landau and proton cyclotron dampings. An immediate consequence of these results is that the damping of turbulence and heating of the protons will arise most likely by the Landau damping and not cyclotron resonances.

In general, for KAWs one has the solar wind flow velocity v_{sw} much larger than the phase velocity ω/k and hence has $f_{sc} \simeq kv_{sw}cos\theta_{kv}/2\pi$ from the frequency transform relation between the plasma-frame frequency (ω) and the satellite-frame frequency $\omega_{sc} = 2\pi f_{sc}$ given by Eq. (5.86). This indicates that the Taylor hypothesis is usually valid for KAWs, that is, KAW fluctuations are approximately "frozen into" the solar wind flow ($\mathbf{v}_{sw}$). Thus, the measured power spectra in the satellite-frame frequency f_{sc} may be interpreted directly as reduced wave number k spectra.

Sahraoui et al. (2010) also calculated the perpendicular and parallel magnetic power spectra of the observed turbulent fluctuations in the satellite-frame frequency f_{sc}. Figure 5.18 shows the results for the interval of $06:14:40-06:25:00$, which are similar to those computed from the other three time intervals, where the parallel (the black line) and perpendicular (the red line) magnetic fluctuations were measured by the flux gate magnetometer for the low-frequency part ($f < 2$ Hz) and by the search-coil magnetometer for the high-frequency part ($f > 2$ Hz) in order to avoid hitting the noise floors of the flux gate magnetometer in the high-frequency part and the search-coil magnetometer in the low-frequency part, and the black dotted line is the in-flight sensitivity floor of the search-coil magnetometer.

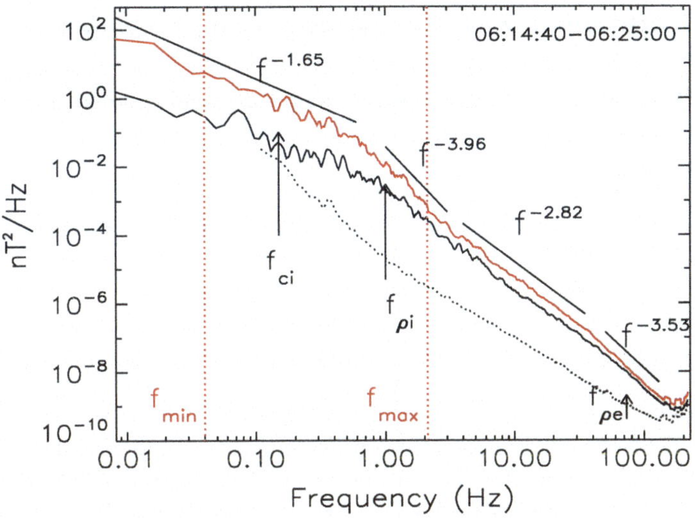

Fig. 5.18 Perpendicular (red curve) and parallel (black curve) magnetic power spectra, where the vertical arrows are the proton gyrofrequency f_{ci} and the "Doppler-shifted" proton (electron) gyroradius ($f_{\rho_{i(e)}} = v_{sw}/2\pi\rho_{i(e)}$) and the vertical dotted lines delimit the interval between f_{min} and f_{max} accessible to analysis by using the k-filtering technique (reprinted the figure with permission from Sahraoui et al., Phys. Rev. Lett. 105, 131101, 2010, copyright 2010 by the American Physical Society)

These spectra presented in Fig. 5.18 show very similar features to those observed by Sahraoui et al. (2009) and Alexandrova et al. (2009), that is, a Kolmogorov scaling law $\propto f^{-1.7}$ in the MHD inertial range of large scales, a breakpoint around the proton gyroscale ($\sim f_{\rho_i}$) (rather than the proton gyrofrequency f_{ci}), a steeper power law spectrum $\propto f^{-2.8}$ in the kinetic inertial range of subproton scales, and then a second breakpoint followed by a steepening around the electron gyroscale ($\sim f_{\rho_e}$, close to the electron inertial length λ_e for $\beta_e = 1$). By comparing these observed spectra with those predicted by the AstroGK simulations for the KAWs turbulence, as shown in Figs. 5.8, 5.9, and 5.10, one can find the striking agreement between the observed and predicted spectra. This strongly proposed that the physical nature of the observed turbulent fluctuations in the kinetic scales is the KAW mode.

Meanwhile, by a careful investigation of these spectra it can be found that the spectra actually steepen more strongly to $\propto f^{-4}$ around the first breakpoint at the proton gyroscale between $f_{sc} \sim 1$–3 Hz, where a transition to subion dispersive cascade occurs. As shown in Fig. 5.14, this stronger steepening can be attributed to the enhanced dissipation at the ion gyroscale (Voitenko et al. 2011), in which the turbulent energy is linearly or nonlinearly damped into protons, while the remaining energy undergoes a dispersive cascade towards smaller scales where it may be dissipated into electron heating.

5.5.3 *Identifications of KAW Magnetic Compressibility*

From Fig. 5.18, in addition, by comparing the parallel and perpendicular magnetic power spectra one can find that the parallel to perpendicular magnetic power spectral ratio increases as the frequency f (i.e., the wave number k) increases. This is consistent with the another important characteristic of KAWs, that is, the magnetic compressibility of KAWs increases with the wave number. Gary and Smith (2009) and Chen and Wu (2011b) analyzed the dependency of the magnetic compressibility parameter,

$$C_{\parallel}(\mathbf{k}) \equiv \frac{|\delta B_{\parallel}(\mathbf{k})|^2}{|\delta \mathbf{B}(\mathbf{k})|^2}, \tag{5.87}$$

on the wave number $\mathbf{k}$ and on the plasma β for different mode waves. The results show that the wave-number dependency of the magnetic compressibility for KAWs is distinctly different from that for whistler waves. In the inertial region of $k_{\perp}\lambda_i \ll 1$ the magnetic compressibility of AWs $C_{\parallel A}(\mathbf{k}) \to 0$ because of the incompressibility of AWs, and in the kinetic region of $k_{\perp}\lambda_i \sim 1$, KAWs have a magnetic compressibility, $C_{\parallel K}(\mathbf{k})$, much smaller than that of whistler waves, $C_{\parallel W}(\mathbf{k})$, that is, $C_{\parallel K}(\mathbf{k}) \ll C_{\parallel W}(\mathbf{k})$. As the perpendicular wave number $k_{\perp}\lambda_i$ increases, however, $C_{\parallel K}(\mathbf{k})$ increases evidently as shown by the middle panel of Fig. 1.3 which is obtained from Eq. (1.30), while $C_{\parallel W}(\mathbf{k})$ is approximately invariable, where the ion inertial length $\lambda_i \sim \rho_i$ for a high-β plasma with $\beta \sim 1$ such as in the solar wind. On the other hand, for a fixed wave number $k_{\perp} \gg k_{\parallel}$, the magnetic compressibility, $C_{\parallel}(\beta)$, is an increasing function of β (or i.e., α_e) for KAWs as shown in the middle panel of Fig. (1.4) given by Eq. (1.30). On the other hand, for whistler waves the magnetic compressibility $C_{\parallel}(\beta)$ nearly is a constant function of β in the kinetic region of $k_{\perp}\rho_i \sim 1$, where $\beta = \beta_i + \beta_e$. Thus, the measurement of the magnetic compressibility of the fluctuations can be used as the distinction between the two models, KAWs or whistler waves (Gary and Smith 2009).

In the observational aspect, based on a database of ACE observations at 1 AU, which was constructed by Hamilton et al. (2008) and consists of 960 intervals spanning the broadest possible range of solar wind conditions including magnetic clouds, Hamilton et al. (2008) analyzed the correlation between the anisotropy of magnetic fluctuations and β_p. Their result indicates that, on average, the magnetic compressibility $C_{\parallel}$ increases with β_p in the range of kinetic scales and that the average tendency of the solar wind fluctuations in the kinetic scales is consistent with the behavior of KAWs and inconsistent with that of quasi-perpendicular magnetosonic or whistler waves. However, a mixture of KAWs and whistler waves can not be ruled out and hence the conclusion is not conclusive one or the other. Therefore, in general, it is of more interest to compare theoretical predictions directly with solar wind observations by plotting the two modes both on the same graph so that quantitative contrasts between them may be made.

Salem et al. (2012) further investigated the physical nature of small-scale turbulent fluctuations in the solar wind using the comparison between the measured magnetic compressibility of fluctuations to the theoretical predictions arising from the models consisting of either KAWs or whistler waves. The data of solar wind fluctuations is from the observation by the Cluster satellite during the interval 03:00:00–04:42:00 UT on 30 January 2003 when the Cluster was travelling in the solar wind. Through the Lorentz transformation, Salem et al. (2012) transform the magnetic compressibility predicted theoretically by the linear modes from the plasma frame into the satellite frame, $\left(|\delta B_\parallel|/|\delta \mathbf{B}_\perp|\right)_{sc}$, so that it may be compared directly to the observational data in situ measured in the satellite frame.

Figure 5.19 presents the predicted magnetic field fluctuation transformed to the satellite frame to predict signature of $|\delta B_\parallel/\delta \mathbf{B}|_{sc}$ for the two wave modes, whistler waves (black/blue) and KAWs (red) with different angles as shown in the figure, where the satellite-frame frequency of the fluctuations f is calculated by $2\pi f = \omega + \mathbf{k}\cdot\mathbf{v}_{sw}$, accounting for the Doppler shift arising from the relative velocity $\mathbf{v}_{sw}$ between the solar wind plasma frame and the satellite frame for a single plane wave with wave vector $\mathbf{k}$ and frequency ω. The frequency $f_b \simeq 0.4$ Hz corresponds the breaking frequency of the power spectrum at the transition between the inertial and dissipative region (Salem et al. 2012). The measured ratio of the parallel to total magnetic fluctuations in the Cluster data, $\delta B_\parallel/\delta B$, is given by the green curve.

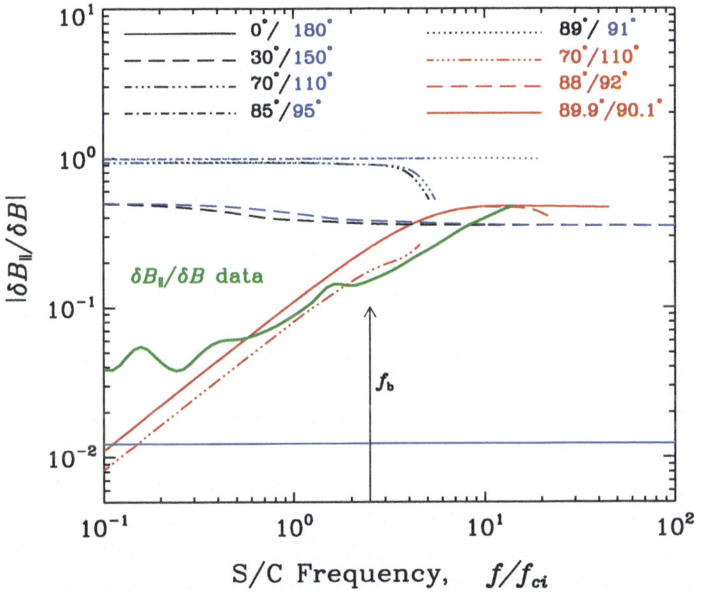

Fig. 5.19 Prediction of $|\delta B_\parallel/\delta \mathbf{B}|_{sc}$ for KAWs (red) or whistler waves (black/blue) with specified angle θ. Cluster flux gate magnetometer measurements up to 2 Hz, or 12 f_{ci}, are shown in green. (from Salem et al. 2012, ©AAS reproduced with permission)

From Fig. 5.19, it is very clear that the measured parallel to total magnetic fluctuation ratio is inconsistent with the predictions by the whistler wave for any angle of the wave vector, but is remarkably good agreement with the prediction for the KAW with a nearly perpendicular wave vector.

Another parameter related to the magnetic compressibility, the magnetic anisotropic ratio of fluctuations (i.e., the ratio of the parallel to perpendicular magnetic fluctuations),

$$\frac{\left(\delta B_{\|}\right)^2}{(\delta B_{\perp})^2} = \frac{C_{\|}}{1 - C_{\|}}, \tag{5.88}$$

also has been used often as a method to identify the physical nature of solar wind turbulent fluctuations near the particle kinetic scales (such as the proton gyroscale $k_{\perp}\rho_p \sim 1$). In particular, the wave-number dependence of the magnetic anisotropic ratio of KAWs, $\left(\delta B_{\|}\right)^2 / (\delta B_{\perp})^2$, has a distinctive feature in the solar wind plasmas with $\beta \sim 1$, that is, the ratio approaches zero in the small wave number limit of $k_{\perp}\rho_p \ll 1$ and increases monotonically to values of order unity when $k_{\perp}\rho_p \sim 1$ (Hollweg 1999; Gary and Smith 2009; TenBarge et al. 2012; Podesta and TenBarge 2012).

Podesta and TenBarge (2012) analyzed 20 high-speed solar wind streams observed by the dual STEREO satellites (denoted by "A" and "B", Acuña et al. 2008) in the ecliptic plane near 1 AU from 2007 through 2011 (see Table 1 of Podesta and TenBarge 2012). In order to facilitate the measurements of the ratio, $\left(\delta B_{\|}\right)^2 / (\delta B_{\perp})^2$, and their comparisons to theory, these streams have been selected during times when the local mean magnetic field $\mathbf{B}_0$ is nearly perpendicular to the local flow velocity of the solar wind $\mathbf{v}_{sw}$. Thus, the quantities $\left(\delta B_{\|}\right)^2$ and $(\delta B_{\perp})^2$ can be measured simply by the average magnetic powers in the $\mathbf{B}_0$ and $\mathbf{v}_{sw} \times \mathbf{B}_0$ directions, respectively.

Figure 5.20 shows a sample of them, which data are from the STEREO-A satellite for the time interval 27 July 2011, 12:00 to 30 July 18:00, 3.25 days, when $v_{sw} \simeq 621$ km/s and $\beta_i \simeq 0.7$. From Fig. 5.20, it can be found that the solar wind measurements (open circles) and the predictions based the Vlasov-Maxwell theory of KAWs (red curve) are good agreement for the high wave number of $k_{\perp}\rho_p > 1$ and both exhibit a monotonic and smooth increase at $k_{\perp}\rho_p \sim 1$ as expected theoretically. Based on the analysis by Podesta and TenBarge (2012), these 20 high-speed solar wind streams observed by the STEREO consistently yield quantitatively similar results that all show a steady increase in the magnetic anisotropic ratio, $\left(\delta B_{\|}\right)^2 / (\delta B_{\perp})^2$, in the neighborhood of $k_{\perp}\rho_p \sim 1$ and are in reasonable agreement with the prediction of KAWs derived from the Vlasov-Maxwell dispersion relation. Therefore, these results can be well interpreted as evidence for the existence of quasi-perpendicular KAWs with $k_{\perp} \gg k_{\|}$ in the fast solar wind at the kinetic scale range near $k_{\perp}\rho_p \sim 1$, where the kinetic physics becomes more important.

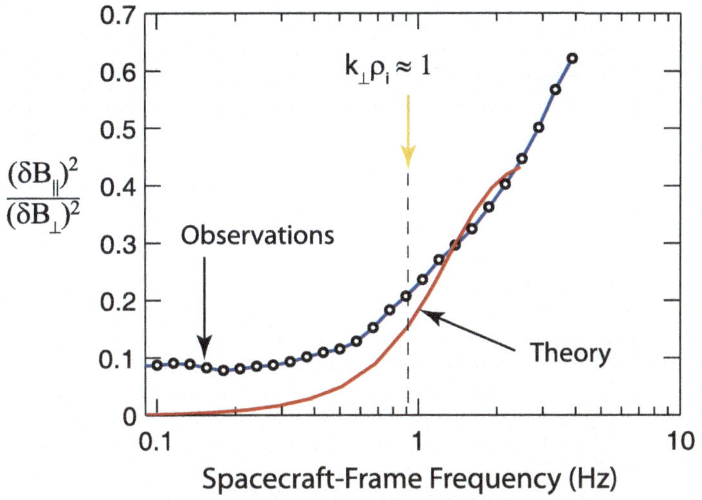

Fig. 5.20 A sample showing comparisons between the theoretical predictions for KAWs (red curve) and solar wind measurements (open circles), including only the quasi-perpendicular data with $84° < \theta_{BV} < 96°$, where θ_{BV} is the angle between the local mean magnetic field $\mathbf{B}_0$ and flow velocity $\mathbf{v}_{sw}$, and the vertical dashed line indicates the wave number $k_\perp \rho_p = k_\perp \rho_i = 1$ (reprinted from Podesta, Sol. Phys., 286, 529–548, 2013, copyright 2013, with permission from Springer)

5.5.4 Identifications of KAW Magnetic Helicity

Besides the magnetic compressibility, the magnetic helicity also is used often to distinctly identify the physical nature of the kinetic-scale fluctuations in solar wind turbulence. In the inertial range the normalized magnetic helicity spectrum σ_m is zero, on average, in the solar wind at 1 AU, while in the kinetic scale range has a distinctive peak near $k\lambda_i = 1$. Figure 5.21 shows a typical sample measured by the STEREO-A during an interval of 4 days from February 13 2008, 08:00 UT to February 17 08:00 UT for an unusually long-lived high-speed stream with the mean flow velocity $v_{sw} \simeq 655$ km/s, proton density $n_p \simeq 2.2\,\text{cm}^{-3}$, temperature $T_p \simeq 1.6 \times 10^5$ K, and the plasma $\beta_p \simeq 0.7$. By use of the Taylor hypothesis, $\omega \simeq k v_{sw}$, the wave number $k\lambda_i = 1$ occurs at 0.7 Hz denoted by the vertical arrow in Fig. 5.21, implying the typical scale of the kinetic turbulence.

From Fig. 5.21, the trace spectrum (upper panel) of the magnetic fluctuations can be fitted well by a power-low spectrum with an index of 1.57 in the inertial range over the satellite-frame frequency from 1 to 100 mHz. When approaching the kinetic scales the trace spectrum breaks evidently and becomes steeper in the kinetic scale range, as expected by the kinetic turbulence theory. On the other hand, from Fig. 5.21, it can be found that the normalized magnetic helicity spectrum σ_m (lower panel) exhibits strong fluctuations with an average value near zero in the inertial range. While in the kinetic scale range immediately after the spectral break the magnetic helicity spectrum remarkably departs from zero and has a clear peak near 1 Hz, a typical

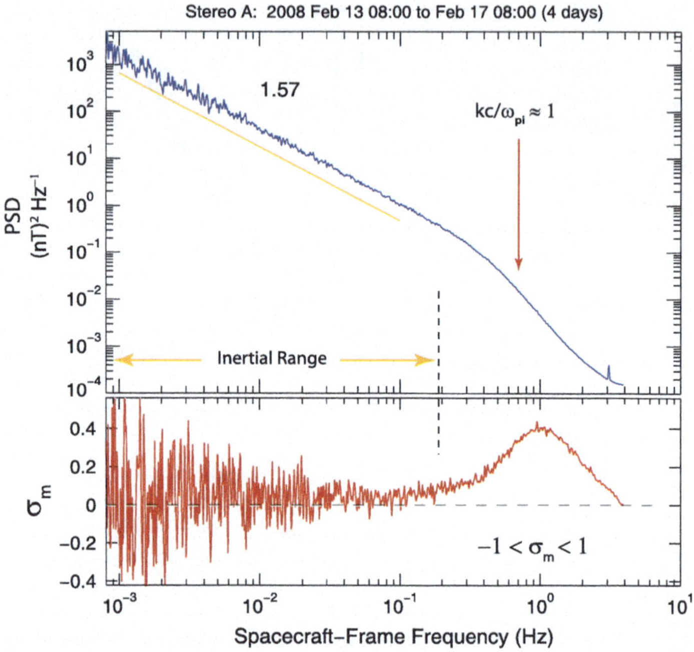

Fig. 5.21 A sample of the trace spectrum (upper panel) and the normalized magnetic helicity spectrum (lower panel) of solar wind magnetic fluctuations near 1 AU, which were measured by the STEREO-A on February 13 2008, 08:00 UT to February 17 08:00 UT (4 days) for an unusually long-lived high-speed stream with the mean parameters $v_{sw} \simeq 655$ km/s, $n_p \simeq 2.2\,\text{cm}^{-3}$, $T_p \simeq 1.6 \times 10^5$ K, and $\beta_p \simeq 0.7$. The spectral slope in the inertial range over the satellite-frame frequency from 1 to 100 mHz is 1.57. By use of the Taylor hypothesis, the wave number $k\lambda_i = 1$ occurs at 0.7 Hz as denoted by the vertical arrow. The normalized magnetic helicity spectrum σ_m exhibits clearly a peak in the kinetic scale range immediately after the spectral break (reprinted from Podesta, Sol. Phys., 286, 529–548, 2013, copyright 2013, with permission from Springer)

kinetic scale. In addition, the fluctuation of the magnetic helicity spectrum also is strongly depressed in the kinetic scale range. Usually this may be explained by waves with a predominantly right-hand sense of polarization propagating away from the Sun (Goldstein et al. 1994; Leamon et al. 1998a). This peak is formed possibly by the ion-cyclotron damping of a cascade of predominantly outward-propagating quasi-parallel Alfvénic ion-cyclotron waves near the spectral breaking point. In consequence, only right-hand polarized quasi-parallel magnetosonic whistler waves cascade through the spectral breaking point into higher wave numbers (Goldstein et al. 1994; Leamon et al. 1998a; Leamon 1998b).

However, the theory and simulation of the plasma turbulence both show that the turbulent energy cascade towards small scales proceeds preferentially in the direction perpendicular to the local mean magnetic field (Goldreich and Sridhar 1995, 1997; Biskamp et al. 1996, 1999; Cho and Vishniac 2000; Maron and Goldreich 2001;

Müller et al. 2003; Cho and Lazarian 2004; Boldyrev 2005, 2006; Howes et al. 2006, 2008a, b; 2011; Schekochihin et al. 2009; Boldyrev and Perez 2012; Boldyrev et al. 2013). This results in that in the kinetic scale range the turbulent energy in quasi-parallel fluctuations would be much less than that in quasi-perpendicular fluctuations, implying that the plasma turbulence in the kinetic scales most likely consists of a cascade of quasi-perpendicular KAWs rather than quasi-parallel whistler waves. Therefore, the ion-cyclotron damping scenario of the formation of the kinetic magnetic helicity spectrum in the kinetic scale range seems unlikely. Howes and Quataert (2010) proposed an alternative interpretation, in which the observed peak in the kinetic magnetic helicity spectrum $\sigma_m(\mathbf{k})$ is caused by quasi-perpendicular KAWs with $k_\perp \gg k_\parallel$. They showed that an anisotropic spectrum of predominantly outward-propagating quasi-perpendicular KAWs, which are also right-hand polarized like the quasi-parallel whistler waves, can produce a magnetic helicity spectrum $\sigma_m(\mathbf{k})$ with a peak in the kinetic scales, which is in reasonable agreement with the observations (Howes and Quataert 2010).

In order to determine further the wave vector direction of the observed magnetic helicity spectrum $\sigma_m(\mathbf{k})$, He et al. (2011, 2012a) and Podesta and Gary (2011) analyzed the look-angle distribution of the observed magnetic helicity spectrum, $\sigma_m(\mathbf{k})$, with respect to the local mean magnetic field. Using the same STEREO-A data as used in Fig. 5.21, two distinct populations of the electromagnetic fluctuations are found at the typical kinetic scales near $k_\perp\rho_i = 1$ as shown in the right panel of Fig. 5.22. One has a left-hand polarization sense with $\sigma_m < 0$ in the satellite frame, denoted by the blue spot near $\theta_{BV} \sim 0$ (i.e., quasi-parallel propagating fluctuations) and the other one has a right-hand polarization sense with $\sigma_m > 0$ in the satellite frame, denoted by the orange and yellow spots located near $\theta_{BV} = 90°$ (i.e., quasi-perpendicular propagating fluctuations). Tentatively, the quasi-parallel fluctuations may be identified as electromagnetic ion-cyclotron waves propagating away from the Sun or magnetosonic whistler waves propagating toward the Sun along the local mean magnetic field, while the quasi-perpendicular waves could be identified as KAWs or oblique whistler waves.

He et al. (2012a) analyzed in detail the hodographs in the local RTN coordinate system (where R points toward the satellite from the Sun center, T is the cross product of the solar rotation axis and R, N completes the right-handed triad) for the fluctuations presented in Fig. 5.22. Figure 5.23 shows two examples of them, which represent the left-hand polarization with $\sigma_m < 0$ (b) and the right-hand polarization with $\sigma_m > 0$ (c). In particular, they found that the major axis of the right-hand polarization ellipses is perpendicular to the local mean magnetic field $\mathbf{B}_0$, implying $\delta B_\perp > \delta B_\parallel$ (He et al. 2012a). This is the typical property of quasi-perpendicular KAWs rather than quasi-perpendicular whistler waves. In fact, for a quasi-perpendicular whistler wave, the major axis of its magnetic polarization ellipse is expected to be aligned with the local $\mathbf{B}_0$, implying significant magnetic compressibility, and the polarization sense turns from right to left handedness as the wave propagation angle (θ_{BV}) increases toward 90°. Therefore, they concluded that, in the kinetic scale range near the break of solar wind turbulence spectra occurring around the proton inertial length (or i.e., the proton gyroradius), the observed right-hand polarization ellipse with orientation

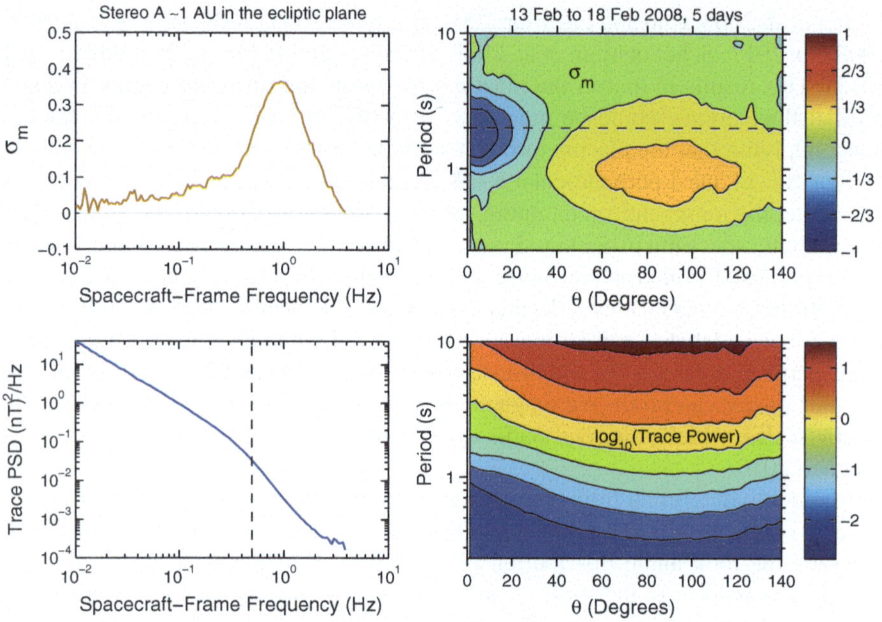

Fig. 5.22 Analysis of a five-day interval of high-speed wind observed in the ecliptic plane near 1 AU by the STEREO-A satellite: Left panel is the magnetic helicity spectrum (upper) and the trace power spectrum (lower) same as that presented in Fig. 5.21 but focus on the kinetic scale range. Right panel shows the look-angle distribution of the magnetic helicity spectrum, where the look angle $\theta = \theta_{BV}$ is the angle between the flow velocity $\mathbf{v}_{sw}$ and the local mean magnetic field $\mathbf{B}_0$ of the solar wind. The vertical (left-lower panel) and horizontal (right-upper panel) dashed lines denote the typical kinetic scale of $k_\perp \rho_i = 1$ (from Podesta and Gary 2011)

perpendicular to the local $\mathbf{B}_0$ represents quasi-perpendicular KAWs (i.e., oblique Alfvénic ion-cyclotron waves) rather than quasi-perpendicular magnetosonic whistler waves, as expected by the theory and simulation of the kinetic turbulence.

In order to further understand the physical nature of the population of the quasi-parallel fluctuations in above the observations (He et al. 2011, 2012a; Podesta and Gary 2011), He et al. (2012b) modeled the measured angular distribution of σ_m by assuming a possible distribution formed by Alfvénic fluctuations in wave vector space and then looked for the best fitting for the observations by adjusting the spectral distributions of the assumed fluctuations. In consequence, a very good agreement between the theoretical model and the observations can be found as shown in Fig. 5.24. Their model shows that the observed two-component angular distribution of the magnetic helicity spectrum ($\sigma_m(k, \theta_{BV})$) can be reproduced well by the superposition of two-component Alfvénic fluctuations, which consists of quasi-perpendicular KAWs as a major component close to $k_\perp$ (i.e., $\theta_{BV} \sim 90°$) and quasi-parallel electromagnetic ion-cyclotron waves as a minor component close to $k_\parallel$ (i.e., $\theta_{BV} \sim 0$).

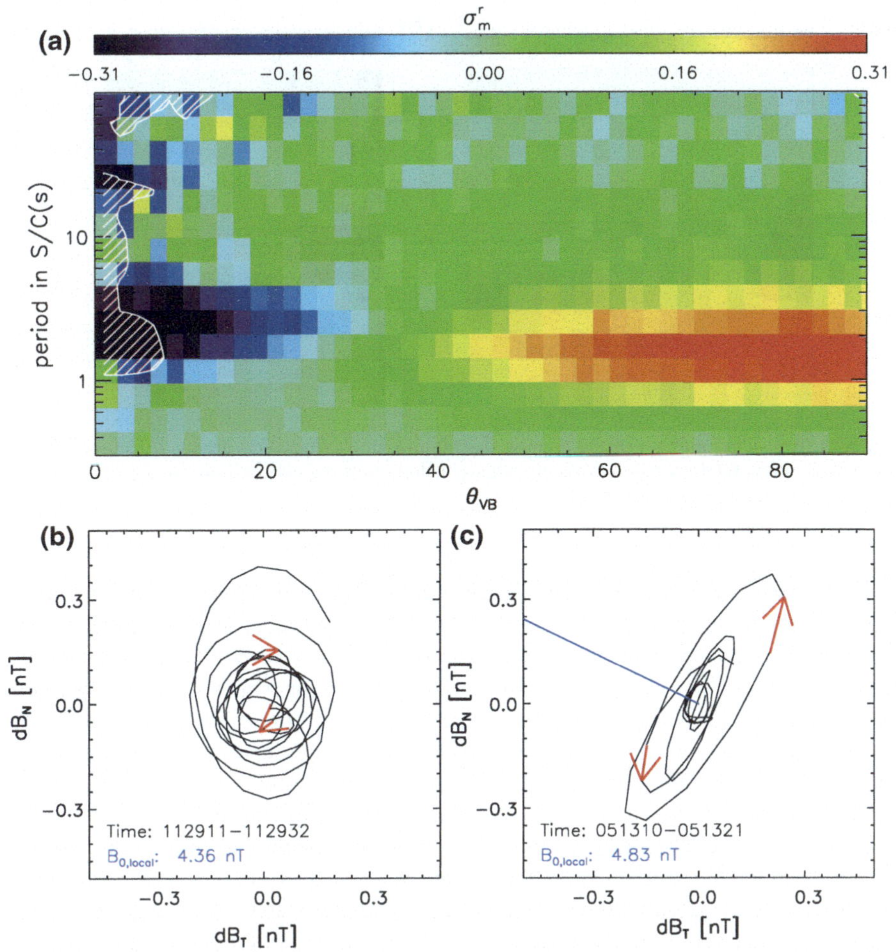

Fig. 5.23 **a** Look-angle (θ_{BV}) distribution of magnetic helicity spectrum (σ_m) derived from the STEREO-A measurements on 13 February 2008 (the same as that in Fig. 5.21, but a shorter interval); **b** An example of the $dB_T - dB_N$ hodograph for negative σ_m in (**a**), which is extracted from a small interval from 11:29:11 to 11:29:32 UT during which the local $\mathbf{B}_0$ is quasi-parallel to the R-direction with $\theta_{BV} < 10°$. Red arrows denote a circle-like left-hand polarization of $dB_T - dB_N$ around the R-direction; **c** One case of the $dB_T - dB_N$ hodograph, which is extracted from an interval from 05:13:10 to 05:13:21 UT when $80° < \theta_{BV} < 90°$. Red arrows denote an ellipse-like right-hand polarization of $dB_T - dB_N$ around the R-direction. Blue line represents the local $\mathbf{B}_0$, which is perpendicular to the major axis of the ellipse, implying $dB_\perp > dB_\parallel$ (from He et al. 2012a, ©AAS reproduced with permission)

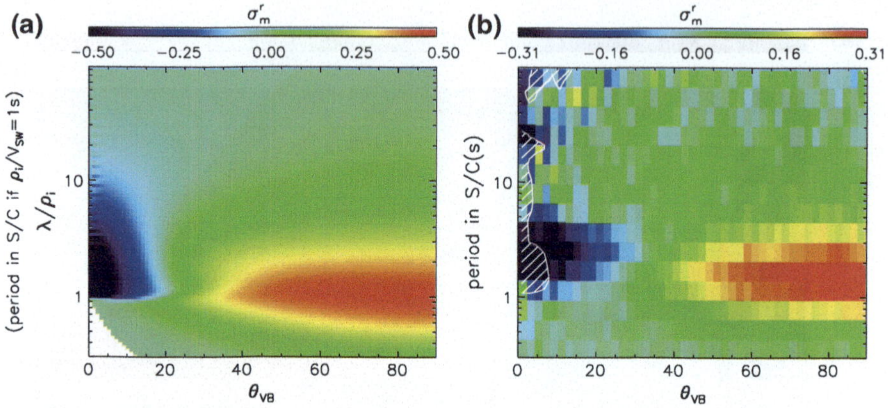

Fig. 5.24 **a** Modeling result of the angular distribution of σ_m based on gradually balanced two-component AWs; **b** Observation of the angular distribution of the two-component σ_m (from He et al. 2012b, ©AAS reproduced with permission)

In an independent but similar modelling study, Klein et al. (2014) obtained similar results. They showed that the observed two-component angular distribution of the magnetic helicity spectrum σ_m can be reproduced by a perpendicular cascade of KAWs and a local parallel non-turbulent ion-cyclotron or whistler waves generated by temperature anisotropy instabilities. By constraining the model free parameters through comparison to in situ data, it is found that, on average, ~95% of the power near the kinetic scales is contained in a perpendicular KAW cascade and that the parallel non-turbulent waves are propagating nearly unidirectionally.

In addition, they also noticed that in order to reproduce the observed diminishing of the magnetic helicity spectrum σ_m at shorter scales of $k_\perp \rho_i > 1$ the balance between outward and inward wave-energy fluxes needs to be reached gradually as the spatial scale decreases. In the observations, however, the diminishing of σ_m occurs at frequencies close to the Nyquist frequency (i.e., the data sampling frequency, He et al. 2011, 2012a; Podesta and Gary 2011; Klein et al. 2014). Also this possibly could influence the diminishing of σ_m in the short scales.

From these observations and modelling investigations of the magnetic helicity spectra, we can include two definite results, one is that in the kinetic scale range of solar wind turbulence the quasi-perpendicular cascade component dominates the magnetic helicity spectrum and contains ~95% of the spectral power, and the other is that the physical nature of the quasi-perpendicular cascade component is KAWs. These results both can be well consistent with the predictions of the theories and simulations of the kinetic turbulence (Boldyrev 2005, 2006; Howes et al. 2006, 2008a, b; 2011; Schekochihin et al. 2009; Boldyrev and Perez 2012; Boldyrev et al. 2013; Grošelj et al. 2018).

References

Abel, I. G., Barnes, M., Cowley, S. C., et al. (2008). Linearized model fokker-planck collision operators for gyrokinetic simulations. I. theory. *Physics of Plasmas*, *15*, 122509.
Acuña, M. H., Curtis, D., Scheifele, J. L., et al. (2008). The STEREO/IMPACT magnetic field experiment. *Space Science Reviews*, *136*, 203–226.
Alexandrova, O., Saur, J., Lacombe, C., et al. (2009). Universality of solar-wind turbulent spectrum from MHD to electron scales. *Physical Review Letters*, *103*, 165003.
Bale, S. D., Kellogg, P. J., Mozer, F. S., et al. (2005). Measurement of the electric fluctuation spectrum of magnetohydrodynamic turbulence. *Physical Review Letters*, *94*, 215002.
Barnes, M., Abel, I. G., Dorland, W., et al. (2009). Linearized model fokker-planck collision operators for gyrokinetic simulations. II. numerical implementation and tests. *Physics of Plasmas*, *16*, 128.
Beresnyak, A., & Lazarian, A. (2006). Polarization intermittency and its influence on MHD turbulence. *The Astrophysical Journal*, *640*, L175–L178.
Biskamp, D. (2003). *Magnetohydrodynamic turbulence*. Cambridge: Cambridge University Press.
Biskamp, D., & Müller, W. C. (2000). Scaling properties of three-dimensional isotropic magnetohydrodynamic turbulence. *Physical Review Letters*, *7*, 4889–4900.
Biskamp, D., Schwarz, E., & Drake, J. F. (1996). Two-dimensional electron magnetohydrodynamic turbulence. *Physical Review Letters*, *76*, 1264–1267.
Biskamp, D., Schwarz, E., Zeiler, A., et al. (1999). Electron magnetohydrodynamic turbulence. *Physics of Plasmas*, *6*, 751–758.
Boldyrev, S. (2005). On the spectrum of magnetohydrodynamic turbulence. *The Astrophysical Journal*, *626*, L37.
Boldyrev, S. (2006). Spectrum of magnetohydrodynamic turbulence. *Physical Review Letters*, *96*, 115002.
Boldyrev, S., & Perez, J. C. (2012). Spectrum of kinetic-Alfvén turbulence. *The Astrophysical Journal Letters*, *758*, L44.
Boldyrev, S., Horaites, K., Xia, Q., & Perez, J. C. (2013). Toward a theory of astrophysical plasma turbulence at subproton scales. *The Astrophysical Journal*, *777*, 41.
Bruno, R., & Carbone, V. (2005). The solar wind as a turbulence laboratory. *Living Reviews in Solar Physics*, *2*, 4.
Bruno, R., & Carbone, V. (2013). The solar wind as a turbulence laboratory. *Living Reviews in Solar Physics*, *10*, 2.
Cerri, S. S., Servidio, S., & Califano, F. (2017). Kinetic cascade in solar-wind turbulence: 3D3V hybrid-kinetic simulations with electron inertia. *The Astrophysical Journal Letters*, *846*, L18.
Chandran, B. D. G., Schekochihin, A. A., & Mallet, A. (2015). Intermittency and alignment in strong RMHD turbulence. *The Astrophysical Journal*, *807*, 39.
Chen, C. H. K., Boldyrev, S., Xia, Q., & Perez, J. C. (2013). Nature of subproton scale turbulence in the solar wind. *Physical Review Letters*, *110*, 225002.
Chen, C. H. K., Horbury, T. S., Schekochihin, A. A., et al. (2010). Anisotropy of solar wind turbulence between ion and electron scales. *Physical Review Letters*, *104*, 255002.
Chen, L., & Wu, D. J. (2011a). Exact solutions of dispersion equation for MHD waves with short-wavelength modification. *Chinese Science Bulletin*, *56*, 955–961.
Chen, L., & Wu, D. J. (2011b). Polarizations of coupling kinetic Alfvén and slow waves. *Physics of Plasmas*, *18*, 072110.
Chen, L., & Zonca, F. (2016). Physics of Alfvén waves and energetic particles in burning plasmas. *Reviews of Modern Physics*, *88*, 015008.
Cho, J., & Lazarian, A. (2004). The anisotropy of electron magnetohydrodynamic turbulence. *The Astrophysical Journal*, *615*, L41–L44.
Cho, J., & Vishniac, E. T. (2000). The anisotropy of MHD Alfvénic turbulence. *The Astrophysical Journal*, *539*, 273–282.

Coleman, P. J, Jr. (1968). Turbulence, viscosity, and dissipation in the solar-wind plasma. *The Astrophysical Journal, 153*, 371–388.
Dobrowolny, M., Mangeney, A., & Veltri, P. (1980). Properties of magnetohydrodynamic turbulence in the solar wind. *Astronomy & Astrophysics, 83*, 26–32.
Fonseca, R. A., Martins, S. F., Silva, L. O., et al. (2008). One-to-one direct modeling of experiments and astrophysical scenario: Pushing the envelope on kinetic plasma simulations. *Plasma Physics and Controlled Fusion, 50*, 124034.
Fonseca, R. A., Silva, L. O., Tsung, F. S., et al. (2002). OSIRIS: A Three-Dimensional, fully relativistic particle in cell code for modeling plasma based accelerators. *Lecture Notes in Computer Science, 2331*, 342–351.
Forman, M. A., Wicks, R. T., & Horbury, T. S. (2011). Detailed fit of critical balance theory to solar wind turbulence measurements. *The Astrophysical Journal, 733*, 76.
Frieman, E. A., & Chen, L. (1982). Nonlinear gyrokinetic equations for low-frequency electromagnetic waves in general plasma equilibria. *Physics of Fluids, 25*, 502.
Frisch, U. (1995). *Turbulence: The legacy of A. N. Kolmogorov*. Cambridge: Cambridge University Press.
Gary, S. P., & Smith, C. W. (2009). Short-wavelength turbulence in the solar wind: Linear theory of whistler and kinetic Alfvén fluctuations. *Journal of Geophysical Research, 114*, A12105.
Goldreich, P., & Sridhar, S. (1995). Toward a theory of interstellar turbulence. 2: Strong Alfvénic turbulence. *The Astrophysical Journal, 438*, 763–775.
Goldreich, P., & Sridhar, S. (1997). Magnetohydrodynamic turbulence revisited. *The Astrophysical Journal, 485*, 680–688.
Goldstein, M. L., Roberts, D. A., & Fitch, C. A. (1994). Properties of the fluctuating magnetic helicity in the inertial and dissipation ranges of solar wind turbulence. *Journal of Geophysical Research, 99*, 11519–11538.
Goldstein, M. L., Wicks, R. T., Perri, S., & Sahraoui, F. (2015). Kinetic scale turbulence and dissipation in the solar wind: Key observational results and future outlook. *Philosophical Transactions of the Royal Society A, 373*, 20140147.
Gollub, J. P., & Swinney, H. L. (1975). Onset of turbulence in a rotating fluid. *Physical Review Letters, 35*, 927–930.
Grappin, R., Frisch, U., Pouquet, A., & Leorat, J. (1982). Alfvénic fluctuations as asymptotic states of MHD turbulence. *Astronomy & Astrophysics, 105*, 6–14.
Grošelj, D., Mallet, A., Loureiro, N. F., & Jenko, F. (2018). Fully kinetic simulation of 3D kinetic Klfvén turbulence. *Physical Review Letters, 120*, 105101.
Hamilton, K., Smith, C. W., Vasquez, B. J., & Leamon, R. J. (2008). Anisotropies and helicities in the solar wind inertial and dissipation ranges at 1 AU. *Journal of Geophysical Research, 113*, A01106.
He, J., Marsch, E., Tu, C., et al. (2011). Possible evidence of Alfvén-cyclotron waves in the angle distribution of magnetic helicity of solar wind turbulence. *The Astrophysical Journal, 731*, 85.
He, J., Tu, C., Marsch, E., & Yao, S. (2012a). Do oblique Alfvén/ion-cyclotron or fast-mode/whistler waves dominant the dissipation of solar wind turbulence near the proton inertial length? *The Astrophysical Journal Letters, 745*, L8.
He, J., Tu, C., Marsch, E., & Yao, S. (2012b). Reproduction of the observed two-component magnetic helicity in solar wind turbulence by a superposition of parallel and oblique Alfvén waves. *The Astrophysical Journal, 749*, 86.
Hollweg, J. V. (1999). Kinetic Alfvén wave revisited. *Journal of Geophysical Research, 104*, 14811–14820.
Hopf, E. (1948). A mathematical example displaying features of turbulence. *Communications on Pure and Applied Mathematics, 1*, 303–322.
Horbury, T. S., Forman, M., & Oughton, S. (2008). Anisotropic scaling of magnetohydrodynamic turbulence. *Physical Review Letters, 101*, 175005.
Howes, G. G., & Quataert, E. (2010). On the interpretation of magnetic helicity signatures in the dissipation range of solar wind turbulence. *The Astrophysical Journal Letters, 709*, L49–L52.

Howes, G. G., Cowley, S. C., Dorland, W., et al. (2006). Astrophysical gyrokinetics: Basic equations and linear theory. *The Astrophysical Journal, 651*, 590–614.

Howes, G. G., Cowley, S. C., Dorland, W., et al. (2008b). A model of turbulence in magnetized plasmas: Implications for the dissipation range in the solar wind. *Journal of Geophysical Research, 113*, A05103.

Howes, G. G., Dorland, W., Cowley, S. C., et al. (2008a). Kinetic simulations of magnetized turbulence in astrophysical plasmas. *Physical Review Letters, 100*, 065004.

Howes, G. G., Tenbarge, J. M., Dorland, W., et al. (2011). Gyrokinetic simulations of solar wind turbulence from ion to electron scales. *Physical Review Letters, 107*, 035004.

Howes, G., Dorland, W., Schekochihin, A., et al. (2010). A weakened cascade model for solar wind turbulence. In *52nd Annual Meeting of the APS Division of Plasma Physics*, 8–12 Nov, APS.

Iroshnikov, R. S. (1963), The turbulence of a conducting fluid in a strong magnetic field, *Astron. Zh. 40*, 742–750. (English Translation: Sov. Astron. 1964, 7, 566).

Kingsep, A. S., Chukbar, K. V., & Yan'kov, V. V. (1990). Electron magnetohydrodynamics. In B. B. Kadomtsev (Ed.), *Reviews of plasma physics* (pp. 243–291). NY: Consultants Bureau.

Kiyani, K. H., Chapman, S. C., Khotyaintsev, Yu V., et al. (2009). Global scale-invariant dissipation in collisionless plasma turbulence. *Physical Review Letters, 103*, 075006.

Klein, K. G., Howes, G. G., TenBarge, J. M., & Podesta, J. J. (2014). Physical interpretation of the angle-dependent magnetic helicity spectrum in the solar wind: The nature of turbulent fluctuations near the proton gyroradius scale. *The Astrophysical Journal, 785*, 138.

Kolmogorov, A. N. (1941). The local structure turbulence in incompressible viscous fluids for very large Reynolds numbers, Dokl. Akad. Nauk. SSSR 30, 301–305. Reprinted in 1991. *Proceedings of the Royal Society of London Series A, 434*, 9–13.

Kraichnan, R. H. (1965). Inertial range spectrum of hyromagnetic turbulence. *Physics of Fluids, 8*, 1385–1387.

López, R. A., Viñas, A. F., Araneda, J. A., & Yoon, P. H. (2017). Kinetic scale structure of low-frequency waves and fluctuations. *The Astrophysical Journal, 845*, 60.

Landau, L. (1944). Stability of tangential discontinuities in compressible fluid. In *Proceedings of the USSR Academy of Sciences* (Vol. 44, pp. 139–141).

Leamon, R. J., Matthaeus, W. H., Smith, C. W., & Wong, H. K. (1998b). Contribution of cyclotron-resonant damping to kinetic dissipation of interplanetary turbulence. *The Astrophysical Journal, 507*, L181–L184.

Leamon, R. J., Smith, C. W., Ness, N. F., et al. (1998a). Observational constraints on the dynamics of the interplanetary magnetic field dissipation range. *Journal of Geophysical Research, 103*, 4775–4787.

Lumley, J. L., & Yaglom, A. M. (2001). A century of turbulence. *Flow, Turbulence and Combustion, 66*, 241–286.

Luo, Q. Y., & Wu, D. J. (2010). Observations of anisotropic scaling of solar wind turbulence. *The Astrophysical Journal Letters, 714*, L138–L141.

Luo, Q. Y., Wu, D. J., & Yang, L. (2011). Measurement of intermittency of anisotropic magnetohydrodynamic turbulence in high-speed solar wind. *The Astrophysical Journal Letters, 733*, L22.

Müller, W. C., Biskamp, D., & Grappin, R. (2003). Statistical anisotropy of magnetohydrodynamic turbulence. *Physical Review E, 67*, 066302.

Maron, J., & Goldreich, P. (2001). Simulations of incompressible magnetohydrodynamic turbulence. *The Astrophysical Journal, 554*, 1175–1196.

Mason, J., Cattaneo, F., & Boldyrev, S. (2006). Dynamic alignment in driven magnetohydrodynamic turbulence. *Physical Review Letters, 97*, 255002.

Matthaeus, W. H., Wan, M., Servidio, S., et al. (2015). Intermittency, nonlinear dynamics and dissipation in the solar wind and astrophysical plasmas. *Philosophical Transactions of the Royal Society A, 373*, 20140154.

Mithaiwala, M., Rudakov, L., Crabtree, C., & Ganguli, G. (2012). Co-existence of whistler waves with kinetic Alfvén wave turbulence for the high-beta solar wind plasma. *Physics of Plasmas*, *19*, 102902.

Ng, C. S., & Bhattacharjee, A. (1996). Interaction of shear-Alfvén wave packets: Implication for weak magnetohydrodynamic turbulence in astrophysical plasmas. *The Astrophysical Journal*, *465*, 845.

Numata, R., Howes, G. G., Tatsuno, T., et al. (2010). AstroGK: Astrophysical gyrokinetics code. *Journal of Computational Physics*, *229*, 9347–9372.

Parker, E. N. (1979). *Cosmical magnetic fields: Their origin and their activity*. New York: Oxford University Press.

Pincon, J. L. (1995). Cluster and the K-filtering. In K.-H. Glassmeier, U. Motschmann, & R. Schmidt (Eds.), *Proceedings of the Cluster Workshops, Data Analysis Tools and Physical Measurements and Mission-Oriented Theory* (p. 87). European Space Agency.

Podesta, J. J. (2009). Dependence of solar-wind power spectra on the direction of the local mean magnetic field. *The Astrophysical Journal*, *698*, 986–999.

Podesta, J. J. (2013). Evidence of kinetic Alfvén waves in the solar wind at 1 AU. *Solar Physics*, *286*, 529–548.

Podesta, J. J., & Gary, S. P. (2011). Magnetic helicity spectrum of solar wind fluctuations as a function of the angle with respect to the local mean magnetic field. *The Astrophysical Journal*, *734*, 15.

Podesta, J. J., & TenBarge, J. M. (2012). Scale dependence of the variance anisotropy near the proton gyroradius scale: Additional evidence for kinetic Alfvén waves in the solar wind at 1 AU. *Journal of Geophysical Research*, *117*, A10106.

Podesta, J. J., Borovsky, J. E., & Gary, S. P. (2010). A kinetic Alfvén wave cascade subject to collisionless damping cannot reach electron scales in the solar wind at 1 AU. *The Astrophysical Journal*, *712*, 685–691.

Pouquet, A., Meneguzzi, M., & Frisch, U. (1986). Growth of correlations in magnetohydrodynamic turbulence. *Physical Review A*, *33*, 4266–4276.

Reynolds, O. (1883). An experimental investigation of the circumstances which determine whether the motion of water shall he direct or sinuous, and of the law of resistance in parallel channels. *Proceedings of the Royal Society*, *35*, 84–99.

Ruelle, D., & Takens, F. (1971). On the nature of turbulence. *Communications in Mathematical Physics*, *23*, 343–344.

Rutherford, P. H., & Frieman, E. A. (1968). Drift instabilities in general magnetic field configurations. *Physics of Fluids*, *11*, 569.

Sahraoui, F., Goldstein, M. L., Belmont, G., et al. (2010). Three dimensional anisotropic k spectra of turbulence at subproton scales in the solar wind. *Physical Review Letters*, *105*, 131101.

Sahraoui, F., Goldstein, M. L., Robert, P., & Khotyaintsev, Yu V. (2009). Evidence of a cascade and dissipation of solar-wind turbulence at the electron gyroscale. *Physical Review Letters*, *102*, 231102.

Salem, C. S., Howes, G. G., Sundkvist, D., et al. (2012). Identification of kinetic Alfvén wave turbulence in the solar wind. *The Astrophysical Journal Letters*, *745*, L9.

Schekochihin, A. A., Cowley, S. C., Dorland, W., et al. (2009). Astrophysical gyrokinetics: Kinetic and fluid turbulent cascades in magnetized weakly collisional plasmas. *The Astrophysical Journal Supplement Series*, *182*, 310–377.

Smale, S. (1967). Differentiable dynamical systems. *Bulletin of the American Mathematical Society*, *73*, 747–817.

Sridhar, S., & Goldreich, P. (1994). Toward a theory of interstellar turbulence 1: Weak Alfvénic turbulence. *The Astrophysical Journal*, *432*, 612–621.

Tatsuno, T., Dorland, W., Schekochihin, A. A., et al. (2009). Nonlinear phase mixing and phase-space cascade of entropy in gyrokinetic plasma turbulence. *Physical Review Letters*, *103*, 015003.

Taylor, G. I. (1938). Production and dissipation of vorticity in a turbulent fluid. *Proceedings of the Royal Society of London Series A*, *164*, 15–23.

TenBarge, J. M., & Howes, G. G. (2012). Evidence of critical balance in kinetic Alfvén wave turbulence simulations. *Physics of Plasmas*, *19*, 419–763.

TenBarge, J. M., Howes, G. G., & Dorland, W. (2013). Collisionless damping at electron scales in solar wind turbulence. *The Astrophysical Journal*, *774*, 1201–1205.

TenBarge, J. M., Podesta, J. J., Klein, K. G., & Howes, G. G. (2012). Interpreting magnetic variance anisotropy measurements in the solar wind. *The Astrophysical Journal*, *753*, 107.

Told, D., Jenko, F., TenBarge, J. M., et al. (2015). Multiscale nature of the dissipation range in gyrokinetic simulations of Alfvénic turbulence. *Physical Review Letters*, *115*, 025003.

Torrence, C., & Compo, G. P. (1998). A practical guide to wavelet analysis. *Bulletin of the American Meteorological Society*, *79*, 61–78.

Tu, C. Y., & Marsch, E. (1995). MHD structures, waves and turbulence in the solar wind: Observations and theories. *Space Science Reviews*, *73*, 1–210.

Voitenko, Yu M, & Keyser, J. D. (2011). Turbulent spectra and spectral kinks in the transition range from MHD to kinetic Alfvén turbulence. *Nonlinear Processes in Geophysics*, *18*, 587–597.

Voitenko, Yu M. (1998a). Three-wave coupling and parametric decay of kinetic Alfvén waves. *Journal of Plasma Physics*, *60*, 497–514.

Voitenko, Yu M. (1998b). Three-wave coupling and weak turbulence of kinetic Alfvén waves. *Journal of Plasma Physics*, *60*, 515–527.

Wicks, R. T., Horbury, T. S., Chen, C. H. K., & Schekochihin, A. A. (2010). Power and spectral index anisotropy of the entire inertial range of turbulence in the fast solar wind. *Monthly Notices of the Royal Astronomical Society*, *407*, L31–L35.

Zhao, J. S., Voitenko, Y. M., Wu, D. J., & Yu, M. Y. (2016). Kinetic Alfvén turbulence below and above ion cyclotron frequency. *Journal of Geophysical Research*, *121*, 5–18.

Zhao, J. S., Wu, D. J., & Lu, J. Y. (2013). Kinetic Alfvén turbulence and parallel electric fields in flare loops. *The Astrophysical Journal*, *767*, 109.

Chapter 6
KAWs in Solar Atmosphere Heating

6.1 Modern View of Solar Coronal Heating Problem

The Sun is our nearest star and the energy released by nuclear reactions near its center is transported by photons inside the inner $\sim$71% of the solar radius ($R_{\odot} \simeq 6.9 \times 10^5$ km), called the radiative zone. Outside this radiative zone, called the convective zone, photons are no longer able to transfer energy efficiently, so convective instabilities set in and vertical flows carry nearly all the excess heat to the solar surface. This visible surface, called the photosphere, is the lowest layer of the solar outer atmosphere, emits almost all the solar light, and lowers its temperature by the radiation. At first sight one may expect the temperature to decrease as we go away from the solar surface, and at first it does do so from $\sim$5,700 K at the surface to a minimum value $\sim$4,200 K at the height $\sim$500 km above the surface. Beyond that, however, it rises slowly through a wider layer $\sim$2,000 km, called the chromosphere, and then rapidly in a narrow transition layer only a few hundreds of km thick, called the transition region, to higher than 10^6 K in the corona, which extends from the transition region into the interplanetary space, where it becomes the solar wind. The variations of the atmospheric temperature, electron and neutral hydrogen densities, and ionization, with height for an average model of the solar outer atmosphere are shown in Fig. 6.1.

Investigating the corona has remained central to astronomy as well as physics, since we can observe it in greater detail than other stars, and can examine numerous plasma processes that are impossible to reproduce in a laboratory. Although so much effort over the last seventy years, the existence of a counter-intuitive high temperature corona above the much cooler photosphere still puzzled solar physicists—it is now well-known as the "solar coronal heating problem". The high temperature of the corona was established, more than seventy years ago, by the Swedish physicist Edlén (1941) who showed that spectral lines observed from the corona were from highly ionized iron between nine and thirteen times. Because of the high degree of ionization of iron in the solar atmosphere, the conclusion was that the temperature had to be around 1 million K. This is surprising, since if the gas in the corona is simply gas

D.-J. Wu and L. Chen, *Kinetic Alfvén Waves in Laboratory, Space, and Astrophysical Plasmas*, Atmosphere, Earth, Ocean & Space,
https://doi.org/10.1007/978-981-13-7989-5_6

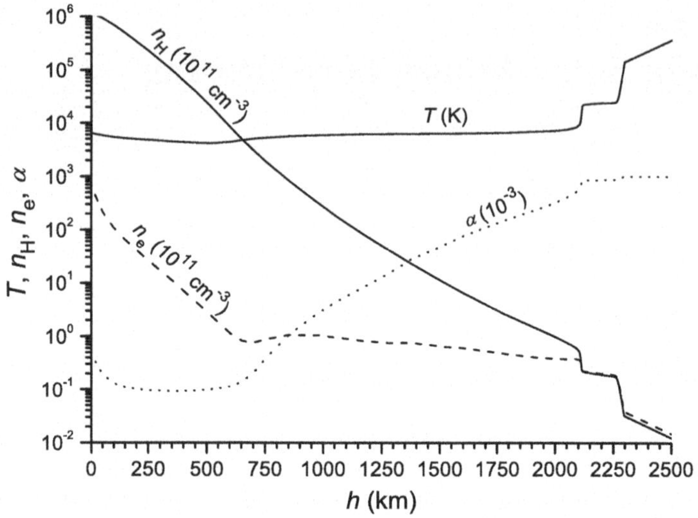

Fig. 6.1 Sketch of the variations of temperature, densities, and ionization with the height, which is calculated based on the VAL average quiet solar atmospheric model given by Vernazza et al. (1981)

from the chromosphere that is advected up into the corona or heated by radiation, the temperature should actually be lower than in the chromosphere, not higher. Moreover, because of radiative losses by EUV and X-ray emission, the corona would just cool off in a matter of hours to days, if the plasma temperature could not be maintained continuously by some heating sources. The existence and maintenance of a hot solar corona have puzzled astronomers ever since it was discovered, but it became clear early on that there must be a heating mechanism at work in the corona, which is not connected to radiation.

The heating of the solar corona has been a fundamental astrophysical issue for more than seventy years. In particular, since the Skylab satellite, the first space station by NASA (the National Aeronautics and Space Administration) of USA (the United States of America), was launched into orbit on May 14, 1973, solar observations from space have revealed the complex and often subtle magnetic-field and plasma interactions throughout the solar atmosphere in unprecedented detail. Skylab carried a white-light coronagraph for the outer corona, an X-ray spectrographic telescope for the corona, an X-ray telescope for the low corona, an EUV spectroheliometer and spectroheliograph, and an UV spectrograph for the chromosphere and transition region. The launch of Skylab initiated a new era of multi-wavelength solar observation from space and signalled a major maturation in space-based solar physics. Some ideas of the enormous productivity of Skylab can be gauged from the fact that the X-ray telescope alone recorded more than 30,000 photographs of the Sun during the nine-month duration of the observations (Bray et al. 1991). In light of the overwhelming observational evidence from the Skylab data analysis, the initial attempts to explain the solar corona as relatively homogeneous and symmetric plasma cannot be sustained, and it first was clearly confirmed that the solar corona is fundamentally structured on a wide range of scales, and extremely active and dynamic and includes intense bright points, active region loops, coronal holes, and coronal mass ejections (CMEs).

The NASA's SMM (Solar Maximum Mission) satellite, which aimed to understand the physics of solar flares and other aspects of solar activity and was launched on the Valentine's Day, 1980 (February 14), not only enriched the picture of coronal structures with observations of spatial structure and flows in the transition region with the temperature range $10^4 < T < 10^6$ K but also further emphasized the great importance of CMEs in the dynamics of the corona and the heliosphere with a massive database of coronagraphic observations (Strong et al. 1999). The structured and dynamic nature of the solar corona was rediscovered more clearly by the Yohkoh ("Sunbeam" in Japanese) satellite launched on August 30, 1991, which was a projection of the ISAS (Institute of Space and Astronautical Science) of Japan and was designed to study solar flares and other transient coronal disturbances. Yokhok provided a full decade of soft X-ray images from the solar disk and from flares. In particular, its soft X-ray telescope came with the great breakthrough in observations of the solar inner corona, revealing for the first time ubiquitous coronal loop structures closely associated with locally close magnetic fields, the geometry and topology of large-scale magnetic field reconfigurations, and magnetic reconnection processes in flares.

The largest complement of instruments for study of the Sun flown since the Skylab is the SOHO (Solar and Heliospheric Observatory) satellite, which was a cooperative effort between NASA of USA and ESA (European Space Agency) and was launched on December 2, 1995. SOHO observes the Sun from inside out to 30 solar radii in the heliosphere and provides a wealth of data on whole dynamic processes in the solar corona. The another NASA's TRACE (Transition Region and Coronal Explorer) satellite, launched on April 2, 1998, was designed to explore quantitatively the connections between fine-scale magnetic fields at the solar surface and the associated plasma structures in the solar outer atmosphere (Aschwanden et al. 2001). TRACE provides coronal observations with high spatial resolution of $\sim 1''$ and has revealed intriguing details about coronal plasma dynamics, coronal plasma heating, and magnetic reconnection processes.

These space-based observations of the solar corona have revolutionized our understanding and appreciation of its surprising beauty and incessant variability, as shown by the cartoon in Fig. 6.2 (Schrijver and Zwaan 2001; Aschwanden et al. 2001). The early physical model of the solar high-temperature corona began simply with the concept of gravitationally stratified layers in the 1950s (the left panel of Fig. 6.2), that is, the photosphere, the chromosphere, and the corona, although the coronal structure, varying through the solar cycle, has been appreciated since the eclipse observations. By the 1980s, it was clear that the basic dynamics of the corona had to take into account the effects of large-scale magnetic structures, especially the physical nature of superradially expanding of solar magnetic flux concentrated in the photosphere into the corona through the chromosphere and the transition region (the middle panel of Fig. 6.2). Since the 1990s, however, our view of the corona

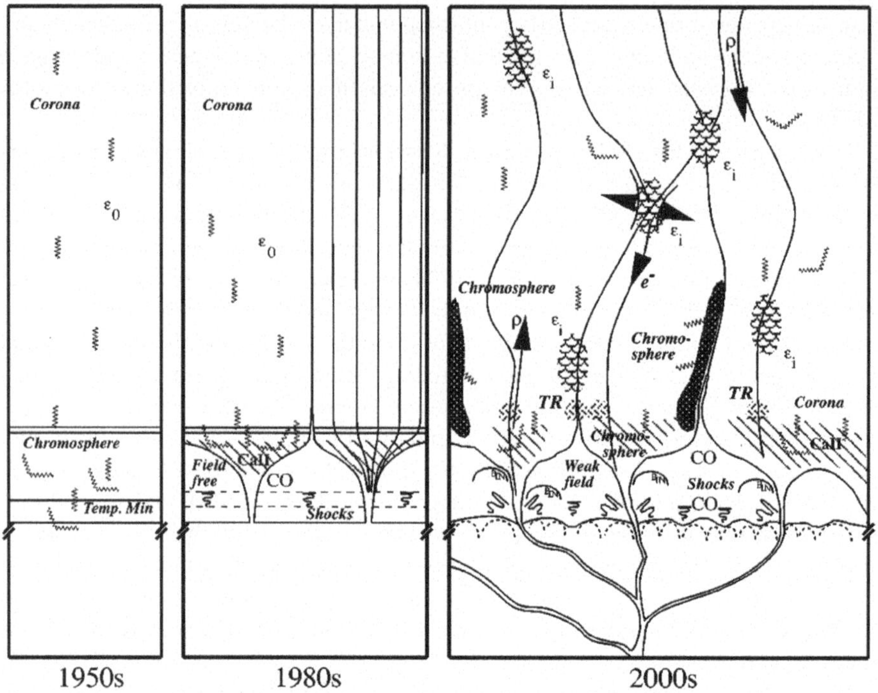

Fig. 6.2 Evolution of the corona cartoon: gravitationally stratified layers in the 1950s (left), vertical flux tubes with chromospheric canopies in the 1980s (middle), and a fully inhomogeneous mixing of photospheric, chromospheric, transition region, and coronal zones by such dynamic processes as heated upflows, cooling downflows, intermittent heating (ε), nonthermal electron beams (e), field line motions and reconnections, emission from hot plasma, absorption and scattering in cool plasma, acoustic waves, and shocks (right) (from Aschwanden et al. 2001)

has been transformed further, in which the conventionally stratified and hydrostatic equilibrium atmosphere is replaced by an intrinsically dynamical, multiscale, and structurized corona (the right panel of Fig. 6.2).

This new view of the corona is further confirmed by latter space-based observations. For instance, the Hinode ("Sunrise" in Japanese) satellite launched in September 2006 has a higher observational resolution of $\sim 0.2''$ and has shown more details about coronal AWs (Erdélyi and Fedun 2007; De Pontieu et al. 2007; Okamoto et al. 2007; Cirtain et al. 2007), small-scale jets (Shibata et al. 2007; Katsukawa et al. 2007), and small-scale magnetic structure and evolution processes (Reale et al. 2007; Sakao et al. 2007; Aulanier et al. 2007; Ichimoto et al. 2007). The SDO (the Solar Dynamics Observatory), the first satellite of the LWS (the loving with a star) program of NASA, was launched on 11 February 2010 and its main goal is to understand the interaction dynamics of solar magnetic fields and plasmas in the solar atmosphere, such as how the magnetic fields are generated and the plasmas are structured by these magnetic fields and how the magneto-plasma interaction energy is stored into local plasma current systems and is released in the corona as solar activities and into the heliosphere as the solar wind (Pesnell 2012).

In fact, the solar corona consists of hot and tenuous low-β plasmas, which are forced to follow solar magnetic field lines and hardly cross the magnetic field lines because of the strong Lorentz force and the insufficiency of the collision (i.e., the high electric conductivity). In consequence, the coronal plasmas are strongly constructed into constantly evolving bright loops or dark filaments by the solar magnetic fields. In active regions of the inner corona, in particular, the magneto-plasma loops, which trace closed magnetic field lines, are the primary structural elements of the solar atmosphere. The non-stationary character of solar plasma magnetic structures manifests itself in various forms of the coronal magnetic loops dynamics (rising motions, oscillations, meandering, twisting; Schrijver et al. 1999; Aschwanden et al. 1999), as well as in the formation, sudden activation and eruption of filaments and prominences. Energetic phenomena, related to these types of magnetic activity, range from tiny transient brightenings (micro-flares) and plasma jets to large-scale flares and CMEs.

Moreover, with observational resolution rising more and more fine-structures in the temporal and spatial have been observed. In fact, up to the diffraction limit of the in being best telescope on board Hinode, the rich dynamical structures on the smallest observable scales imply that the spatial scale of elementary heating processes is still unresolved (Thomas and Weiss 2004; de Wijn et al. 2009). We believe that the elementary structures are well beyond what can be observed even in the foreseeable future as well as today. In conclusion, the solar outer atmosphere is revealed to be an inhomogeneous, intricate, dynamic system, subtle plasma and magnetic-field interactions occur over a wide range of spatial and temporal scales and create rich coronal plasma structures. These plasma structures are shaped strongly by solar magnetic fields and perform various dynamic processes such as heated upflows, cooling down-flows, small-scale jets, ubiquitously propagating MHD and plasma waves, intermittent and transient heating, nonthermal electron beams, eruptive electromagnetic emissions, magnetic flux transports and reconnections.

These space-based high-resolution observations not only revealed a new inhomogeneous and dynamic corona, but also brought us to a modern view of the solar coronal heating problem. In the hot and expanding solar corona the total energy loss consists of three main forms, which are the electromagnetic wave radiation, the heat conduction, and the solar wind outflow. In order to maintain the coronal plasmas at the high-temperature state it is necessary to supply continuously the energy into these coronal plasmas to compensate their energy losses. Based on the modern view of the solar coronal heating problem, we can disintegrate the coronal heating problem into three gradated basic questions about the heating energy, which are necessary to answer for any theoretical model of coronal heating mechanisms, as follows:

The first is the generation mechanism of the heating energy. It is very evident that the heating energy impossibly comes from the cosmic environment surrounding the Sun because those possible cosmic energy sources are either too faraway or too weak to supply energy to heat the solar corona. Thus, the heating energy must come from the Sun itself, that is, from the photosphere as the solar surface or directly from the convective zone under the solar surface;

The second is the transport mechanism of the heating energy. There are three main ways to transport energy fluxes from the solar surface, through the cooler and denser photosphere and chromosphere, into the hotter and tenuous corona. One is the radiation photons from the photosphere, which, however, hardly contribute to heat coronal plasmas because these radiation photons nearly travel entirely transparently through the tenuous corona directly into the sky. Other one is MHD turbulent waves driven by the convection under the photosphere, which propagate and carry wave energy flux into the corona along solar magnetic fields. And another one is field-aligned currents associated with local electric circuits of current-carrying loops, which either are carried by upwards emerging magnetic flux tubes from under the photosphere or are driven by the continuous photospheric motions at the loop footpoints, such as often observed rotation of sunspots;

The third is the transform mechanism of the heating energy. After the heating energy in the forms of waves or currents is transported into the corona, it is necessary that these wave energy or current energy are locally dissipated into the particle kinetic energy of coronal plasmas at the proper efficiency that can match to the local energy loss rate before they are continued to transport to else places, such as into the interplanetary space with the solar wind.

The answer of the first question is evident, that is, the energy of heating coronal plasmas must originate from the convective zone or the photosphere. On the second question it is not controversial either that the turbulent MHD waves (mainly AWs, see discussion below) and local electric circuits both can effectively transport the heating energy into the solar corona. As for the third question, so far it still is an open and focus issue in the study of the coronal heating problem, which is well known as the wave-energy dissipation mechanism (also called AC mechanism) or as the magnetic-energy release mechanism (also called DC mechanism). In principle, the wave energy of propagating in the corona can be dissipated into the kinetic energy of coronal plasma particles through the classical collisional resistance and viscosity or via the kinetic wave-particle interaction. While the magnetic energy associated with some local electric circuits usually is first gradually accumulated and stored in the electric circuits through increasing currents or electric potentials in the circuits and then explosively released via the so-called magnetic field reconnection as solar active phenomena when the currents or potentials approaching to some critical values of the circuit instability. However, in these two types of the energy transformation from the wave energy and the magnetic energy into the kinetic energy of plasma particles there are still some critical and basic important problems, for instance, what is the microphysical process for the wave energy dissipation or the magnetic energy release.

It is believed extensively that the wave-energy (AC) and magnetic-energy (DC) heating mechanisms are responsible for the steady and transient coronal heating processes, respectively. The former contributes mainly to the essential heating of the global high-temperature corona, while the latter mostly corresponds to eruptive heating phenomena of coronal plasmas associated with solar active phenomena such as solar flares. In this chapter we focus our concerns mainly on the former, that is, the wave energy heating of solar coronal plasmas.

Practically, of the three basic modes of MHD waves, AW, fast, and slow magnetosonic modes, AW is unique one that can propagate into the corona nearly without the dissipation. The fast magnetosonic wave can propagate nearly isotropically and hence can carry energy across the local magnetic field. In particular, during propagation the fast waves are refracted into regions of low-Alfvén velocity (Habbal et al. 1979). This can possibly result in one disadvantageous situation of the fast waves in the coronal heating problem, that is, these waves approaching the transition region from the chromosphere are heavily refracted and even are totally internally reflected so that they may never enter the corona. For example, the Brewster angle is only 6° if the ratio of the Alfvén velocities above and below the transition region is 10 (Hollweg 1981).

On the other hand, the slow waves can propagate only at the directions nearly parallel to the magnetic field for the weak ($v_s \gg v_A$, i.e., $\beta \gg 1$) and strong ($v_s \ll v_A$, i.e., $\beta \ll 1$) field cases, while the allowed propagating directions locate within a cone of half angle less than 30° for the intermediate case of $v_s \sim v_A$ (i.e., $\beta \sim 1$). Thus, similar to the acoustic waves, as the density decreases drastically in the upper-chromosphere and the transition region the wave amplitude increases rapidly and easy develops into a shock, which is dissipated quickly in a thin shocked layer. In consequence, also the slow waves originating from the photosphere are very difficult to transport directly the wave energy into the corona due to the strong shocked dissipation in the chromosphere.

Finally, AWs seem to remain as unique viable option to transport the wave energy into the corona. AWs propagate and transport energy exactly along the magnetic field at the Alfvén velocity, and hence never show evanescence or total internal reflection and hardly form shocks either. This is qualitatively agreement with the magnetic correlation of the coronal heating shown in observations. In fact, it is evident that the structurized and dynamic nature of the solar corona should be attributed to solar magnetic fields. The solar corona consists of tenuous and fully ionized plasmas, which are forced to follow the magnetic fields and cannot cross the magnetic fields due to the serious impediment of the strong Lorentz forces. Therefore, the corona is structured into bright loops and dark filaments along the magnetic fields. Nearly uniform brightness along these loops and filaments and clearly sharp boundaries across them imply an approximately constant density and temperature along them but large transverse gradients near their boundaries. The magnetically inhomogeneous feature of the brightness distribution indicates that the coronal heating mechanism has evidently magnetic nature, that is, a magnetically correlation process.

The first heating mechanism involving the role of magnetic fields was proposed by Alfvén (1947), in which the ohmic dissipation of AW currents was proposed to be responsible for the heating of coronal plasmas. Subsequent investigations by many authors (e.g., Piddington 1956; Cowling 1958; Osterbrock 1961) showed that the direct dissipation of AWs due to the classical collisional resistance and viscosity is neither adequate to balance the empirical chromospheric radiative cooling rate (Vernazza et al. 1981) nor effectively to dissipate the energy of AWs in the chromosphere (Narain and Ulmschneider 1990, 1996). This indicates that AWs, if they do exist in below, can pass through the cooler and denser photosphere and chromosphere nearly without dissipation and hence could carry their wave energy into the hotter and tenuous corona along the solar magnetic fields.

In the photosphere, the fluid movements are indeed strong enough to overcome the Lorentz forces, and hence the photospheric fluid turbulence is able to pump energy into photospheric magnetic fields via braiding the magnetic field lines and to transport the energy to the corona along the magnetic fields in the form of AWs. By a simple dimensional argument, Parker (1979, 1983) showed that the flux of the wave energy from the photosphere is much more than that sufficiently to heat and maintain the corona at the observed high temperature. Moreover, in virtue of the recent high-resolution observations aboard the Hinode satellite, the identified evidence for AWs in the corona and the chromosphere has been eventually discovered (De Pontieu et al. 2007; Okamoto et al. 2007; Cirtain et al. 2007; Tomczyk et al. 2007). However, there still is a key difficulty that is how to convert the wave energy of these AWs into the kinetic energy of plasma particles in the corona, because the direct dissipation of the AWs becomes so low in the corona that the AWs pass unhindered through the nearly collisionless coronal plasmas and directly into the interplanetary space.

In order to increase the dissipation rate of AWs in various coronal plasma environments, many more promising dissipation mechanisms of AWs have been proposed, such as distinct resonant heating in coronal loops (Hollweg 1978, 1981; Schwartz et al. 1984), turbulent heating due to Kelvin-Helmholtz instabilities (Hollweg 1984; Heyvaerts and Priest 1983), phase mixing caused by Alfvén velocity gradients (Heyvaerts and Priest 1983; Browning and Priest 1984; Abdelatif 1987), spatial resonance (i.e., resonant mode conversion, Ionson 1978; Lee and Roberts 1986; Hollweg 1987), nonlinear mode coupling (Chin and Wentzel 1972; Uchida and Kaburaki 1974; Wentzel 1974, 1976), and kinetic Landau damping (Stéfant 1970; Hollweg 1971). In the microphysical scenario, the competition among these mechanisms may be attributed to the presence of smaller scale AWs because the dissipation rate of AWs, in general, is proportional to the squaring wave number k^2. However, for the heating rate due to the AW dissipation to be able to balance the energy loss rate of coronal plasmas, small-scale AWs have so small scales close to the particle kinetic scales for typical coronal plasmas, such as the ion gyroradius (ρ_i) or the electron inertial length (λ_e), and much less than the mean free length of coronal plasmas. In these particle kinetic scales AWs become dispersive KAWs and the physics of their wave-particle interaction, especially the kinetic dissipation of the wave energy and the plasma particle heating, is totally different from that of nondispersive AWs as shown in Chap. 1.

Although the coronal plasma heating how to perform still is an unsolved problem, even becomes more complex, the modern coronal heating study has achieved some significant progresses in the comprehensive understanding of heating mechanisms.

Parametric dependence of heating mechanism: First, the observations of the inhomogeneity in the coronal brightness structions indicate that the coronal heating mechanism must be strongly localized, that is, sensitively dependent on local plasma parameters;

Magnetic correlation of heating mechanism: Second, the obviously magnetic correlation of these coronal brightness structures in the observations implies that the heating mechanism should be magnetic correlation, such as the transport and dissipation of AWs or KAWs;

Kinetic characteristics of heating mechanism: Third, the coronal heating mechanism should be a kinetic process due to the wave-particle interaction of occurring at the particle kinetic scales of coronal plasmas because the direct dissipation of AWs due to the classic collisional resistance and viscosity is much too low to balance the local energy loss of coronal plasmas.

It is evident that one of the most natural as well as the most promising candidates for the coronal heating mechanism is the kinetic dissipation of KAWs due to the wave-particle interaction because it can simultaneously have the above three characteristics (see the relevant discussions in Chap. 1). This chapter tries to present a comprehensive discussion of the application of KAWs to the coronal heating problem, also including the chromospheric heating and the solar wind heating. After possible generation mechanisms and observational evidence of KAWs in the solar atmosphere are briefly introduced in Sect. 6.2, the rest sections from Sects. 6.3–6.6 discuss the applications of KAWs to inhomogeneous heating of plasmas in the upper chromosphere, the inner solar corona, the extended solar corona, and the interplanetary solar wind, respectively.

6.2 Generation and Observation of KAWs in Solar Atmosphere

6.2.1 Generation of KAWs by Solar AW Turbulence

The theoretical studies and in situ satellite measurements presented in previous chapters have shown that KAWs, small-scale dispersive AWs, play an important role in the auroral electron acceleration in the magnetosphere-ionosphere coupling dynamics (Chap. 3), in the anomalous particle transport crossing the boundary layers in the solar wind-magnetosphere interaction (Chap. 4), and in the solar wind kinetic turbulence (Chap. 5). As professor Hannes Alfvén believed that theories of cosmic phenomena must also agree with known results from laboratory experiments (Fälthammar 1995; Gekelman 1999), these study results based on in situ measurements by satellites in space plasmas near the earth also must be helpful for us to understand physical processes of remote astrophysical plasmas. In particular in the case of the solar atmosphere, although significant differences, there are strongly analogical phenomena between the terrestrial thermosphere-ionosphere-magnetosphere coupling system and the solar photosphere-chromosphere-corona coupling system. For example, the terrestrial auroras and the solar flares have the common nature driven by magnetic reconnection; the auroral particle energization and the coronal plasma heating have the similar wave-particle interaction processes in the microphysics mechanism; there is the analogical variation features of plasma parameters with height in their outer atmospheres and the alike electrodynamical coupling among their various atmospheric layers due to the vertical magnetic fields; and there is the commonable magnetic driving nature of the atmospheric plasma dynamics.

Up to now, we have not yet explored in situ the solar atmosphere although the Voyager 1 and 2 have gotten the external interstellar space outside the heliosphere. The best resolvable scale of the remote observations of the solar atmosphere still is much larger than the characteristic scale of KAWs, that is, the local kinetic scales of plasma particles in the solar atmosphere. Therefore, it still is impossible to identify directly KAWs in the solar atmosphere, as least, in the foreseeable future. Despite all this, we have good reason to believe that KAWs can ubiquitously exist in the solar atmosphere, just like they exist ubiquitously in space plasmas nearby the Earth. We believe that the most important and most common generation mechanism for KAWs in the solar atmosphere is the transferring of the AWs originating from under the photosphere towards small-scale AWs during their outward propagation along solar magnetic fields.

The theory (Parker 1979, 1983) and observation (De Pontieu et al. 2007; Okamoto et al. 2007; Cirtain et al. 2007; Tomczyk et al. 2007) both show that AWs driven by the fluid turbulence under the photosphere can carry the wave-energy flux into the corona, which is sufficient to heat the solar corona as well as to accelerate the solar wind. Many theoretical works have suggested that the large-scale AWs originating from the solar surface can be converted into smaller and smaller-scale AWs and form a continuous Kolmogorov-like AW turbulence spectrum as they propagate outward and enter the corona along solar magnetic fields through Kelvin-Helmholtz instabilities (Hollweg 1984; Heyvaerts and Priest 1983), the phase mixing (Heyvaerts and Priest 1983; Browning and Priest 1984; Abdelatif 1987), the resonant mode conversion (Ionson 1978; Lee and Roberts 1986; Hollweg 1987), and nonlinear mode coupling (Stéfant 1970; Hollweg 1971; Chin and Wentzel 1972; Uchida and Kaburaki 1974; Wentzel 1974, 1976; Voitenko et al. 2003; Shukla et al. 2004; Zhao et al. 2011), which can be caused by ubiquitous inhomogeneities in the corona.

The Goldreish-Sridhar theory of MHD AW turbulence in an interstellar medium predicts that the turbulent cascading process of AWs has strong anisotropy, in which the wave energy cascades primarily in the cross-field preference with $k_\perp \gg k_\parallel$ (Sridhar and Goldreich 1994; Goldreich and Sridhar 1995, 1997). In result, the large-scale AWs are transformed inevitably into KAWs with small cross-field scales (Howes et al. 2006; Schekochihin et al. 2009). Moreover, as shown in Sects. 5.4 and 5.5, numerical simulations of magneto-plasma turbulence (Cho and Vishniac 2000; Maron and Goldreich 2001; Cho et al. 2002; Müller et al. 2003) and in situ measurements of turbulence in the solar wind (Leamon et al. 1998; Horbury et al. 2008; Podesta 2009; Wicks et al. 2010; Luo and Wu 2010) both further confirmed this anisotropic cascading model of AW turbulence. In the solar atmosphere, similarly, it should be reasonably predictable that the ultimate production of the converted AWs towards small scales must be KAWs, which is dissipated effectively into the kinetic energy of plasma particles by the wave-particle interaction.

Therefore, the most natural as well as the most important generation mechanism of KAWs in the solar atmosphere is the continuous conversion of the AWs originating from the solar surface towards small-scale AWs during their outward propagation. Although we have no directly observed evidence due to the absence of in situ measurements at present, we can reasonably believe that KAWs can exist ubiquitously in the solar atmosphere, just like they ubiquitously exist in space plasmas near the Earth.

On the other hand, as shown in Sect. 6.1, the solar atmosphere is in natural a magneto-plasma coupling system far from dynamical as well as thermal equilibrium, in particular, the solar corona consists of magnetically structurized inhomogeneous plasmas, in which there are various free energies that can drive plasma instabilities to lead to the growth of KAWs, such as the field-aligned current instability presented in Sect. 1.4 and the field-aligned density striation instability presented in Sect. 1.5.

6.2.2 Generation of KAWs by Solar Field-Aligned Currents

As pointed out by Benz (Benz 2002): "No smoke without a fire, and no magnetic field without a current somewhere in the plasma permeated by the magnetic field." Therefore, a description in terms of currents often is physically more interesting than that in terms of magnetic fields (Alfvén and Carlqvist 1967). Complex electrodynamics of the coronal magneto-plasma loops, together with action of some subphotospheric dynamo mechanisms, turn the majority of the coronal loops into current-carrying systems (Khodachenko et al. 2009). Various active phenomena of solar coronal plasmas are closely related to the dynamics of these current-carrying loops and their interaction via the magnetic field and currents (Beaufume et al. 1992; Canfield et al. 1993; Gary and Demoulin 1995; Lee et al. 1998; Zaitsev et al. 1998; Tan et al. 2006; Tan 2007). In solar observations, however, the observable parameter is the magnetic field, and electric currents may be inferred only indirectly. For example, the Ampere's law intimately ties the magnetic field to the current density. Thus, measurements of vector magnetic fields on the photosphere, at least, can provide the information of the vertical electric current flowing from below the photosphere into the corona via the calculation of the vertical component of $\nabla \times \mathbf{B}$.

Since the systematic measurements of the transverse Zeeman effect became available in 1960s, many authors have diagnosed the coronal electric current via the measurements of vector magnetic fields. For example, by using this method Severny (1964) mapped vertical currents of the order of $\sim 10^{11}$ A in small regions ($<10^7$ m in diameter) neighborhood sunspots. Canfield et al. (1993) obtained the total current strength of 6.6–8.8 $\times 10^{12}$ A in solar activity area. Gary and Demoulin (1995) used vector magnetograms observed by Marshall Flight Center to derive the total current strength $\sim 10^{13}$ A in some activity areas. Based on high resolution measurements of vector magnetic field by big bear Lake solar physics observatory, Tan et al. (2006) deduced the total current strength $\sim 3.6 \times 10^{12}$ A.

These currents flowing along the coronal loops presumably arise in the subphotospheric layer due to stresses exerted by the fluid motions at the loop footpoints. As the coronal loop emerges as a magnetic flux tube in the corona, it experiences a transition from a high-β medium in the photosphere to a low-β medium in the corona. This implies that the photospheric motions at the loop footpoints act as the driver of the coronal dynamics via the torsion of the magnetic field lines of the flux tube into the corona (Title et al. 1989). Thus, the continuous photospheric motions at the footpoints can keep driving a field-aligned current along the coronal loop as a continual process (Rosner et al. 1978). As argued by Spicer (1991): "a field-aligned current can be a discharging current of charges accumulated in a photospheric condenser in connection with convection cell motions".

In a series of works, Chen et al. (2011, 2013), Chen and Wu (2012) investigated systematically the generation of KAWs excited by the field-aligned current instability and found that, as shown in Sect. 1.4, the plasma pressure parameter β can significantly influence the excitation of KAWs by the current instabilities because the threshold condition of the current instability, that is, the ratio of the current drift velocity ($v_D \equiv j_\parallel/en_0$) to the local Alfvén velocity (v_A), $v_D/v_A > v_D^c/v_A$, sensitively depends on the local plasma pressure parameter β, where $j_\parallel$ is the field-aligned current density and n_0 is the local ambient electron density. For example, for the low-β case of $\beta < 2Q \ll 1$ the threshold condition $v_D^c/v_A > 1$ requires $v_D^c > v_A > v_{T_e}$ (see Eq. (1.61), Chen et al. 2013). This indicates that in a low-β plasma with $\beta < 2Q$ KAWs could be hardly excited directly by the field-aligned current instability, instead of, the Langmuir wave should be excited preferentially at a higher growth rate when $v_D > v_{T_e}$. In the intermediate-β case of $2Q < \beta \ll 1$ (i.e., $v_A < v_{T_e}$), the field-aligned current instability of KAWs could have a lower threshold drift velocity $v_D^c < v_A$ (Chen and Wu 2012). While in a high-β plasma (Chen et al. 2011) KAWs can be generated easier by the united field-aligned current and temperature anisotropy instability (see Figs. 1.8 and 1.9).

The plasma pressure parameter β undoubtedly has a complex spatial distribution and temporal variation because of the complexity of magneto-plasma structures in the solar atmosphere. Figure 6.3 shows the average distribution and variation range of the plasma pressure parameter β in the solar atmosphere. From Fig. 6.3, the plasma pressure parameter $\beta\sim$ or $>$ 1 in the photosphere and lower chromosphere ($R < 10^3$ km) as well as the extended solar corona and the solar wind ($R > 10^5$ km). However, from the upper chromosphere through the inner corona (i.e., $10^3 < R < 10^5$ km) the magnetic pressure generally dominates well the static gas pressure (i.e., $\beta < 1$), where strongly dynamical processes of local plasma current systems and related solar activities perform continuously due to interactions between magnetic fields and plasmas.

For the case of current-carrying loops, which exist ubiquitously in the solar corona and have the plasma pressure parameter β of ranging typically $1 \gg \beta \gg 2Q$, it is believed extensively that the coronal electric currents deduced from vector magnetograms usually are concentrated in a set of current-carrying loops centered on a neutral line of active regions (Hagyard 1989; Canfield et al. 1993; Melrose et al. 1991; Tan et al. 2006). This implies that the current of current-carrying loops

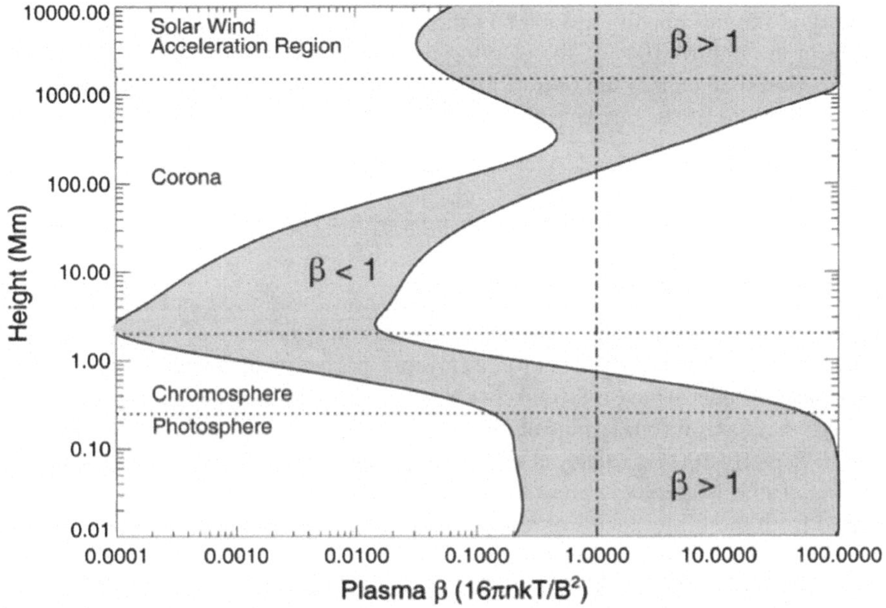

Fig. 6.3 Average distribution and variation range of the plasma pressure parameter β in the solar atmosphere: the boundaries of the average β regime (denoted by grey region) are given by assuming photospheric magnetic fields, 100 G (right) and 2500 G (left), which represent typical magnetic field strengths in solar quiet and active regions, respectively. The magnetic pressure generally dominates the static gas pressure (i.e., $\beta < 1$) from the upper chromosphere through the inner corona (i.e., $10^3 < R < 10^5$ km). As with all plots of physical quantities against height, a broad spatial and temporal average is implied (from Aschwanden et al. 2001)

actually is confined to current channels in these coronal loops. For instance, the so-called skin effect results in the current to stay at a narrow current-shell (Beaufume et al. 1992) or a thin current-rope (Tan 2007) in an individual loop. Thus, the cross-section of the real current channel in a loop probably is much less than the cross-section of the loop, that is, the so-called filling factor for the coronal electric current density in an individual coronal loop

$$f \equiv \left(\frac{d}{D}\right)^2 \ll 1, \tag{6.1}$$

where d and D are the cross-section scales of the real current channel and the loop, respectively. In consequence, the real field-aligned drift velocity of current-carrying electrons relative to ions in the current channel of a current-carrying coronal loop can be estimated from the observed current intensity I:

$$V_D = \frac{1}{f}\frac{4I}{en\pi D^2} \simeq \frac{I_{12}}{n_{15}D_6^2} \times \frac{10}{f}\ \mathrm{km/s}, \tag{6.2}$$

where f is the current filling factor in the coronal loop, I_{12}, n_{15}, and D_6 are the current in the unit of 10^{12} A, the electron density in the unit of $10^{15}\,\mathrm{m}^{-3}$, and the cross-section diameter in the unit of 10^6 m in the coronal loop, respectively. This leads to the ratio of the current drift velocity to the Alfvén velocity in the coronal loop

$$\frac{V_D}{v_A} = \frac{I_{12}}{D_6^2\sqrt{n_{15}}}\frac{1}{7Bf}, \tag{6.3}$$

where the Alfvén velocity $v_A \simeq 70B/\sqrt{n_{15}}$ km/s has been used and B is the local magnetic field strength (in the unit of G) in the coronal loop.

Based on the observations of coronal currents, for a typical current-carrying loop in the solar corona, we have total current $I \sim 0.1\text{–}10 \times 10^{12}$ A, cross-section diameter $D \sim 10^6$ m, electron (or i.e. plasma) number density $n \sim 10^{15}\,\mathrm{m}^{-3}$, magnetic field strength $B \sim 10\text{–}100$ G (Zaitsev et al. 1998; Tan 2007; Malovichko 2008; Aschwanden et al. 1999; Nakariakov and Ofman 2001). The current filling factor f can be inferred from the filling factor of hot plasmas in the coronal loop, to be $f \sim 10^{-1}$-10^{-4} (Martens et al. 1985; Beaufume et al. 1992). From Eq. (6.3) the field-aligned drift velocity of electrons V_D can plausibly be estimated as $V_D/v_A \sim 0.1$ to 10. The results show that KAWs can be excited effectively by the field-aligned current instability at a high growth rate $\gamma \sim 0.01\text{–}0.1\,\omega_{ci} \sim 10^3\text{–}10^4\,\mathrm{s}^{-1}$ in the ubiquitous current-carrying loops in the solar corona (Chen and Wu 2012).

Zhao et al. (2013) further investigated the spectral structure of KAW turbulence in solar coronal loops. Their results show that the spectra of the magnetic and electric field fluctuation display an evident transition at near the electron inertial length scale and the turbulence cascades mainly toward the direction perpendicular to the local unperturbed magnetic field at the small scale smaller than the electron inertial length, as expected by the theory of KAW turbulence in the Chap. 5. In particular, the parallel electric field increases as the turbulent scale decreases and may be much larger than the Dreicer field by a factor of $\sim 10^2\text{–}10^4$ when the turbulent scale approaches the electron inertial length.

6.2.3 Generation of KAWs by Solar Field-Aligned Striations

Other one of important free energy sources for driving the instability growth of KAWs is the ubiquitous density inhomogeneity in the solar atmosphere (Wu and Chen 2013; Chen et al. 2015). The solar corona consists of hot and tenuous low-β plasmas, which are forced to follow solar magnetic field lines and hardly cross the magnetic field lines because of the magnetized characteristic due to the Lorentz force. In consequence, the coronal plasmas are strongly constructed into constantly evolving bright loops or dark filaments by the solar magnetic fields, as shown in modern space-based observations. Therefore, these field-aligned filamentous structures, called the field-aligned density striations, are characterized by a strongly anisotropical density distribution, in which the parallel length scales of inhomogeneity are much larger than the perpendicular inhomogeneity scales and the most common perpendicular scales are in the orders of the kinetic scales of plasma particles, such as the ion gyroradius and the electron inertial length (Wu 2012).

Field-aligned density striation is one of the most ubiquitous inhomogeneity phenomena in magnetized plasmas such as space plasmas in the corona and the aurora as well as magnetically confined plasmas in laboratory. For example, high-resolution observations have now given an impressive image of the solar corona as a rapidly evolving dynamic plasma where energetic phenomena occur mainly on very small scales that are characterized by density striations (Golub et al. 1990; Aschwanden et al. 2001). On the other hand, auroras and associated plasma processes have various classes of spatial and temporal scales, and change from whole auroras in hundreds of kilometer to microscale structures (such as narrow discrete arcs, auroral curls and folds, and flickering auroras) in hundreds of meter, which is comparable to the ion gyroradius (Hallinan and Davis 1970; Mcfadden et al. 1990; Borovsky and Suszcynsky 2013; Trondsen and Cogger 1997, 1998). Particle simulation by Tsiklauri (2011, 2012) also showed that KAWs can be effectively excited by the transverse density inhomogeneity and that the parallel and perpendicular electric fields of the excited KAWs can efficiently lead to the field-aligned acceleration of electrons and the cross-field heating of ions, respectively, as expected by the theory of KAWs.

The similar phenomena of spontaneously producing KAWs in field-aligned density striations also have been observed in laboratory plasmas, as shown in Sect. 2.4. Motivated by the need to understand the dynamics of waves generated spontaneously in field-aligned density striations, Maggs and Morales (1996, 1997) designed and performed a laboratory experiment in LAPD at UCLA, in which a controlled field-aligned density depletion was produced by selective coating of an emissive oxide coated cathode and density and magnetic fluctuations generated spontaneously in the density striation were measured. They identified the nature of the fluctuations as the drift AW for $\beta > 2Q$ and as broadband magnetic shear AW turbulence for $\beta < 2Q$, as shown by Fig. 2.6. The driving source for the fluctuations is the cross-field density (Fig. 2.6) and temperature (Fig. 2.7) gradients in the edge of the density striations, which have typical scales close to the electron inertial length.

Based on the two-fluid model of magnetized plasmas, Wu and Chen (2013) and Chen et al. (2015) investigated the dispersion relation and the instability excitation of KAWs in a magneto-plasma with density striation structures. Their results show that the KAWs become instable in the presence of the density striation and the corresponding instability excitation has a maximal growth rate at the perpendicular wavelength close to the spatial scale of the density gradient perpendicular to the magnetic field (see Sect. 1.5), as shown by the laboratory experiment (Maggs and Morales 1996, 1997) and the particle simulation (Tsiklauri 2011, 2012). For some typical solar coronal plasma parameters, the growth rate of the KAW instability can reach $\sim 1/3$ of the wave frequency, implying a very effective growing of KAWs.

On the other hand, the KAW growth rate excited by the field-aligned density striation instability increases as the plasma pressure parameter β decreases, and hence the field-aligned density striation instability has a higher growth rate in a low-β plasma of $\beta < 2Q$ (i.e., $\alpha_e < 1$) and the growth rate rapidly falls below the effectively exciting level when $\beta > 2Q$ (i.e., $\alpha_e > 1$) or $\lambda_e k_x > 1$ (see Figs. 1.10 and 1.11). This is obviously different from the field-aligned current instability, which leads to a more effective generation of KAWs in plasmas with $\beta > 2Q$ but hardly an effective excitation of KAWs in a low-β plasma of $\beta < 2Q$. This indicates that the field-aligned density striation can effectively produce KAWs with a perpendicular wavelength larger than the electron inertial length in the inertial parametric regime of $\alpha_e < 1$, as shown by observations in auroral plasmas.

The so-called Bennett pinching effect can establish a close relation between these two excitation mechanisms for KAWs, the field-aligned electric current and density striation instabilities. In fact, the Bennett pinching may lead to the concentration of field-aligned currents and hence to the simultaneous increase of the local current density and the local plasma density. Thus, in a plasma region with a local plasma pressure parameter $\beta > 2Q$ the increasing current density can effectively generate KAWs by the field-aligned current instability. While in a plasma region with a local parameter $\beta < 2Q$, the increasing plasma density leads to the formation of field-aligned density striations and may effectively produce KAWs by the field-aligned density striation instability.

Besides ubiquitous field-aligned currents and striations, fast electron and ion beams presented often in the solar atmosphere, especially during solar activities, also may excite effectively the instability growth of KAWs. Voitenko (1996, 1998) studied the excitation of KAWs by proton-beams produced in a flare and proposed that the KAWs are possibly responsible for the flaring loop heating. Chen et al. (2014) investigated the growth rate of KAWs excited by fast electron beams in an intermediate-β plasma (i.e., $1 \gg \beta \gg 2Q$), which is the case of typical solar coronal plasmas. Their results show that the maximal growth rate of KAWs depends sensitively on the feature velocity of fast electron beams, v_b, and the most favorable beam velocity occurs between $8v_A < v_b < 10v_A$. However, the maximal relative growth rate of KAWs by fast electron beams, in general, is only $\sim$1% of the wave frequency, and hence possibly is not important excitation source of KAWs in the solar atmosphere.

6.2.4 Possible Solar Observation Evidence of KAWs

In general, KAWs and associated processes are very difficult to be observed directly because of the limitation of observational resolution of telescopes. However, Wu et al. (2007) reported a novel kind of fine structures in solar radio bursts, called "Solar Microwave Drifting Spike" (SMDS), which were detected by the Solar Broadband Radio Spectrometer (SBRS) of the National Astronomical Observatories of China (NAOC) with high temporal ($\sim$1 ms) and spectral ($\sim$0.3%) resolutions (Fu et al. 1995). Their analysis suggested that these SMDSs probably are produced by a group of SKAWs, in which the electrons in the SKAWs are accelerated by the SKAW electric fields to tens of keV and trapped within the SKAW potential well. It is these

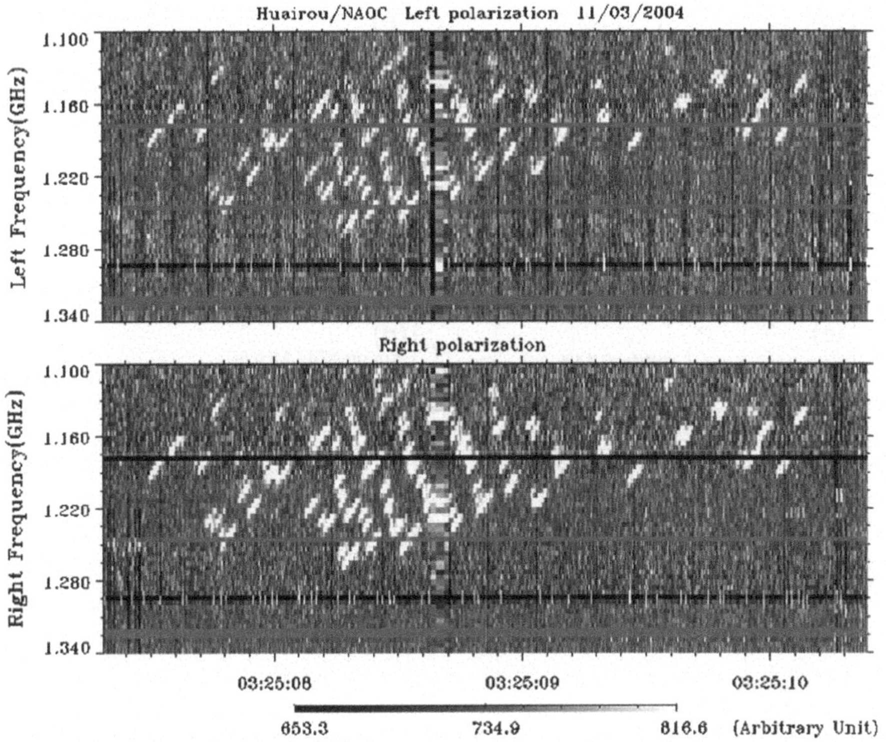

Fig. 6.4 Dynamic spectra of SMDSs observed by SBRS/NAOC on November 3, 2004. Top: Left-hand polarized component. Bottom: Right-hand polarized component (from Wu et al. 2007, ©AAS reproduced with permission)

trapped electrons that trigger the SMDSs, whose frequency drifts are attributed to the SKAW propagation along the magnetic field.

In order to deeply analyze the fine structures of solar microwave bursts a fast scanning mode of SBRS with high temporal (1.25 ms) and spectral (4 MHz) resolutions had worked in the frequency band of 1.10–1.34 GHz for six months (from 2004 October to 2005 March). Figure 6.4 shows a group of SMDSs observed on November 3, 2004, where top and bottom panels are the dynamical spectra of their left and right polarization components, respectively. This group consists of 65 individual SMDSs during from 03:25:7.5 UT to 03:25:10.2 UT in the frequency range between 1.112 and 1.268 GHz, which appeared just at the beginning of a major radio burst associated with a flare in the active region NOAA 10696 localized at N09E45 on the solar disk.

The SMDSs are characterized typically by a short lifetime (∼tens of ms), a narrow bandwidth (relative bandwidth $<1\%$), and an intermediate frequency drift rate ($\sim$ a few hundred of MHz/s). Figure 6.5 presents their observable features, where the horizontal coordinates are the center times of the SMDSs in second after 03:25 UT.

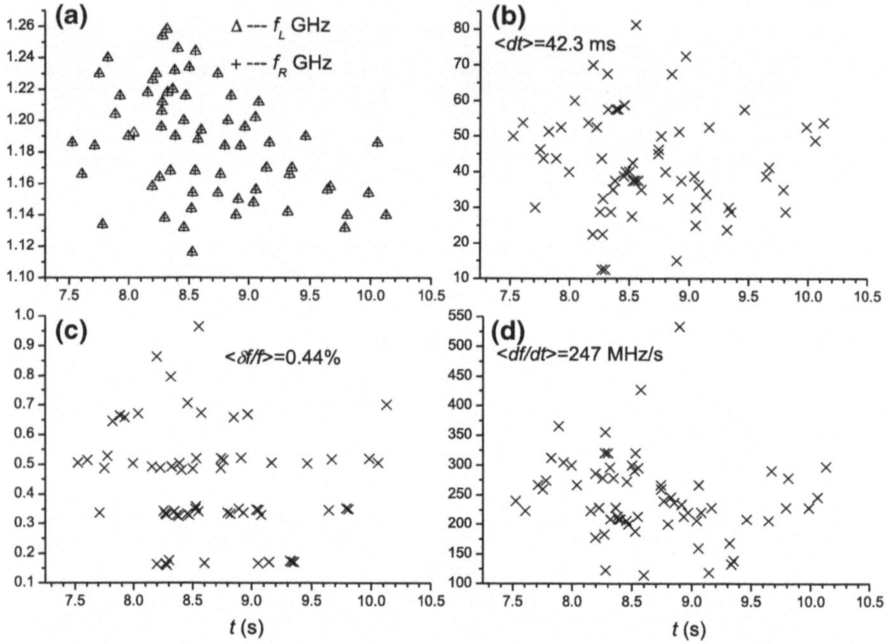

Fig. 6.5 Characteristic parameters of the SMDSs. **a** Their central frequencies (in GHz), where triangles and plus signs represent their left and right polarization components, respectively. **b** Their lifetimes (in ms). **c** Relative bandwidths (in percent). **d** Frequency drifting rates (in MHz/s) (from Wu et al. 2007, ©AAS reproduced with permission)

Figure 6.5a presents their center frequencies in GHz, where the triangle ("Δ") and cross ("+") represent the left and right polarization components, respectively. From Fig. 6.5a, all SMDSs have the left and right polarizations in pairs but the very low polarization degrees even lower than the observational resolution (10%). Figure 6.5b shows their lifetimes (dt), which are in the order of several tens of ms and have an average lifetime of 42.3 ms. Their relative bandwidths ($\delta f/f$) are given in Fig. 6.5c, which all are below 1% and have an average bandwidth of 0.44%. Finally, Fig. 6.5d displays their frequency drifting rates, which are in the order of a few hundreds of MHz/s and have an average drifting rate of 247 MHz/s. In comparison to solar microwave bursts reported previously, the intermediate drifting rates of the SMDSs are similar to that of the fiber bursts (Bernold and Treumann 1983; Aurass and Kliem 1992), but their lifetimes and frequency drifting ranges are much less than that of the fiber burst. On the other hand, their short lifetimes and narrow frequency drifting ranges are similar to that of the millisecond radio spikes, which are believed to be generated by the electron cyclotron maser emission (ECME; Benz 1986; Messmer and Benz 2000; Fleishman et al. 2003), but their frequency drifting rates are much less than that of the millisecond spikes.

The narrow bandwidth imposes fairly stringent constraints on the microphysics that is responsible for the generation of the SMDSs because the theoretical bandwidth does not exceed the observed bandwidth. The observed bandwidth may be a superposition of various factors including the natural bandwidth of the emission line and its broadening due to the inhomogeneity of the emission source, the turbulent motion of the emission, and the scattering of the emission during its propagating from the source to the observer. The natural bandwidth of ECMEs is typically in the range 0.1–0.4% for the plasma environments in the solar corona (Fleishman 2004). In particular, the observed bandwidth constraints the possible sizes of the emitting sources and the upper limit of sizes of individual SMDS sources may be estimated by $l < (\delta f/f)L$, where L is the scale of magnetic inhomogeneity in the source region (Benz 1986). Taking $L \sim 1{,}000$ km, one has a size of individual SMDS source less than 10 km in the present observation.

According to the model of solar radio active regions (Zhao 1995), the height of the local source for cyclotron emissions at frequency f (in units of GHz), H, can be estimated by

$$H = D\left[\left(\frac{11.2B_0}{f}\right)^{1/3} - 1\right], \tag{6.4}$$

where the magnetic field has been assumed to be a dipole field with a depth D under the photosphere and a strength B_0 (in units of kG) on the photosphere. Thus, the moving velocity of the source can be calculated by

$$v_d \equiv \frac{dH}{dt} = \frac{D}{3}\left(\frac{11.2B_0}{f}\right)^{1/3}\frac{1}{f}\frac{df}{dt}, \tag{6.5}$$

where df/dt is the frequency drifting rate of the source. Taking the depth $D \sim 3.5 \times 10^4$ km and the magnetic strength $B_0 \sim 1.5$ kG, we have the mean height of the emitting sources of the SMDSs $H_s \sim 5 \times 10^4$ km and the mean moving velocity of the sources $v_{sd} \sim 6 \times 10^3$ km/s, where the mean emitting frequency $f_s \simeq 1.186$ GHz and frequency drift rate $df/dt \simeq 247$ MHz/s have been used. The mean magnetic field of the source regions can be estimated by the mean emitting frequency ($f_s = 1.186$ GHz) as $B_s \sim 100$ G.

The energetic electrons that trigger ECME (Benz 1986; Messmer and Benz 2000; Fleishman et al. 2003) have typical energy range from a few tens to one hundred keV. This implies that the electron velocity ranges from several tens to one hundred 10^3 km/s, much larger than the drift velocity of the source $v_{sd} \sim 6 \times 10^3$ km/s, above inferred by the observed frequency drifting rate of the source. On the other hand, the solar coronal plasma has a typical density $\sim 10^9$ cm^{-3} at the altitude of the source ($H_s \sim 5 \times 10^4$ km), implying a local Alfvén velocity $v_A \sim 7 \times 10^3$ km/s in the source region, which is close to the source drifting velocity v_{sd}. This indicates that the sources of these SMDSs are moving with the local Alfvén velocity, when the ambient plasma density $n_0 \simeq 1.3 \times 10^9$ cm^{-3}. This result proposes that the SMDSs can be explained in the SKAW scenario.

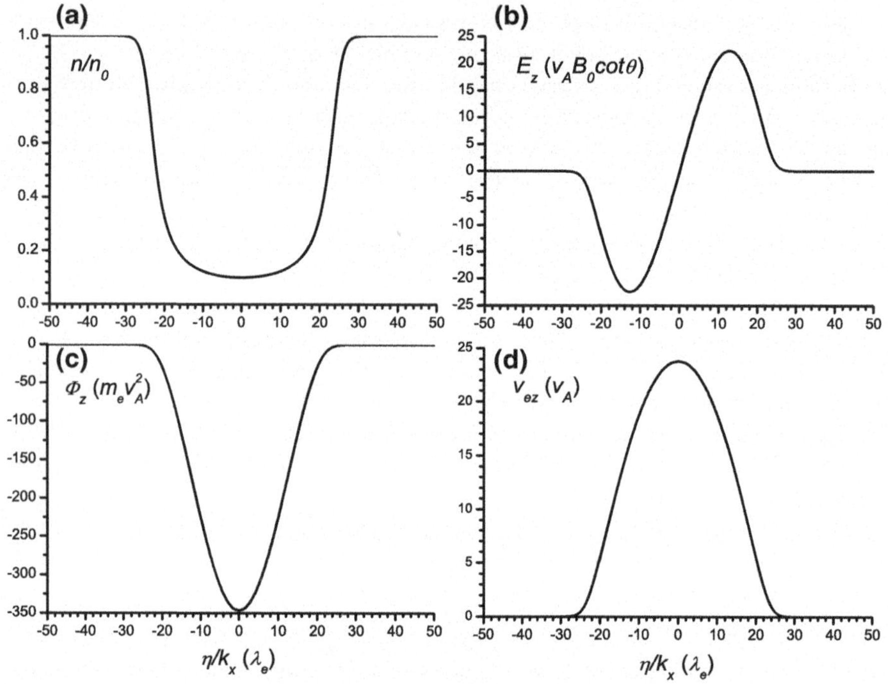

Fig. 6.6 Structure of SKAWs. **a** Dip density SKAW with amplitude 0.9. **b** Distributions of the parallel electric field. **c** Electric potential. **d** Electron velocity in the SKAW (from Wu et al. 2007, ©AAS reproduced with permission)

Figure 6.6 shows an example of SKAWs with a density amplitude 0.9, where panels (a), (b), (c), and (d) are the distributions of the density (n_e), the electric field (E_z), the electric potential (Φ_z), and the electron velocity (v_{ez}), respectively, in the SKAW, which propagates at the local Alfvén velocity v_A along the magnetic field. From Fig. 6.6, the SKAW has a cross-field size $\sim 50\lambda_e$. In particular, the electrons trapped in the SKAW by the potential well can be self-consistently accelerated by the SKAW electric field to a velocity $\sim 20v_A \simeq 0.4c$ km/s, that is, with an energy ~ 50 keV, which is the typical energy of the energetic electrons triggering solar microwave bursts.

In recent works (Wu et al. 2002; Yoon et al. 2002; Chen et al. 2002) an electron beam-maser instability based on ECME has been proposed to explain solar radio bursts, in which the streaming electrons possess a distribution with a perpendicular population inversion; such distribution may be formed by the pitch-angle scattering of the streaming electrons by ubiquitous turbulent waves in solar coronal plasmas. Observations of SKAWs by satellites in space plasmas show them to be frequently accompanied by various turbulence (e.g., Wahlund et al. 1994a). Therefore, we believe that the electrons trapped within SKAWs can be scattered by the accompanying turbulence and form an inverted velocity distribution in the perpendicular velocity space. It is these trapped electrons in the SKAWs that directly trigger the observed SMDSs through their ECME. Thus, the frequency drifts of the observed SMDSs naturally are attributed to the propagation of the SKAWs along the magnetic field. This is different from that of solar type III bursts, which is directly due to the motion of the triggering electrons. Also, the combination Alfvén solitons with electron cyclotron maser emission was ever proposed to explain solar intermediate drift bursts, where the solitons just modulate the emission induced by pre-existing energetic electrons (Benz and Mann 1998; Treumann et al. 1990; Dabrowski et al. 2005). In the present scenario, however, the emitting electrons are produced self-consistently by the SKAWs.

The SKAW lifetime may be estimated approximately by the inverse of the KAW damping rate, $\gamma \sim 0.5\nu_e k_\perp^2 \lambda_e^2$ (Voitenko and Goossens 2000), where the electron collision frequency $\nu_e = 2.91 \times 10^{-6} n_e \ln \Lambda T_e^{-3/2}$ and $\ln \Lambda$ is the Coulomb logarithm. Taking the typical temperature $T \sim 100$ eV for coronal plasmas, one has $\ln \Lambda \sim 20$ and $\nu_e \sim 100$ Hz. On the other hand, the wavelength of linear KAWs may be estimated by the characteristic scale of the inhomogeneity in the SKAWs, $\lambda_\perp \sim 10\lambda_e$ in Fig. 6.6a. In consequence, the damping rate $\gamma \sim 20$ Hz, that is, the SKAW lifetime $\tau \sim \gamma^{-1} \sim 50$ ms, which is consistent with the observed SMDS lifetime.

The typical field-aligned size of the SKAWs can be estimated by the average relative bandwidth $\delta f/f \sim 0.44\%$ of the SMDSs, as $l \sim 0.44\% L \sim 4$ km, based on the emitting source size $l \sim (\delta f/f)L$ (Benz 1986), where the scale of the magnetic inhomogeneity in the source region $L \sim 1,000$ km has been used. Obviously, the field-aligned size of the SKAWs is much larger than their cross-field size $\sim 50\lambda_e \sim 10$ m (see Fig. 6.6). This indicates that the SKAWs propagate quasi-perpendicular to the magnetic field (Wu and Chao 2004b). Finally, the frequency of the SKAWs $\sim v_A/l \sim 1.5 \times 10^3$ Hz, which is much less than the local ion gyrofrequency $f_{ci} \sim 1.5 \times 10^5$ Hz as expected by the SKAW theory.

In observation, Wu et al. (2007) for the first time presented probably observational evidence of SKAWs in the solar corona. They reported a novel kind of fine structures in solar microwave drifting spikes (SMDSs), which were detected by SBRS (the Solar Broadband Radio Spectrometer) of NAOC (the National Astronomical Observatories of China) with high temporal ($\sim$1 ms) and spectral ($\sim$0.3%) resolutions. These SMDSs are typically characterized by a short lifetime ($\sim$tens ms), a narrow bandwidth (a relative bandwidth $<1\%$), and an intermediate frequency drift rate ($\sim$ a few hundreds of MHz/s). Their analysis suggests that these SMDSs are probably produced by a group of SKAWs, in which the electrons trapped in the potential of the SKAWs are field-aligned accelerated by the SKAW electric fields to tens of keV. It is these trapped electrons in the SKAWs that trigger the SMDSs. The narrow bandwidth of the SMDSs implies the small size of the emitting sources, that is, the SKAWs, their short lifetime can be explained by the dissipation time of the SKAWs in the local plasma environment, while the frequency drifts of the SMDSs are attributed to the propagation of the SKAWs along the magnetic field at the local Alfvén velocity.

6.3 Upper Chromospheric Heating by Joule Dissipation of KAWs

6.3.1 Sunspot Upper-Chromospheric Heating Problem

Sunspots are dark magnetic structures on the solar surface because their strong magnetic fields inhibit the normal energy transport by the convection below the solar photosphere, so that their temperature is lower than the surrounding quiet photosphere by ~2000 K. However, the temperature increases more quickly with heights in a sunspot than in the quiet solar atmosphere outside the sunspot. In result, the sunspot chromosphere and corona have higher temperature than the surrounding chromosphere and corona of the quiet solar atmosphere (Avrett 1981; Maltby et al. 1986).

Based on the spectral profiles of Hα, Hβ, CaII H, K and infrared (λ8498, λ8542, λ8662) lines observed with the 13.5 m vacuum spectrometer of the McMath telescope at Kitt Peak Observatory on July 1, 1985 and on the non-LTE (local thermal equilibrium) radiative transfer calculation, Ding and Fang (1989, 1991) obtained the semiempirical atmospheric models of a sunspot penumbra and umbra, respectively. Figure 6.7 depicts the results for their umbra model (solid lines), where the panels (from the top down) are the temperature variation with heights, the electron and total hydrogen atom (ionized and neutral) number densities, and the ionization estimated simply by $\alpha = n_e/(1.2n_H)$. Here, the factor of 1.2 is because the total electron number is contributed by the electrons from other heavy ions as well as the ionized hydrogen. For comparison, the corresponding atmospheric parameters in the VAL (Vernazza, Avrett, and Loeser) quiet solar model given by Vernazza et al. (1981) are plotted in the dashed lines in the figure.

From Fig. 6.7 one can see that the sunspot atmosphere has height variation very similar to that of the VAL quiet solar atmosphere, except the sunspot atmosphere varies more quickly through the chromosphere to the corona after the temperature minimum region at ~500 km, where there is interface between the photosphere and the chromosphere. In particular, from Fig. 6.7, it is worth to notice that there is a higher ionization besides a high magnetic field strength in the sunspot upper chromosphere. It can be expected that there is some heating mechanism in the sunspot upper chromosphere, which must be associated with plasma processes as well as magnetic fields because there is a higher ionization as well as a higher magnetic field strength in the sunspot upper chromosphere.

What causes the heating of the upper solar atmospheres has long been an outstanding problem in solar physics (Ulmschneider 1991). Three types of processes have been proposed to explain the chromospheric heating: (1) heating by upward-propagating acoustic waves generated in the convection zone (Carlsson and Stein 1992); (2) resistive dissipation of direct currents $\mathbf{j} = \nabla \times \mathbf{B}/\mu_0$ (Rabin and Moore 1984; Goodman 2004); and (3) local ohmic heating driven by magnetic reconnection (Parker 1988). Recent results based on observations of the TRACE satellite, however, show that acoustic waves do not carry enough energy to sustain the heating rate of the

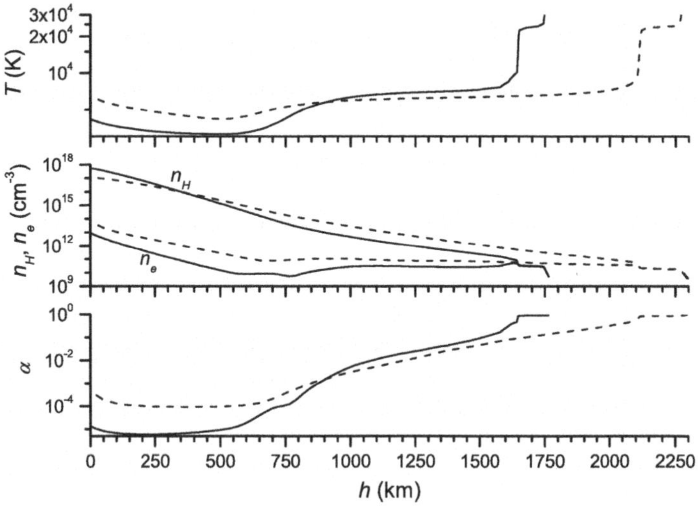

Fig. 6.7 Semi-empirical atmospheric model of a sunspot: panels (from the top down) are temperature, electron and atom densities, and ionization, where dashed lines represent those of the average quiet solar atmosphere (from Wu and Fang 2007, ©AAS reproduced with permission)

internetwork chromosphere in solar non-magnetic areas (Fossum and Carlsson 2005). Meanwhile, Socas-Navarro (2005) analyzed three-dimensional vector currents and temperatures observed in a sunspot from the photosphere to the chromosphere and concluded that resistive current dissipation can not provide the dominant source to heat the sunspot chromosphere in that case. In addition, since magnetic reconnection is associated with strong current sheets, reconnection also may not be important, at least in the relatively simple magnetic topology of sunspots. The fact that the temperature increases more quickly in a sunspot with stronger magnetic fields and the sunspot chromosphere has a higher temperature implies that magnetic waves such as AWs may play an important role in the chromosphere (Socas-Navarro 2005; Jefferies et al. 2006).

De Pontieu et al. (2001) studied the chromospheric damping of MHD AWs due to ion-neutral collisions for various solar structures, such as active region plage, quiet sun, and sunspots, and found that the ion-neutral damping of AWs is the most effective for the sunspot case. Their results show that the maximum damping always occurs around the temperature minimum region of the solar atmosphere, where is the interface between the photosphere and the chromosphere. In particular, the ion-neutral damping can contribute significantly to the energy balance in the region between 500 and 1000 km. In the upper-chromosphere above 1000 km, however, the heating rate due to the ion-neutral damping quickly decreases, and hence is much lower than the local radiation cooling rate (De Pontieu et al. 2001).

6.3.2 Joule Dissipation of KAWs by Collision Resistance

Current of KAWs is contributed mainly by the field-aligned motion of electrons in the wave. Therefore, when the electron collision frequency is enough high in comparison with the wave frequency the wave energy is effectively dissipated by the electron collision damping, that is, so-called the Joule dissipation. Low solar atmospheres in the photosphere and chromosphere consist of partly ionized plasmas, in which the ionization increases with heights from lower than 10^{-5} in the photosphere to close 1 (i.e., nearly fully ionized) at the top chromosphere, as showed by the dashed line in Fig. 6.7. In such partly ionized plasmas, the electron collision frequency is contributed by the electron-ion collision (Coulomb collision due to the long-range Coulomb force) and the electron-atom collision (neutral atom collision due to the short-range molecule force). For the Coulomb collision between electrons and ions, we have

$$\nu_{ei} = \frac{8 \ln \Lambda}{3\pi^2} \frac{\omega_{pe}}{N_D} \simeq 110 n_e T^{-3/2} \tag{6.6}$$

where $\ln \Lambda = 10$ has been used and n_e and T are in units of cm^{-3} and K, respectively. While for the electron-atom collision, we have

$$\nu_{ea} = \frac{4}{\pi}\sqrt{\frac{T}{m_e}}\sigma_a n_a \simeq 3.86 \times 10^{-10} \frac{1-\alpha}{\alpha} n_e T^{1/2} \tag{6.7}$$

where n_a and α are the atom number density and the ionization, respectively, and the atom collision cross-section $\sigma_a = 4.4 \times 10^{-16}\ \mathrm{cm}^{-3}$ and the ionization relation $n_a = (1-\alpha)\, n_e/\alpha$ have been used.

For the sunspot atmospheric model presented in Fig. 6.7 the electron collision frequencies by ions and atoms are showed by solid and dashed lines in Fig. 6.8, respectively. From Fig. 6.8, it is clear that the electron-atom collision frequency ν_{ea} decreases monotonously with increasing height from the photosphere through the upper chromosphere, as expected. While the Coulomb collision frequency decreases in the photosphere and the lower chromosphere below the height ~800 km, then slightly increases and maintains almost constant in the upper chromosphere until the height ~1650 km, and above this drastically decreases after entering the transition region and the corona because the great density decreases and temperature increases. In particular, the Coulomb collision frequency ν_{ei} becomes larger than the neutral collision frequency ν_{ea} and dominates the collisional electric resistance in the upper chromosphere because of the increase of the ionization, where the ionization is considerably larger than 10^{-3}.

Including the electron collision effect, the continuity and field-aligned motion equations of electrons can be written as

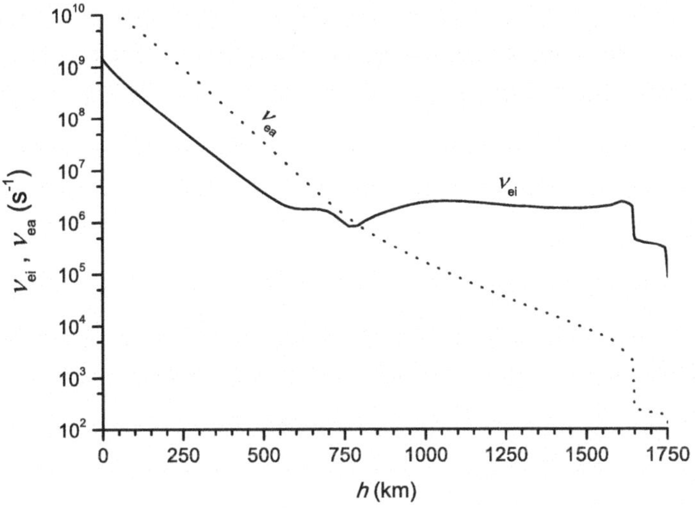

Fig. 6.8 Electron collision frequencies by ions (solid line) and by neutral hydrogen atoms (dashed line) (from Wu 2012)

$$\begin{aligned}\frac{\partial n_e}{\partial t}+\frac{\partial\left(n_e v_{ez}\right)}{\partial z}&=0\\ \frac{\partial v_{ez}}{\partial t}+v_{ez}\frac{\partial v_{ez}}{\partial z}&=-\frac{e}{m_e}E_z-\nu_{ei}v_{ez}.\end{aligned} \tag{6.8}$$

For KAW with frequency ω and parallel wavenumber k_z, Eq. (6.8) leads to

$$\begin{aligned}v_{ez}&=\frac{\omega}{k_z}\delta n\\ E_z&=i\frac{m_e\omega}{e}\frac{\omega+i\nu_{ei}}{k_z}\delta n,\end{aligned} \tag{6.9}$$

where $\delta n=\delta n_e/n_e$ is the relative density fluctuation. In result, the Joule heating rate due to the classical collisional resistance of KAWs, including the Coulomb and neutral collisions, can be given by

$$\begin{aligned}Q_{HK}&\simeq\frac{1}{2}\text{Re}(j_zE_z^*)=\frac{1}{2}m_en_e\nu_{ei}\left(\frac{\omega}{k_z}\right)^2(\delta n)^2\\ &\simeq\frac{1}{2}m_en_e\nu_{ei}v_A^2(\delta n)^2\\ &\simeq 9.3\times10^{-4}n_e^2n_H^{-1}T^{-3/2}B_0^2\delta n_K^2,\end{aligned} \tag{6.10}$$

where n_H is the total atmospheric density (atoms and ions), δn_K denotes the relative density fluctuation for KAWs, the prefixion "Re" denotes the real part of complex, the superscript "$*$" denotes the complex conjugate, and the current $j_z \simeq -en_e v_{ez}$ has been used. In addition, the parallel phase speed of KAWs has been approximated by the local Alfvén velocity, i.e., $\omega/k_{\parallel} \simeq v_A$. This is an enough good approximation for the weakly dispersive case of $k_{\perp}\lambda_e$ and $k_{\perp}\rho_i \ll 1$.

For the sake of comparison, here we present the heating rate by acoustic waves, which is extensively believed to be responsible for the heating mechanism of the photosphere and low chromosphere. In the scenario of the acoustic wave heating, the damping rate of acoustic waves caused by the classical collision resistance is given by $\gamma_S = -c_s\lambda_c k^2/6$, where c_s is the acoustic speed, λ_c the collision mean free path, and $k = 2\pi/\lambda$ the wavenumber. Since the heating energy directly comes from the dissipated regime of turbulent spectra of acoustic waves, which have a wavelength comparable to the mean free path, we can estimate the heating rate by acoustic waves as

$$Q_{HS} = \frac{1}{12} m_H n_H c_s^2 \nu_c \delta n_S^2 \simeq 2.27 \times 10^{-28} n_H^2 T^{3/2} \delta n_S^2, \qquad (6.11)$$

where m_H and ν_c are the mass and the collision frequency of hydrogen atoms, δn_S is the relative density fluctuation for acoustic waves, and the relation, $\delta v = c_s \delta n_S$, has been used for the perturbed velocity in the acoustic waves.

To compare Eqs. (6.10) and (6.11), we can find that their local heating rates depend on the local atmospheric density and temperature parameters, $n_{H(e)}$ and T, in considerably different form. In particular, the KAW heating rate strongly depends on the magnetic field B_0 in the form proportional to its square, that is, this is a strongly magnetic correlation heating mechanism.

6.3.3 *Upper-Chromospheric Heating by KAW Joule Dissipation*

In the heating problem of the solar atmosphere, the main role of heating mechanisms is to balance the energy loss due to the radiative cooling. Figure 6.9 shows the radiative cooling rate, R_L, of the sunspot atmosphere in the sample presented by Ding and Fang (1991), where the dashed line is calculated based on the semiempirical model presented in Fig. 6.7. Similar to the VAL quiet solar model (Vernazza et al. 1981) and to the sunspot penumbra model (Ding and Fang 1989), a negative cooling rate round the temperature minimum region (~400–600 km) in the sunspot umbra model seems to be unavoidable in semiempirical atmospheric modelling too. One of possible explanations for this negative cooling rate is that some ignored spectral lines should be included to balance this negative cooling rate (Noyes and Avrett 1987). To avoid this negative cooling rate problem, we may fit numerically the calculated cooling rate R_L by an analytical expression Q_L as follows:

$$Q_L = n_H n_e Z(h)\Lambda(T). \qquad (6.12)$$

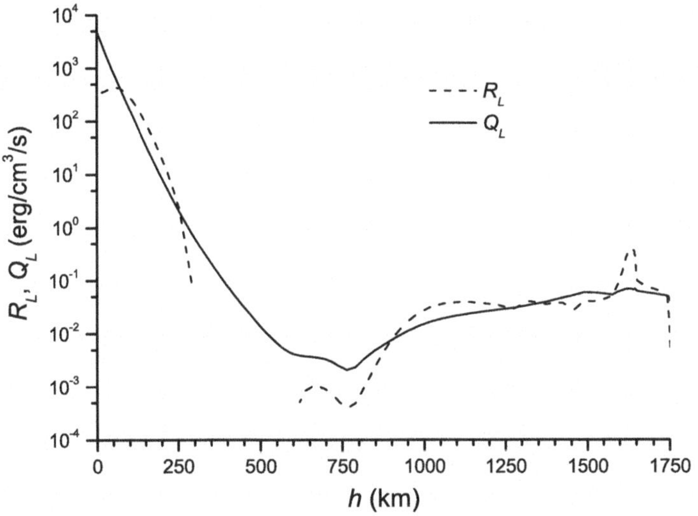

Fig. 6.9 Calculated radiative cooling rate (dashed line) and the analytically fitting repression (solid line) (from Wu and Fang 2007, ©AAS reproduced with permission)

Here, the derivation of the functions $Z(h)$ and $\Lambda(T)$ is based on reproducing, as well as possible, the detailed non-LTE calculations outside the temperature minimum region. Following Gan and Fang (1990), by trial and error, we obtain:

$$\Lambda(T) = 1.547 \times 10^{-29} T^{3/2} \tag{6.13}$$

and

$$Z(h) = \begin{cases} 10^{0.00363h-5.445} + 2.3738 \times 10^{-4} e^{-h/163} & \text{for } h < 1500 \text{ km} \\ 1 & \text{for } h > 1500 \text{ km} \end{cases} \tag{6.14}$$

The result is presented in Fig. 6.9 by the solid line (i.e., Q_L). It reproduces the numerical results considerably well, in particular, in the upper-chromosphere above 800 km, except for a small hump at the chromospheric top.

Here we further constrain our work on the sunspot atmosphere below 1750 km, where the height variation of the magnetic field may be ignored. To balance the energy loss due to the radiative cooling at the rate Q_L, the required wave amplitudes in the density fluctuation forms, δn_K and δn_S, can be obtained from Eqs. (6.10) and (6.11), respectively. Figure 6.10 shows the required δn_S (dotted line) and δn_K (dashed line) when the radiative cooling is balanced by the acoustic wave heating (i.e., $Q_L = Q_{HS}$) and by the KAW heating (i.e., $Q_L = Q_{HK}$), respectively, where the magnetic field $B_0 = 10^3$ G in the sunspot umbra has been used (Ding and Fang 1989).

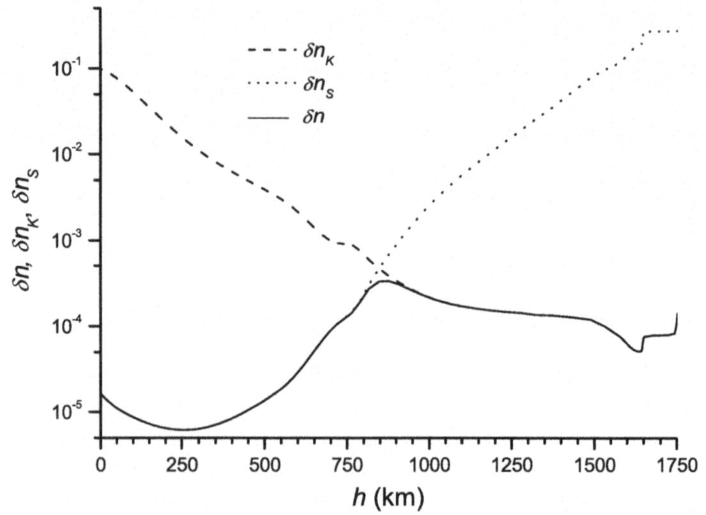

Fig. 6.10 Relative density amplitudes required by the KAW heating (dashed line), by the acoustic wave heating (dotted line), and by the KAW and acoustic wave combining heat (solid line) (from Wu and Fang 2007, ©AAS reproduced with permission)

From Fig. 6.10, one finds that stronger and stronger acoustic waves (i.e., δn_S) are required to balance the cooling rate with increasing height. This indicates that the transferring efficiency of the acoustic wave energy into the kinetic energy of particles is lower and lower with increasing height because the collision frequency becomes lower and lower when the atom density decreases with height. On the other hand, however, the plasma density determined by the electron density slightly increases and then maintains almost a constant in the chromosphere after reaching its minimum in the temperature minimum region (see Fig. 6.7). In consequence, the ionization rapidly increases in the chromosphere, in particular, the Coulomb collision frequency ν_{ei} does not decrease with height. This implies that the transferring efficiency of the energy of KAWs into the kinetic energy of plasma particles is higher and higher with increasing height, and hence the KAW heating becomes more and more important in comparison to the acoustic wave heating with increasing height in the chromosphere. The dashed line in Fig. 6.10 clearly shows that the density amplitude δn_K required by the KAW heating is much lower than that required by the acoustic wave heating to balance the radiative cooling in the sunspot chromosphere above the height of ~850 km. This indicates that the KAW heating dominates the heating mechanism in the upper chromosphere above ~850 km although the acoustic wave heating is possibly more important in the photosphere and the lower chromosphere below ~850 km.

When combining the acoustic wave heating and the KAW heating to balance the radiative cooling, the required density amplitude is in a lower level ($\sim 10^{-5}$) in the photosphere, then increases in the temperature minimum region, and reaches a maximum of $\sim 3 \times 10^{-4}$ at ~850 km, as showed by δn (solid line) in Fig. 6.10. In particular, the required density amplitude maintains at a relatively low level $\sim 10^{-4}$ in the upper chromosphere above 850 km, where the heating mechanism is dominated by the KAW heating.

The necessary KAW energy flux can be estimated as $\sim 3 \times 10^5$ erg/cm^2/s by the cooling rate $\sim$0.03 erg/cm^3/s (see Fig. 6.9) and the radiative scale height $\sim$100 km. On the other hand, the AW flux generated by convection at the photosphere can be estimated by

$$F_A = \frac{B_0^2}{2\mu_0} v_A \left(\frac{\delta v}{v_A}\right)^2, \tag{6.15}$$

where δv is fluid convection velocity. To take $\delta v \sim 1$ km/s and $v_A \sim 10$ km/s at the photosphere, one has $F_A \sim 4 \times 10^8$ erg/cm^2/s. The required wave energy flux by the KAW heating in the sunspot upper chromosphere is only about one in a thousand of the AW energy flux supplied by convection below the photosphere. This indicates that the required wave energy flux by the KAW heating mechanism is very easy to be satisfied.

The above results demonstrated that the KAW heating dominates the heating mechanism in the sunspot upper chromosphere, in which the heating is attributed to the Joule dissipation of KAWs due to the Coulomb collision because of the increase of the ionization there. Although analysis here is for a special sample of sunspot atmospheres, we believe that the conclusion is commonly valid for other magnetic atmospheres, for example, for the quiet solar atmosphere. In fact, as showed in Fig. 6.7 the sunspot atmosphere has height variation very similar to that of the VAL quiet solar atmosphere, except that the sunspot atmosphere varies more quickly from the photosphere through the corona. In particular, similar ionization increasing with height occurs in the quiet solar atmosphere too.

Therefore, it is reasonable to conclude that the KAW heating caused by the ohmic dissipation of KAWs dominates the heating mechanism in the upper chromospheres of the quiet solar atmosphere as well as in the sunspot upper chromospheres. The difference between the two atmospheric models lies in only that the KAW heating becomes the dominating mechanism at different heights, in dependence on the magnetic field strength of the models. That is, the atmosphere with a stronger magnetic field has a lower upper-chromosphere, where the heating mechanism is dominated by the KAW heating, because the relative importance of the KAW heating more quickly increases with heights when the magnetic field is stronger.

6.4 Inner Coronal Heating by Landau Damping of KAWs

6.4.1 Heating Problems of Coronal Loops and Plumes

The most impressive image of the solar corona presented by the current generation space-based observation is its surprising complexity and richness both in structure and dynamics. The hot plasma with temperatures higher than 10^6 K in the corona emits mainly at soft X-ray bands with typical wavelengths of a few tens Å, which dominates the radiative cooling process of the hot coronal plasmas. Observations show that these soft X-ray emissions tend to be concentrated in some looplike structures, called coronal loops, in solar active regions with closed magnetic topology and raylike structures, called coronal plumes, in coronal holes with open magnetic topology (Golub and Pasachoff 1997). It is widely believed in Skylab observations, early on, that these coronal loops correspond to magnetic flux tubes whose footpoints are rooted in the photosphere and looptops extend into the inner corona at the heliocentric distance $R < 1.2R_\odot$ (Bray et al. 1991). However, later Yohkoh soft X-ray observations with a higher spatial resolution showed clearly that these bright loops are not confined within isolated magnetic flux tubes, but are located in some weaker magnetic field parts of the ambient magnetic field (Porter and Klimchuk 1995).

By use of a comparison between the Yohkoh soft X-ray data and ground-based Hα observations, Fang et al. (1997) presented a typical example of enhanced soft X-ray coronal loops and further confirmed the above result. They showed the soft X-ray images of an active region in the corona, obtained by the Yohkoh soft X-ray telescope (Tsuneta et al. 1991) in 1994 May, which present several looplike structures that start from the sunspot and connect regions with an opposite magnetic polarity (see Figs. 1 and 2 of Fang et al. 1997). Also they analyzed the positions and shapes of the sunspot from the two-dimensional spectral observations in Hα by the solar tower at the Nanjing University in Nanjing of China and calculated potential field lines using the magnetograms from the Huairou Station of NAO (National Astronomical Observatories) in Beijing of China. Their result showed clearly that the field lines can match well with the looplike structures, except for the small twisted structure at the central part of the sunspot region (see Fig. 4 of Fang et al. 1997). This further confirms the previous result that the bright coronal plasma loops are associated closely with some magnetic flux tubes connecting two areas of opposite magnetic polarity, and that the whole structure of loops is dominated by the ambient sunspot magnetic field (Bray et al. 1991; Webb and Zirin 1981). From more detailed analyses, however, they obtained an interesting result that the plasma in the magnetic flux tubes is not heated at an uniform way in the ambient magnetic field, in which the parts near the weaker regions tend to be brighter than the parts near the stronger field regions. This well is consistent with the discoveries by Porter and Klimchuk (1995).

Soft X-ray brightened loops are ubiquitous structures in solar active regions. Since they were discovered, the physical nature of their heating mechanism has been a puzzled problem (Bray et al. 1991). Significantly different from the case of the partly ionized upper chromosphere, where the classical collisional resistance can provide an effective heating mechanism as shown in last section Sect. 6.3, the solar corona consists of fully ionized high-temperature and tenuous plasmas, in which the classical Joule dissipation of KAWs does hardly work because the great density decrease and temperature increase both lessen heavily the classical collision frequency. For example, the typical Coulomb collision frequency is well below 10^2 Hz, which is much less than the ion gyrofrequency $\sim 10^5$ Hz for typical parameters of coronal loops in solar active regions, the density $n_e \sim 10^9$ cm^{-3}, the temperature $T \sim 10^6$ K, and the magnetic field $B \sim 100$ G. The result presented by Fang et al. (1997)

could have some important implications for us to understand the physical nature of coronal loop heating mechanisms, which indicates that the heating efficiency is associated with the magnetic field in some more complex ways.

Coronal plumes are other kind of nonuniform coronal structures, which are embedded in relatively uniform solar coronal holes with open magnetic fields. The solar coronal holes are the darkest and least active parts in the solar atmosphere, which are associated with open solar magnetic fields directly extending into the interplanetary space and hence are main source regions of high-speed solar wind streams (Cranmer 2009). Coronal plumes are a common feature in solar coronal holes and are observed as long, bright ray structures that extend nearly radially from the photosphere to a distance larger than $10R_\odot$. These structures are thought to be denser than the surrounding media in the coronal hole. From white-light eclipse photographs and extreme ultraviolet observations (Saito 1958, 1965; Bohlin et al. 1975; Ahmad and Withbroe 1977; Widing and Feldman 1992; Walker et al. 1993), the electron density in the coronal plume is estimated to be ~3–5 times higher than that in the surrounding media in the coronal hole. Using the CDS (the Coronal Diagnostic Spectrometer) on board the SOHO satellite, a detailed analysis by Young et al. (1999) showed even a higher coronal plume density ~10 times that of the background. For example, the typical electron densities of the coronal plume and the interplume region in some limb coronal holes can be estimated as 10^9–10^{10} cm^{-3} and $\sim 10^8$ cm^{-3}, respectively (Walker et al. 1993; Young et al. 1999). While their temperatures have no evident difference and both are typically in the range $(1\text{–}1.5) \times 10^6$ K (Walker et al. 1993; Young et al. 1999).

Based on simultaneous data from the SOHO satellite, and some other space-based as well as ground-based observatories, De Forest et al. (1997) clearly showed that the coronal plumes are associated with small-scale ($\sim 10^3$ km) magnetic flux concentrations in the chromosphere, and that the strongest flux concentrations in the coronal hole have a radial field strength in the order ~100 G. This permits one to understand the geometry of the lowest portion of the coronal plume in terms of the magnetic field configuration: taking typically a density of $\sim 10^{10}$ cm^{-3} and temperature of $\sim 10^6$ K at the base of a coronal plume (Walker et al. 1993), one has the plasma pressure parameter $\beta \sim 4 \times 10^{-3} \ll 1$, implying that the plume geometry should be close to the shape of a force-free field (De Forest et al. 1997). Although no measure of magnetic fields is available, the low-β plasma condition in coronal holes is commonly accepted.

Another fundamental observational result is that the structures of coronal plumes are quite stable and can last for tens hours or days. This indicates that they can be modeled as radial steady-flow structures with transverse (perpendicular to the radial direction) pressure balance. It seems reasonable to assume that the coronal plumes are rooted in open magnetic flux concentration regions. Supporting this idea is the observations of a supper-radial expansion of coronal plumes near their base in the range 1–1.2 $R_\odot$ (Saito 1965; Ahmad and Withbroe 1977). What is observed is obviously a density behavior, but if the coronal plume is in equilibrium, then it must be threaded by diverging field lines with increasing height (Ahmad and Withbroe 1977). Therefore, in some early theories of coronal plumes (Newkirk and Harvey

1968; Suess 1982) potential fields were often used to model the coronal plume geometry. This has also been explicitly proven later by a work of Del Zanna et al. (1997). They showed that the flow has negligible effect on geometric spreading of the coronal plume.

As noticed by some authors (e.g., Wang 1994), however, an additional heating in the coronal plumes is necessary to balance the extra radiative cooling of these brightening and dense coronal plumes. Obviously, this needs a structured (or i.e. nonuniform) heating mechanism too. In fact, for the case of equal temperature in the coronal plume and in the surrounding interplume media in the hole, the radiative cooling rate in the plume is much higher than that in the surrounding interplume media by a factor of $\sim$10–100, since the radiative cooling rate is proportional to the density square (McClymont and Canfield 1983). On the other hand, the observed fact that some small-scale transient brightening structures in the coronal plumes propagate coherently outward at the local Alfvén velocity along the radial axis of the coronal plumes indicates that there is some relationship between the additional heating process and Alfvénic fluctuations (De Forest et al. 1997).

The observations of these coronal loops and plumes both require an inhomogeneous heating mechanism that depends sensitively on the local magneto-plasma parameters so that there is a higher heating efficiency in these bright soft X-ray structures. It was proposed by Wu and Fang (1999, 2003) that in a nearly collisionless plasma such as the coronal plasmas, the kinetic Landau damping of KAWs due to the wave-particle interaction could be one of the most suitable candidates for such heating mechanism with the sensitively parametric dependence, because the so-called thermal resonant condition of KAWs due to the wave-particle interaction near the electron thermal velocity could play an important and crucial role in determining the heating efficiency.

6.4.2 *Landau Damping of KAWs*

Although the classic Coulomb collision between particles can not provide an effective dissipation mechanism in a collisionless plasma, the wave-particle interaction can lead to the dissipation of the wave energy if only the proper resonant condition is satisfied, for instance, the so-called thermal resonant condition, in which the phase velocity of the wave matches to the thermal velocity of particles. The wave energy dissipation due to the resonant wave-particle interaction is called the Landau damping. Hasegawa and Chen (1975, 1976) investigated first the ion Landau damping of KAWs due to the resonant interaction between ions and KAWs when $v_{T_e} \gg v_A$. Employing the gyrokinetic equation for electrons, Wu and Fang (1999) further studied the electron Landau damping of KAWs due to the resonant interaction between electrons and KAWs when $v_{T_e} \sim v_A$ and found that the heating rate due to the electron Landau damping of KAWs has a maximal value when the electron thermal resonant condition, $v_{T_e} = v_A$, is satisfied. Thus, they proposed that this

KAW heating mechanism by the electron Landau damping can probably provide a reasonable explanation for the structured nonuniform heating phenomena, which are ubiquitous in the solar corona.

Since the Landau damping is a pure kinetic process, in general, this should be treated by employing the exact Vlasov equation. However, for the electron kinetics in the low frequency waves such as KAWs with frequency much lower than the ion gyrofrequency, the one-dimensional drift kinetic equation along the unperturbed magnetic field (see e.g., Hasegawa and Sato 1989) is an enough good approximation, which can greatly simplify the mathematical treatment of the problem. Practically, in the drift kinetic description the motion of electrons may be treated as one dimension field-aligned motion, v_{ez}, which contributes the majority of field-aligned currents of KAWs. While the electrons and ions have the exact same cross-field electric drift velocity, $\mathbf{v}_E = \mathbf{E} \times \mathbf{B}/B^2$, which does not contribute to the net current $\mathbf{j}$.

The field-aligned drift kinetic equation of electrons in the case of KAWs can be written as follows (Hasegawa and Uberoi 1982; Wu and Fang 1999):

$$\left(\frac{\partial}{\partial t} + v_{ez}\frac{\partial}{\partial z} - \frac{eE_z}{m_e}\frac{\partial}{\partial v_{ez}}\right) f_e = 0, \tag{6.16}$$

where $f_e = f_e^{(0)} + f_e^{(1)}$ is the distribution function of electrons in the velocity space, and $f_e^{(0)}$ and $f_e^{(1)}$ are the equilibrium and perturbed distributions, respectively. Assuming the equilibrium distribution of electrons to be the Maxwellian distribution and the ambient plasma density to be homogeneous, i.e.,

$$f_e^{(0)} = \frac{n_0}{\sqrt{2\pi} v_{T_e}} e^{-v_{ez}^2/2v_{T_e}^2}, \tag{6.17}$$

the linear perturbed distribution function $f_e^{(1)}$ can be expressed as follows (Hasegawa and Uberoi 1982; Wu and Fang 1999):

$$f_e^{(1)} = i\frac{en_0E_z}{\sqrt{2\pi} m_e k_z v_{T_e}^3} \frac{v_{ez}e^{-v_{ez}^2/2v_{T_e}^2}}{v_{ez} - \omega/k_z}. \tag{6.18}$$

From this Eq. (6.18), the field-aligned current carried by electrons can be given by

$$\begin{aligned} j_{ez} &= -e\int_{-\infty}^{\infty} v_{ez} f_e^{(1)} dv_{ez} \\ &= -i\sqrt{\frac{2}{\pi}}\frac{n_0e^2E_z}{m_ek_zv_{T_e}}\left[P\oint_{-\infty}^{\infty}\frac{x^2e^{-x^2}}{x-x_0}dx + i\pi x_0^2e^{-x_0^2}\right], \end{aligned} \tag{6.19}$$

where $x = v_{ez}/\sqrt{2}v_{T_e}$, $x_0 = \omega/\sqrt{2}k_z v_{T_e}$, and "$P\oint$" denotes the principal value integral. The second term in the square brackets above originates from the singular integral at the singular point of $x = x_0$ and represents the Landau damping effect due to the wave-particle resonant interaction. The Plemelj formula (Fried and Conte 1961) for an arbitrary complex function $G(x)$

$$\int_{-\infty}^{\infty} \frac{G(x)}{x - x_0} dx = P \oint_{-\infty}^{\infty} \frac{G(x)}{x - x_0} dx + i\pi G(x_0) \tag{6.20}$$

has been used to derive the second identity of Eq. (6.19).

The heating rate by the electron Landau damping may be calculated by the Joule heating of the field-aligned current, j_{ez}, and it reads:

$$Q_H = \frac{1}{2}\text{Re}\left(j_{ez}E_z^*\right) = \omega x_0 e^{-x_0^2} \frac{\sqrt{\pi}}{k_z^2\lambda_D^2} U_{E_z}, \tag{6.21}$$

where $U_{E_z} \equiv \epsilon_0 E_z E_z^*/2 = \epsilon_0 |E_z|^2/2$ is the energy density of the parallel component E_z of the perturbed electric field. For the weakly dispersive case of $k_x^2\lambda_e^2$ and $k_x^2\rho_i^2 \ll 1$, from the polarization relations of KAWs in Eqs. (1.32) and (1.33) we have

$$U_{E_z} \simeq \frac{k_z^2}{k_x^2}\delta_\perp^2 U_{E_x} \simeq \frac{k_z^2}{k_x^2}\frac{v_A^2}{c^2}\delta_\perp^2 U_B \tag{6.22}$$

and $\omega/k_\parallel \simeq v_A$, where U_{E_x} and U_B are the energy densities of the perpendicular component E_x of the perturbed electric field and the perturbed magnetic field, respectively, and $\delta_\perp \equiv k_x^2\rho_s^2$ for $v_A < v_{T_e}$ or $\equiv k_x^2\lambda_e^2$ for $v_A > v_{T_e}$ is a small parameter. Substituting this relation into Eq. (6.21) and by use of the equality $\lambda_D = (v_A/c)\rho_i$, the heating rate by the Landau damping Q_H of Eq. (6.21) can be written as

$$Q_H = \omega\sqrt{\pi} x_0 e^{-x_0^2} \delta_\perp U_B. \tag{6.23}$$

Following Hasegawa and Chen (1976), the fractional damping rate can be expressed by

$$\frac{\gamma}{\omega} = \sqrt{\pi} x_0 e^{-x_0^2} = \sqrt{\frac{\pi}{2}}\frac{B_0}{B_R} e^{-B_0^2/2B_R^2}, \tag{6.24}$$

where B_0 is the ambient magnetic field strength, and

$$B_R = \sqrt{\mu_0 n_0 T_e m_i/m_e} \simeq 1.922 \times 10^{-4}\sqrt{n_0 T_e}\ (\text{G}) \tag{6.25}$$

is the "resonant" magnetic field strength at the resonant point $v_A = v_{T_e}$. In the last expression above, n_0 and T_e are in units of cm^{-3} and eV, respectively. It is easy to find that the result of Hasegawa and Chen (1976) can be obtained anew in the limit of $v_A \ll v_{T_e}$.

Figure 6.11 shows the fractional damping rate versus the ratio of the Alfvén velocity v_A (i.e., the phase velocity of waves) to the electron thermal velocity v_{T_e}. It is straightforward to find that the effective dissipation of waves by the Landau damping

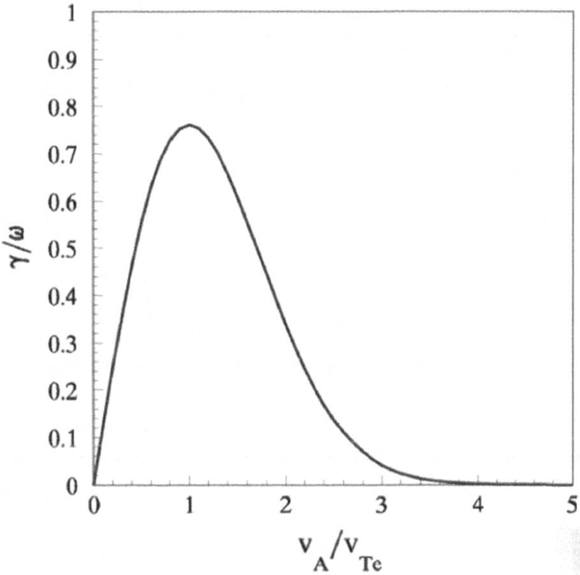

Fig. 6.11 The fractional damping rate γ/ω versus the velocity ratio v_A/v_{T_e} (from Wu and Fang 1999, ©AAS reproduced with permission)

mechanism occurs in a properly wide range of v_A/v_{T_e}, i.e., 0.1–2.5, and the fractional damping rate, (γ/ω), reaches its maximum $\simeq$0.76 at $v_A = v_{T_e}$. This is consistent with the physical nature of the wave-particle resonant interaction of the Landau damping mechanism, where the major of thermal electrons in the Maxwellian distribution is resonant with the wave, called the thermal resonant condition.

6.4.3 Coronal Loop Heating by KAW Landau Damping

From Eq. (6.23), one can find that the heating ratio Q_H depends on the ambient magnetic field (B_0) as well as the wave energy density ($U_B = B_y^2/2\mu_0$) of KAWs. However, it is very difficult to determine the exact spectrum and the wave energy density of KAWs or AWs in the solar atmosphere. In convenient for the comparison between theory and observation, we assume that the wave field B_y depends on the ambient field B_0 in the following form

$$B_y = C_B B_0^{\alpha}, \tag{6.26}$$

where C_B and α are constant parameters. Thus, the heating rate Q_H in Eq. (6.23) can be rewritten as follows:

$$Q_H = Q_H^M \left(\frac{e}{2\alpha+1}\right)^{\alpha+0.5} \left(\frac{B_0}{B_R}\right)^{2\alpha+1} e^{-B_0^2/2B_R^2}, \tag{6.27}$$

where

$$Q_H^M = \omega\sqrt{\frac{\pi}{2}}\left(\frac{2\alpha+1}{e}\right)^{\alpha+0.5}\delta_\perp\frac{\left(C_B B_R^\alpha\right)^2}{2\mu_0} \quad (6.28)$$

is the maximal heating rate, occurring at the ambient magnetic field $B_0 = B_M$,

$$B_M = \sqrt{2\alpha+1}B_R = 1.922\times10^{-4}\sqrt{(2\alpha+1)\,n_0 T_e}\ (\mathrm{G}). \quad (6.29)$$

For the case of the wave field B_y independent of the ambient field B_0, i.e., $\alpha = 0$, the maximal heating rate occurs at the same position as the most effective damping rate, that is, at the wave-electron resonant point $B_M = B_R$. As the dependence of the wave field on the ambient field (i.e., α increases), the maximal heating rate moves to the region where the ambient field is stronger.

Figure 6.12 plots the normalized heating rate Q_H/Q_H^R against the ambient magnetic field strength B_0 for three different values of the parameter $\alpha = 0, 0.5$, and 1, where the panel (a) shows the case of the loop top, the panel (b) shows the case of the loop leg, and Q_H^R is defined by

$$Q_H^R \equiv \omega\sqrt{\frac{\pi}{2}}\delta_\perp\frac{\left(C_B B_R^\alpha\right)^2}{2\mu_0} = \left(\frac{e}{2\alpha+1}\right)^{\alpha+0.5} Q_H^M. \quad (6.30)$$

In fact, Fig. 6.12 also shows distribution of the heating rate function Q_H across the ambient magnetic field, i.e., the sunspot magnetic field, where the maximal strength at the sunspot center has been taken as $B_0 = 1000$ G for the example of Fang et al. (1997). From Fig. 6.12 it can be found clearly that the maximal heating rate, implying the brightest parts of the loop, does not come from the sunspot center with the strongest magnetic field, but from the region closer the boundary of the sunspot, where there is weaker magnetic field. This is consistent with the soft X-ray observations of coronal loops (Porter and Klimchuk 1995; Fang et al. 1997). Moreover, as inferred above from Eq. (6.29) the loop center (i.e., the brightest part of the loop) will be closer the sunspot center (i.e., the center of ambient magnetic field) for a larger α value because of stronger dependence of the wave field on the ambient field.

Table 6.1 lists the physical parameters associated with the coronal loop presented by Fang et al. (1997), where the first and second rows correspond to the top region and leg region of the loops, respectively. The condition of the energy equilibrium between the heating rate and the total energy loss rate means

$$Q_H = Q_L + \nabla\cdot\mathbf{q}, \quad (6.31)$$

where the radiative loss rate function Q_L around temperature $T_e \sim 3\times10^6$ K can be calculated by (Nagai 1980)

$$Q_L \simeq 4.645\times10^{-19} n_e^2 T_e^{-0.604}\ \mathrm{erg\ cm^{-3}s^{-1}} \quad (6.32)$$

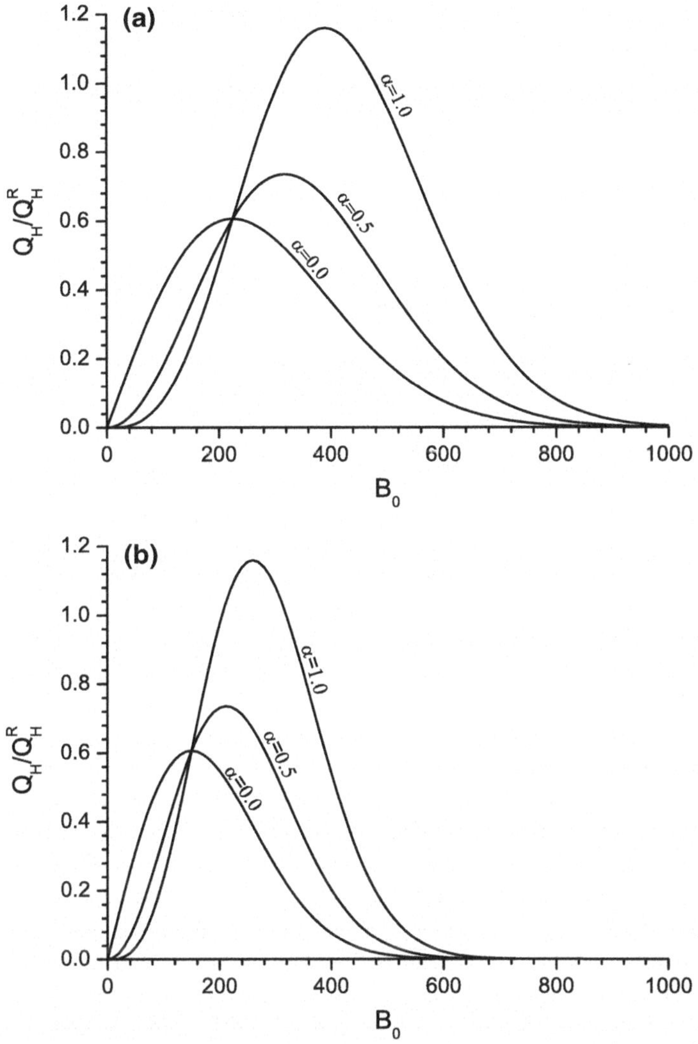

Fig. 6.12 The normalized heating rate Q_H/Q_H^R versus the ambient magnetic field strength B_0 for the parameter $\alpha = 0$, 0.5, and 1, where (**a**) is for the case of the loop top, and (**b**) is for the case of the loop leg (from Wu and Fang 1999, ©AAS reproduced with permission)

with temperature in units of K and the heat conduction loss rate $\nabla \cdot \mathbf{q}$ can be estimated by

$$\nabla \cdot \mathbf{q} = \nabla \cdot \left(C_0 T_e^{5/2} \nabla T_e\right) \simeq C_0 T_e^{7/2} (2/L)^2 \tag{6.33}$$

with the constant coefficient $C_0 = 2 \times 10^{-6}$ and the loop length L in units of cm. For the sample presented by Fang et al. (1997), we have $\nabla \cdot \mathbf{q} \approx 3.1 \times 10^{-3}$ erg · cm^{-3}s^{-1}.

Table 6.1 Parameters of coronal loops given by Fang et al. (1997) (from Wu and Fang 1999, ©AAS reproduced with permission)

Region	n_e	T_e	B_R	B_M	Q_H^M	$Q_L + \nabla \cdot \mathbf{q}$
	$10^9/\mathrm{cm}^3$	10^6K	G	G	10^{-3}erg/cm^3/s	10^{-3}erg/cm^3/s
Loop top	5.3	3.0	225	266	4.7	4.7
Loop legs	3.5	2.0	150	177	4.0	4.0

The consistence in modeling requires

$$\frac{(Q_H^M)_{\mathrm{top}}}{(Q_H^M)_{\mathrm{leg}}} = \frac{(Q_L + \nabla \cdot \mathbf{q})_{\mathrm{top}}}{(Q_L + \nabla \cdot \mathbf{q})_{\mathrm{leg}}}. \tag{6.34}$$

Combining Eq. (6.28) this leads to the parameter $\alpha \simeq 0.2$. Taking the wave period $\sim 10^{-3}$ s we have the wave frequency $\omega \simeq 6.3 \times 10^3$ s^{-1}, which is much lower than the ion gyrofrequency $\omega_{ci} \sim 10^6$ s^{-1}. If further taking the small parameter $\delta_\perp \simeq 10^{-4}$, the wave field strength required by this heating mechanism can be estimated by combining Eq. (6.31) and the parameters given in Table 6.1. In result, we have $B_y = 0.48$ G, that is, the relative fluctuation $\sim$0.18% at the top region and $B_y = 0.44$ G, that is, the relative fluctuation $\sim$0.25% at the leg region. This indicates that the energy loss of the bright loops can be balanced by the KAW heating due to the Landau damping of KAWs with a very low relative magnetic fluctuation $\sim$0.2%, which is easy to be satisfied.

Neglecting the variation of the ambient magnetic field strength along the magnetic flux tube, from Table 6.1 and Fig. 6.12 it can be found that the region with the maximum heating rate, and hence with the maximum radiative cooling rate (i.e., the brightest region of the loop) is nearby $B_0 = B_M^{\mathrm{top}} = 266$ G for the top of the loop and nearby $B_0 = B_M^{\mathrm{leg}} = 177$ G for the legs of the loop. On the other hand, the ambient sunspot magnetic field strength varies in the plane perpendicular to the flux tube of the loop from $B_0 \sim 100$ G at the outer boundary of the sunspot to $B_0 \sim 900$ G at the central region of the sunspot (Fang et al. 1997). This indicates that the brightest region of the coronal loop is located in the outer region of the sunspot, and the top brightest region is slightly nearer to the center of the sunspot than the leg brightest regions. In the meantime, comparing the panels (a) and (b), one can find that the effective heating region at the loop top is wider than that at the loop legs. These results are, qualitatively at least, consistent with the soft X-ray observations of coronal loops (Porter and Klimchuk 1995; Fang et al. 1997). Moreover, from Eqs. (6.25) and (6.28), we have

$$\frac{(Q_H^M)_{\mathrm{top}}}{(Q_H^M)_{\mathrm{leg}}} = \left(\frac{B_R^{\mathrm{top}}}{B_R^{\mathrm{leg}}}\right)^{2\alpha} = \left(\frac{n_e^{\mathrm{top}} T_e^{\mathrm{top}}}{n_e^{\mathrm{leg}} T_e^{\mathrm{leg}}}\right)^{\alpha}. \tag{6.35}$$

The observation shows, in general, $n_e^{\rm top}T_e^{\rm top} > n_e^{\rm leg}T_e^{\rm leg}$. Thus, provided that $\alpha > 0$, the tops of loops are brighter than their legs, as showed by solar soft X-ray observations (Porter and Klimchuk 1995; Kano and Tsuneta 1995, 1996; Fang et al. 1997).

In consideration of that coronal loops are ubiquitous magneto-plasma structures in the solar corona, especially in solar active regions of the inner solar corona lower than the solar atmospheric height $h \sim 0.2R_\odot$, the heating mechanism of coronal loops can dominate reasonably the heating of the whole solar corona, although it has typically nonuniform features. The KAW heating caused by the ohmic dissipation of KAWs due to the classic Coulomb collision, which is the dominating heating mechanism for upper chromospheres as demonstrated in the last section, can not effectively work in the solar coronal heating because of the collisionless nature of hot and thin coronal plasmas. The KAW heating caused by the Landau damping of KAWs due to the wave-particle interaction, however, probably becomes one of the most important coronal heating mechanisms, which is responsible for the nonuniform heating of coronal loops confined in close magnetic fields of sunspots, as proposed by Wu and Fang (1999).

In comparison, these two heating mechanisms both are caused directly by the dissipation of KAWs and have the strongly magnetic correlationship, that is, their heating rates both sensitively depend on the local magnetic field strength. However, also there is remarkable difference between the two heating mechanisms. The heating mechanism due to the Joule dissipation of KAWs is effective only in a collision plasma because the Joule dissipation of KAWs is caused directly by the Coulomb collision between particles, but the heating mechanism due to the Landau damping of KAWs can play an important role in a collisionless plasma because the Landau damping of KAWs is attributed to the wave-particle interaction. On the other hand, the heating rate by the Joule dissipation of KAWs is directly proportional to the square of the ambient magnetic field strength, while the heating rate by the Landau damping of KAWs depends on the local magnetic field in a more complex manner that is determined by the wave-particle resonant condition. Therefore, the latter usually leads to the radiative brightness of heated plasmas having a more complex spatial distribution, as depicted by the soft X-ray observations of the solar corona.

6.4.4 Empirical Model of Coronal Hole and Plumes

Different from the case of coronal loops, which distribute mainly in solar active regions of the inner corona below the height $h \sim 0.2R_\odot$ above the solar surface and in which the hot plasmas are controlled mainly by closed magnetic fields of sunspots, coronal plumes present in solar coronal holes, in which hot plasmas are guided by open magnetic fields and can extend to the height $h \sim 10R_\odot$, much larger than the height of the coronal loops. Solar coronal holes are the darkest and least active parts in the solar atmosphere and associated with open solar magnetic fields (Cranmer 2009). Coronal plumes are a common feature in solar coronal holes and are observed as long ray structures that extend nearly radially from the photosphere to the heliocentric distance larger than, $R > 10R_\odot$.

Both theory and observation of the solar corona and the solar wind show that the main features of coronal plume structures can be described by three regions, distinguished in structure and dynamics with the radial distance from the Sun. The first one is the "root" region of the plume near the base of the plume in the lower corona (from the solar surface at $R = 1R_\odot$ to $R \sim 1.2R_\odot$), where the magnetic field in the coronal hole concentrates mainly in some small open flux regions at the solar surface, then rapidly spreads with the height in the form of superradial expansion, and finally fills nearly uniformly the coronal hole at the top of this region. The plumes in the coronal hole are believed to be rooted in these open flux concentration regions. The body plasma flow in the hole (including the plumes), however, is almost static in this region (De Forest et al. 1997).

The sequent "acceleration" region of the plume is located from $R \sim 1.2R_\odot$ to $R \sim 10R_\odot$, where the main body of the plume, as a thin and long bright ray in the hole, appears in white-light observations. The body flow in the hole (including the plumes) is accelerated to the interplanetary solar wind velocity of several hundreds of km/s in this region, where the hole has an approximately uniform magnetic field and a low-β plasma. The major energy of AWs, which originate from turbulent motions in the solar surface and propagate along the open field lines, is dissipated in this region through heating the coronal plasma and accelerating the solar wind flow.

The third region is the "mixing" region of the plume. In fact, from $R \sim 10R_\odot$, the coronal plasma in the hole flows into the broad interplanetary space in the form of the high velocity solar wind. With the increase of the plasma β parameter as well as the flow velocity, however, the dense structure of the plume in the hole could become more unstable in this region (Andries et al. 2000; Andries and Goossens 2001). This leads to the plume mixing with interplume plasmas and the disruption of the plume structure. If assuming the constant-velocity solar wind in this region, both the magnetic field and plasma are nearly radial expansion in the form of a square inverse proportional to the heliocentric distance according to the requirement of the mass conservation.

In order to explain comprehensively the heating and acceleration of the solar wind, theories must incorporate detailed empirical knowledge of the physical conditions in the extended corona in the range $R \sim 1.2$–$10\ R_\odot$ where the coronal plasmas are heated and accelerated into the solar wind. The electron density plays a major role in the estimation of plasma parameters, such as the plasma β parameter, the Alfvén velocity, and the collision frequency. In the inner solar atmosphere, the electron density can be derived using various atmospheric models, which are constrained by observed spectral lines. In the extended corona, the electron density usually is calculated based on polarization brightness measurements of this region. However, in comparing the two sets of density, it has long been suspected that the electron density in the extended corona might be overestimated by a factor $\sim$10, due to line-of-sight effects in the measurements of the polarized brightness (Koutchmy 1977; Esser and Sasselov 1999; Teriaca et al. 2003).

Combining the estimation of the electron densities in the higher corona and in the solar wind, which are derived from observations from 13.8 MHz to a few kHz by the radio experiment WAVES aboard the WIND satellite (Leblanc et al. 1998), Wu and Fang (2003) presented an empirical model for the electron density distribution with the heliocentric distance in plumes. In the surrounding coronal hole, however, the coronal plasma has a much lower density by a factor $\sim$10 (Walker et al. 1993; Young et al. 1999). According to the empirical model of coronal plumes presented by Wu and Fang (2003), the radial variation of the plume density and the surrounding plasma density (Wu and Yang 2007) in the coronal hole can be modeled by

$$n_p = \frac{10^{10}}{e^{50(r-1)}} + \left(\frac{680}{r^6} + \frac{35}{r^4} + 2.8\right)\frac{10^5}{r^2}\ (\mathrm{cm}^{-3}) \tag{6.36}$$

and

$$n_b = \frac{n_p}{1 + 9e^{-(r-1)^2/100}}, \tag{6.37}$$

respectively, where $r \equiv R/R_\odot$ is the heliocentric distance in units of the solar radius $R_\odot$, and the denominator in Eq. (6.37) models the density in the hole lower than that in the plume by a factor $\sim$10 in the extended coronal region $\sim$1.2–10 $R_\odot$. This model has a density $\sim 10^7$ cm^{-3} in the coronal base at 1.1 $R_\odot$, which is in agreement with the maximum density derived by the lower solar atmospheric models (Koutchmy 1977; Esser and Sasselov 1999; Teriaca et al. 2003) and a density of a few electrons per cubic centimeter at 1 AU, which is a typical value of the solar wind density in satellite in situ measurements. The result is presented in panel (a) of Fig. 6.13, where the plume density is plotted by dashed line, the surrounding plasma density in the hole is plotted by solid line, and they superpose after $R > 20R_\odot$.

Based on the observations of both FeIX and FeXII emission lines, the temperature in the lower corona is estimated to be in the range of $(1–1.5) \times 10^6$ K. On the other hand, some numerical two-fluid models of high-speed solar wind streams associated with coronal holes showed (Hu et al. 1997) that the radial distribution of the electron temperature can be characterized by a quickly increasing from $\sim 5 \times 10^5$ K to $\sim 1.5 \times 10^6$ K within less than $3R_\odot$, then a slow decrease by a factor of $\sim$10 in a radial distance of about a few tens of $R_\odot$, followed by a slower decrease in interplanetary space. Following Wu and Fang (2003), this radial distribution of the temperature can be modeled empirically by

$$T_e = 4.2 \times 10^5 r e^{1-r/3} + 1.5 \times 10^5 \left(\frac{3}{r}\right)^{0.3}\ (\mathrm{K}). \tag{6.38}$$

The result is shown in panel (b) of Fig. 6.13.

Although there are few observations of the magnetic field available for the corona, the low-β condition is commonly supposed to hold in the extended corona of a coronal hole. An early empirical formula that fits the Faraday rotation data in the extended corona was derived by Mariani and Neubauer (1990) as $6r^{-3} + 1.18r^{-2}$ G.

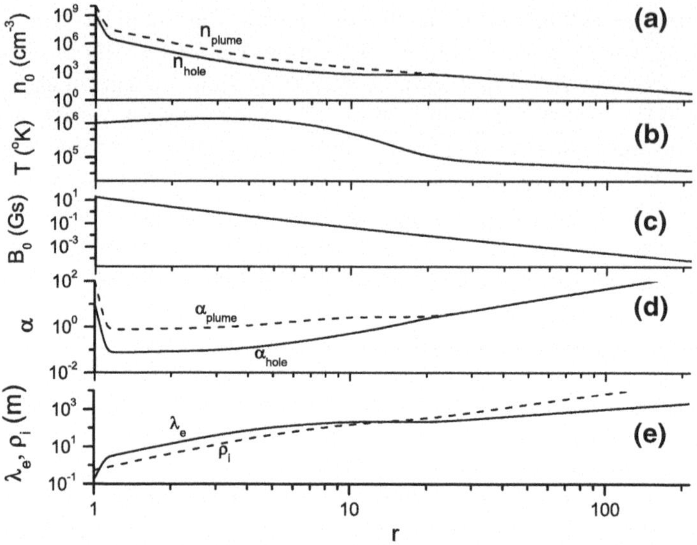

Fig. 6.13 Empirical model of background plasma parameters for a coronal hole (from Wu and Yang 2007, ©AAS reproduced with permission)

In extrapolating this formula to 1 AU, however, one obtains an interplanetary magnetic field of ∼2 nT, which is obviously less than the in situ measured value of ∼5 nT by a factor of ∼2.5. Since the measured Faraday rotation is directly proportional to the product $n_e B_r$ (Volland et al. 1977) and the electron density in the extended corona could have been overestimated, a larger magnetic field could be expected from the Faraday rotation. Following Wu and Fang (2003), the radial variation of the magnetic field in coronal hole may be fitted by a similar formula but enlarging it by a factor of 2.5 to fit the correct magnetic field value at 1 AU, that is,

$$B_0 = \frac{15}{r^3} + \frac{3}{r^2} \text{ (G)}. \tag{6.39}$$

The result is shown in panel (c) of Fig. 6.13.

The electron thermal resonant parameter $\alpha = v_{T_e}/v_A$ for the Landau damping of KAWs is showed in panel (d) of Fig. 6.13, where the dashed line is for the case in the plume and the solid line is for the case in the surrounding plasma in the hole. Finally, panel (e) of Fig. 6.13 presents the comparison between the electron inertial length (solid line) and the ion gyroradius (dashed line) in the hole. The result indicates that dispersion of KAWs is dominated by the electron inertial effect (i.e., $\lambda_e > \rho_i$) and by the finite ion gyroradius effect (i.e., $\lambda_e < \rho_i$) in the regions $R < 10R_\odot$ and $R > 20R_\odot$, respectively, and the transition region is between $10R_\odot < R < 20R_\odot$.

On the other hand, the stability of coronal plumes indicates that it can be modeled as a radial steady flow with the transverse pressure balance. Because of the low-β plasma in the hole, the effect of the transverse gradients in the plasma density on the geometry of the plume main body is so little (Suess et al. 1998) that the geometry can be modeled as a radial (i.e., truncated conical) column with a length $\sim 10R_\odot$ and a circular cross-section with a radius $\sim 0.1R_\odot$. It is embedded in the center of the coronal hole magnetized by the radial field $\mathbf{B} = B_0\hat{\mathbf{z}}$, where $\hat{\mathbf{z}}$ is the unit vector along the solar radial direction.

For the sake of simplification, we consider a single plume embedded in the center of a coronal hole. In a local cylindrical polar coordinate system (ρ, ϕ, z) with the z-axis at the center of the hole along the solar radial direction, the axial symmetrical density distribution in the cross-section of the hole can be modeled by the function (Wu and Fang 2003)

$$n = \left(n_p - n_b\right) e^{-\rho^2/\rho_P^2} + n_b, \tag{6.40}$$

where n_p is the electron number density on the plume axis (i.e., $\rho = 0$), which is the maximum of the density, estimated qualitatively as 2 times the observed average plume density; n_b is the electron number density in the surrounding media at the boundary of the hole of $\rho \to \infty$, which is approximately the average ambient density, and ρ and ρ_P are the transverse distance off the plume axis and the characteristic cross-section radius of the plume, respectively.

Assuming pressure balance and a uniform temperature distribution in the cross-section of the hole, the distribution of the magnetic field in the cross-section can be derived as

$$B_z = B_0\left(1 - be^{-\rho^2/\rho_P^2}\right), \tag{6.41}$$

where

$$b = \left(1 - \frac{n_b}{n_p}\right)\frac{\beta}{2} \tag{6.42}$$

is a small modification to balance the density non-uniformity due to the dense plume in the hole, and $\beta \sim 2\beta_e$ is the plasma pressure parameter. For a low-β plasma environment such as coronal holes, the small modification parameter, $b \ll 1$, indicates an approximately uniform ambient magnetic field B_0 in the cross-section of the hole.

6.4.5 Coronal Plume Heating by KAW Landau Damping

Although coronal holes are quiet and uniform in comparison to active regions, it is clear that the plume heating in the hole still has relation to a structured heating mechanism. The plume is denser than the surrounding plasma in the hole, and hence has a higher radiative cooling rate Q_L. According to McClymont and Canfield (1983), for a wider range of variation in temperature above 10^5 K, the plasma radiative cooling rate can be analytically fitted by

$$Q_L = 10^{-16} n^2 T_e^{-1}\ \mathrm{erg \cdot cm^{-3} s^{-1}}. \tag{6.43}$$

To fuel the bright plume, a higher heating rate Q_H is needed in the plume than in the surrounding plasma because the cooling rate Q_L is proportional to the square of the density.

The fact that the small-scale structures in the plume brightness propagate coherently outward at the Alfvén speed (De Forest et al. 1997) probably indicates that there is some heating mechanism associated with Alfvénic fluctuations in the plume. Wu and Fang (2003) proposed that the Landau damping of KAWs due to the electron thermal resonance could be responsible for the heating mechanism in the brightening plumes. From the empirical atmospheric model shown in Fig. 6.13, it can be found that the resonant condition of $v_{T_e} = v_A$ (i.e., $\alpha = 1$, see the dashed line in the panel (d) of Fig. 6.13) can be well satisfied in the main body of the plume, which has a long radial extension between 1.1 and 10 $R_\odot$. In the local cross-section of the hole, assuming that the resonant condition is satisfied at the plume axis ($\rho = 0$), the cross-section distribution of the heating rate (Q_H) by the Landau damping of KAWs can be calculated from Eq. (6.23). By use of the cross-section distribution of density in Eq. (6.40), one has

$$Q_H = \sqrt{\frac{n_b}{n}} e^{(n/n_b - 1)/2} Q_b, \tag{6.44}$$

where

$$Q_b = \omega \sqrt{\frac{\pi}{2}} \sqrt{\frac{n_0}{n_b}} e^{-n_0/2n_b} \delta_\perp U_B \tag{6.45}$$

represents the heating rate in the ambient plasma surrounding the plume in the hole, $n_0 = n_p$, and the magnetic field B_0 is adopted as a constant in the cross-section because of $b \ll 1$ in Eq. (6.41). Figure 6.14 shows the cross-section distributions of the heating rate, Q_H, for the density ratio of the plume axis to the hole plasma surrounding the plume, $n_0/n_b = 4$, 6, 8, and 10 (from the top down), where the dashed lines are the cross-section distributions of the density.

From Fig. 6.14 it can be found that the additional heating to fuel the bright dense plume can be self-consistently produced by the Landau damping of KAWs in the dense plume, because of the electron thermal resonance there. Moreover, the extra KAW heating rate increases remarkably with the plume to ambient density ratio, n_0/n_b. In general, the density ratio (n_0/n_b) is a function of the heliocentric distance R. In particular, the disappearance of the plume appearance in the mixing region can be attributed to the disappearance of the high density ratio with increasing R, due to some instability (Andries et al. 2000; Andries and Goossens 2001). As shown in panel (a) of Fig. 6.15 (also see the dashed line in the panel (d) of Fig. 6.13), the plume deviates evidently from the resonant condition of $\alpha = 1$ at the mixing region when $R > 10R_\odot$, and this results in that the KAW heating mechanism does not work in the mixing region.

Combining Eqs. (6.23) and (6.24), we have $Q_H = \gamma \epsilon U_B$, where the term ϵU_B represents the part of the wave energy that can be directly converted into the kinetic energy of plasma particles, that is, the plasma heating, due to the wave-particle interaction. In general, this is done directly by the electric field in the wave. Assuming

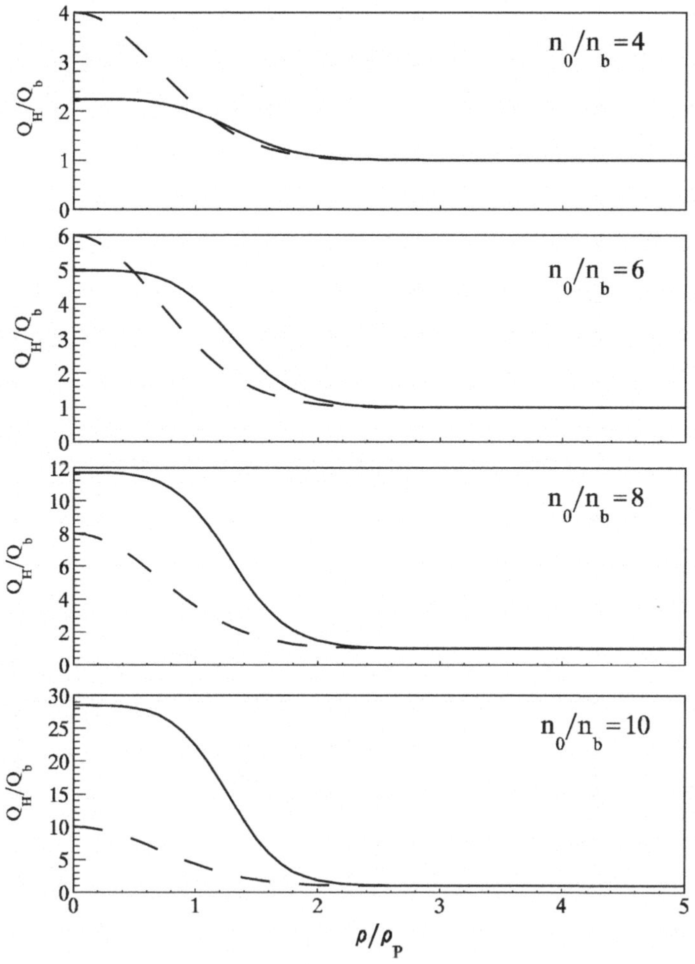

Fig. 6.14 The cross-section distributions of the heating rate, Q_H, by KAWs in the local cross-section of the surrounding coronal hole for the ratio of the density at the plume axis to that in the surrounding hole, $n_0/n_b = 4, 6, 8$, and 10, from the top down. The dashed lines are responsible for the distributions of the corresponding densities in the local cross-section (from Wu and Fang 2003, ©AAS reproduced with permission)

$\epsilon U_B \sim U_E$, one has $\epsilon \sim U_E/U_B \simeq v_A^2/c^2$ for KAWs. Panel (b) of Fig. 6.15 shows the radial profile of the parameter ϵ, and it has the maximum in the main body of the plume, implying that the Landau damping of KAWs has the highest efficiency for the plasma heating in the plume body.

Panel (c) of Fig. 6.15 plots the radiative cooling rate Q_L given by Eq. (6.43) (McClymont and Canfield 1983) at the plume axis versus the heliocentric distance. Further assuming the local frequency of KAWs $\omega \sim 0.01\omega_{ci}$, which is typical for KAWs measured in space plasmas (Louarn et al. 1994; Wahlund et al. 1994a; Volwerk

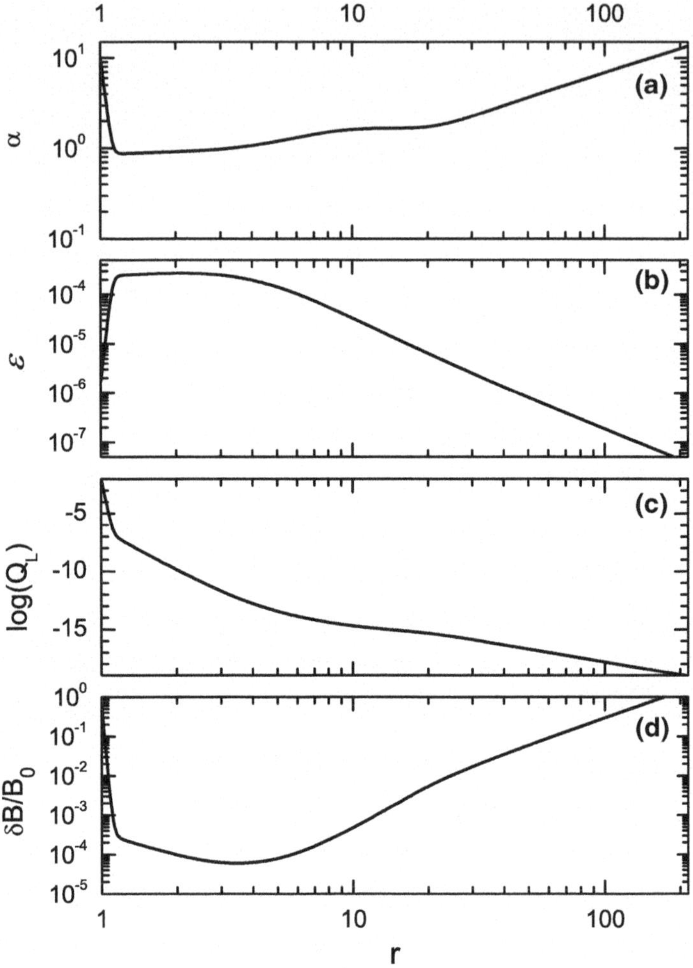

Fig. 6.15 Radial profiles of **a** the resonant parameter α, **b** the dissipative parameter of KAWs ϵ, **c** the radiative cooling rate Q_L in the form of the common logarithm, and **d** the KAW strength required by the KAW heating mechanism (from Wu and Fang 2003, ©AAS reproduced with permission)

et al. 1996; Wu et al. 1996a, b, 1997; Huang et al. 1997; Stasiewicz et al. 1997, 1998, 2000b; Bellan and Stasiewicz 1998; Chaston et al. 1999), the required KAW strength can be estimated from the balance $Q_H = Q_L$. Panel (d) of Fig. 6.15 shows the radial variation of the required wave strength in the relative fluctuation magnetic field, $\delta B/B_0$, which is well below 10^{-3} in the main body of the plume between 1.1 and 10 $R_\odot$. This implies that it is not difficult for the KAWs to supply energy enough to fuel the bright dense plume.

6.5 Extended Solar Coronal Heating by Nonlinear KAWs

6.5.1 Anomalous Energization Phenomena of Coronal Ions

One of the major unsolved problems in solar physics is the determination of the heating process of the magnetically open corona and the corresponding acceleration process of the fast solar wind. It is extensively believed that the extended solar corona of ranging in $\sim 1.2 - 10 R_{\odot}$ is the main acceleration region, in which coronal plasmas are guided by open magnetic fields and accelerated to the solar wind velocity of hundreds of km/s into the interplanetary space. The complexity of the extended coronal plasma is a result of its decreasing density with height that leads to a transition from a collisionally dominated plasma to one that is nearly collisionless. As a result, every ion species tends to have its own unique temperature, its own velocity distribution departure from the Maxwellian, and its own outflow velocity. However, it is a great difficulty to gain the physical information associated with the acceleration processes, because the hydrogen gas has been fully ionized into free electrons and protons and hence there is only some spectral lines of minor heavy ions to be available for spectral line observations, which is main method of inferring the physical situation and process in the source region.

The goals of the UVCS/SOHO, which is used for spectroscopic observations of the extended solar corona, are to provide detailed empirical descriptions of the extended solar corona, and to use the descriptions to gain information about the physical processes controlling the large-scale and small-scale features of the extended solar corona and the acceleration and composition of the solar wind near the Sun.

The most surprising result in its observations (Kohl et al. 1997, 1998; Esser et al. 1999; Cranmer et al. 1999) is that the effective temperatures of heavy ions increase with their masses in the solar corona at $\sim$1.5–5 $R_{\odot}$ from the solar center. For example, the measured spectral line widths of the heavy ions O^{5+} and Mg^{9+} imply their effective temperatures $T^{\mathrm{eff}} \sim 10^8$ K, much higher than that of protons ($\sim 10^6$ K) (see Fig. 6.16). In particular, the effective temperatures of these ions are strongly anisotropic, with $T_{\perp}^{\mathrm{eff}} \gg T_{\parallel}^{\mathrm{eff}}$, where "$\perp$" and "$\parallel$" are directions with respect to the ambient magnetic field $\mathbf{B}_0$ in the solar corona. Also the observations show that both the ion temperature anisotropy and the ion-proton temperature ratio increase rapidly from $\sim$1 at the lower corona below 1.3 $R_{\odot}$ to their maximums above $\sim 4 R_{\odot}$ with heliocentric distance (see Fig. 6.16). These results indicate that these ions must have experienced an anisotropic (mainly preferential cross-field) and mass-charge dependent energization (but neither mass proportional nor mass-to-charge proportional: Esser et al. 1999).

A popular explanation for the ion cross-field energization is based on the ion-cyclotron damping of high-frequency Alfvén waves injected into the base of the solar corona (Hu and Habbal 1999; Hu et al. 2000; Ofman et al. 2002; Xie et al. 2004), in which the wave dissipation range is formed by the high-frequency waves with frequencies close to the local ion gyrofrequency (Xie et al. 2004). The ion-cyclotron heating scheme, however, is not free of difficulties. Not only is the origin of the high-

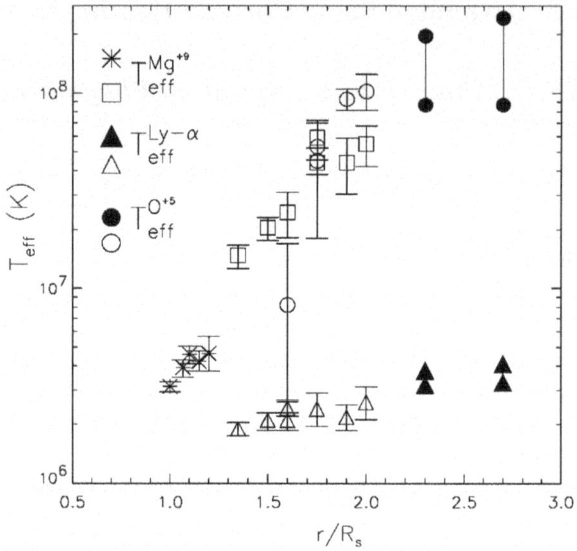

Fig. 6.16 Effective temperatures derived from the measured line widths: open and solid symbols for O^{+5} and Ly-α are from Refs. Esser et al. (1999) and Kohl et al. (1997), respectively, where upper/lower symbols represent upper/lower limits for these observations, asterisks and squares for Mg^{+9} are from Hassler et al. (1990) and Esser et al. (1999), respectively (from Esser et al. 1999)

frequency waves at the coronal base unclear, but their ability to propagate undamped throughout the corona to distances of several solar radii is also doubtful (Voitenko and Goossens 2004, 2005a, 2006). Although the exact spectral composition of the wave flux is unknown, most of its energy should reside in the low-frequency waves. Another one possibility proposed by Lee and Wu (2000) is the ions heated and accelerated by fast shocks that originate in the chromosphere by magnetic reconnection in small scales.

It has been considered that the study of plasmas near Earth can help understanding astrophysical plasmas. In particular, although significant differences there are strongly commonable phenomena in the auroral plasma and the solar plasma. Recent in situ measurements by satellites show that similar particle energization processes are rather common in the auroral plasma (Wygant et al. 2002; Chaston et al. 2003, 2004, 2005). It is especially remarkable that the cross-field ion energization as well as the field-aligned electron energization in the auroral plasma are mainly associated with KAWs (Stasiewicz et al. 2000a), but not with ion-cyclotron waves. Despite of the low-frequency, the perpendicular wavelength of KAWs can become comparable to microkinetic length scales of plasma particles, such as the ion gyroradius ρ_i or the electron inertial length λ_e. In this case, the electron and ion motions decouple from each other, so that a space charge is built up and electrostatic electric fields are generated. The field-aligned component of wave electric fields, $E_\parallel$, makes KAWs available for the field-aligned energization of electrons that drive bright aurora as shown in Chap. 3 (Wu and Chao 2003, 2004a). On the other hand, the short perpendicular wavelength makes KAWs almost electrostatic in the sense that the wave electric field is almost parallel to the wave vector, that is, $|\mathbf{k} \cdot \mathbf{E}| \gg |\mathbf{k} \times \mathbf{E}|$, and hence $|E_\perp| \gg |E_\parallel|$. The large cross-field electric field $E_\perp$ makes KAWs capable of playing an important role in the cross-field energization of ions (Voitenko and Goossens 2004, 2006; Wu and Yang 2006, 2007; Chaston et al. 2004).

Large-scale AWs are one of most popular low-frequency electromagnetic fluctuations in the lower solar atmosphere. Some recent theoretical works suggest that they can be converted into small-scale KAWs when propagating outward due to the inhomogeneity, the phase-mixing, the parametric instability, the wave-wave coupling, etc. (Voitenko 1998; Voitenko and Goossens 2002, 2005b, 2006; Voitenko et al. 2003; Shukla and Sharma 2001; Shukla et al. 2004). Also numerous observations suggest that the high effective temperature of the heavy ions has close relation to the flux of Alfvén waves and that a significant part of this flux is dissipated at heliocentric distances $\geq 1.5R_\odot$ (Esser et al. 1999; Cranmer and van Ballegooijen 2005). In particular, in a work Wu et al. (2007) first presented probably observational evidence of SKAWs in the solar corona (see Sect. 6.2). They reported a novel kind of fine structures in solar radio bursts, SMDS, which are typically characterized by a short lifetime ($\sim$tens ms), a narrow bandwidth (relative bandwidth $<1\%$), and an intermediate frequency drift rate ($\sim$ a few hundreds of MHz/s) and suggested that these SMDS events can be self-consistently produced by a group of SKAWs, in which electrons are field-aligned accelerated by the electric fields of the SKAWs to tens of keV and trapped within their potential wells. It is these trapped electrons that trigger these SMDS events.

Recently, Voitenko and Goossens (2004, 2006) have proposed that the non-adiabatic ion acceleration by KAW trains can provide an alternative to the ion-cyclotron explanation for the intense transverse heating of coronal heavy ions observed by UVCS/SOHO at 1.5–4 $R_\odot$. Alternatively, Wu and Yang (2006, 2007) studied the interaction of heavy ions with nonlinear solitary KAWs (i.e., SKAWs) in a low-β plasma and argued that, for the case of minor heavy ions as in the extended corona, SKAWs can produce a preferential cross-field energization for the heavy ions due to the large cross-field electric field $E_\perp$.

6.5.2 Interaction of Minor Heavy Ions with SKAWs

Assuming the ambient magnetic field $\mathbf{B}_0$ in the coronal hole is uniform and along the z axis (i.e. the radial direction), the interaction of heavy ions with KAWs can be described by following equations (Wu and Yang 2006, 2007): (i) the continuity equations of electrons, protons, and ions

$$\begin{aligned}
&\frac{d}{d\eta}\left[n_e\left(k_z v_{ez}-M\right)\right]=0,\\
&\frac{d}{d\eta}\left[n_p\left(k_x v_{px}-M\right)\right]=0,\\
&\frac{d}{d\eta}\left[n_i\left(k_x v_{ix}-M\right)\right]=0,
\end{aligned}\tag{6.46}$$

(ii) the parallel momentum equation of electrons

$$E_z = (M - k_z v_{ez}) \frac{dv_{ez}}{d\eta} - \alpha^2 k_z \frac{d \ln n_e}{d\eta}, \tag{6.47}$$

(iii) the electric polarization drifts for protons and ions

$$v_{px} = -M \frac{dE_x}{d\eta}, \; v_{ix} = -A_i M \frac{dE_x}{d\eta} = A_i v_{px}, \tag{6.48}$$

(iv) the Faraday and Ampere laws for the electric and magnetic fields

$$\frac{d}{d\eta}(k_z E_x - k_x E_z) = M \frac{dB_y}{d\eta}, \; k_x \frac{dB_y}{d\eta} = -n_e v_{ez}, \tag{6.49}$$

and (v) the charge neutrality condition

$$n_e = n_p + Z_i n_i, \tag{6.50}$$

where Z_i is the ion charge in units of the proton charge e and A_i is the ion mass-charge ratio in units of the proton mass-charge ratio m_p/e.

By using the localized boundary conditions for SKAWs (see Sect. 3.4): $n \equiv n_e = n_0 = 1$, $n_p = n_{p0} = 1 - Z_i n_{i0}$, $n_i = n_{i0}$, $v_{ez} = v_{p(i)x} = d/d\eta = 0$ for $\eta \to \pm\infty$, where n_{i0} is the ion abundance ratio in the background plasma (i.e. the unperturbed ion number density in units of the unperturbed electron density), integrating the above equation system of Eqs. (6.46)–(6.49) and limiting the attention on the low-β case of $\alpha < 1$, one can obtain the nonlinear Sagdeev equation of the plasma density in SKAWs as follows (Wu and Yang 2006, 2007):

$$\frac{1}{2}\left(\frac{dn_e}{d\eta}\right)^2 + K(n_e) = 0, \tag{6.51}$$

where the Sagdeev potential $K(n_e)$ is expressed by

$$K(n_e) = -\frac{n_e^4}{k_x^2 M_z^2}\left[\frac{(n_e - 1)^2}{2}\left(M_z^2 - \frac{n_e + 2}{3n_e}\right) - n_e^2 I_1(n_e)\right], \tag{6.52}$$

and $I_1(n_e)$ is a modification term for the Sagdeev potential $K(n_e)$, due to the presence of the second ion species with density n_i. Similar with the case in the electron-proton plasma (Hasegawa and Mima 1976; Wu et al. 1995, 1996b), also the solution of the Sagdeev equation (6.51) is a SKAW. Based on the existential criterion for SKAWs (Wu et al. 1996b), one has $K(n_e) < 0$ for n_e between 1 and n_{em} and $K(n_e) = 0$ at $n_e = 1$ and $n_e = n_{em}$, where n_{em} is the minimal value for the electron density (i.e. the plasma density) distribution in the SKAW, and hence the plasma density amplitude of the SKAWs $\delta n_{em} = 1 - n_{em}$. The parallel phase speed $M_z = M/k_z$ can be given by the condition $K(n_{em}) = 0$ in the following form of the nonlinear dispersion relation:

$$M_z^2 = \frac{2 + n_{em}}{3n_{em}} + \frac{2n_{em}^2 I_1(n_{em})}{(1 - n_{em})^2}. \tag{6.53}$$

According to the element abundance of the solar corona the coronal plasma consists of mainly electrons and protons (i.e. ionized hydrogen atoms). Other heavy elements in the corona have only very low abundance less than 10^{-3}, except for helium, which is ~5% (i.e., $n_{\mathrm{He}} \sim 4.7 \times 10^{-2}$), for example, $n_{\mathrm{O}} \sim 8.55 \times 10^{-4}$, $n_{\mathrm{Mg}} \sim 1.74 \times 10^{-4}$, $n_{\mathrm{Fe}} \sim 1.32 \times 10^{-4}$ (Reames 1994). Since the ion abundance must be less than the corresponding element abundance, there are only minor heavy ion species in the corona, that is, $n_{i0} \ll 1$. In this minor ion approximation of $n_{i0} \ll 1$ the modification term $I_1(n_e)$ can be represented as follows:

$$\begin{aligned}\frac{I_1(n_e)}{N_{i0}} = {} & \frac{2A_i^2 - 9A_i + 6}{6A_i^3} + \frac{5(A_i - 1) - (A_i - 1)^2 n_i/n_{i0}}{6A_i^3 n_e} \\ & + \frac{1 - (A_i - 1) n_i/n_{i0}}{6A_i^2 n_e^2} - \frac{1 - n_i/n_{i0}}{3A_i n_e^3} - \frac{(A_i - 1)^2}{A_i^4} \ln \frac{n_i}{n_{i0}},\end{aligned} \tag{6.54}$$

where the parameter $N_{i0} = (A_i - 1) Z_i n_{i0}$ and the ion density n_i is given by

$$\frac{n_i}{n_{i0}} = \frac{n_e}{A_i - (A_i - 1) n_e}. \tag{6.55}$$

Table 6.2 lists the mass in units of the proton mass m_p, the charge Z_i in units of the proton charge e, and mass-charge ratio A_i in units of m_p/e for some heavy ions that are often present in the UVCS observations.

Let the perturbed density $\delta n_e \equiv n_e - 1 \propto \exp(i\eta)$, the linearization result of the Sagdeev equation (6.51), by use of Eqs. (6.52), (6.54), and (6.55), presents the linear dispersion relation:

$$M_z^2 = \frac{1 - N_{i0}}{1 + \lambda_e^2 k_x^2}. \tag{6.56}$$

In particular, we obtain the well-known linear dispersion relation of KAWs in the low-β case, $M_z^2 = 1/(1 + \lambda_e^2 k_x^2)$, when $N_{i0} \to 0$.

Table 6.2 Parameters of heavy ions (from Wu and Yang 2007, ©AAS reproduced with permission)

Species	H^{1+}	He^{2+}	O^{5+}	Mg^{9+}	Fe^{11+}
m_i/m_p	1	4	16	24.305	55.847
Z_i	1	2	5	9	11
A_i	1	2	3.2	2.7	5.077

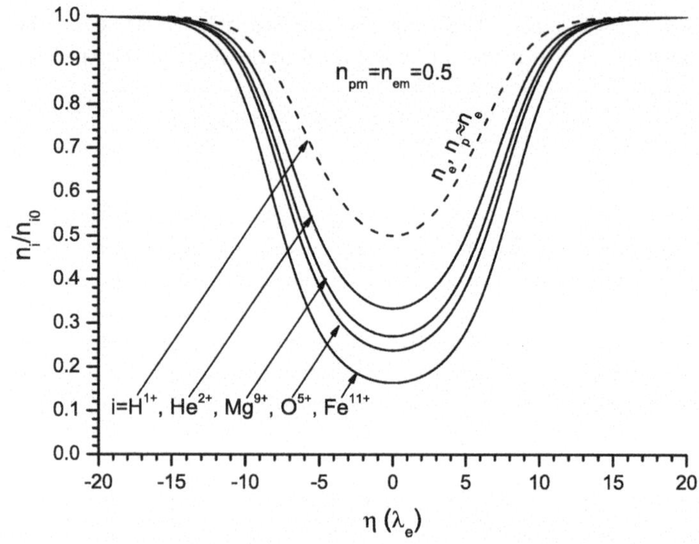

Fig. 6.17 Distributions of densities in a solitary nonlinear wavelet of KAWs for the He^{2+}, Mg^{9+}, O^{5+}, and Fe^{11+} ions. Dashed line is the density distribution of the background plasma (i.e. protons and electrons) (from Wu and Yang 2007, ©AAS reproduced with permission)

Based on the solution of Eq. (6.51), by use of Eq. (6.55), Fig. 6.17 shows the density distributions of the ions (He^{2+}, O^{5+}, Mg^{9+}, and Fe^{11+}) in the SKAW, where the dashed line is the local plasma density fluctuation with an amplitude $N_{em} \equiv 1 - n_{em} = 0.5$. Figure 6.18 displays distributions of the perturbed electric and magnetic fields in the SKAW, where the perturbed fields have been normalized by their amplitudes (i.e., their maximums). The amplitudes of the perturbed fields (E_{xm}, E_{zm}, and B_{ym}) versus the density amplitude ($N_{em} \equiv 1 - n_{em}$) are plotted in Fig. 6.19.

From Fig. 6.17, the solitary nonlinear wavelet of KAWs has a typical width of $\sim 20\lambda_e$ in the cross-field direction. The local density distribution of the ions has a dip-soliton profile similar to that of electrons and deepens with the ion mass-charge ratio A_i, where the proton density $n_{H^{1+}} \simeq n_e$ because of the assumption of minor ions ($n_i \ll n_e$) and hence $n_p \simeq n_e$.

From Figs. 6.18 and 6.19, the perturbed electric and magnetic fields of the SKAW have a typical kink structure and their amplitudes increase with the density amplitude. In particular, the cross-field component of the perturbed electric field E_{xm} is much larger than the field-aligned component E_{zm} by a factor $\sim 10^2$ for weak SKAWs with $N_{em} < 30\%$ or by a factor ~ 10 for strong SKAWs with $N_{em} > 70\%$.

Now let us consider the ion energization in a SKAW, which structure is presented in the last section. In the field-aligned direction, the energization of charged particles is caused simply by the direct acceleration of the parallel electric field E_z, in result $m_i v_{i\parallel}^2/2 = |q_i\phi_z|$, where ϕ_z is the field-aligned electric potential drop of the SKAW.

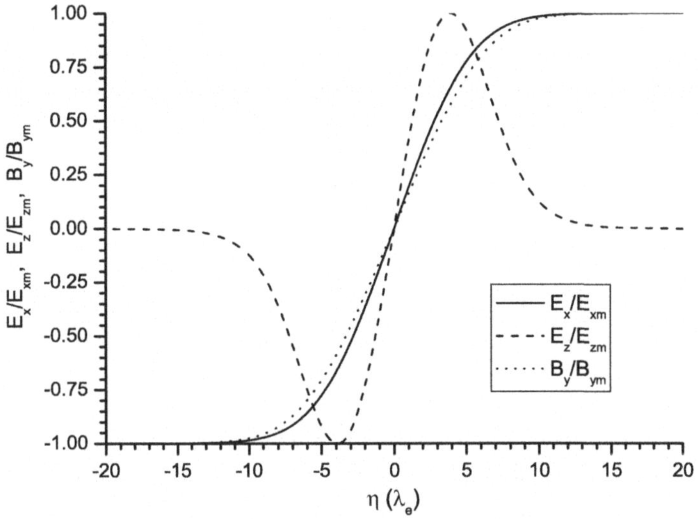

Fig. 6.18 Distributions of perturbed electric and magnetic fields in a solitary nonlinear wavelet of KAWs, where solid, dashed, and dotted lines represent E_x, E_z and B_y, respectively (from Wu and Yang 2007, ©AAS reproduced with permission)

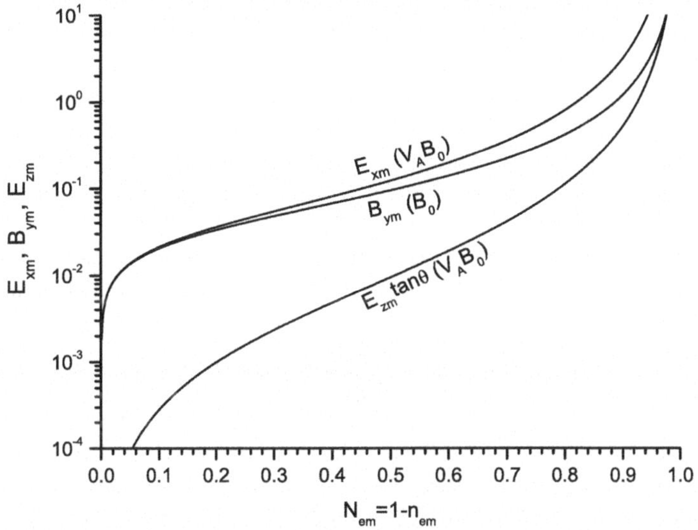

Fig. 6.19 Plots of amplitudes of perturbed electric and magnetic fields versus the density amplitude of the background plasma (from Wu and Yang 2007, ©AAS reproduced with permission)

This indicates that the field-aligned energy of the particles depends on the field-aligned electric potential ϕ_z and is directly proportional to the particle charge q_i, but independent of its mass. On the other hand, in the cross-field direction, the ions cannot be accelerated directly by the perpendicular field E_x because of their Larmor cyclotron motion with a gyrofrequency much higher than the wave frequency. Thus, the ions accelerated by E_x in the first half period of one ion-cyclotron period will be decelerated by E_x in the following half period. Consequently, the cross-field ion energy sensitively depends on the ion gyrofrequency and hence on the ion mass and charge.

From Eq. (6.46), the field-aligned electron velocity and the cross-field proton and ion velocities in the SKAW can be obtained as follows:

$$\begin{aligned} v_{ez} &= (1 - 1/n_e)\, M_z, \\ v_{px} &= \left(1 - n_{p0}/n_p\right) M_z \cot\theta \simeq v_{ez}\cot\theta, \\ v_{ix} &= (1 - n_{i0}/n_i)\, M_z \cot\theta \simeq A_i v_{px}, \end{aligned} \tag{6.57}$$

where the boundary conditions v_{ez}, v_{px}, $v_{ix} = 0$, $n_e = 1$, $n_p = n_{p0}$, $n_i = n_{i0}$ for $\eta \to \infty$ have been used, the ion density n_i is given by Eq. (6.55), θ is the wave propagating angle, and $\cot\theta = k_z/k_x$. Figure 6.20 shows the velocity distributions of the ions (He^{2+}, O^{5+}, Mg^{9+}, and Fe^{11+}) in the SKAW, where $\theta = 89°$ has been used because of the quasi-perpendicular propagation of the SKAW, that is, $k_z \ll k_x$. For the sake of comparison, the proton velocity is presented by the dashed line in Fig. 6.20. It is clear that the ion velocity increases with the ion mass-charge ratio A_i. In fact, from the last one equality of Eq. (6.57) the ion velocity is directly proportional to A_i. In particular, it is worth noting that the ion velocities reach maximum at the center of the SKAW and decrease to zero at its boundaries. This implies that the ion motions are strictly restricted within the local structure of the SAKW with the typical size $\sim$ a few tens λ_e.

Based on the physics of the field-aligned energization of charged particles described above, from Eq. (6.57) we have the field-aligned ion energy in the SKAW

$$\varepsilon_{i\|} = Z_i \varepsilon_{e\|} = \left(\frac{1}{n_e} - 1\right)^2 M_z^2 Z_i \varepsilon_A, \tag{6.58}$$

where $\varepsilon_{e\|}$ is the field-aligned electron energy and $\varepsilon_A \equiv m_e v_A^2/2$. This indicates that the field-aligned energy distribution of ions depends on the density n_e in the sense that it increases as n_e decreases from zero at the boundary ($n_e = 1$) to the maximal energy

$$\varepsilon_{i\|m} = \left(\frac{1}{n_{em}} - 1\right)^2 M_z^2 Z_i \varepsilon_A. \tag{6.59}$$

at the center ($n_e = n_{em} < 1$). Thus, the average field-aligned ion energy can be estimated as Wu and Yang (2007)

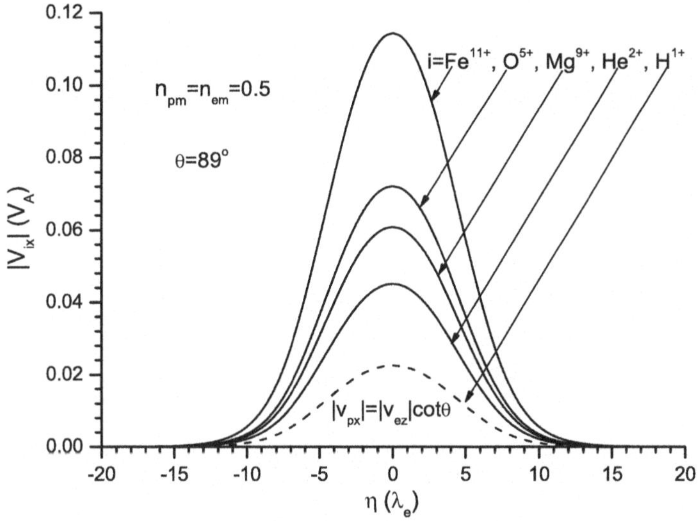

Fig. 6.20 Distributions of velocities in a solitary nonlinear wavelet of KAWs for the He^{2+}, Mg^{9+}, O^{5+}, and Fe^{11+} ions. Dashed line is the velocity distribution of protons (from Wu and Yang 2007, ©AAS reproduced with permission)

$$\langle \varepsilon_{i\parallel} \rangle = Z_i \langle \varepsilon_{e\parallel} \rangle = \left[1 + \frac{(1-n_{em})^2 (1-2.5n_{em})}{3n_{em}^2 (1-n_{em}+n_{em}\ln n_{em})} \right] M_z^2 Z_i \varepsilon_A. \quad (6.60)$$

For the cross-field ion energy, we have, from Eq. (6.57),

$$\varepsilon_{i\perp} \equiv \frac{1}{2} m_i v_{ix}^2 = \left(1 - \frac{n_{i0}}{n_i} \right)^2 \varepsilon_{i0}, \quad (6.61)$$

where $\varepsilon_{i0} = A_i Z_i M_z^2 (\varepsilon_A/Q) \cot^2\theta$. Similarly, we can estimate the average cross-field ion energy as follows (Wu and Yang 2007):

$$\langle \varepsilon_{i\perp} \rangle = \left[1 + \frac{A_i^2 (1-n_{em})^2 [A_i (1-n_{em}) - 1.5n_{em}]}{3n_{em}^2 [A_i (1-n_{em}) + n_{em} \ln (A_i/n_{em} - A_i + 1)]} \right] \varepsilon_{i0}. \quad (6.62)$$

Figure 6.21 plots the average cross-field energies of the He^{2+}, O^{5+}, Mg^{9+}, and Fe^{11+} ions versus the density amplitude $\delta n_{em} \equiv N_{em}$. As expected the ion transverse energy increases rapidly with δn_{em} (the SKAW strength). Figure 6.22 presents the anisotropic ratio of the ion energization $\delta_i \equiv \langle \varepsilon_{i\perp} \rangle / \langle \varepsilon_{i\parallel} \rangle$ (solid lines) and the ion-proton energy ratio $\gamma_i \equiv \langle \varepsilon_{i\perp} \rangle / \langle \varepsilon_{p\perp} \rangle$ (dashed lines). Unlike the ion energies, the ion anisotropic ratio δ_i and the ion-proton energy ratio γ_i do not sensitively depend on the density amplitude δn_{em} of the SKAW. From Fig. 6.22 the ion anisotropy (δ_i) increases with the ion mass-charge ratio A_i, but not with a simply proportional relationship. In specialty, we have $\delta_i \sim 4$ for He^{2+}, ~ 10 for Mg^{9+}, ~ 17 for O^{5+}, and ~ 65 for

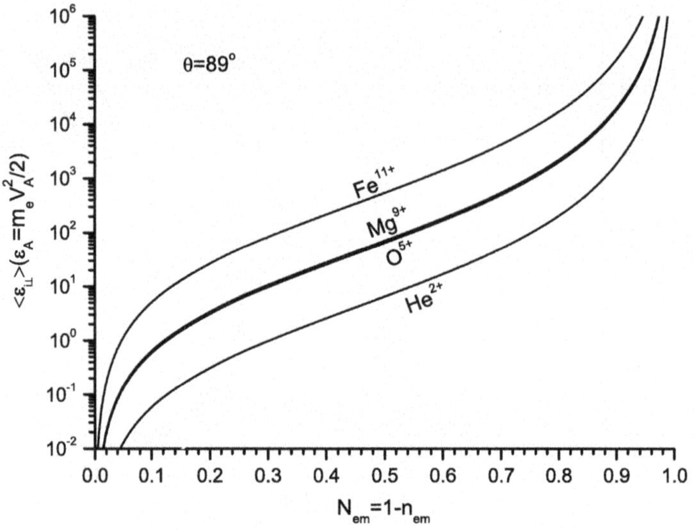

Fig. 6.21 Plots of average cross-field energies of the He^{2+}, Mg^{9+}, O^{5+}, and Fe^{11+} ions versus the density amplitude of the background plasma (from Wu and Yang 2007, ©AAS reproduced with permission)

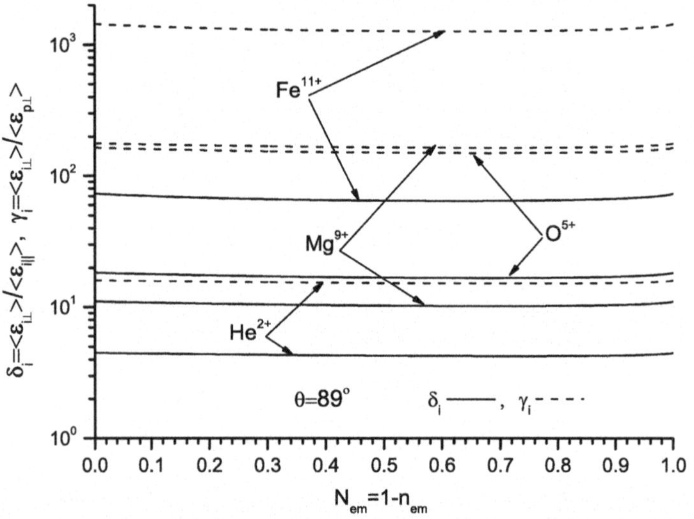

Fig. 6.22 Plots of temperature anisotropic parameter (solid lines) of the ions and temperature ratio (dashed lines) of the ions to protons versus the density amplitude of the background plasma (from Wu and Yang 2007, ©AAS reproduced with permission)

Fe^{11+}, and hence approximately have $\delta_i \sim 0.5A_i^3$, proportional to the cube of the ion mass-charge ratio. On the other hand, for the ion-proton energy ratio γ_i, we have $\gamma_i \sim 15$ for He^{2+}, $\sim$160 for Mg^{9+}, $\sim$150 for O^{5+}, and $\sim$1300 for Fe^{11+}, and hence we approximately have $\gamma_i \sim 0.9A_i^3 Z_i$, proportional to the product of the cube of the ion mass-charge ratio and the ion charge, or i.e. the product of the square of the ion mass-charge ratio and the ion mass.

6.5.3 Energization of Minor Heavy Ions by SKAWs

In the case of coronal holes, we can imagine that AWs originating in the lower solar atmosphere are converted into KAWs when they propagate outwards and enter the corona. The KAWs develop into solitary nonlinear wavelets (i.e., SKAWs) in the extended coronal region because they are not damped resonantly by the thermal electrons in the ambient plasma. These solitary wavelets have a small cross-field scale $\sim$ a few tens λ_e and wander randomly in the extended coronal region. In particular, they can lead to the anisotropic energization of the ions, as shown in the last section, and the motion of the energized ions trapped in the wavelets manifests itself as a random motion or turbulence. In other words, the energized ions have anisotropic effective temperatures, $\langle\varepsilon_{i\perp}\rangle/k_B$ and $\langle\varepsilon_{i\parallel}\rangle/k_B$ (where k_B is the Boltzmann constant).

On the other hand, since the plasma density decreases in the solar corona, and hence the electron inertial length increases with increasing heliocentric distance, the ions in larger heliocentric distance will be energized by the wavelets with larger cross-field scale. This implies the variation of the ion energization with heliocentric distance is determined by the spectrum distribution of KAWs in the extended corona. It is rather difficult, however, to calculate exactly the turbulent spectrum of KAWs in the extended corona. The anisotropic cascading model of MHD turbulence proposed by Goldreich and Sridhar (1995, 1997) predicts that the one-dimensional energy spectrum is proportional to $k_\perp^{-5/3}$: an anisotropic Kolmogorov spectrum. Boldyrev (2005) further showed that this anisotropic spectrum can change, depending on the strength of the ambient magnetic field, from $\delta B_k^2 \propto k_\perp^{-5/3}$ to $\delta B_k^2 \propto k_\perp^{-3/2}$ as the ambient field increased from $B_0^2 < \delta B^2$ to $B_0^2 > \delta B^2$. Taking the feature width of solitary wavelets in the wave-vector space as $\Delta k \sim k$, its amplitude can be estimated by $\delta B_m^2 \propto k_\perp^{-2/3} \propto \lambda_\perp^{2/3}$ for the weak field case or by $\propto k_\perp^{-1/2} \propto \lambda_\perp^{1/2}$ for the strong field case. Its density amplitude N_{em} can be calculated from the amplitude relation between the field and density presented in Fig. 6.19. As a result, the average field-aligned and cross-field ion energies, $\langle\varepsilon_{i\parallel}\rangle$ and $\langle\varepsilon_{i\perp}\rangle$, can be obtained from Eqs. (6.60) and (6.62), respectively.

Consequently, total effective temperature of the ions in the extended corona consists of the anisotropic random motion caused by the KAW wavelets and the kinetic temperature of the ambient plasma, T_0, which is assumed to be isotropic. Thus, we have

$$T_{i\parallel}^{\mathrm{eff}} = T_0 + \frac{\langle\varepsilon_{i\parallel}\rangle}{k_B} \tag{6.63}$$

and

$$T_{i\perp}^{\text{eff}} = T_0 + \frac{\langle \varepsilon_{i\perp} \rangle}{k_B}. \tag{6.64}$$

Figures 6.23 and 6.24, respectively, show the variations of the ion temperature anisotropy

$$\delta_i \equiv \frac{T_{i\perp}^{\text{eff}}}{T_{i\parallel}^{\text{eff}}} = \frac{1 + \langle \varepsilon_{i\perp} \rangle / k_B T_0}{1 + \langle \varepsilon_{i\parallel} \rangle / k_B T_0} \tag{6.65}$$

and the ion-proton temperature ratio

$$\gamma_i \equiv \frac{T_{i\perp}^{\text{eff}}}{T_{p\perp}^{\text{eff}}} = \frac{1 + \langle \varepsilon_{i\perp} \rangle / k_B T_0}{1 + \langle \varepsilon_{p\perp} \rangle / k_B T_0}, \tag{6.66}$$

with heliocentric distance, where the typical cross-field width of the KAW wavelets, $\sim 10\lambda_e$, has been used, an anisotropic spectrum $\propto k_\perp^{-3/2}$ has been assumed in the coronal base $R \sim 1R_\odot$, because of the strong field there, and the perturbed field in the long wavelength $\lambda_0 \sim 1000$ km has been estimated by $\delta B_{\lambda_0} \sim 0.01 B_0$, based on observations in the coronal base. In Figs. 6.23 and 6.24, the solid and dashed lines represent the cases of the KAW spectra $\propto k_\perp^{-5/3}$ for a weak field and $\propto k_\perp^{-3/2}$ for a strong field in the extended corona, respectively. It is clear that the ion energization does not sensitively depend on the exact form of the spectra, except that the rapid increase of the ion temperature anisotropy (δ_i) and the ion-proton temperature ratio (γ_i) occurs in larger heliocentric distances for a harder spectrum.

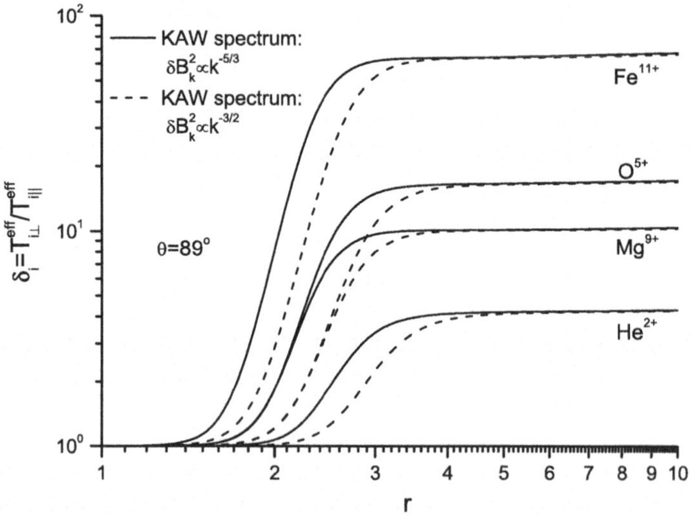

Fig. 6.23 Variations of temperature anisotropic parameter of the ions with heliocentric distance (from Wu and Yang 2007, ©AAS reproduced with permission)

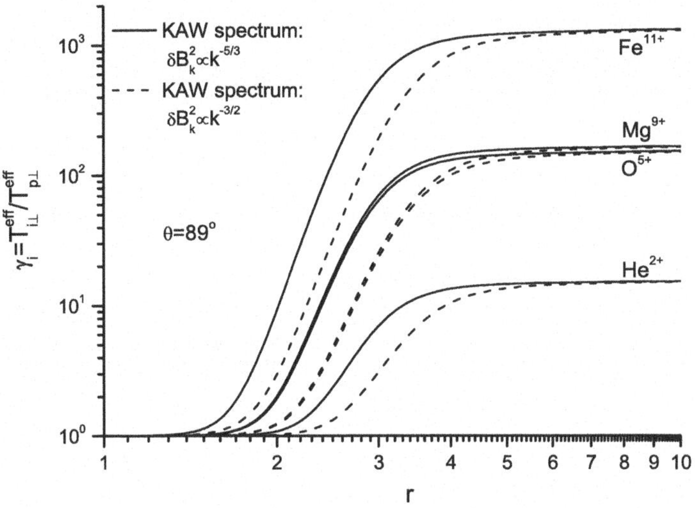

Fig. 6.24 Variations of temperature ratio of the ions to protons with heliocentric distance (from Wu and Yang (2007), ©AAS reproduced with permission)

From Fig. 6.23, we can see that the anisotropy of the effective ion temperature rapidly increases with heliocentric distance in $R \sim 1.5$–$3R_\odot$ as the ion energization by KAWs increases. In the lower corona of $R < 1.3R_\odot$, the effective ion temperature is dominated by the kinetic temperature of the ambient plasma, and hence is almost isotropic. In the higher corona of $R > 4R_\odot$, the anisotropy approaches the saturation value $\langle\varepsilon_{i\perp}\rangle/\langle\varepsilon_{i\parallel}\rangle$ (see Fig. 6.22), since the ion energization by KAWs dominates the effective temperature, that is, $\langle\varepsilon_{i\perp}\rangle \gg k_B T_0$. From Fig. 6.23 we have the saturation values of the anisotropy $\delta_i \sim 4$ for the He^{2+} ions, ~10 for the Mg^{9+} ions, ~17 for the O^{5+} ions, and ~65 for the Fe^{11+} ions, and hence we have an approximate empirical relation for the maximal anisotropy of the effective ion temperature in the extended corona hole as follows:

$$\delta_{im} \sim 0.5A_i^3. \tag{6.67}$$

On the other hand, from Fig. 6.24, it can be found that the ion-proton temperature ratio has a similar behavior in the radial variation. The temperature ratio γ_i rapidly increases with heliocentric distance in $R \sim 1.5$–$4R_\odot$ as the ion energization by KAWs increases. In the lower corona of $R < 1.3R_\odot$, one has $\gamma_i \sim 1$, since the effective temperature is dominated by the isotropic ambient kinetic temperature (i.e. $k_B T_0 \gg \langle\varepsilon_{i\perp}\rangle$). In the higher corona of $R > 5R_\odot$, the temperature ratio also approaches the saturation value $\langle\varepsilon_{i\perp}\rangle/\langle\varepsilon_{p\perp}\rangle$ (see Fig. 6.22), since the ion energization by KAWs dominates the effective temperature. From Fig. 6.24, we have the saturation values of the ratio $\gamma_{im} \sim 15$ for the He^{2+} ions, ~150 for the O^{5+} ions, ~160 for the Mg^{9+} ions, and ~1300 for the Fe^{11+} ions, and hence we have an approximate empirical relation for the maximal ion-proton temperature ratio in the extended corona as follows:

$$\gamma_{im} \sim 0.9 A_i^3 Z_i. \tag{6.68}$$

In comparison to the UVCS/SOHO observation, it is easy to find that the results presented by Figs. 6.23 and 6.24 can in reason and self-consistently explain the observed ion temperature anisotropy and the ion-proton temperature ratio, their dependence on the ion mass and charge, and the variation with heliocentric distance. The extended corona is an important region in coronal holes, where coronal plasmas are heated and accelerated into the solar wind. The spectroscopic measurements by UVCS/SOHO provided very important information and stringent constraints on the coronal plasma physics in this extended region (Kohl et al. 1997, 1998; Esser et al. 1999; Cranmer et al. 1999; also see Fig. 6.16):

1. The effective temperatures of minor heavy ions (i.e., O^{5+}, Mg^{9+}, and Fe^{11+}) have strongly anisotropy with $\delta_i = T_{i\perp}^{\rm eff}/T_{i\parallel}^{\rm eff} \gg 1$;
2. These heavy ions have an effective temperature much larger than that of protons, and the ion-proton temperature ratio, $\gamma_i = T_{i\perp}^{\rm eff}/T_{p\perp}^{\rm eff}$, is much larger than their mass ratio;
3. The variations of the parameters δ_i and γ_i both are characterized by a rapid increase with heliocentric distance in 1.5–3.5 $R_\odot$, from $\simeq$1 in the lower corona below 1.3 $R_\odot$ to the saturation values in the higher corona above 4 $R_\odot$.

Coronal holes have been identified as source regions of the fast solar wind. Although AWs activity has been detected in the solar corona by remote sensing, and in the solar wind by in situ satellite measurements, it still is unknown how to convert the AWs energy into the kinetic energy of coronal plasma particles and the flow energy of the solar wind (Cranmer et al. 1999; Cranmer and van Ballegooijen 2005; Trimble and Aschwanden 2005; Voitenko and Goossens 2006; Del Sarto et al. 2006; Isenberg and Vasquez 2007; Markovskii et al. 2008, 2009; Cranmer 2009). The observations of the anisotropic and mass-charge dependent energization of minor heavy ions in the extended corona by UVCS/SOHO (Kohl et al. 1997, 1998; Esser et al. 1999; Cranmer et al. 1999) provide us an opportunity to inspect various microphysics mechanisms. In particular, these observations emphasize the importance of particle kinetics due to wave-particle interactions (Marsch 2006). The resonant interaction between ions and parallel-propagating ion cyclotron waves has been proposed to be responsible for the heating and acceleration of the fast solar wind, in which the preferential effects naturally arise from the ability of minor ions to simultaneously resonate with several modes in a spectrum of inward- and outward-propagating waves in the way equivalent to a second-order Fermi acceleration of these thermal particles (Isenberg and Vasquez 2007), while protons can encounter only a single resonance for a given particle parallel speed. Markovskii et al. (2008, 2009) pointed out that the injecting energy in the limit of cyclotron-resonant quasi-linear diffusion is insufficient to account for the observed acceleration. Also they further investigated the role of nonlinear processes in the resonant absorption of the waves, and found that, although the nonlinearity is particularly strong when the intensity of antisunward-propagating waves is comparable to that of sunward waves, the nonlinear wave damping operating alone can not be responsible for the cross-field energization of ions because it only causes their field-aligned diffusion.

The results presented in Figs. 6.23 and 6.24 (Wu and Yang 2006, 2007) can well reproduce above these substaintial characteristics of the anomalous energization phenomena of minor heavy ions in the extended corona. In the scenario described by Wu and Yang (2006, 2007), AWs that originate the MHD turbulence in the lower solar atmosphere are converted into KAWs and develop into solitary wavelets in the extended corona, which can be modeled by SKAWs, when they propagate outward and enter into the corona along open solar magnetic fields. It is these KAW wavelets that lead to the anisotropic and mass-dependent energization of the heavy ions.

The observational effective temperatures, derived from the spectral line widths of ions measured by UVCS/SOHO, consist of the kinetic temperature (T_0), which originates in the kinetic heating in the lower corona and is isotropic and uniform for all ion species because of collisionally dominated plasma, and the turbulent temperature ($\langle\varepsilon_i\rangle/k_B$) that is caused by the KAW wavelets. Their characteristic wavelength ($\sim$ a few tens of λ_e) increases with the heliocentric distance R (see Fig. 6.13e). Therefore, the corresponding turbulent energy spectrum intensifies as the heliocentric distance increases. As a result, in the lower corona below 1.3 $R_\odot$, the kinetic temperature T_0 dominates the total effective temperature (i.e., $T_0 \gg \langle\varepsilon_i\rangle/k_B$) due to the weak KAW wavelets in the lower corona and we have both δ_i and $\gamma_i \simeq 1$. In the extended corona of $R \gtrsim 1.3R_\odot$, the strength of the KAW wavelets and hence the turbulent temperature increase with the heliocentric distance so rapidly that the turbulent temperature becomes dominating one, and this results in the observed anisotropy and mass-charge dependence of the heavy-ion anomalous energization (i.e., $\delta_i \gg 1$ and $\gamma_i \gg 1$) in the extended corona. In particular, we notice that in the higher corona above $4R_\odot$, both δ_i and γ_i approach their saturation values, that is, their maximums: $\delta_{im} = \langle\epsilon_{i\perp}\rangle/\langle\epsilon_{i\parallel}\rangle \simeq 0.5A_i^3$ and $\gamma_{im} = \langle\epsilon_{i\perp}\rangle/\langle\epsilon_{p\perp}\rangle \simeq 0.9A_i^3 Z_i$.

6.6 Interplanetary Solar Wind Heating by Turbulent KAWs

The solar wind is the extension of the solar high-temperature expansion corona. The first dynamical model of the solar wind was proposed by Parker (1958). It has been believed that the acceleration of the solar wind from a hydrostatic atmosphere to an outward fluid at several hundreds km/s (i.e., the solar wind) occurs mainly in the extended coronal region within $\sim 10R_\odot$, although the exact acceleration mechanism is still unclear. After this acceleration region ($R > 10R_\odot$), the solar wind, in general, is considered as the spherical and adiabatic expansion. Assuming a constant velocity solar wind with a density $n \propto R^{-2}$, on average, the solar wind temperature T decreases with the heliocentric distance R at the power-law way of $T \propto n^{-2/3} \propto R^{-4/3}$, called the adiabatic cooling dependence.

Gazis and Lazarus (1982) first measured the radial variation of the solar wind temperature. They compared the measurements of the solar wind proton temperature measured by the Voyager-1 satellite when it traveled from the heliocentric distance $R \sim 1$ AU to $\sim$10 AU with the temperatures measured by the IMP-8 satellite in the Earth orbit (i.e., $\sim$1 AU). Figure 6.25 shows their results, where the upper series of boxes are average solar wind temperatures (in units of $T = 5.6 \times 10^4$ K) at $R \sim 1$ AU (i.e., the Earth orbit), which were obtained by the IMP-8 satellite, and the scalar temperature is calculated by taking the average of the second moments of the reduced proton distribution functions measured by the three main sensors. In Fig. 6.25, the lower series of boxes present the ratioes of temperatures measured by the Voyager-1 satellite, where in order to ensure the same solar wind period explored by the two satellites, the start and stop times of each Voyager-1 averaging period are time-shifted from the start and stop times of each IMP-8 averaging period by taking account of the solar wind travel time and the difference in longitude of the two satellites. The straight line in Fig. 6.25 presents the adiabatic cooling dependence, that is, $T \propto R^{-4/3}$. In addition, each rectangle represents an average over two solar rotations, the width of the rectangle represents the motion of the Voyager-1 satellite during the averaging period, and the height of each rectangle represents the standard deviation of the temperature measurements over the averaging period.

From Fig. 6.25, one can find that the solar wind temperature decreases with the distance R evidently slower than that predicted by the adiabatic law of $R^{-4/3}$, decreasing only as $T \propto R^{-0.7} \sim R^{-2/3}$. This implies that there should be some heating process to complement the adiabatic expansion cooling of the solar wind. In other words, the solar wind also requires an additional heating mechanism during its propagating in the interplanetary space, and this is so-called the solar wind heating problem.

Gazis (1984) proposed that the solar wind heating possibly results from an additional heat flux due to heat conduction Hollweg (1974, 1976), for instance, the interaction between fast and slow solar wind flows which converts bulk kinetic energy into local thermal energy in the interaction regions via forming shocks. In general, shocks do not contribute significant heating (in the sense of entropy generation) in the inner heliosphere, although they may possibly make important contributions beyond a few AU (Goldstein and Jokipii 1977; Whang et al. 1989, 1990). The electron temperature varies only very slowly with R due to a strong heat conduction from the Sun, but the coupling between electrons and protons outside the corona is probably too weak to provide significant proton heating (Marsch et al. 1983). Another major candidate for the proton heating mechanism is so-called the turbulent heating, in which the turbulent cascade of solar wind Alfvénic fluctuations towards small scales transfers the turbulent energy into the kinetic scales of plasma particles and is dissipated into particle kinetic energy (Tu et al. 1984, 1989; Tu 1988).

Verma et al. (1995) further investigated in detail the possibility of the dissipation of the solar wind turbulence as the solar wind heating mechanism. By use of the generalized Kolmogorov-like (for isotropic non-conductive fluid) and Iroshnikov-Kraichanan-like (for anisotropic MHD) turbulent spectra and taking account of the suppression effect of the cross helicity (H_c) on the dissipation, where H_c is defined by

$$H_c \equiv \frac{1}{2} \int (\mathbf{v} \cdot \mathbf{B})\, d^3\mathbf{x}, \tag{6.69}$$

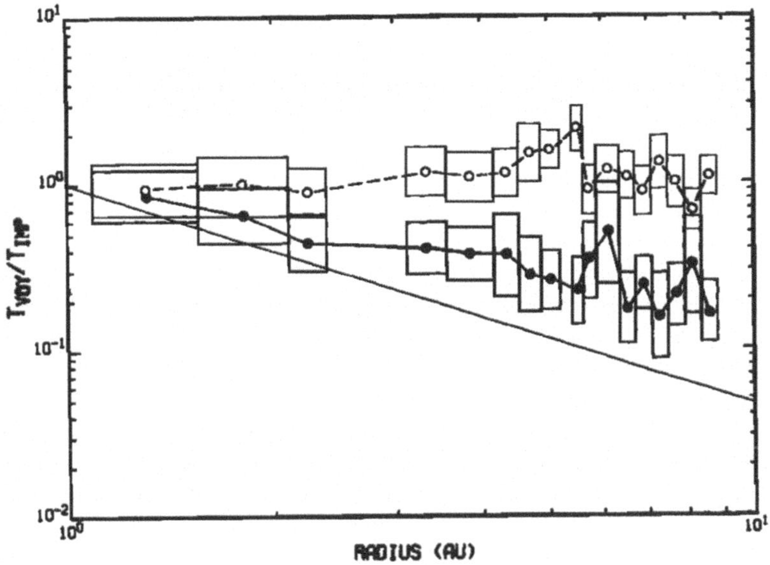

Fig. 6.25 Average proton temperature of the solar wind is plotted as a function of heliocentric radius. The upper series of boxes are temperatures (in units of $T = 5.6 \times 10^4$ K) obtained by the IMP-8 satellite in the Earth orbit (i.e., $R \sim 1$ AU). The lower series of boxes are the ratioes of temperatures obtained by the Voyager-1 satellite over the same period of the solar wind to the IMP-8 measured values. The straight line varies as $R^{-4/3}$. Each rectangle represents an average over two solar rotations. The width of the rectangle represents the motion of the Voyager-1 satellite during the averaging period. The height of each rectangle represents the standard deviation of the temperature measurements over the averaging period (from Gazis and Lazarus 1982)

they calculated the turbulent dissipation rate of AWs and the resulting radial evolution of temperature in the solar wind. Their results showed that the turbulent dissipation predicted by the isotropic Kolmogorov-like turbulence model could lead to a sufficient heating rate to explain the measured slower-than-adiabatic radial dependence of the solar wind temperature, and while the anisotropic Iroshnikov-Kraichanan-like turbulence yields less turbulent heating rate because of its higher cross helicity and seems inadequate to explain the observed temperature radial dependence (Verma et al. 1995). However, Verma et al. (1995) also pointed out that the Kolmogorov-like turbulent heating can likely dominate only in the inner heliosphere and may possibly be only one of competitive heating mechanisms with, such as, heating by shocks and the assimilation of interstellar pickup ions in the outer heliosphere.

Early observations indicated that the solar wind contains MHD waves (mainly AWs, Unti and Neugebauer 1968; Belcher and Davis 1971) as well as fluid turbulence (Coleman 1966, 1968; Matthaeus and Goldstein 1982). In particular, Coleman (1968) inspired a lively debate about whether or not solar wind fluctuations are best described as non-interacting MHD waves or as MHD turbulent phenomena, and this debate has continued to the present day (Ofman 2010; Goldstein et al. 2015;

Matthaeus et al. 2015). One of the important issues causing the debate is the observation that the magnetic and velocity fluctuations in the solar wind are often nearly aligned each other, that is, to satisfy the so-called Alfvén relation $\delta\mathbf{v} = \mp\delta\mathbf{b}$, where $\delta\mathbf{b} \equiv \delta\mathbf{B}v_A/B_0$. For the totally aligned case that the Alfvén relation is exactly satisfied, the nonlinear terms in the ideal (i.e. zero dissipation) incompressible MHD equations will vanish entirely and the Alfvén relation is an exact solution of "nonlinear" AWs (Matthaeus et al. 2015). In result, there can not be any actively evolving turbulence. However, on the other hand, as pointed out by Coleman (1966, 1968), the power spectra of the solar wind magnetic fluctuation resemble those of the incompressible Navier-Stokes turbulence, in which the shape of the spectra can be well described by the Kolmogorov theory (Kolmogorov 1941).

The cross helicity given by Eq. (6.69) describes quantitatively the alignment degree between the magnetic field and fluid-velocity field fluctuations. In general, a high cross helicity (i.e., a high alignment degree) tends to reduce the nonlinear terms in the MHD equations. By using data from the Helios 1 and 2 satellites together with data from the Voyager-1 and -2 spacecraft, which ranges the heliosphere from 0.3 to 20 AU, Roberts et al. (1987a, b) further analyzed the "waves" versus "turbulence" scenarios for describing the solar wind fluctuation and concluded that both points of view had varying amounts of validity. Meanwhile, they further confirmed that for highly Alfvénic fluctuations characterized by a high alignment degree between velocity and magnetic fluctuations (i.e., the maximal cross helicity), which tends to reduce the nonlinear terms in the MHD equations, there is little evidence of turbulent evolution driven by the nonlinear cascade of waves, especially in fast solar wind out to 1 AU (Roberts et al. 1987a, b). This indicates that the solar wind fluctuations consist of two different populations, one is nonlinear waves or intermittent structures such as shocks and current sheets, and other one is fully developed turbulence which has a typical Kolmogorov-like power spectrum in the inertial regime.

On the other hand, the results of Roberts et al. (1987a, b) also show that, on average, the alignment degree (i.e., the cross helicity) tends to decrease with heliocentric distance, and moreover the decline of the alignment degree, in general, is faster for the small-scale fluctuations and slower for the large-scale ones. In addition, the decrease of the alignment degree tends to be rapider and larger when the fluctuations encounter some strong structures, such as shocks, current sheets, and co-rotating interaction regions. This suggests that the nonlinear interaction induced by the strong structures possibly triggers and enhances the nonlinear coupling among Alfvénic fluctuations.

The solar wind is typically a collisionless plasma environment, as shown in Chap. 5, the turbulent dissipation occurs in the kinetic regime at the kinetic scales of plasma particles. Moreover, the kinetic dissipation of the solar wind turbulence transfers the turbulent wave energy into the kinetic energy of ions and electrons, that is, heating ions and electrons, in different scales because of their different kinetic scales. For a steady-state turbulent cascade process, the energy dissipated rate at these small kinetic scales should match to the energy injected rate in large scales. Therefore, the solar wind heating rate (i.e., the energy dissipated rate of the solar wind turbulence) can be estimated by the energy transferring rate in the solar wind turbulent cascade (MacBride et al. 2008; Stawarz et al. 2009), which may be derived from the third-order moment of the solar wind fluctuations (Podesta et al. 2009; Coburn et al. 2012). Based on the ten-years solar wind data during 1998–2007 by the ACE (Advanced Composition Explorer) satellite at the earth L1 point, MacBride et al. (2008) showed that the derived turbulent energy injection and dissipation rates are consistent with the local solar wind heating rate inferred from the local temperature gradient. The turbulent cascade even may deposit more energy than that required by the observed proton heating and the excess energy could cascade and transfer into the smaller electron scales to heat electrons by the Landau damping (Leamon et al. 1999; Stawarz et al. 2009; Sahraoui et al. 2010).

By use of the third-order moment method (Smith et al. 2009), Stawarz et al. (2010) found that for the solar wind fluctuations with large cross-helicity states the energy cascade rate of the dominant outward-propagating component is surprisingly negative, implying that the high-alignment state dominates a significant back-transfer (i.e., inverse cascade) energy from small-scale to large-scale structures in the inertial range. However, the physical mechanism of causing such a back-transfer process is unclear, and moreover also the inverse cascade phenomena seem to conflict with the observed temperature of the solar wind near 1 AU. The analyses based on the third-order moment imply the assumption that the fluctuations are incompressible, but the solar wind fluctuation is not exactly incompressible. In addition, Podesta (2010) also criticized the work by Smith et al. (2009) and pointed out that they possibly have underestimated seriously the statistical errors inherent in their use of the third-order moment because the slow convergence of the third-order moment to its statistical mean may lead to rather large uncertainty in the measurements of third-order moments (Podesta et al. 2009).

In the present in situ measurements of space plasmas, it often is possible that the same observational data result in different conclusions using different analytical methods or technologies. For example, using the solar wind data observed by the ACE satellite from 1998/02/04 to 2009/12/23 during it was at the L1 point, Osman et al. (2011) measured the solar wind heating, such as the electron temperature, the ion temperature, and the electron heat flux, and compared with the local dissipation density estimated from MHD simulations. They found that the enhancements of the solar wind heating in the in situ measurements and of the dissipation in the MHD simulations both are associated with coherent structures that can be identified, by use of the so-called PVI (partial variance increments) method (see e.g., Matthaeus et al. 2015), as current sheets generated by MHD turbulence and also are possible sources of inhomogeneity in the solar wind and intermittency in MHD turbulence. Thus, they concluded that significant inhomogeneous heating can occur in the solar wind, which is closely connected with current sheets, that is, intermittent structures generated by MHD turbulence.

Using nearly same ACE solar wind data from 1998 to 2010, Borovsky and Denton (2011) identified a total of 194,070 strong current sheets by a large rotation angle ($\phi > 45°$) of the solar-wind magnetic-field direction in 128 s. They compared the proton temperature, proton specific entropy, and electron temperature at each current sheet with the same quantities in the plasmas adjacent to the current sheet. By contrast with the results above presented by Osman et al. (2011), Borovsky and Denton (2011) concluded that, statistically, the plasma at the current sheets is not hotter or of higher entropy than the plasmas just outside the current sheets, implying that there is no evidence for the significant localized heating of the solar-wind protons or electrons at strong current sheets. They found, however, that current sheets are more prevalent in hotter solar wind plasmas because more current sheets are counted in the hotter fast solar wind than in the slow solar wind (Borovsky and Denton 2011). Obviously, more further investigations are needed to clarify possible roles of current sheets or intermittent structures in the solar wind heating.

The relationship between the two populations of the solar wind fluctuations, "nonlinear intermittent structures" and "fully developed turbulence", has been an issue in the solar wind turbulence. Their contributions to the solar wind heating still are an open problem. Different from the case of the coherent intermittent structures within which the specific kinetic mechanisms are still unclear (Osman et al. 2011), however, in the context of the turbulent heating by the solar wind turbulence the solar wind heating mechanism should be closely associated with the kinetic dissipation of the solar wind turbulence due to the wave-particle interaction of KAWs in the kinetic scales of plasma particles, which dissipates the turbulent energy of the solar wind turbulence into the kinetic energy of plasma particles (Howes et al. 2006; Schekochihin et al. 2009).

As presented in Chap. 5, extensive investigations from theoretical analysis, numerical simulations, and in situ measurements all show that the power spectra of the solar wind turbulence are dominated by AWs in large MHD scales and by KAWs in small kinetic scales and have evidently two spectral breakpoints nearly at the ion and electron gyroradius, which can be attributed partly to the kinetic dissipations by ions and electrons, respectively, and partly to the formation of turbulent intermittent structures by the kinetic alignment and anisotropy. While the KAW turbulence is natural and inevitable result of the anisotropic turbulent cascade of the solar wind turbulence from large-scale MHD turbulence to small-scale kinetic turbulence in the dynamical evolution of the solar wind turbulence. Therefore, it can be expected reasonably that the KAW turbulence can play an important role in the solar wind heating, like KAWs in the coronal heating problem.

In summary, based on KAWs, we can depict an unified physical scenario for the plasma heating problem of the solar atmosphere from the partly ionized upper chromosphere to the fully ionized solar corona and solar wind. In the nearly neutral photosphere and lower chromosphere the atmospheric heating, if need, still is dominated by the acoustic wave heating due to the neutral atomic collision dissipation as expected by the conventional heating theory, which has little relation to magnetic fields as well as plasmas.

As the temperature and hence the ionization increases with heights, the Joule heating due to the classic electron-ion Coulomb collision resistance becomes the main heating mechanism in the partly ionized upper chromosphere, where the ionization, in general, is larger than 10^{-3}, and the plasma processes dominate the neutral gas kinetics. The heating efficiency of this KAW Joule heating mechanism is directly proportional to the strength of local magnetic fields, implying that the upper chromospheric heating has a strong magnetic dependence so that upper chromospheric regions with stronger magnetic field look brighter, as shown in the observations of upper chromospheres of sunspots and the chromospheric flocculi associated with small-scale magnetic flux tubes in quiet solar atmospheric regions.

In the corona over the transition region, however, the Joule dissipation of KAWs is no longer an effective heating mechanism because the classic Coulomb collision frequency, $\nu_{ei} \propto nT^{-3/2}$, rapidly decreases as the temperature increases as well as the density decreases when coming into high-temperature and tenuous coronal plasmas. In such nearly collisionless coronal plasmas, the Landau damping of KAWs due to the kinetic dissipation caused by the wave-particle interaction of KAWs can become a more effective heating mechanism. However, the heating efficiency of the kinetic dissipation mechanism of KAWs sensitively depends on the local plasma parameters in the way that is determined by the wave-particle resonant condition. This leads to that the heating efficiency and hence the radiative brightness have more complex spatial distributions in the corona. This parametric dependent and hence inhomogeneous heating mechanism also leads to the formation of complex coronal plasma structures, which are characterized by field-aligned filaments such as coronal loops and coronal plumes because the anisotropy of low-β plasmas in the corona tends the plasmas to be much more uniform along magnetic fields than across magnetic fields, as displayed in based-space high resolution observations.

In fact, the plasma heating consists of two processes, one is the energization of plasma particles, which leads to the increase of particle kinetic energies, and other one is the thermalization of plasma particles, that is, the randomization of particle kinetic energies. In the upper chromospheric heating by the KAW Joule dissipation, the energization and thermalization of particles are performed by current-carrying electric fields and collisions, respectively. For the inner coronal heating by the KAW Landau damping, the particle energization is generated by the wave-particle interaction, in which the field energy of KAWs is transferred into the kinetic energy of the resonant particles, and the thermalization of particles may be realized through the collisions between the resonant particles and other non-resonant particles or the particle scattering by turbulent waves, because the collision frequency may be higher than the energy loss rate of the resonant particles although much lower than the resonant interaction rate.

In the more tenuous extended corona, however, the minor heavy ions energized by the nonlinear KAW wavelets (i.e., SKAWs) can effectively transfer their kinetic energies into other particles because the collisions are so rare that they may conserve their kinetic energy as their effective temperature when are trapped within SKAWs till these SKAWs are mixed or transferred into the solar wind turbulence. The solar wind heating is the natural and inevitable result of the kinetic dissipation in the kinetic scale range when the solar wind turbulence cascades towards smaller scales.

References

Abdelatif, T. E. (1987). Heating of coronal loops by phase-mixid shear Alfvén waves. *The Astrophysical Journal, 322*, 494–502.

Ahmad, I. A., & Withbroe, G. L. (1977). EUV analysis of polar plumes. *Solar Physics, 53*, 397–408.

Alfvén, H. (1947). Granulation, magneto-hydrodynamic waves and the heating of the solar corona. *Monthly Notices of the Royal Astronomical Society, 107*, 211–219.

Alfvén, H., & Carlqvist, P. (1967). Currents in the solar atmosphere and a theory of solar flares. *Solar Physics, 1*, 220.

Andries, J., & Goossens, M. (2001). Kelvin-Helmholtz instabilities and resonant flow instabilities for a coronal plume model with plasma pressure. *Astronomy & Astrophysics, 368*, 1083–1094.

Andries, J., Tirry, W. J., & Goossens, M. (2000). Modified Kelvin-Helmholtz instabilities and resonant flow instabilities in a one-dimensional coronal plume model: Results for plasma $\beta = 0$. *The Astrophysical Journal, 531*, 561–570.

Aschwanden, M. J., Fletcher, L., Schrijver, C. J., & Alexander, A. D. (1999). Coronal loop oscillations observed with the transition region and coronal explorer. *The Astrophysical Journal, 520*, 880–894.

Aschwanden, M. J., Poland, A. I., & Rabin, D. M. (2001). The new solar corona. *Annual Review of Astronomy and Astrophysics, 39*, 175–210.

Aulanier, G., Golub, L., Deluca, E. E., et al. (2007). Slipping magnetic reconnection in coronal loops. *Science, 318*, 1588–1591.

Aurass, H., & Kliem, B. (1992). Fiber bursts in type IV DM radio continua as a signature of coronal current sheet dynamics. *Solar Physics, 141*, 371–379.

Avrett, E. H. (1981). Reference model atmospheric calculation-The sunspot sunspot model. In Cram & Thomas, (Ed.), *The Physics of Sunspots* (pp. 235–255). New Mexico: Sacramento Peak Observatory.

Beaufume, P., Coppi, B., & Golub, L. (1992). Coronal loops-current-based heating processes. *The Astrophysical Journal, 393*, 396–408.

Belcher, J. W., & Davis, L. (1971). Large-amplitude Alfvén wave in the interplanetary medium. *Journal of Geophysical Research, 76*, 3534–3563.

Bellan, P. M., & Stasiewicz, K. (1998). Fine-scale cavitation of ionospheric plasma caused by inertial Alfvén wave ponderomoive force. *Physical Review Letters, 80*, 3523–3526.

Benz, A. O. (1986). Millisecond radio spikes. *Solar Physics, 104*, 99–110.

Benz, A. O. (2002). *Plasma astrophysics, astrophysics and space science library* (2nd ed., p. 279). Dordrecht: Kluwer.

Benz, A. O., & Mann, G. (1998). Intermediate drift bursts and the coronal magnetic field. *Astronomy and Astrophysics, 333*, 1034–1042.

Bernold, T. X., & Treumann, R. A. (1983). The fiber fine structure during solar type IV radio bursts: Observations and theory of radiation in presence of localized whistler turbulence. *The Astrophysical Journal, 264*, 677–688.

Bohlin, J. D., Sheeley, N. R., & Tousey, R. (1975). Structure of the sun's polar cap at wavelength 240–600 Å. In *Space Research XV* (pp. 651–656).

Boldyrev, S. (2005). On the spectrum of magnetohydrodynamic turbulence. *The Astrophysical Journal Letters, 626*, L37.

Borovsky, J. E., & Denton, M. H. (2011). No evidence for heating of the solar wind at strong current sheets. *The Astrophysical Journal Letters, 739*, L61.

Borovsky, J. E. & Suszcynsky, D. M. (2013), Optical measurements of the fine structure of auroral arcs. In Lysak, R. L. (ed.), *Auroral Plasma Dynamics*. Geophysical Monograph Series (vol. 80, pp. 25–30), Washington, D. C.

Bray, R. J., Cram, L. E., Durrant, C. J., & Loughhead, R. E. (1991). *Plasma loops in the solar corona*. Cambridge: Cambridge University Press.

Browning, P. K., & Priest, E. R. (1984). Kelvin-Helmholtz instability of a phased-mixed Alfvén wave. *Astronomy and Astrophysics, 131*, 283–290.

Canfield, R. C., de La Beaujardiere, J.-F., Fan, Y. H., et al. (1993). The morphology of flare phenomena, magnetic fields, and electric currents in active regions. I-Introduction and methods. *The Astrophysical Journal, 411*, 362–369.
Carlsson, M., & Stein, R. F. (1992). Non-LTE radiating acoustic shocks and CA II K2V bright points. *The Astrophysical Journal, 397*, L59–L62.
Chaston, C. C., Bonnell, J. W., Carlson, C. W., et al. (2003). Properties of small-scale Alfvén waves and accelerated electrons from FAST. *Journal of Geophysical Research: Space Physics, 108*, 8003–8019.
Chaston, C. C., Bonnell, J. W., Carlson, C. W., et al. (2004). Auroral ion acceleration in dispersive Alfvén waves. *Journal of Geophysical Research: Space Physics, 109*, A04205.
Chaston, C. C., Peticolas, L. M., Carlson, C. W., et al. (2005). Energy deposition by Alfvén waves into the dayside auroral oval: Cluster and FAST observations. *Journal of Geophysical Research: Space Physics, 110*, A02211.
Chaston, C. C., Carlson, C. W., Peria, W. J., et al. (1999). FAST observations of inertial Alfvén waves in the dayside aurora. *Geophysical Research Letters, 26*, 647–650.
Chen, L., & Wu, D. J. (2012). Kinetic Alfvén wave instability driven by field-aligned currents in solar coronal loops. *The Astrophysical Journal, 754*, 123.
Chen, L., Wu, D. J., & Hua, Y. P. (2011). Kinetic Alfvén wave instability driven by a field-aligned current in high-β plasmas. *Physical Review E, 84*, 046406.
Chen, L., Wu, D. J., & Huang, J. (2013). Kinetic Alfvén wave instability driven by field-aligned currents in a low-β plasma. *Journal of Geophysical Research: Space Physics, 118*, 2951.
Chen, L., Wu, D. J., Zhao, G. Q., et al. (2014). Excitation of kinetic Alfvén waves by fast electron beams. *The Astrophysical Journal, 793*, 13.
Chen, L., Wu, D. J., Zhao, G. Q., et al. (2015). A possible mechanism for the formation of filamentous structures in magnetoplasmas by kinetic Alfvén waves. *Journal of Geophysical Research: Space Physics, 120*, 61–69.
Chen, Y. P., Zhou, G. C., Yoon, P. H., & Wu, C. S. (2002). A beam-maser instability: Direct amplification of radiation. *Physics of Plasmas, 9*, 2816–2821.
Chin, Y. C., & Wentzel, D. G. (1972). Nonlinear dissipation of Alfvén waves. *Astrophysics and Space Science, 16*, 465–477.
Cho, J., & Vishniac, E. T. (2000). The anisotropy of MHD Alfvénic turbulence. *The Astrophysical Journal, 539*, 273–282.
Cho, J., Lazarian, A., & Vishniac, E. T. (2002). Simulation of magnetohydrodynamic turbulence in a strong magnetized medium. *The Astrophysical Journal, 564*, 291–301.
Cirtain, J., Golub, L., Lundquist, L., et al. (2007). Evidence for Alfvén waves in solar X-ray jets. *Science, 318*, 1580–1582.
Coburn, J. T., Smith, C. W., Vasquez, B. J., et al. (2012). The turbulent cascade and proton heating in the solar wind during solar minimum. *The Astrophysical Journal, 754*.
Coleman, P. J. (1966). Variations in the interplanetary magnetic field: Mariner 2: 1. Observed properties. *Journal of Geophysical Research, 71*, 5509–5531.
Coleman, P. J. (1968). Turbulence, viscossity, and dissipation in the solar wind plasma. *The Astrophysical Journal, 153*, 371–388.
Cowling, T. G. (1958). Solar electrodynamics. *Proceedings of the International Astronomical Union, 6*, 105.
Cranmer, S. R. (2009). Coronal holes. *Living Reviews in Solar Physics, 6*(3), 1–64.
Cranmer, S. R., & van Ballegooijen, A. A. (2005). On the generation, propagation, and reflection of Alfvén waves from the solar photosphere to the distant heliosphere. *The Astrophysical Journal Supplement Series, 156*, 265–293.
Cranmer, S. R., Field, G. B., & Kohl, J. L. (1999). Spectroscopic constraints on models of ion cyclotron resonance heating in the polar solar corona and high-speed solar wind. *The Astrophysical Journal, 518*, 937–947.
Dabrowski, B. P., Rudawy, P., Falewicz, R., et al. (2005). Millisecond radio spikes in decimetre band and their related active solar phenomena. *Astronomy & Astrophysics, 434*, 1139–1153.

De Forest, C. E., Hoeksema, J. T., Gurman, J. B., et al. (1997). Polar plume anatomy: Results of a coordinated observation. *Solar Physics*, *175*, 393–410.
De Pontieu, B., Martens, P. C. H., & Hudson, H. S. (2001). Chromospheric damping of Alfvén waves. *The Astrophysical Journal*, *558*, 859–871.
De Pontieu, B., McIntosh, S., Carlsson, M., et al. (2007). Chromospheric Alfvénic waves strong enough to power the solar wind. *Science*, *318*, 1574–1577.
de Wijn, A. G., Stenflo, J. O., Solanki, S. K., & Tsuneta, S. (2009). Small-scale solar magnetic fields. *Space Science Reviews*, *144*, 275–315.
Del Sarto, D., Califano, F., & Pegoraro, F. (2006). Electron parallel compressibility in the nonlinear development of two-dimensional collisionless magnetohydrodynamic reconnection. *Modern Physics Letters B*, *20*, 931–961.
Del Zanna, L., Hood, A. W., & Longbottom, A. W. (1997). An MHD model for solar coronal plumes. *Astronomy and Astrophysics*, *318*, 963–969.
Ding, M. D., & Fang, C. (1989). A semi-empirical model of sunspot penumbra. *Astronomy and Astrophysics*, *225*, 204–212.
Ding, M. D., & Fang, C. (1991). A semi-empirical model of sunspot umbra. *Chinese Astronomy and Astrophysics*, *15*, 28–36.
Edlén, B. (1941). An attemp to identify the emission lines in the spectrum of the solar corona. *Arkiv för Matematik, Astronomi och Fysik*, *28B*, 1–4.
Erdélyi, R., & Fedun, V. (2007). Are there Alfvén waves in the solar atmosphere? *Science*, *318*, 1572–1574.
Esser, R., & Sasselov, D. (1999). On the disagreement between atmospheric and coronal electron densities. *The Astrophysical Journal Letters*, *521*, L145–L148.
Esser, R., Fineschi, S., Dobrzycka, D., et al. (1999). Plasma properties in coronal holes drived from measurements of minor ion spectral lines and polarized white light intensity. *The Astrophysical Journal*, *510*, L63–L67.
Fälthammar, C. G. (1995). In memoriam: Hannes Alfvén. *Astrophysics and Space Science*, *234*, 173–175.
Fang, C., Tang, Y. H., Ding, M. D., et al. (1997). Coronal loops above sunspot region. *Solar Physics*, *176*, 267–277.
Fleishman, G. D. (2004). Natural spectral bandwidth of electron cyclotron maser emission. *Astronomy Letters*, *30*, 603–614.
Fleishman, G. D., Gary, D. E., & Nita, G. M. (2003). Decimetric spike bursts versus microwave continuum. *The Astrophysical Journal*, *593*, 571–580.
Fossum, A., & Carlsson, M. (2005). High-frequency acoustic waves are not sufficient to heat the solar chromosphere. *Nature*, *435*, 919–921.
Fried, B. D., & Conte, S. D. (1961). *The plasma dispersion function*. New York: Academic Press.
Fu, Q. J., Qin, Z. H., Ji, H. R., & Pei, L. B. (1995). A broadband spectrometer for decimeter and microwave radio bursts. *Solar Physics*, *160*, 97–103.
Gan, W. Q., & Fang, C. (1990). A hydrodynamic model of the gradual phase of the solar flare loop. *The Astrophysical Journal*, *358*, 328–337.
Gary, G. A., & Demoulin, P. (1995). Reduction, analysis, and properties of electric current systems in solar active regions. *The Astrophysical Journal*, *445*, 982.
Gazis, P. R. (1984). Observations of plasma bulk parameters and the energy balance of the solar wind between 1 and 10 AU. *Journal of Geophysical Research: Space Physics*, *89*, 775–785.
Gazis, P. R., & Lazarus, A. J. (1982). Voyager observations of solar wind proton temperature: 1–10 AU. *Geophysical Research Letters*, *9*, 431–434.
Gekelman, W. (1999). Review of laboratory experiments on Alfvén waves and their relationship to space observations. *Journal of Geophysical Research: Space Physics*, *104*, 14417–14435.
Goldreich, P., & Sridhar, S. (1995). Towards a theory of interstellar turbulence. II. Strong Alfvénic turbulence. *The Astrophysical Journal*, *438*, 763–775.
Goldreich, P., & Sridhar, S. (1997). Magnetohydrodynamic turbulence revisited. *The Astrophysical Journal*, *485*, 680–688.

Goldstein, B. E., & Jokipii, J. R. (1977). Variation of solar wind parameters. *Journal of Geophysical Research*, *82*, 1095.

Goldstein, M. L., Wicks, R. T., Perri, S., & Sahraoui, F. (2015). Kinetic scale turbulence and dissipation in the solar wind: Key observational results and future outlook. *Philosophical Transactions of the Royal Society A: Mathematical, Physical and Engineering Sciences*, *373*, 20140147.

Golub, L., & Pasachoff, J. M. (1997). *The solar corona*. Cambridge: Cambridge University.

Golub, L., Herant, M., Kalata, K., et al. (1990). Sub-arcsecond observations of the solar X-ray corona. *Nature*, *344*, 842–844.

Goodman, M. L. (2004). On the creation of the chromospheres of solar type stars. *Astronomy & Astrophysics*, *424*, 691–712.

Habbal, S. R., Leer, E., & Holzer, T. E. (1979). Heating of coronal loops by fast mode MHD waves. *Solar Physics*, *64*, 287–301.

Hagyard, M. J. (1989). Observed nonpotential magnetic fields and the inferred flow of electric currents at a location of repeated flaring. *Solar Physics*, *115*, 107–124.

Hallinan, T. J., & Davis, T. N. (1970). Small-scale auroral arc distortions. *Planetary and Space Science*, *18*, 1735–1736.

Hasegawa, A., & Chen, L. (1975). Kinetic process of plasma heating due to Alfvén wave excitation. *Physical Review Letters*, *35*, 370–373.

Hasegawa, A., & Chen, L. (1976). Kinetic process of plasma heating by resonant mode conversion of Alfvén wave. *The Physics of Fluids*, *19*, 1924–1934.

Hasegawa, A., & Mima, K. (1976). Exact solitary Alfvén wave. *Physical Review Letters*, *37*, 690–693.

Hasegawa, A. & Sato, T. (1989). *Space plasma physics: I-stationary processes*. Physics and Chemistry in Space (Vol. 16, pp. 181). Berlin: Springer.

Hasegawa, A., & Uberoi, C. (1982). *The Alfvén waves*. Technical Information Center, US Department of Energy: Oak Ridge.

Hassler, D. M., Rottman, G. J., Shoub, E. C., & Holzer, T. E. (1990). Line broadening of MG X 609 and 625 A coronal emission lines observed above the solar limb. *The Astrophysical Journal Letter*, *348*, L77–L80.

Heyvaerts, J., & Priest, E. R. (1983). Coronal heating by phase-mixed shear Alfvén waves. *Astronomy and Astrophysics*, *117*, 220–234.

Hollweg, J. V. (1971). Nonlinear landau damping of Alfvén waves. *Physical Review Letters*, *27*, 1349–1352.

Hollweg, J. V. (1974). On electron heat conduction in the solar wind. *Journal of Geophysical Research*, *79*, 3845–3850.

Hollweg, J. V. (1976). Collisionless electron heat conduction in the solar wind. *Journal of Geophysical Research*, *81*, 1649–1658.

Hollweg, J. V. (1978). Alfvén waves in the solar atmosphere. *Solar Physics*, *56*, 305–333.

Hollweg, J. V. (1981). Alfvén waves in the solar atmosphere. II-Open and closed magnetic flux tubes. *Solar Physics*, *70*, 25–66.

Hollweg, J. V. (1984). Alfvénic resonant cavities in the solar atmosphere: Simple aspects. *Solar Physics*, *91*, 269–288.

Hollweg, J. V. (1987). Resonance absorption of magnetohydrodynamic surface waves physical discussion. *The Astrophysical Journal*, *312*, 880–885.

Horbury, T. S., Forman, M., & Oughton, S. (2008). Anisotropic scaling of magnetohydrodynamic turbulence. *Physical Review Letters*, *101*, 175005.

Howes, G. G., Cowley, S. C., Dorland, W., et al. (2006). Astrophysical gyrokinetics: Basic equations and linear theory. *The Astrophysical Journal*, *651*, 590–614.

Hu, Y. Q., & Habbal, S. R. (1999). Resonant acceleration and heating of solar wind ions by dispersive ion cyclotron waves. *Journal of Geophysical Research: Space Physics*, *104*, 17045–17056.

Hu, Y. Q., Esser, R., & Habbal, S. R. (1997). A fast solar wind model with anisotropic proton temperature. *Journal of Geophysical Research: Space Physics*, *102*, 14661–14676.

Hu, Y. Q., Esser, R., & Habbal, S. R. (2000). A four-fluid turbulence-driven solar wind model for preferential acceleration and heating of heavy ions. *Journal of Geophysical Research: Space Physics*, *105*, 5093–5111.

Huang, G. L., Wang, D. Y., Wu, D. J., et al. (1997). The eigenmode of solitary kinetic Alfvén waves by Freja satellite. *Journal of Geophysical Research: Space Physics*, *102*, 7217–7224.

Ichimoto, K., Suematsu, Y., Tsuneta, S., et al. (2007). Twisting motions of sunspot penumbral filaments. *Science*, *318*, 1597–1599.

Ionson, J. A. (1978). Resonant absorption of Alfvénic surface waves and the heating of solar coronal loops. *The Astrophysical Journal*, *226*, 650–673.

Isenberg, P. A., & Vasquez, B. J. (2007). Preferential perpendicular heating of coronal hole minor ions by the Fermi mechanism. *The Astrophysical Journal*, *668*, 546–556.

Jefferies, S. M., McIntosh, S. W., Armstrong, J. D., et al. (2006). Magnetoacoustic portals and the basal heating of the solar chromosphere. *The Astrophysical Journal*, *648*, L151–L155.

Kano, R., & Tsuneta, S. (1995). Scaling law of solar coronal loops obtained with YOHKOH. *The Astrophysical Journal*, *454*, 934–944.

Kano, R., & Tsuneta, S. (1996). Temperature distributions and energy scaling law of solar coronal loops obtained with YOHKOH. *Publications of the Astronomical Society of Japan*, *48*, 535–543.

Katsukawa, Y., Berger, T. E., Ichimoto, K., et al. (2007). Small-scale jetlike features in penumbral chromospheres. *Science*, *318*, 1594–1597.

Khodachenko, M. L., Zaitsev, V. V., Kislyakov, A. G., & Stepanov, A. V. (2009). Equivalent electric circuit models of coronal magnetic loops and related oscillatory phenomena on the sun. *Space Science Reviews*, *149*, 83.

Kohl, J. K., Noci, G., Antonucci, E., et al. (1997). First results from the SOHO ultravoilet coronagraph spectrometer. *Solar Physics*, *175*, 613–644.

Kohl, J. L., Noci, G., Antonucci, E., et al. (1998). UVCS/SOHO empirical determinations of anisotropic velocity distributions in the solar corona. *The Astrophysical Journal Letters*, *501*, L127–L131.

Kolmogorov, A. N. (1941). The local structure turbulence in incompressible viscous fluids for very large Reynolds numbers, Dokl. Akad. Nauk. SSSR 30, 301–305. Reprinted in 1991. *Proceedings of the Royal Society of London. Series A: Mathematical and Physical Sciences*, *434*, 9–13.

Koutchmy, S. (1977). Study of the June 30, 1973 trans-polar coronal hole. *Solar Physics*, *51*, 399–407.

Leamon, R. J., Ness, N. F., Smith, C. W., & Wong, H. K. (1999). Dynamics of the dissipation range for solar wind magnetic fluctuations. *AIP Conference Proceedings*, *471*, 469.

Leamon, R. J., Smith, C. W., Ness, N. F., et al. (1998). Observational constraints on the dynamics of the interplanetary magnetic field dissipation range. *Journal of Geophysical Research: Space Physics*, *103*, 4775–4787.

Leblanc, Y., Dulk, G., & Bougeret, J. (1998). Tracing the electron density from the corona to 1 AU. *Solar Physics*, *183*, 165–180.

Lee, J., McClymont, A. N., Mikić, Z., et al. (1998). Coronal currents, magnetic fields, and heating in a solar active region. *The Astrophysical Journal*, *501*, 853.

Lee, L. C., & Wu, B. H. (2000). Heating and acceleration of protons and minor ions by fast shocks in the solar corona. *The Astrophysical Journal*, *535*, 1014–1026.

Lee, M. A., & Roberts, B. (1986). On the behavior of hydromagnetic surface waves. *The Astrophysical Journal*, *301*, 430–439.

Louarn, P., Wahlund, J. E., Chust, T., et al. (1994). Observations of kinetic Alfvén waves by the Freja spacecraft. *Geophysical Research Letters*, *21*, 1847–1850.

Luo, Q. Y., & Wu, D. J. (2010). Observations of anisotropic scaling of solar wind turbulence. *The Astrophysical Journal Letters*, *714*, L138–L141.

MacBride, B. T., Smith, C. W., & Forman, M. A. (2008). The turbulent cascade at 1 AU: Energy transfer and the third-order scaling for MHD. *The Astrophysical Journal*, *679*, 1644.

Maggs, J. E., & Morales, G. J. (1996). Magnetic fluctuations associated field-aligned striations. *Geophysical Research Letters*, *23*, 633–636.

Maggs, J. E., & Morales, G. J. (1997). Fluctuations associated a filamentary density depletion. *Physics of Plasmas*, *4*, 290–299.

Malovichko, P. P. (2008). Stability of magnetic configurations in the solar atmosphere under temperature anisotropy conditions. *Kinematics and Physics of Celestial Bodies*, *24*, 236–241.

Maltby, P., Avrett, E. H., Carlsson, M., et al. (1986). A new sunspot umbral model and its variation with the solar cycle. *The Astrophysical Journal*, *306*, 284–303.

Mariani, F., & Neubauer, F. M. (1990). The interplanetary magnetic field. In Schwenn & Marsch (Ed.), *Physics of the Inner Heliosphere* (pp. 183–206). New York: Springer.

Markovskii, S. A., Vasquez, B. J., & Hollweg, J. V. (2009). Proton heating by nonlinear field-aligned Alfvén waves in solar coronal holes. *The Astrophysical Journal*, *695*, 1413–1420.

Markovskii, S. A., Vasquez, B. J., & Smith, C. W. (2008). Statistical analysis of the high-frequency spectral break of the solar wind turbulence at 1 AU. *The Astrophysical Journal*, *675*, 1576–1583.

Marsch, E. (2006). Kinetic physics of the solar corona and solar wind. *Living Reviews in Solar Physics*, *3*(1), 1–100.

Marsch, E., Rosenbauer, H., & Schwenn, R. (1983). On the equation of state of solar wind ions derived from Helios measurements. *Journal of Geophysical Research: Space Physics*, *88*, 2982–2992.

Maron, J., & Goldreich, P. (2001). Simulations of incompressible magnetohydrodynamic turbulence. *The Astrophysical Journal*, *554*, 1175–1196.

Martens, P. C. H., Van Den Oord, G. H. J., & Hoyng, P. (1985). Observations of steady anomalous magnetic heating in thin current sheets. *Solar Physics*, *96*, 253–275.

Matthaeus, W. H., & Goldstein, M. L. (1982). Stationarity of magnetaohydrodynamic fluctuations in the solar wind. *Journal of Geophysical Research: Space Physics*, *87*, 10347–10354.

Matthaeus, W. H., Wan, M., Servidio, S., et al. (2015). Intermittency, nonlinear dynamics and dissipation in the solar wind and astrophysical plasmas. *Philosophical Transactions of the Royal Society A: Mathematical, Physical and Engineering Sciences*, *373*, 20140154.

McClymont, A. N., & Canfield, R. C. (1983). Flare loop radiative hydrodynamics-Part two-Thermal stability of empirical models. *The Astrophysical Journal*, *265*, 497–506.

Mcfadden, J. P., Carlson, C. W., & Boehm, M. H. (1990). Structure of an energetic narrow discrete arc. *Journal of Geophysical Research: Space Physics*, *95*, 6533–6547.

Melrose, D. B. (1991). Neutralized and unneutralized current patterns in the solar corona. *The Astrophysical Journal*, *381*, 306–312.

Messmer, P., & Benz, A. O. (2000). The minimum bandwidth of narrowband spikes in solar flare decimetric radio waves. *Astronomy & Astrophysics*, *354*, 287–295.

Müller, W. C., Biskamp, D., & Grappin, R. (2003). Statistical anisotropy of magnetohydrodynamic turbulence. *Physical Review E*, *67*, 066302.

Nagai, F. (1980). A model of hot loops associated with solar flares I-Gasdynamics in the loops. *Solar Physics*, *68*, 351–379.

Nakariakov, V. M., & Ofman, L. (2001). Determination of the coronal magnetic field by coronal loop oscillations. *Astronomy & Astrophysics*, *372*, L53–L56.

Narain, U., & Ulmschneider, P. (1990). Chromospheric and coronal heating mechanisms. *Space Science Reviews*, *54*, 377–445.

Narain, U., & Ulmschneider, P. (1996). Chromospheric and coronal heating mechanisms II. *Space Science Reviews*, *75*, 453–509.

Newkirk, G., & Harvey, J. (1968). Coronal polar plumes. *Solar Physics*, *3*, 321–343.

Noyes, R. W., & Avrett, E. H. (1987). The solar chromosphere. In Dalgarno & Layzer (Eds.), *Spectroscopy of astrophysical plasmas* (p. 125). Cambridge: Canbridge University Press.

Ofman, L. (2010). Wave modeling of the solar wind. *Living Reviews in Solar Physics*, *7*, 1–48.

Ofman, L., Gary, S. P., & Viñas, A. (2002). Resonant heating and acceleration of ions in coronal holes driven by cyclotron resonant spectra. *Journal of Geophysical Research: Space Physics*, *107*, 1461–1470.

Okamoto, T., Tsuneta, S., Berger, T., et al. (2007). Coronal transverse magnetohydrodynamic waves in a solar prominence. *Science*, *318*, 1577–1580.

Osman, K. T., Matthaeus, W. H., Greco, A., & Servidio, S. (2011). Evidence for inhomogeneous heating in the solar wind. *The Astrophysical Journal Letters*, *727*, L11.
Osterbrock, D. E. (1961). The heating of the solar chromosphere, plages, and corona by magnetohydrodynamic waves. *The Astrophysical Journal*, *134*, 347.
Parker, E. N. (1958). Dynamics of the interplanetary gas and magnetic fields. *The Astrophysical Journal*, *128*, 664–676.
Parker, E. N. (1979). *Cosmical magnetic fields: Their origin and their activity*. Oxford: Clarendon Press.
Parker, E. N. (1983). Magnetic neutral sheets in evolving fields-part two-formation of the solar corona. *The Astrophysical Journal*, *264*, 642–647.
Parker, E. N. (1988). Nanoflares and the solar X-ray corona. *The Astrophysical Journal*, *330*, 474–479.
Pesnell, W. D. (2012). The solar dynamics observatory (SDO). *Solar Physics*, *275*, 3–15.
Piddington, J. H. (1956). Solar atmospheric heating by hydromagnetic waves. *Monthly Notices of the Royal Astronomical Society*, *116*, 314.
Podesta, J. J. (2009). Dependence of solar-wind power spectra on the direction of the local mean magnetic field. *The Astrophysical Journal*, *698*, 986–999.
Podesta, J. J. (2010). Comment on "Turbulent cascade at 1 AU in high cross-helicity flows". *Physical Review Letters*, *104*, 169001.
Podesta, J. J., Forman, M. A., Smith, C. W., et al. (2009). Accurate estimation of third-order moments from turbulence measurements. *Nonlinear Processes in Geophysics*, *16*, 99–110.
Porter, L. J., & Klimchuk, J. A. (1995). Soft X-ray loops and coronal heating. *The Astrophysical Journal*, *454*, 499–511.
Rabin, D., & Moore, R. (1984). Heating the sun's lower transition region with fine-scale electric currents. *The Astrophysical Journal*, *285*, 359–367.
Reale, F., Parenti, S., Reeves, K. K., et al. (2007). Fine thermal structure of a coronal active region. *Science*, *318*, 1582–1585.
Reames, D. V. (1994). Coronal element abundances derived from solar energetic particles. *Advances in Space Research*, *14*, 177–180.
Roberts, D. A., Goldstein, M. L., Klein, L. W., & Mathaeus, W. H. (1987a). Origin and evolution of fluctuations in the solar wind: Helios observations and Helios-Voyager comparisons. *Journal of Geophysical Research: Space Physics*, *92*, 12023–12035.
Roberts, D. A., Klein, L. W., Goldstein, M. L., & Mathaeus, W. H. (1987b). The nature and the evolution of magnetohydrodynamic fluctuations in the solar wind: Voyager observations. *Journal of Geophysical Research: Space Physics*, *92*, 11021–11040.
Rosner, R., Golub, L., Coppi, B., & Vaiana, G. S. (1978). Heating of coronal plasma by anomalous current dissipation. *The Astrophysical Journal*, *222*, 317–332.
Sahraoui, F., Goldstein, M. L., Belmont, G., et al. (2010). Three dimensional anisotropic k spectra of turbulence at subproton scales in the solar wind. *Physical Review Letters*, *105*, 131101.
Saito, K. (1958). Polar rays of the solar corona. *Publications of the Astronomical Society of Japan*, *10*, 49–78.
Saito, K. (1965). Polar rays of the solar corona. II. *Publications of the Astronomical Society of Japan*, *17*, 1–26.
Sakao, T., Kano, R., Narukage, N., et al. (2007). Continuous plasma outflows from the edge of a solar active region as a possible source of solar wind. *Science*, *318*, 1585–1588.
Schekochihin, A. A., Cowley, S. C., Dorland, W., et al. (2009). Astrophysical gyrokinetics: Kinetic and fluid turbulent cascades in magnetized weakly collisional plasmas. *The Astrophysical Journal Supplement*, *182*, 310–377.
Schrijver, C. J., & Zwaan, C. (2001). Solar and stellar magnetic activity: The solar dynamo. *Physics Today*, *54*, 54–56.
Schrijver, C. J., Title, A. M., Berger, T. E., et al. (1999). A new view of the solar outer atmosphere by the transition region and coronal explorer. *Solar Physics*, *187*, 261–302.

Schwartz, S. J., Cally, P. S., & Bel, N. (1984). Chromospheric and coronal Alfvénic oscillations in non-vertical magnetic fields. *Solar Physics*, *92*, 81–98.
Severny, A. (1964). Solar magnetic fields. *Space Science Reviews*, *3*, 451–486.
Shibata, K., Nakamura, T., Matsumoto, T., et al. (2007). Chromospheric anemone jets as evidence of ubiquitous reconnection. *Science*, *318*, 1591–1594.
Shukla, A., & Sharma, R. P. (2001). Transient filaments formation by nonlinear kinetic Alfvén waves and its effect on solar wind turbulence and coronal heating. *Physics of Plasmas*, *8*, 3759–3765.
Shukla, A., Sharma, R. P., & Malik, M. (2004). Filamentation of Alfvén waves associated with transverse perturbation. *Physics of Plasmas*, *11*, 2068–2074.
Smith, C. W., Stawarz, J. E., Vasquez, B. J., et al. (2009). Turbulent cascade at 1 AU in high cross-helicity flows. *Physical Review Letters*, *103*, 201101.
Socas-Navarro, H. (2005). Are electric currents heating the magnetic chromosphere? *The Astrophysical Journal Letters*, *633*, L57–L60.
Spicer, D. S. (1991). In P. Ulmschneider, E. R. Priest & R. Rosner (Eds.), *Mechanisms of chromospheric and coronal heating* (Vol. 547). Heidelberg: Springer.
Sridhar, S., & Goldreich, P. (1994). Towards a theory of interstellar turbulence. I. weak Alfvénic turbulence. *Astrophysical Journal*, *432*, 612–621.
Stasiewicz, K., Bellan, P., Chaston, C., et al. (2000a). Small scale Alfvénic structure in the aurora. *Space Science Reviews*, *92*, 423–533.
Stasiewicz, K., Gustafsson, G., Marklund, G., et al. (1997). Cavity resonators and Alfvén resonance cones observed on Freja. *Journal of Geophysical Research: Space Physics*, *102*, 2565–2575.
Stasiewicz, K., Holmgren, G., & Zanetti, L. (1998). Density depletions and current singularities observed by Freja. *Journal of Geophysical Research: Space Physics*, *103*, 4251–4260.
Stasiewicz, K., Khotyaintsev, Y., Berthomier, M., & Wahlund, J. E. (2000b). Identification of widespread turbulence of dispersive Alfvén waves. *Geophysical Research Letters*, *27*, 173–176.
Stawarz, J. E., Smith, C. W., Vasquez, B. J., et al. (2009). The turbulent cascade and proton heating in the solar wind at 1 AU. *The Astrophysical Journal*, *697*, 1119–1127.
Stawarz, J. E., Smith, C. W., Vasquez, B. J., et al. (2010). The turbulent cascade for high cross-helicity states at 1 AU. *The Astrophysical Journal*, *713*, 920.
Stéfant, R. J. (1970). Alfvén wave damping from finite gyroradius coupling to the ion acoustic mode. *The Physics of Fluids*, *13*, 440–450.
Strong, K. T., Saba, J. L. R., Haisch, B. M., & Schmelz, J. T. (1999). *The many faces of the sun: A summary of the results from NASA's solar maximum mission*. New York: Springer.
Suess, S. T. (1982). Polar coronal plumes. *Solar Physics*, *75*, 145–159.
Suess, S. T., Poletto, G., Wang, A. H., et al. (1998). The geometric spreading of coronal plumes and coronal holes. *Solar Physics*, *180*, 231–246.
Tan, B. L. (2007). Distribution of electric current in solar plasma loops. *Advances in Space Research*, *39*, 1826–1830.
Tan, B. L., Ji, H. S., Huang, G. L., et al. (2006). Evolution of electric currents associated with two M-class flares. *Solar Physics*, *239*, 137–148.
Teriaca, L., Poletto, G., Romoli, M., & Biesecker, D. (2003). The nascent solar wind: Origin and acceleration. *The Astrophysical Journal*, *588*, 566–577.
Thomas, J. H., & Weiss, N. O. (2004). Fine structure in sunspots. *Annual Review of Astronomy and Astrophysics*, *42*, 517–548.
Title, A. M., Tarbel, T. D., Topka, K. P., et al. (1989). Statistical properties of solar granulation derived from the SOUP instrument on Spacelab 2. *The Astrophysical Journal*, *336*, 475–494.
Tomczyk, S., McIntosh, S. W., Keil, S. L., et al. (2007). Alfvén waves in the solar corona. *Science*, *317*, 1192–1196.
Treumann, R. A., Güdel, M., & Benz, A. O. (1990). Alfvén wave solitons and solar intermediate drift bursts. *Astronomy & Astrophysics*, *236*, 242–249.
Trimble, V., & Aschwanden, M. (2005). Astrophysics in 2004. *Publications of the Astronomical Society of the Pacific*, *117*, 311–394.

Trondsen, T. S., & Cogger, L. L. (1997). High-resolution television observations of black aurora. *Journal of Geophysical Research: Space Physics*, *102*, 363.
Trondsen, T. S., & Cogger, L. L. (1998). A survey of small-scale spatially periodic distortions of auroral forms. *Journal of Geophysical Research: Space Physics*, *103*, 9405–9415.
Tsiklauri, D. (2011). Particle acceleration by circularly and elliptically polarised dispersive Alfvén waves in a transversely inhomogeneous plasma in the inertial and kinetic regimes. *Physics of Plasmas*, *18*, 092903.
Tsiklauri, D. (2012). Three dimensional particle-in-cell simulation of particle acceleration by circularly polarised inertial Alfvén waves in a transversely inhomogeneous plasma. *Physics of Plasmas*, *19*, 082903.
Tsuneta, S., Acton, L., Bruner, M., et al. (1991). The soft X-ray telescope for the SOLAR-A mission. *Solar Physics*, *136*, 37–67.
Tu, C. Y. (1988). The damping of interplanetary Alfvénic fluctuations and the heating of the solar wind. *Journal of Geophysical Research: Space Physics*, *93*, 7–20.
Tu, C. Y., Pu, Z. Y., & Wei, F. S. (1984). The power spectrum of interplanetary Alfvénic fluctuations: Derivation of the governing equation and its solution. *Journal of Geophysical Research: Space Physics*, *89*, 9695–9702.
Tu, C. Y., Roberts, D. A., & Goldstein, M. L. (1989). Spectral evolution and cascade constant of solar wind Alfvénic turbulence. *Journal of Geophysical Research: Space Physics*, *94*, 13575–13578.
Uchida, Y., & Kaburaki, O. (1974). Excess heating of corona and chromosphere above magnetic regions by non-linear Alfvén waves. *Solar Physics*, *35*, 451–466.
Ulmschneider, P. (1991). Acoustic heating. In Priest Ulmschneider & Rosner, (Eds.), *Mechanisms of chromospheric and coronal heating* (pp. 328–343). Berlin, Germany: Springer.
Unti, T. W. J., & Neugebauer, M. (1968). Alfvén waves in the solar wind. *Physics of Fluids*, *11*, 563–568.
Verma, M. K., Roberts, D. A., & Goldstein, M. L. (1995). Turbulent heating and temperature evolution in the solar wind plasma. *Journal of Geophysical Research: Space Physics*, *100*, 19839.
Vernazza, J. E., Avrett, E. H., & Loeser, R. (1981). Structure of the solar chromosphere III-Models of the EUV brightness components of the quiet-sun. *The Astrophysical Journal Supplement Series*, *45*, 635–725.
Voitenko, Y. (1996). Flare loops heating by the 0.1–1.0 MeV proton beams. *Solar Physics*, *168*, 219–222.
Voitenko, Y. (1998). Excitation of kinetic Alfvén waves in a flaring loop. *Solar Physics*, *182*, 411–430.
Voitenko, Y., & Goossens, M. (2000). Nonlinear decay of phase-mixed Alfvén waves in the solar wind. *Astronomy & Astrophysics*, *357*, 1073–1085.
Voitenko, Y., & Goossens, M. (2002). Nonlinear excitation of small-scale Alfvén waves by fast waves and plasma heating in the solar atmosphere. *Solar Physics*, *209*, 37–60.
Voitenko, Y., & Goossens, M. (2004). Cross-field heating of coronal ions by low-frequency kinetic Alfvén waves. *The Astrophysical Journal Letters*, *605*, L149–L152.
Voitenko, Y., & Goossens, M. (2005a). Cross-scale nonlinear coupling and plasma energization by Alfvén waves. *Physical Review Letters*, *94*, 135003.
Voitenko, Y. & Goossens, M. (2005b), Nonlinear coupling of Alfvén waves with widely different cross-field wavelengths in space plasmas. *Journal of Geophysical Research: Space Physics*, *110* A10S01.
Voitenko, Y., & Goossens, M. (2006). Energization of plasma species by intermittent kinetic Alfvén waves. *Space Science Reviews*, *122*, 255–270.
Voitenko, Y., Goossens, M., Sirenko, O., & Chian, A. (2003). Nonlinear excitation of kinetic Alfvén waves and whistler waves by electron beam-driven Langmuir waves in the solar corona. *Astronomy & Astrophysics*, *409*, 331–345.
Volland, H., Bird, M., Levy, G., et al. (1977). Helios-1 Faraday-rotation experiment-results and interpretations of solar occultations in 1975. *Journal of Geophysics Zeitschrift Geophysik*, *42*, 659–672.

Volwerk, M., Louarn, P., Chust, T., et al. (1996). Solitary kinetic Alfvén waves: A study of the Poynting flux. *Journal of Geophysical Research: Space Physics, 101*, 13335–13343.

Wahlund, J. E., Louarn, P., Chust, T., et al. (1994a). On ion-acoustic turbulence and the nonlinear evolution of kinetic Alfvén waves in aurora. *Geophysical Research Letters, 21*, 1831–1834.

Walker, A. B. C., De Forest, C. E., Hoover, R. B., & Barbee, T. W. (1993). Thermal and density structure of polar plumes. *Solar Physics, 148*, 239–252.

Wang, Y. M. (1994). Polar plumes and the solar wind. *The Astrophysical Journal, 435*, L153–L156.

Webb, D. F., & Zirin, H. (1981). Coronal loops and active region structure. *Solar Physics, 69*, 99–118.

Wentzel, D. G. (1974). Coronal heating by Alfvén waves. *Solar Physics, 39*, 129–140.

Wentzel, D. G. (1976). Coronal heating by Alfvén waves, II. *Solar Physics, 50*, 343–360.

Whang, Y. C., Behannon, K. W., Burlaga, L. F., & Zhang, S. (1989). Thermodynamic properties of the hellospheric plasma. *Journal of Geophysical Research: Space Physics, 94*, 2345.

Whang, Y. C., Liu, S., & Burlaga, L. F. (1990). Shock heating of the solar wind plasma. *Journal of Geophysical Research: Space Physics, 95*, 18769–18780.

Wicks, R. T., Horbury, T. S., Chen, C. H. K., & Schekochihin, A. A. (2010). Power and spectral index anisotropy of the entire inertial range of turbulence in the fast solar wind. *Monthly Notices of the Royal Astronomical Society: Letters, 407*, L31–L35.

Widing, K. G., & Feldman, U. (1992). Element abundances and plasma properties in a coronal polar plume. *The Astrophysical Journal, 392*, 715–721.

Wu, C. S., Wang, C. B., Yoon, P. H., et al. (2002). Generation of type III solar radio bursts in the low corona by direct amplification. *The Astrophysical Journal, 575*, 1094–1103.

Wu, D. J. (2012). *Kinetic Alfvén wave: Theory, experiment and application*. Beijing: Science Press.

Wu, D. J., & Chao, J. K. (2003). Auroral electron acceleration by dissipative solitary kinetic Alfvén waves. *Physics of Plasmas, 10*, 3787–3789.

Wu, D. J., & Chao, J. K. (2004a). Model of auroral electron acceleration by dissipative solitary kinetic Alfvén wave. *Journal of Geophysical Research: Space Physics, 109*, A06211.

Wu, D. J., & Chao, J. K. (2004b). Recent progress in nonlinear kinetic Alfvén waves. *Nonlinear Processes in Geophysics, 11*, 631–645.

Wu, D. J., & Chen, L. (2013). Excitation of kinetic Alfvén waves by density striation in magneto-plasmas. *The Astrophysical Journal, 771*, 3.

Wu, D. J., & Fang, C. (1999). Two-fluid motion of plasma in Alfvén waves and heating of solar coronal loops. *The Astrophysical Journal, 511*, 958–964.

Wu, D. J., & Fang, C. (2003). Coronal plume heating and kinetic dissipation of kinetic Alfvén waves. *The Astrophysical Journal, 596*, 656–662.

Wu, D. J., & Fang, C. (2007). Sunspot chromospheric heating by kinetic Alfvén waves. *The Astrophysical Journal Letters, 659*, L181–L184.

Wu, D. J., & Yang, L. (2006). Anisotropic and mass-dependent energization of heavy ions by kinetic Alfvén waves. *Astronomy & Astrophysics, 452*, L7–L10.

Wu, D. J., & Yang, L. (2007). Nonlinear interaction of minor heavy ions and kinetic Alfvén waves and their anisotropic energization in coronal holes. *The Astrophysical Journal, 659*, 1693–1701.

Wu, D. J., Huang, G. L., Wang, D. Y., & Fälthammar, C. G. (1996b). Solitary kinetic Alfvén waves in the two-fluid model. *Physics of Plasmas, 3*, 2879–2884.

Wu, D. J., Huang, G. L., & Wang, D. Y. (1996a). Dipole density solitons and solitary dipole vortices in an inhomogeneous space plasma. *Physical Review Letters, 77*, 4346–4349.

Wu, D. J., Huang, J., Tang, J. F., & Yan, Y. H. (2007). Solar microwave drifting spikes and solitary kinetic Alfvén waves. *The Astrophysical Journal Letters, 665*, L171–L174.

Wu, D. J., Wang, D. Y., & Fälthammar, C. G. (1995). An analytical solution of finite-amplitude solitary kinetic Alfvén waves. *Physics of Plasmas, 2*, 4476–4481.

Wu, D. J., Wang, D. Y., & Huang, G. L. (1997). Two dimensional solitary kinetic Alfvén waves and dipole vortex structures. *Physics of Plasmas, 4*, 611–617.

Wygant, J. R., Keiling, A., Cattell, C. A., et al. (2002). Evidence for kinetic Alfvén waves and parallel electron energization at 4–6 R_E altitudes in the plasma sheet boundary layer. *Journal of Geophysical Research: Space Physics, 107*, 1201–1215.
Xie, H., Ofman, L., & Viñas, A. (2004). Multiple ions resonant heating and acceleration by Alfvén/cyclotron fluctuations in the corona and the solar wind. *Journal of Geophysical Research: Space Physics, 109*, A08103.
Yoon, P. H., Wu, C. S., & Wang, C. B. (2002). Generation of type III solar radio bursts in the low corona by direct amplification. II. Further numerical study. *The Astrophysical Journal, 576*, 552–560.
Young, P. R., Klimchuk, J. A., & Mason, H. E. (1999). Temperature and density in a polar plume-measurements from CDS/SOHO. *Astronomy and Astrophysics, 350*, 286–301.
Zaitsev, V. V., Stepanov, A. V., Urpo, S., & Pohjolainen, S. (1998). LRC-circuit analog of current-carrying magnetic loop: Diagnostics of electric parameters. *Astronomy and Astrophysics, 337*, 887–896.
Zhao, J. S., Wu, D. J., & Lu, J. Y. (2011). Kinetic Alfvén waves excited by oblique MHD Alfvén waves in coronal holes. *The Astrophysical Journal, 735*, 114.
Zhao, J. S., Wu, D. J., & Lu, J. Y. (2013). Kinetic Alfvén turbulence and parallel electric fields in flare loops. *The Astrophysical Journal, 767*, 109.
Zhao, R. Y. (1995). A model of solar (radio) active regions. *Astrophysics and Space Science, 234*, 125–137.

Chapter 7
KAWs in Extrasolar Astrophysical Plasmas

In previous chapters we comprehensively introduced KAWs in space and solar plasmas, including their observational identifications and applications to related particle kinetic phenomena, such as the auroral energetic electron acceleration in auroral magnetospheric plasmas, the anomalous particle transport in the magnetopause, the kinetic turbulence in the solar wind, and the solar coronal heating problem. The solar and solar-system plasma, especially the solar-terrestrial coupling system, is a natural laboratory for plasma astrophysics. It is extensively believed that the results based on this natural laboratory may be extrapolated and applied to extrasolar astrophysical plasmas. In the last chapter of this monograph, we will introduce possible applications of KAWs to extrasolar astrophysical phenomena. First, observation of a uniform and universal big power-law spectrum for cosmic plasma turbulence over a huge scale range from cosmological to kinetic scales and its implication to KAW turbulence in cosmic plasmas are introduced in Sect. 7.1. Then, the heating problem of cosmic hot gases and its possible associations with KAWs are presented in Sect. 7.2 (stellar coronae) and Sect. 7.3 (galactic halos). Finally, KAWs and their possible roles in the reacceleration of energetic electrons and the generation of electric currents in extragalactic extended radio sources are discussed in Sect. 7.4.

7.1 Plasma Turbulence from Cosmological to Kinetic Scales

One of major discoveries of modern astronomy is that vast cosmic space, from interplanetary and interstellar to intergalactic space, is filled by various dispersion interstellar medium (ISM) and intracluster medium (ICM), such as stellar coronae and winds, supernova remnants, molecular clouds, intergalactic hot gases and its galactic halos, and so on. While one of the most important developments in the field of ISM dynamics is that it has widely recognized that most processes and structures of ISM

D.-J. Wu and L. Chen, *Kinetic Alfvén Waves in Laboratory, Space, and Astrophysical Plasmas*, Atmosphere, Earth, Ocean & Space,
https://doi.org/10.1007/978-981-13-7989-5_7

are strongly affected by turbulence. An initial connection between ISM dynamics and fluid turbulence can be found in some early descriptions for ISM (von Weizsäcker 1951), in which ISM consists of cloudy objects with a hierarchical structures formed in interacting shocks by supersonic turbulence that is stirred on the largest scale by differential galactic rotation and dissipated on small scales by atomic viscosity. For instance, it was noticed that rms (root mean square) differences in emission-line velocities of the Orion nebula increased with projected separation as a power law with a power index α between 0.25 and 0.5, implying that ISM within the Orion nebula may be described as a fluid turbulence formed by the Kolmogorov energy cascade (von Hoerner 1951), and for the latter, as well known, the corresponding power-law index is $\alpha = 1/3$.

However, these early proposals regarding pervasive turbulence had not been believed extensively in the field of ISM community, until observations of interstellar scintillation at radio wave lengths resulted in the wider recognization of ubiquitous turbulence in ISM. When radio waves from radio pulsars or compact radio sources propagate through ISM, electron density fluctuations in the ionized ISM can scatter the radio waves and give rise to scintillation of these radio sources (Rickett 1970, 1990; Narayan 1992). The nature of this phenomenon was first understood by Scheuer (1968), who proposed that the observed pulsar variability in its radio radiation could be caused due to that some ionized inhomogeneous blob structures (irregularities) scatter the radio waves and estimated the characteristic scale of these ionized blobs lying in a wide range $\sim 10^7$–10^{11} m. The scale of these plasma blobs is the same scale at which cosmic rays excite magnetic turbulence by streaming instabilities (Wentzel 1968), but much less than the typical scale of interstellar clouds $\sim 10^{18}$ m. The observations also showed that these plasma irregularities seem to have a correlation in the wide scale range and possibly to be related to turbulence (Salpeter 1969; Rickett 1970; Little and Matheson 1973).

Lee and Jokipii (1976) further proposed that these plasma irregularities have a turbulent origin and their power spectrum follows an isotropic Kolmogorov-like spectrum. In particular, they conjectured that this power-law spectrum can have much wider extending range, from the small-scale plasma irregularities at scales $\sim 10^{11}$ m < 1 AU to the large-scale interstellar clouds at scales $\sim 10^{17}$ m > 1 pc (Lee and Jokipii 1976), although there was little understanding of the physical connection between these small-scale plasma irregularities and the larger-scale cloud motions in ISM. Soon afterwards, based on various observations of spectral lines from many molecular clouds (also including young stars) Larson (1979, 1981) further extended this ISM turbulent spectrum to larger scales over a scale range from < 1 pc to > 10 kpc. He found that there are well power-law correlations between the velocity dispersion inferred from linewidths and the molecular cloud sizes, which is similar to the Kolmogorov scaling law for subsonic turbulence. Therefore, he further proposed that the power-law spectrum of the velocity dispersion can be formed through a common hierarchical process of ISM turbulent motions in which smaller-scale motions are produced by the turbulent cascade of larger-scale ones (Larson 1979, 1981).

In the aspect of large-scale interstellar cloud motions, Larson's work was soon followed by more homogeneous observations that showed similar correlations (Myers 1983; Dame et al. 1986; Solomon et al. 1987), implying that these motions were believed to be turbulent because of their power-law nature. On the other hand, in the aspect of small-scale plasma irregularities, further supports came from the dispersion measurements of compact extragalactic radio sources as well as pulsars (Rickett et al. 1984; Cordes et al. 1985; Phillips and Wolszczan 1991). Another important contribution to the recognization of the universal existence of the ISM turbulence is the surprising discovery of a power spectrum for widespread HI emission that was comparable to the Kolmogorov power spectrum for velocity in incompressible turbulence (Crovisier and Dickey 1983). In consequence, based on many independent observations from small-scale plasma irregularities to large-scale interstellar clouds, a universal power-law spectrum of plasma density fluctuations is well established, which follows the Kolmogorov scaling law with the three-dimensional power-law index $\alpha = 11/3$ and spans over 12 orders of magnitude (Lee and Jokipii 1976; Armstrong et al. 1981, 1995).

In a recent work, using in situ measurements from the Voyager 1 satellite, Lee and Lee (2019) further extended this universal turbulent spectrum from the ordinary inertial-scale range down to the kinetic-scale region. The Voyager 1 satellite crossed the heliopause at 122 AU on 2012 August 25 and reached 142 AU away from the Sun in 2018 (Gurnett et al. 2013). The electron density can be inferred by identifying the local electron Langmuir frequency from electrostatic wave measurements because the plasma science experiment instrument onboard the Voyager 1 satellite was shut in 2007. By use of the Lomb-Scargle periodogram, they calculated the power spectral density of electron density fluctuations measured by the Voyager 1 satellite during its interstellar journey from 2012 September 3 to 2016 October 25. In order to reduce the inaccuracy of the calculation, they further separated the density profiles into five groups and conducted the Lomb-Scargle periodogram on each of them to obtain the spectral density functions (SDFs) for different scales.

Figure 7.1 shows their results, where the purple dots are the SDF of the scale $\sim$50–500 m, the green dots are the SDF of the scale $\sim$10–1000 km, the blue dots are the SDF of the scale from $\sim$0.05–2.5 AU, and the red dots are the SDF of the scale $\sim$2.8–15 AU, which are obtained by the complete density profile. In addition, the fine grey dots and lines in Fig. 7.1 show the composite spectra obtained by Armstrong et al. (1981, 1995) and Chepurnov and Lazarian (2010), respectively. Armstrong et al. (1981, 1995) constructed a composite electron density spectrum over the scale of $10^{6.5}$–10^{13} m based on several studies of interstellar scattering and over the scale of 10^{16}–10^{18} m from interstellar velocity fluctuations, rotation measures of magnetic variations, and the gradient scale of interstellar clouds. Chepurnov and Lazarian (2010) obtained a more precise estimation of power density at the scale of 10^{16}–10^{18} m based on the data of Wisconsin Hα Mapper.

These SDFs of various scales are combined to construct the composite spectrum over scales from $\sim$50 to $\sim 10^{18}$ m, over 16 orders of magnitude, as shown in Fig. 7.1. For typical values of magnetic field $\sim$0.5 nT, electron density $n_e \sim 0.09$ cm^{-3}, and temperature $T \sim 100$ eV in the local ISM (Gurnett et al. 2013), we have the particle

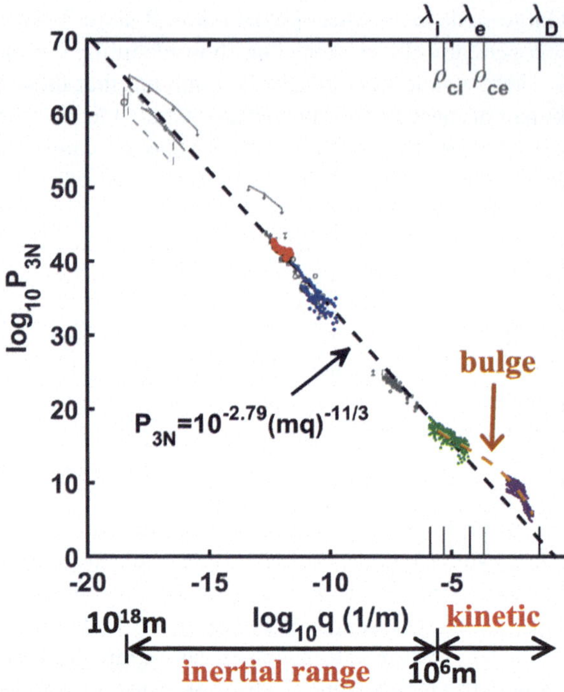

Fig. 7.1 Composite spectrum (red, blue, green and purple dots) obtained from the in situ measurements by the Voyager 1 satellite outside the heliopause in the interstellar plasma. The black dashed line is the best-fitting line for the inertial part of these data by the isotropic Kolmogorov spectrum with spectral index $\alpha = 11/3$. The dashed orange line shows the presence of a bulge in the kinetic scale range. The spectral densities obtained from earlier remote sensing methods are shown as fine grey dots and lines (from Lee and Lee 2019)

kinetic scales as follows: the ion inertial length $\lambda_i \sim 7.6 \times 10^5$ m, the ion gyroradius $\rho_i \sim 2.0 \times 10^5$ m, the electron inertial length $\lambda_e \sim 1.8 \times 10^4$ m, the electron gyroradius $\rho_e \sim 4.8 \times 10^3$ m, and the Debye length $\lambda_D \sim 25$ m, as marked in Fig. 7.1. Therefore, by combining the in situ measurements by the Voyager 1 satellite in the local ISM (Lee and Lee 2019) and the earlier ground remote observations (Armstrong et al. 1981, 1995; Chepurnov and Lazarian 2010), we now have the turbulent spectrum that covers not only a wide Kolmogorov inertial range (10^6–10^{18} m) but also a complete kinetic range down to the Debye length (10^6 m to 50 m). The black dashed line in Fig. 7.1 represents the best-fitting line for the inertial part of these data by the isotropic Kolmogorov spectrum with spectral index $\alpha = 11/3$. While the orange dashed line in Fig. 7.1 shows the presence of a bulge in the kinetic scale range.

What is the reason for a so big universal power-law spectrum uniformly extending from ~100 pc of cosmological scales down to ~50 m of microscopic kinetic scales? This is indeed a mysterious and interesting problem in modern astronomical communities. Moreover, the observed ISM turbulence of originating from electron density fluctuations in ionized ISM unexpectedly has the same dynamically spectral parameter (i.e., $\alpha \simeq 11/3$) with the Kolmogorov theory for neutral fluid turbulence, although the μG-order magnetic field ubiquitously exist in ISM (see e.g., Beck 2001). In principle, in the pc-order large scales the dynamics of the interstellar clouds is dominated by the gravity and the electromagnetic force can play an important role in the dynamics of the plasma irregularities in the AU-order small scales. Why can do different dynamical drivers lead to the same dynamically turbulent spectrum?

Neglecting the gravity but taking account of the effect of magnetic fields, Higdon (1984) proposed a model for strongly anisotropic MHD turbulence to describe quantitatively the turbulent origin of small-scale irregular plasma density fluctuations in ionized ISM observed by Armstrong et al. (1981). In his model, the observed density fluctuations are interpreted to be two-dimensional isobaric entropy variations with oppositely directed gradients in temperature and density projected transversely to the local magnetic field. The observed distribution of the electron density variations is produced by the convection and distortion of these dynamically passive entropy fluctuations by turbulent velocity and magnetic field fluctuations. These small-scale turbulent velocity and magnetic field fluctuations are distributed anisotropically and concentrated transversely to the local magnetic field. In particular, the transverse transport coefficients are dramatically decreased in the presence of magnetic fields, implying that the low dissipation due to the magnetic field would allow the energy-conserving inertial range to extend into smaller scales than the inertial range of isotropic turbulent flows without magnetic fields. This may remove the difficulty that the dissipative scale is possibly larger than the smallest scale observed by Armstrong et al. (1981).

Assuming a Kolmogorov turbulent spectrum for the velocity and magnetic field fluctuations, Montgomery et al. (1987) also claimed that the same Kolmogorov spectrum may be obtained for the density fluctuations by the nearly incompressible MHD model. In their model, the linearized small fluctuations in the density are given by the pressure variations, which may be calculated from the velocity and magnetic field fluctuations by the incompressible MHD equation.

As shown in Chap. 5, an early self-consistent nonlinear theory of incompressible MHD turbulence is the Iroshnikov-Kraichnan theory of the isotropic cascade of AWs towards small scales, but predicts a power-law spectrum with an one-dimensional spectral index $\alpha = 3/2$ different from $\alpha = 5/3$ for the Kolmogorov spectrum. Sridhar and Goldreich (1994) and Goldreich and Sridhar (1995, 1997) further developed the Iroshnikov-Kraichnan theory for the weak, strong, and intermediate Alfvénic turbulence in the inertial scale range, that is, the well-known Goldreich-Sridhar theory. The most important characteristic of the Goldreich-Sridhar theory is to predict that the anisotropic cascade of AW turbulence towards small scales performs preferentially along the direction perpendicular to the local magnetic field. The Goldreich-Sridhar theory has been widely accepted as the standard theory for the MHD turbulence in ISM because the anisotropic cascade of AW turbulence predicted by this theory has widely been confirmed not only by a series of numerical simulations (Cho and Vishniac 2000; Maron and Goldreich 2001; Müller et al. 2003) but also by a large number of observations (Horbury et al. 2008; Podesta 2009; Wicks et al. 2010; Luo and Wu 2010).

The most straightforward result of the anisotropic cascade of AW turbulence by the standard Goldreich-Sridhar theory is that as the anisotropic cascade proceeds towards small scales, the AW turbulence in the inertial scale range naturally and inevitably transfers into the KAW turbulence in the kinetic scale region (see e.g., Howes et al. 2006; Schekochihin et al. 2009; and Chap. 5 for more details). In consequence, as shown in Chap. 5, the turbulent spectrum has the two breakpoints nearly at the ion and electron gyroscales due to the effects of the dissipation, dispersion, and alignment of KAWs in small scales, so that the kinetic turbulent spectrum deviates evidently the inertial Kolmogorov spectrum. For instance, as shown in Figs. 5.9, 5.10, and 5.12, in the kinetic inertial region that covers the kinetic scale range from the ion gyroscale $\sim\rho_i$ to the electron gyroscale $\sim\rho_e$, the magnetic field power spectrum steepens, but the electric field spectrum becomes more flattened, implying that the two kinetic spectra both obviously deviate from the inertial Kolmogorov spectrum.

Also the kinetic-scale spectrum obtained by Lee and Lee (2019) exhibits a similar spectral transition behavior from the inertial-scale spectrum to the kinetic-scale spectrum. The SDFs of the green and purple dots in Fig. 7.1 cover well the ion and electron kinetic scales, and the orange dashed line represents the fitting spectrum in the kinetic scale region. One can clearly find that in the kinetic scale range, the kinetic fitting spectrum represented by the orange dashed line obviously deviates away from the Kolmogorov spectrum represented by the black dashed line, exhibits an enhanced intensity, and forms an evident bulge, as illustrated in Fig. 7.1. Lee and Lee (2019) suggested that the shocks of solar origin propagating in the local ISM possibly could be responsible for the enhanced bulge in the kinetic scale range of the SDFs in Fig. 7.1. Several shock/foreshock events have been identified during 2012 October to 2014 December (Gurnett et al. 2015; Kim et al. 2017). Various plasma waves, including KAWs, mirror waves, ion and electron cyclotron waves, and Langmuir waves, can be excited as shocks propagate in magnetized plasma (Tsurutani et al. 1982; Gurnett et al. 2015; Lee 2017). While the excitation of these waves may cause the enhancement of electron density fluctuations in the kinetic range.

In general, ISM shocks can not exist universally like the ISM turbulence with the universal Kolmogorov spectrum. While, in fact, the universal big-range spectrum with the Kolmogrov spectrum in the inertial scale and the enhanced around the kinetic scale can possibly exist widespreadly in ISM. For instance, the enhanced bulge of the turbulent spectrum in the kinetic scale region also has been observed in the solar wind (Neugebauer 1975; Kellogg and Horbury 2005; Alexandrova et al. 2008; Sahraoui et al. 2009; Chen et al. 2012; Šafránková et al. 2015). In the kinetic scales KAWs and whistler waves both can contribute to the enhancement of plasma density fluctuations. However, as shown by discussions in Chap. 5, KAWs are evidently better and more proper candidates than whistler waves for the wave mode of the solar wind fluctuations in the kinetic scale range. The compressibility of KAWs, which increases as the scale decreases (see Figs. 5.19, 5.20, and the relevant discussions), can cause active density fluctuations and hence lead to an enhanced density spectral power in the kinetic scales (Chandran et al. 2009; Chen et al. 2012; Šafránková et al. 2015). Thus, the interstellar kinetic turbulent spectrum obtained by Lee and Lee (2019) based on the in situ measurements of the Voyager 1 satellite in the local ISM provides new evidence for the existence of KAWs in ionized ISM.

Therefore, we can in reason believe that KAWs, just like AWs, can ubiquitously exist in various dispersion ISM. These widespread KAWs in ISM can play an important role in kinetic processes of the ionized ISM, such as the heating of stellar coronae, coronae of accretion disks around compact objects, interstellar and intergalactic hot gases, and galactic hot halos, as well as the acceleration of energetic electron beams in planetary magnetospheres and the reacceleration of energetic electrons in extragalactic radio jets.

7.2 KAW Heating of Stellar and Accretion Disk Coronae

In the viewpoint of astronomy, our sun is only an ordinary star of skyful stars and lies in the main sequence region for middle-age stars in the H-R (Hertzsprung-Russel) diagram of the stellar evolution. Like the sun has a high-temperature corona, the stars ubiquitously possess theirselves coronae, called stellar coronae. These coronae consist of high-temperature hot plasmas at temperatures exceeding 10^6 K and emit mainly X-ray photons as stellar X-ray sources, like the solar corona. The first stellar X-ray source is the binary system Capella, which was identified by Catura et al. (1975) as the optical counterpart of a soft X-ray source detected occasionally during a rocket flight in 1974. They estimated the X-ray luminosity at 10^{31} erg/s, much higher than the solar X-ray luminosity by four orders of magnitude, and the plasma electron temperature at about 8×10^6 K, again several times higher than the solar coronal temperature. This result was confirmed by Mewe et al. (1975) from X-ray observations onboard the ANS (Astronomical Netherlands Satellite) launched in August 1974. Moreover, they were the first to interpret the soft X-rays as solar-like coronal emission at an enhanced level. Around the same time, Heise et al. (1975) monitored the first stellar coronal X-ray flares on the stars YZ CMi and UV Cet with the X-ray observations by the ANS and one of the flares was recorded simultaneously with an optical burst. Since that tens of thousands of stellar X-ray sources have been discovered by a series of based-space X-ray observatories.

The optical counterparts of these stellar X-ray sources abound among all types of stars, across nearly the whole H-R diagram and most stages of stellar evolution (Vaiana et al. 1981). Figure 7.2, based on about 2000 X-ray sources with identified optical stars, shows their basic features known from an optical H-R diagram, including the loci of these stars in the optically determined absolute magnitude M_V and the color index $B - V$, where the size of the circles characterizes their X-ray luminosity $\log L_X$ as indicated in the panel at lower left and the vertical dashed line denotes the so-called corona vs wind dividing line that separates coronal giants and supergiants to its left from stars with massive winds to its right (Güdel 2004).

The widespread occurrence of the stellar X-ray coronae along the main sequence utterly refutes the acoustic heating model for the stellar coronae because only stars in a rather narrow range of spectral types would have enough "acoustic flux" to heat

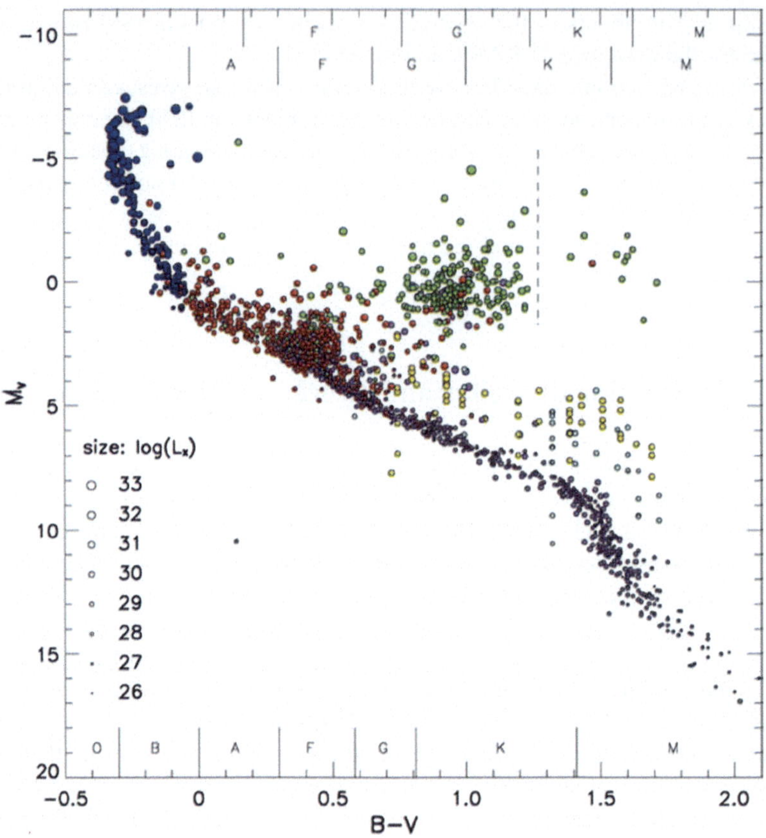

Fig. 7.2 Hertzsprung-Russell diagram based on about 2000 X-ray detected stars extracted from the catalogs by Berghöfer et al. (1996) (blue), Hünsch et al. (1998a, b) (green and red, respectively), and Hünsch et al. (1999) (pink). Where missing, distances from the Hipparcos catalog (Perryman et al. 1997) were used to calculate the relevant parameters. The low-mass pre-main sequence stars are taken from studies of the Chamaeleon I dark cloud (Alcalá et al. 1997; Lawson et al. 1996; yellow and cyan, respectively) and are representative of other star formation regions. The size of the circles characterizes log L_X as indicated in the panel at lower left. The ranges for the spectral classes are given at the top with the upper row for supergiants and lower row for giants, and at the bottom for main-sequence stars (reprinted from Güdel, Astron. Astrophys. Rev., 12, 71–237, 2004, copyright 2004, with permission from Springer)

a high-temperature corona (Walter et al. 1980). For stars with convective envelopes (e.g., main sequence stars of spectral type F and later), it seems reasonable that these stars also have strong magnetic fields driven by the convective dynamo and a magnetically confined corona. Thus, the effect of magnetically coronal heating similar to the case of the solar coronal heating could offer a much better explanation of the observations than the acoustic wave heating. In fact, for the case of the solar corona there is the very close relationship between magnetic activity and coronal

X-ray emissions. Based on the simultaneous soft X-ray data from the Yohkoh soft X-ray telescope and magnetic field data from the Kitt Peak Solar Observatory during the period of 1991–2001, Benevolenskaya et al. (2002) investigated the correlation between the soft X-ray intensity of coronal loops and their photospheric magnetic flux and the solar activity cycle variation of this correlation. Their results are shown in Fig. 7.3, where panel (a) is for the declining period of the solar activity cycle (from 1991 November 11 to 1996 July 25) and (b) for the rising period of the solar activity cycle (from 1996 June 28 to 2001 March 13). In Fig. 7.3, different subperiods are marked by different colors, which are coded in terms of Carrington rotation (CR) numbers (Benevolenskaya et al. 2002).

From Fig. 7.3, one can clearly find a very evident power-law correlation between the soft X-ray intensity of coronal loops (I_X) and their photospheric magnetic field ($B_\parallel$), which has a very high correlation coefficient $\simeq$0.9 and a power-law index, on average, close to 2, that is, $I_X \propto B_\parallel^2$ (Benevolenskaya et al. 2002). This implies that the energy of heating the coronal loops is transported by AWs from the photosphere into the corona along these loops, in which the loop magnetic field plays a role of magnetic channel for AWs propagating. The AWs of propagating into the corona are transferred into small-scale KAWs via the turbulent cascade or the resonant mode conversion, and then the KAWs eventually are dissipated inhomogeneously to heat coronal plasmas and to compensate their radiative cooling, as shown in Sect. 6.4. In addition, the power-law relations exhibited in Fig. 7.3a, b both have an evident spectral breakpoint around $\log|B_\parallel| \sim 1.5$, implying that the magnetic dependence of the X-ray radiance is significantly steeper for the weak magnetic field than the strong field. This different dependence could possibly be caused by the drastic decrease of the mean level of X-ray intensity during the low-activity phases of the solar cycle.

In particular, the correlationship between the X-ray radiative luminosity and the magnetic flux can be extended smoothly to the case of stellar X-ray radiations and hence possibly is a universal relationship. Figure 7.4 shows the power-law relation between the X-ray spectral radiance (L_X) and the magnetic flux Φ for different cases of the Sun as well as stellar objects, including solar quiet regions (dots), active regions (diamonds), X-ray bright points (squares), and solar disk averages (pluses) from the Sun as well as types G, K, and M dwarfs (crosses) and T Tauri stars (circles) from active stars, where the solid line represents the power-law approximation $L_X \propto \Phi^{1.15}$ of combined data set (Pevtsov et al. 2003). Among different subsets, the power-law index α in the form of $L_X \propto \Phi^\alpha$ can vary between $\alpha \sim 1$ and $\alpha \sim 2$, and the L_X-Φ dependence for the subset of solar disk averages also shows a similar spectral breakpoint (i.e., "knee" in Fig. 7.4). However, the change in the instrumental response also is a possible reason for this spectral breakpoint because different sections of the SXT entrance filters failed over the period from 1992 November 13 to 1998 January 24, which could increase the response of the telescope at long wavelengths (i.e., lower temperatures), as pointed out by Pevtsov et al. (2003). In any case, on average, we can find that the L_X-Φ dependence of these stars is very well consistent with the linear extrapolation from the solar results in a wide range over 12 orders of magnitude. This indicates again that the same mechanisms for the solar coronal heating also heat the coronae of these stars.

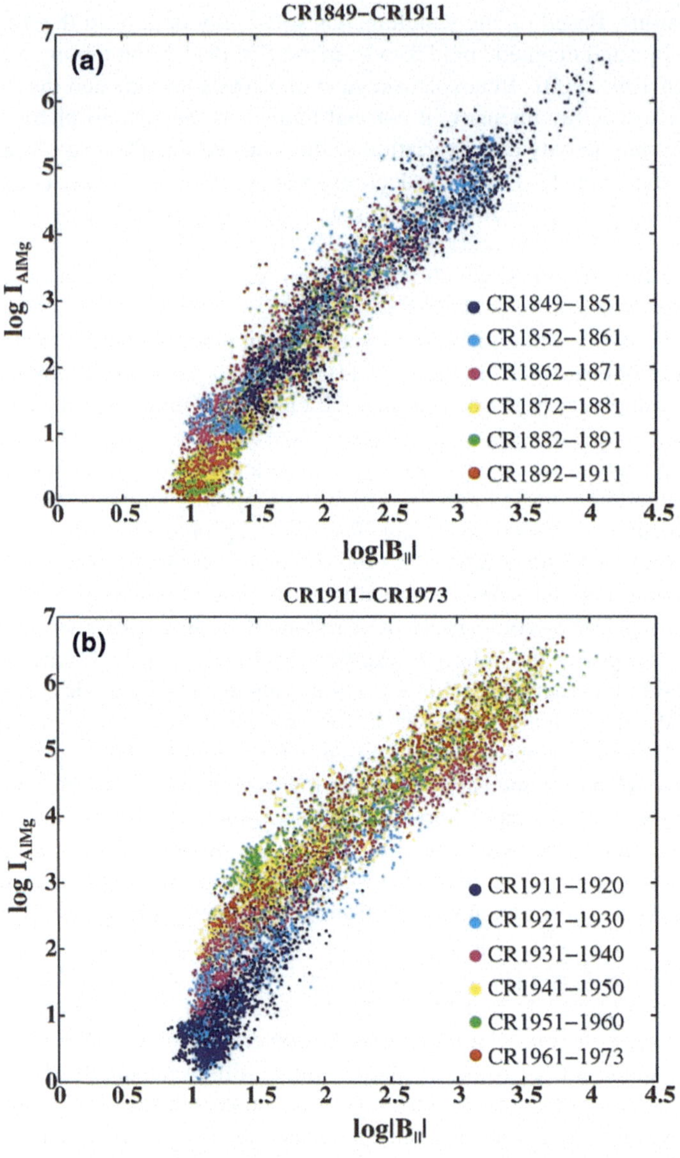

Fig. 7.3 Scatter plots of the soft X-ray intensity from SXT (Yohkoh) data in the AlMg filter as a function of the magnetic flux (from Kitt Peak Observatory) in the natural logarithmic scale for the latitudinal zone between +55° and −55° for two periods: **a** 1991 November 11 to 1996 July 25 for the declining period of the solar activity cycle and **b** 1996 June 28 to 2001 March 13 for the rising period of the solar activity cycle. Different subperiods are marked by different colors and the color-coding in terms of Carrington rotation (CR) numbers is shown in the figures (from Benevolenskaya et al. 2002)

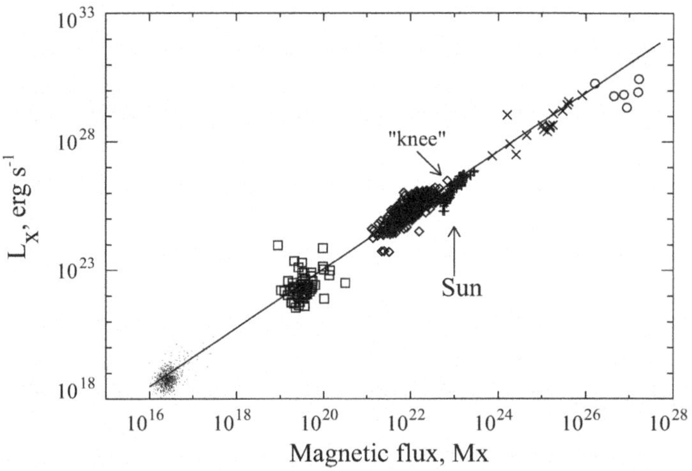

Fig. 7.4 X-ray spectral radiance L_X vs total unsigned magnetic flux for solar and stellar objects. Dots: Quiet Sun. Squares: X-ray bright points. Diamonds: Solar active regions. Pluses: Solar disk averages. Crosses: G, K, and M dwarfs. Circles: T Tauri stars. Solid line: Power-law approximation $L_X \propto \Phi^{1.15}$ of combined data set (from Pevtsov et al., 2003, ©AAS reproduced with permission)

On the other hand, the hypothesis that the X-ray stars have a magnetically confined corona similar with the solar corona also is further demonstrated by the observations of numerous stellar flares (Pallavicini et al. 1990; Haisch et al. 1991), because the stellar as well as solar flares are extensively believed to be driven by the eruptive release of magnetic energy via the magnetic reconnection or electric current instability. Flare stars constitute about 10% of the stars in the Galaxy. Flares, solar and stellar flares, usually are understood as a physical process, in which the catastrophic releasing magnetic energy leads to particle acceleration and electromagnetic radiation, although the magnetic energy release and conversion have never been directly observed. Similar to solar flares, stellar flares produce radiations in multi-waveband ranges from meter-wave range to γ rays. The most important wavelength regions, however, from which we have learned diagnostically on stellar coronae, include the radio (decimetre to centimetre) range and the X-ray domain. The former is produced by the magnetic cyclotron or synchrotron radiation of energetic electrons in local magnetic field and hence sensitive to accelerated electrons in magnetic fields. While the latter is attributed to the Coulomb bremsstrahlung of thermal or nonthermal energetic electrons and hence independent of the local magnetic field. The two wavebands, in general, both can be observed simultaneously and are related each other. For example, White et al. (1978) related a soft X-ray flare on HR 1099 with a simultaneous radio burst.

In particular, the luminosity ratio of the X-ray to radio radiations for stellar X-ray stars seems also to exhibit a very well consistency with that for solar flares. Benz and Güdel (1994) compared the X-ray to radio luminosity ratio of solar flares with that of various active late-type stars, as shown in Fig. 7.5. From Fig. 7.5, the solar

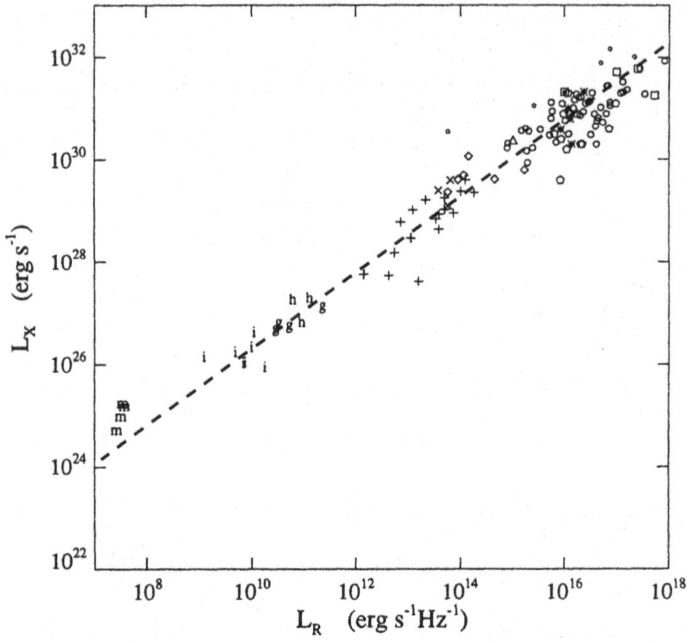

Fig. 7.5 Comparison of soft X-ray (L_X) versus microwave (L_R) luminosities of solar flares and active corona stars of different classes, where the symbols are, respectively, m: solar microflares; i: intermediate impulsive solar flares; h: gradual solar flares with dominating large impulsive phase; g: pure gradual solar flares; $+$: dM(e) stars; $\times$: dK(e) stars; $\Diamond$: BY Dra binaries; $\bigcirc$: RS CVn binaries; $\circ$ RS CVn binaries with two giants; $\triangle$: AB Dor; $*$: Algol-type binaries; $\square$: FK Com stars; *pentagon*: post-T Tau stars (from Benz and Güdel 1994)

flares, except for the four microflares denoted by m, can well connect smoothly and linearly to the active stellar coronae, covering about eight orders of magnitude in the $L_X - L_R$ plot. The four microflares were observed on same day with a narrow microwave beam by the 100-m radio telescope at Effelsberg, which might not be centered on the source and lead to the underestimate of their radio luminosity (Benz and Güdel 1994). Other one possibility for the relative weakness of microwave in microflares is because of a lower microwave productivity of microflares due to their lower energetic level. Apart from the microflares, the X-ray to radio luminosity ratio of solar flares and that of different type active stars is compatible with a linear correlation, that is, $L_R \propto L_X$ (Güdel and Benz 1993; Benz and Güdel 1994).

In fact, the similar luminosity correlationship between X-ray and radio radiations can be well satisfied in a much larger scale range. Falcke et al. (2004) in detail analyzed and compared X-ray and radio data for some compact sources, mainly including the candidates of stellar-mass and supermassive black holes, for example, XRBs (X-ray binaries, possible stellar-mass black holes) and AGN (active galactic nuclei, possible supermassive black holes). The standard model for these compact sources is the so-called accretion disk-jet system, in which the strong gravity of the central black hole can attract very efficiently media from surrounding environment and form an accretion disk at a Kepler-like motion around the black hole and a large jet originating from the central region and extending to a far distance along the axis of the accretion disk. For the disk-jet system with a low accreting rate lower than the critical accreting rate ($\sim$a few percent of the Eddington accreting rate, Maccarone 2003) the radiation from the jet can possibly dominate that from the disk. When the axis directs the line of sight, in particular, the jet appears a compact source, which is optically thin and thick for X-ray and radio emissions, respectively.

Figure 7.6 shows the correlation between the radio (L_R) and X-ray (L_X) luminosities for XRBs and AGN, where in order to remove the dependence of the luminosity on the mass (Falcke and Biermann 1995) and the Doppler boosting effect (Ghisellini et al. 1993), the proper scaling corrections have been made (Falcke et al. 2004). For Sgr A* (the Galactic central black hole with a middle mass $\sim 10^6 M_\odot$, Schödel et al. 2002), the quiescent and flare luminosities both (Baganoff et al. 2001) are presented in Fig. 7.6. Other samples in Fig. 7.6 are GX339-4: a representative XRB with a mass $\sim 6M_\odot$ and hence a typical candidate of stellar-mass black hole (Corbel et al. 2003; Gallo et al. 2003); LLAGN: some low-luminosity AGN (Nagar et al. 2000; Terashima and Wilson 2003); FRI: FR I radio galaxies (Spinrad et al. 1985; Chiaberge et al. 1999); and XBL and RBL: X-ray and radio selected BL Lac objects (Sambruna et al. 1996). These samples include low mass stellar-order black holes ($\sim 6M_\odot$), middling mass galactic-order black holes ($\sim 10^6 M_\odot$), and supermassive AGN-order black holes ($\sim 10^9 M_\odot$) and their distributions in the X-ray and radio luminosities both cover very wide scale ranges over about 20 and 15 orders of magnitude, respectively. The results presented in Fig. 7.6 show that for all these different type sources, their radio and X-ray luminosities can be unified smoothly and fall on a common $L_R - L_X$ correlation. This suggests that the radio and X-ray emissions have not only a common driving energy source (i.e., the gravitational energy) but also should have a similar producing mechanism for these different types of objects. In particular, the magnetic field, which is one of the most essential elements in the disk-jet model, can play an important role in the relevant energy transport and transform processes from the initial gravitational energy released by accreting flows around the black hole into the eventual radiative energy in X-ray and radio wavebands.

Like the magnetically confined coronae surrounding normal stars, in fact, a magnetically confined corona possibly is formed around a black hole. For the case of a high accreting rate above the critical accreting rate (Maccarone 2003), the radiation originating from the accretion disk can possibly dominate that from the jet (Falcke et al. 2004). In general, the accretion-disk radiation consists of two components, one is a soft blackbody-like X-ray component at low energies (<10 keV) and other one is a hard power-law-like γ-ray component at high energies up to several hundreds keV (Tanaka and Lewin 1995). The low-energy X-ray component is usually attributed to the emission from an optically thick, geometrically thin cold accretion disk, and the high-energy γ-ray component is attributed to an optically thin, geometrically thick hot corona in either a plane parallel to the disk or with a spherical geometry above the disk (Shakura and Sunyaev 1973; Liang 1998). Motivated by the solar atmospheric structure, Zhang et al. (2000) proposed a three-layered atmospheric structure for the

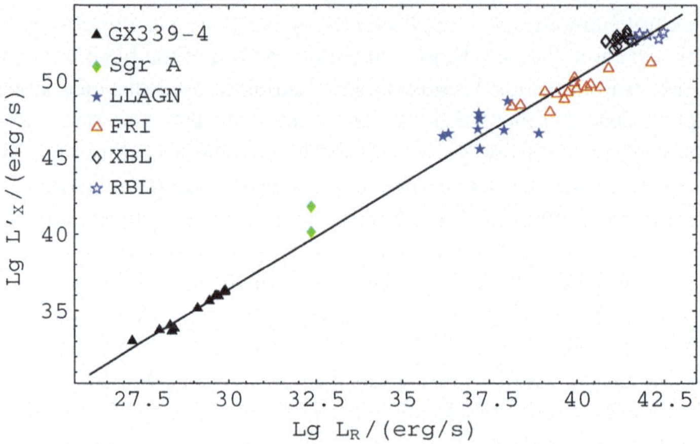

Fig. 7.6 Radio and X-ray correlation for XRBs and AGN, where the samples are: $GX339-4$: a XRB with a mass $\sim 6M_{\odot}$ as a typical candidate of stellar-mass black hole; $Sgr\ A^*$: the Galactic central black hole with a mass $\sim 10^6 M_{\odot}$; $LLAGN$: low-luminosity AGN; FRI: FR I radio galaxies; XBL: X-ray selected BL Lac objects; RBL: radio selected BL Lac objects (from Falcke et al. 2004)

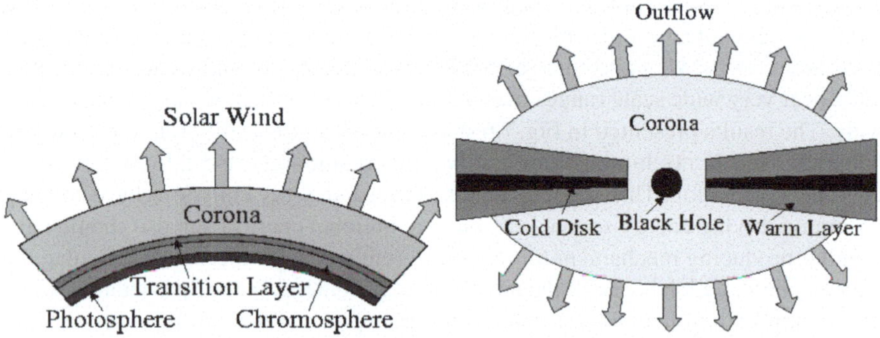

Fig. 7.7 Schematic diagrams of the solar atmosphere and accretion disk structure. The temperatures in the solar atmosphere are approximately: 6×10^3 K (photosphere), 3×10^4 K (chromosphere), and 2×10^6 K (corona), respectively. For the black hole disk atmosphere, the corresponding temperatures are approximately 500 times higher: 3×10^6 K (cold disk), 1.5×10^7 K (warm layer), and 1×10^9 K (hot corona), respectively (from Zhang et al. 2000)

accretion disks around a black hole. In their model, above the cold and optically thick dense disk with a temperature 0.2–0.5 keV, there are a warm layer with a temperature of 1.0–1.5 keV and an optical depth around 10 and a much hotter, optically thin tenuous and extensive corona above the warm layer, with a temperature of 100 keV or higher and an optical depth around unity, which are comparable with the solar photosphere, chromosphere, and corona, respectively, as shown in Fig. 7.7.

In particular, Zhang et al. (2000) suggested that the structural similarity between the accretion disk atmosphere and the solar atmosphere implies that similar physical processes may be operating in these different systems. For instance, the empirically-invoked viscosity for the disks might have originated from the same dynamo processes operating on the sun (Hawley et al. 1999). Like the solar photospheric convective turbulence, the accreting flow turbulence in the cold disk can cause enhanced AWs propagating upwards through the warm layer into the hot corona, and eventually provide energy sources for the heating of the hot corona. The differential rotation of accreting flows in the cold disk may twist magnetic fields and lead to the formation of magnetic loops, and then these magnetic loops emerge into the hot corona and trigger magnetic flares or "shoots" above the accretion disk, which may be responsible for the observed X-ray variability in compact sources, such as XRBs and AGN (Haardt et al. 1997; Di Matteo et al. 1999). For example, the magnetic reconnection model, which is widely believed to be a possible triggering mechanism for solar flares and coronal mass ejections, was used by Dai et al. (2006) to explain the X-ray re-brightening in the X-ray afterglow of a short γ-ray burst, which can be caused by a differentially rotating young neutron star formed from the merger of binary neutron stars.

In addition, the fact that the temperatures of the three layers are higher in the accretion disks by an approximately same factor $\sim$500 than the corresponding layers in the sun also indicates that the similar magnetic activities in the accretion disks are responsible for powering the upper atmosphere, especially coronal activities in both cases (Zhang et al. 2000). In addition, by analogy, the outflows (sometimes the jets also are taken as "collimated" outflows) also can originate from the coronal activities above the accretion disk, just like the solar wind, including coronal mass ejections, is driven out by strong solar coronal activities. It is worth noting that disk coronae powered by magnetic flares are also believed to exist in the accretion disks around supermassive black holes (Haardt et al. 1997). Therefore, it is plausible conjecture that similar physical processes in the solar atmosphere also may work in disk-jet systems with different properties and scales, such as in the Galactic center and AGN.

The Sun is a natural astrophysical laboratory because not only this nearest star can provide us the clearest observations and the most detailed information, but also those astrophysical phenomena occurring on the Sun can take place in many other astrophysical objects with enormously different scales. One of the most distinctive characteristics of plasma physics is the plasma scaling law, that is, plasmas in different environments and sizes can have similar dynamical processes and perform similar plasma physical phenomena when they have similar dimensionless scaling parameters, such as the plasma kinetic to magnetic pressure ratio β. Therefore, these stellar and accretion disk coronae may be made up of various solar-like features, and hence our knowledge, which has been established to describe the plasma physical processes in the active solar corona, can be greatly helpful for us to understand these extrasolar coronal phenomena, including their heating as well as flares. In particular, the heating mechanisms for the solar corona by KAWs, which are presented in Chap. 6, can provide the most promising explanation for the heating of the stellar and accretion disk coronae.

7.3 KAW Heating of Galactic Halos and Intracluster Medium

Although overwhelmingly major constituents of the universe are made of plasmas, the gravity force seems still to have an overwhelmingly advantage and to enjoy a dominant position in the formation and evolution of the universe structures. At the present stage of cosmical evolution, our universe has a hierarchical structure, from planets and stars to galaxies and galaxy clusters, which all are local gravity centers in the universe. The gas clouds out of which galaxies and galaxy clusters form were heated by the energy released during their initial gravitational collapse. Some of the gas then cooled to form the stellar objects readily observed today. In the case of ordinary galaxies much of the gas cooled rapidly, but in massive galaxies and clusters the gas cooled more slowly and a quasi-hydrostatic atmosphere formed. The mass of uncooled hot gas exceeds that in visible stars in clusters of galaxies. The temperature of the atmosphere is close to the local virial temperature (typically $T(r) = G\mu m_H M(r)/r \sim 10^7$–$10^8$ K, where G is the gravitational constant, m_H is the hydrogen mass, $\mu \simeq 0.61$ is the mean molecular weight of the ionized plasma, $M(r)$ is the total mass within the local radius r) and is directly observable only in the X-ray waveband as extended X-ray sources with X-ray luminosities of $L_x \sim 10^{43}$–10^{45} ergs/s. This hot atmosphere at a high temperature $\sim 10^7$–10^8 K, called intracluster medium (ICM), consists of very tenuous plasmas and fills the space between the galaxies within the cluster.

As the largest clearly define objects in the universe, the clusters of galaxies have sizes of several Mpc and masses up to $10^{15} M_\odot$ and are recognizable in photographs as distinct concentrations of galaxies centered on one or more brightest cluster members. Correspondingly, the ICM also exhibits some interesting astrophysical superlatives, for example, the ICM is the hottest thermal equilibrium plasma that we can study in detail, their gravitational potentials give rise to the largest effect of light deflection with deflection angles exceeding half an arcmin and produce the most spectacular gravitational lensing effects (Hattori et al. 1999), the hot plasma cloud of the ICM casts the darkest shadows onto the cosmic microwave background through the Sunyaev-Zeldovich effect (Birkinshaw 1999), and the merger of galaxy clusters produces the largest energy release in the universe after the big bang itself with energies up to orders of 10^{63} erg (Feretti et al. 2002).

The Coma and Perseus clusters were among the first clusters to be identified as X-ray sources by the first X-ray satellite, the Uhuru satellite in the early 1970s (Giacconi et al. 1971; Gursky et al. 1971; Forman et al. 1972). By the mid-1970s, at least 40 clusters of galaxies were identified as extended and powerful X-ray sources (Gursky and Schwartz 1977). Figure 7.8 shows a composite image of the Coma cluster of galaxies, where an optical image from the Palomar Sky Survey showing the galaxy distribution of the Coma cluster is superposed in grey scale on top of an X-ray image from the ROSAT All-Sky Survey with X-ray brightness coded in red color. It can be clearly recognized that the X-ray image displays the cluster as one connected entity and there is an evident concentration center in its X-ray image, which is possibly

Fig. 7.8 A composite image of the Coma galaxy cluster with X-rays in the ROSAT All-Sky Survey (underlaying red color) and the optically visible galaxy distribution in the Palomar Sky Survey Image (galaxy and stellar images superposed in grey, from Böhringer and Werner (2010))

related to two supergiant galaxies lied in the center (as shown by the optical image in Fig. 7.8), the elliptical galaxy NGC4889 and the spiral galaxy NGC4874. This illustrates the fact that galaxy clusters indeed are well defined one of fundamental structures in the universe.

The plasma of ICM is very tenuous, with densities of 10^{-5}–10^{-1} cm^{-3} from the cluster outskirts to the densest regions of cool core clusters. In general, all created photons leave the ICM plasma due to its low density. Thus no radiative transfer calculation is necessary for the interpretation of the X-ray spectra of the ICM. This means in particular that the spectrum we observe from a galaxy cluster provides the information of the entire ICM, which is different from stellar spectra that provide information on merely a very thin layer on the stellar surface, that is, the stellar photosphere. This is one of reasons for the cluster X-ray spectra can be so informative and straightforward to interpret. Therefore, although having large sizes, the ICM can, in general, be treated as an optically thin coronal plasma in ionization equilibrium and their radiations are produced mainly by thermal bremsstrahlung emission of hot electrons in the X-ray band (Sarazin 1988). The thermal origin for the X-ray emission has been confirmed in the Coma, Perseus, and Virgo clusters with the detection of the collisionally excited, 6–7 keV Fe-K emission feature by the Ariel 5 (Mitchell et al. 1976) and OSO-8 observatories (Serlemitsos et al. 1977).

At temperatures below $\sim 3 \times 10^7$ K, however, the X-ray emission is increasingly dominated by the recombination lines of iron, oxygen, silicon, and other elements (Sarazin 1988). The emission from these heavy elements, in particular, the iron K lines at 6–7 keV and the iron L lines below 1 keV in cooler plasmas, significantly alters the shape of the spectrum. This and the exponential decline in emission at high energies permit the temperature and metallicity of the ICM to be measured accurately with modern X-ray telescopes (McNamara and Nulsen 2007). In general, the X-ray telescopes used to study ICM and clusters have commonly employed proportional counters or CCDs (charge coupled devices) as detectors, which are sensitive to photons with energies spanning the range 0.1–10 keV, and hence can well match to the thermal radiation from the ICM. Thus, the ICM density can be estimated from the X-ray surface brightness of the ICM because the radiation power of the thermal bremsstrahlung emission due to Coulomb collisions is proportional to the square of the density (McNamara and Nulsen 2007).

The surface brightness of the ICM (I_x) declines with the distance from the center (r) approximately as $I_x \propto r^{-3}$ at large distances. Despite the rapid decline in I_x, the relatively low X-ray background permits X-ray emission to be traced to very large radii, making it an excellent probe of gas temperature, metallicity, and mass throughout much of the volume of a cluster. Surface brightness profiles have traditionally been characterized using the isothermal "β-profile", that is,

$$I_x \propto \left(1 + \frac{r^2}{r_c^2}\right)^{-3\beta+1/2}, \tag{7.1}$$

where r_c is the radius of the central core of the ICM, the parameter β is the ratio of the energy per unit mass in galaxies to that in the ICM, and generally $\beta \simeq 2/3$ for relaxed bright clusters (Cavaliere and Fusco-Femiano 1976; Branduardi-Raymont et al. 1981; Forman and Jones 1982). This model can provide a reasonably good fit to the surface brightness profiles of clusters at intermediate radii. In the central regions of some clusters, where the temperature declines and the density rises rapidly, the fit is poorer. At large radii, the observed surface brightness profiles steepen below the β-profile (Vikhlinin et al. 2006). For an isothermal ICM (or with temperature $T > 2$ keV), the electron density profile for the β-model ICM

$$n_e(r) = n_0 \left(1 + \frac{r^2}{r_c^2}\right)^{-3\beta/2}, \tag{7.2}$$

where n_0 is the central electron density. However, more generally, the density and temperature profiles are determined by "deprojection" (Fabian et al. 1981). In general, the ICM as the atmosphere of the central dominant galaxies of the galaxy cluster may be represented as a series of shells of uniform density, temperature, and composition and its properties can be determined by fitting X-ray spectra extracted from corresponding annular regions (Ettori 2000; Pointecouteau et al. 2004).

Faraday rotation measurements of background radio galaxies and radio galaxies within clusters have revealed that the ICM is threaded with magnetic fields (Carilli and Taylor 2002; Kronberg 2003; Govoni and Feretti 2004; Vallée 2004). The Coma cluster first was detected to have magnetic fields of orders of a few μG (Kronberg 2003). In the cores of clusters with cooling flows, magnetic field strengths could possibly reach an order of tens μG inferred from Faraday rotation measures (Clarke et al. 2001). These magnetic fields in the ICM can possibly be formed by the extrapolation and extension of radio galaxies and/or quasars in the galaxy cluster (Clarke et al. 2001; Furlanetto and Loeb 2001; Kronberg et al. 2001) or by the primordial fields that have been amplified over time by ICM turbulence (Carilli and Taylor 2002).

Except occasionally in the inner several kpc or so, in general, the magnetic fields seem not to be dynamically important in the ICM because the ratio of kinetic to magnetic pressure (i.e., the plasma β) is typically tens, much larger than one. However, the magnetic fields, especially their structures can significantly influence, even entirely control the particle transport processes in the ICM because the mean free path of the Coulomb collisions still is much larger than the kinetic scales of particles (Tribble 1989; Govoni 2006). For example, Vikhlinin et al. (2001) observed remnants with size $\sim$3 kpc of the ISM in the dominant galaxies (i.e., NGC4874 and NGC4889) at the center of the Coma cluster and found that the thermal conduction must be suppressed there by a factor of 30–100 from the classical value (Braginskii 1965), in order for these remnants to survive in thermal evaporation.

The temperature of ICM, T_X, can be associated with the spectral energy distribution of the thermal bremsstrahlung for the Coulomb collisions between electrons and ions, which may be given by (e.g., Gronenschild and Mewe 1978)

$$F(\nu, T_X) = C_0 \frac{n_e n_i Z^2}{\sqrt{T_X}} g_{ff}(Z, T_X, \nu) \exp\left(-h\nu/T_X\right), \tag{7.3}$$

where C_0 is a constant of consisting of the basic physical constants, n_e and n_i are the electron and ion density, respectively, Z is the charge number of the ion, and g_{ff} is the gaunt factor, a quantity close to unity. The most evident spectral signature of T_X is the sharp cut-off of the emission spectrum at high energies, due to the exponential term with the argument $-h\nu/T_X$. However, the precise measurement of the temperature distribution of the ICM requires the X-ray telescope with sufficient spectral and angular resolution. The advanced X-ray observatories Chandra and XMM-Newton launched in 1999 now can routinely provide localized measurements of the ICM temperature (Vikhlinin et al. 2006; Pratt et al. 2007). The observed results show that there are two types of the ICM in temperature distribution: one is the cooling core ICM with a dense core, in which the ICM has a temperature profile decreasing toward the center in the core regions and other one is non-cooling core ICM, which has a moderate central densities (typically below 10^{-2} cm^{-3}) and a flat or even slightly increasing temperature profile toward the cluster center.

In general, the non-cooling core ICM has usually central cooling times exceeding the Hubble time. For the cooling core ICM with high densities in their cores, however, they usually have shorter central cooling times, typically one or two orders of magnitude smaller than the Hubble time, because the higher central density by factors of 10 or more leads to the surface brightness of the ICM central core rising dramatically above a β-model, by factors of up to 100, and reaching values of 10^{45} ergs/s, more than 10% of the clusters total luminosity. If this radiative loss is uncompensated by heating, the central dense hot plasma would lose its kinetic energy and cool on a time scale of $t_c = (\gamma - 1)^{-1}\, p/n_e n_H \Lambda(T) < 10^9$ year (Silk 1976; Cowie and Binney 1977; Fabian and Nulsen 1977; Mathews and Bregman 1978), where p is the pressure, $\Lambda(T)$ is the cooling function, and γ is the ratio of specific heats. As the central core radiates, it is compressed by the surrounding ICM hot plasmas and flows inward. Moreover, the cooling time further decreases as the density increases, so that the temperature of the central core drops rapidly to $<10^4$ K, and eventually, the cooled central core condenses on to the central dominant galaxy of the cluster. The condensing central core is replenished by hot ICM lying above, leading to a steady, long-lived, pressure-driven inward flow at a rate of up to 1000 $M_\odot$ per year (Fabian 1994).

However, the observed cooling rate seems significantly lower than the expectation of the above cooling flow model, in which the cooling flow is approximately steady and the power radiated from the steady flow equals the sum of the enthalpy carried into it and the gravitational energy dissipated within it. This discrepancy also implies that the cooling flow is not condensing at the predicted rates, and hence that radiation losses are being replenished. In particular, the advanced high-sensitive XMM-Newton observations failed to confirm the previous estimates strength of low-energy spectral lines, including the Fe L features, at $\sim$1 keV from cooling flow clusters (Cowie 1981; Tamura et al. 2001; Peterson et al. 2001, 2003). The failure to detect the low energy X-ray lines at the expected levels also indicates that the canonical cooling rates were overestimated by an order of magnitude or more (Peterson and Fabian 2006).

Since at the same time high angular resolution Chandra images showed signs of strong interaction of the central AGN with the ambient ICM in many cooling core regions (David et al. 2001; Nulsen et al. 2002; Fabian et al. 2005), the most probable solution to the absence of strong cooling was soon believed to be the heating of cooling core regions by the central AGN, which harbors at the center of the dominant galaxy of the cluster and is engined by the central engine of the AGN that consists a massive black hole and its accretion disk. As the change from the cooling flow paradigm to an AGN heating scenario seems to be now widely accepted, the interest has shifted to the question: how is the ICM actually heated by the interaction between the ICM and AGN?

One of the first best studied cases is that of the Perseus cluster with the central dominant galaxy NGC 1275, which, as the brightest X-ray source in the sky, has now been observed with the Chandra observatory for total 900 ks (1 ks$= 10^3$ s) deep exposure, as proposed by Fabian et al. (2005), providing the most detailed picture of the cooling core in cluster central region. Figure 7.9 shows a multi-color image of the

Fig. 7.9 A composite image of the central region of the Perseus cluster produced from the Chandra observations in three energy bands 0.3–1.2 (red), 1.2–2 (green), and 2–7 (blue) keV. An image smoothed on a scale of 10 arcsec (with 80% normalization) has been subtracted from the image to highlight regions of strong density contrast in order to bring out fainter features lost in the high-intensity range of raw images (from Fabian et al. 2006)

central region of the Perseus cluster from the Chandra deep exposure images in three energy bands 0.3–1.2 (red), 1.2–2 (green), and 2–7 (blue) keV (Fabian et al. 2006). In Fig. 7.9, the central bright part of the image shows two inner cavities containing the active radio lobes and two outer cavities, called ghost cavities, which were initially interpreted as aging radio relics that had broken free from the jets originating AGN of NGC 1275 and had risen 20–30 kpc into the ICM of the cluster (McNamara et al. 2001; Fabian et al. 2002), and later known they are filled with low-frequency radio emission and may be connected by tunnels back to the AGN (Clarke et al. 2005; Wise et al. 2007). These cavities are surrounded by a series of nearly concentric "ripples", which can be more clearly brought out by the unsharp masking processing of the image (Fabian et al. 2006). A more detailed analysis of the density and temperature variations across the ripples implies typical pressure variations associated with the ripples with amplitudes of about $\pm(5-10)$ %, which are interpreted as sound waves or very weak shock waves in the innermost region (Fabian et al. 2006). The blue structure to the north of the AGN is caused by absorption in the falling high-velocity system, projected at least 60 kpc in front of the AGN of NGC1275 (Gillmon et al. 2004).

Fig. 7.10 Chandra X-ray image with 500 ks exposure for M87 in the energy band 0.5–1.0 keV of showing structures in the ICM associated with the AGN outburst. Several small cavities and two large eastern and southwestern arms with a series of filaments and loop-like structures are visible (from Forman et al. 2007, ©AAS reproduced with permission)

The supergiant elliptical galaxy M87 (NGC 4486) is one of the central dominant galaxies of the Virgo cluster. As the nearest (a distance ~18 Mpc, or the redshift $z \simeq 0.0042$) cooling core cluster and the strong radio galaxy (Virgo A), also M87 is one of the most suitable systems for studying the interaction between an AGN and the ICM surrounding it through relativistic radio jets (Böhringer et al. 1995; Churazov et al. 2001), and hence has been extensively investigated. The central engine of M87 has a supermassive black hole with a mass $3.2 \times 10^9 M_\odot$ (Harms et al. 1994; Ford et al. 1994; Macchetto et al. 1997) and its well-studied jet (Sparks et al. 1996; Perlman et al. 2001; Marshall et al. 2002; Harris et al. 2003). In X-rays, M87 is the second brightest extragalactic source (after the Perseus cluster), and the X-ray emission is dominated by thermal radiation from the ICM at temperature ~2 keV (Gorenstein et al. 1977; Fabricant and Gorenstein 1983; Böhringer et al. 2001; Matsushita et al. 2002; Belsole et al. 2001; Molendi 2002). By use of a Chandra observation with 40 ks exposure, Young et al. (2002) reported X-ray cavities and edges in the surface brightness profile.

Figure 7.10 shows the X-ray image of M87 by the Chandra with deep exposure of 500 ks in the soft energy band of 0.5–1.0 keV, in which a lot of complex structures are clearly visible in both two large-scale arms, especially the cavities and loop-like structures in the broader eastern arm and filaments in the long southwestern arm. These structures form a complex web of resolved loop-like and filamentary structures. By the combination of the observations of Chandra with 100 and 500 ks exposure (Forman et al. 2005, 2007) and XMM-Newton with 120 ks exposure (Simionescu et al. 2007, 2008), these complexes in M87 have been investigated more detailedly. The loop-like structures in the broad eastern arm can be interpreted as a series of cavities at different evolutionary stages as they rise in the atmosphere of M87. While the long southwestern arm appears to be composed of several intertwined filaments, which are very soft and are not apparent at energies above 2 keV (Forman et al. 2005, 2007).

By combining the hard emission (3.5–7.5 keV) observation, Forman et al. (2005; 2007) suggested further that the enhanced-emission edges or rings at $\sim$14 and $\sim$17 kpc were likely shock fronts rising into the atmosphere of M87, that is, the ICM of the Virgo cluster, which were possibly driven by the AGN outbursts that occurred $1\text{–}2 \times 10^7$ years ago. The temperature in the ring rises from 2.0 to 2.4 keV, implying the shock Mach number of $M \sim 1.2$ and the shock velocity $v \sim 880$ km/s. At the shock, the density jump is 1.33, which yields a Mach number of 1.22, consistent with that derived from the temperature jump too. Their results also show that outside the shock fronts there are two regions of lower pressure in the northeastern and in the southwestern, which are identified as radio lobes filled with relativistic plasma emerging from the central AGN, but not contributing to the X-ray emission.

The giant galaxy MS0735.6+7421, hosting a radio source 4C+74.13, also is a central dominant galaxy in a cooling core cluster, similar to M87, but at a much larger distance of $\sim$716 Mpc (the redshift $z \simeq 0.216$). McNamara et al. (2005) reported the discovery of giant cavities of a diameter $\sim$200 kpc and shock fronts in the giant galaxy cluster MS0735.6+7421 caused by an interaction between a radio source and the ICM surrounding it. The average pressure surrounding the cavities is $p \simeq 6 \times 10^{-11}$ erg/cm^3. The work required to inflate each cavity against this pressure is $pV \sim 10^{61}$ erg, where V is the volume of the cavity. The enthalpy (free energy) of the cavities can approach $\sim$4pV per cavity, depending on the equation of state of the gas filling them, giving a value of about 8×10^{61} erg, exceeding that in M87 by more than four orders of magnitude.

Figure 7.11 presents the Hubble Space Telescope visual image of the MS0735.6+7421 cluster superposed with the Chandra X-ray image (blue) and a radio image from the Very Large Array at a frequency of 330 MHz (red). The X-ray image in Fig. 7.11 shows an enormous pair of cavities, each roughly 200 kpc in diameter, that are filled with radio emission. This indicates that the radio jets emanating from the AGN at the center core have been inflating the cavities for 10^8 years with an average power of $\sim \times 10^{46}$ erg/s since the AGN outbursted and the supermassive black hole in the AGN grew by at least $< 3 \times 10^8 M_\odot$ during the outburst. The cavities and radio source are bounded by a giant weak shock front, as contrast, the central dominant galaxy at the center of the optical image, one of the largest galaxies in the universe, is dwarfed by the giant cavities and the radio source.

Another example of giant-scale cavities and shocks with an order of magnitude similar to the MS0735.6+7421 cluster is the Hydra A at a distance of the redshift $z = 0.055$. In the combined Chandra and XMM-Newton study of the central region of the galaxy cluster associated with the radio source Hydra A, a similar front diagnostics

Fig. 7.11 Hubble Space Telescope visual image of the MS0735.6+7421 cluster superposed with the Chandra X-ray image (blue) and a radio image from the Very Large Array at a frequency of 330 MHz (red) (from McNamara and Nulsen 2007)

has also been performed by Nulsen et al. (2005) and Simionescu et al. (2009), in which the shock feature has a much larger radial extent of 200–300 kpc, implying a much more powerful by about two orders of magnitude than that in M87. Then, a deep XMM-Newton observation provides higher photon statistics that allows a more detailed analysis of the temperature and density (Simionescu et al. 2009). Modeling the shock in three dimensions and fitting the projected model to the data implies a Mach number of the best fitting model of about 1.2–1.3 (Simionescu et al. 2009). While simple one-dimensional modeling of the evolution of the shock driven by the energy input of the radio lobes as well as detailed three-dimensional hydrodynamical simulations may provide further estimates for the age and the total energy of the driven shock, and the results lead to the age of $\sim 2 \times 10^7$ years and the energy of 10^{61} erg (Nulsen et al. 2005; Simionescu et al. 2009).

In galaxy clusters and groups, or galaxies with central AGN, in principle, cavities (or bubbles) produced by the interaction between the ICM or hot halo and the relativistic radio jets emanating from the AGN outbursts, may be very common structures in their X-ray images of the ICM, although they are very difficult to detect because of large distances (McNamara and Nulsen 2007; Böhringer and Werner 2010). These cavities vary enormously in size, from diameters smaller than 1 kpc like those in

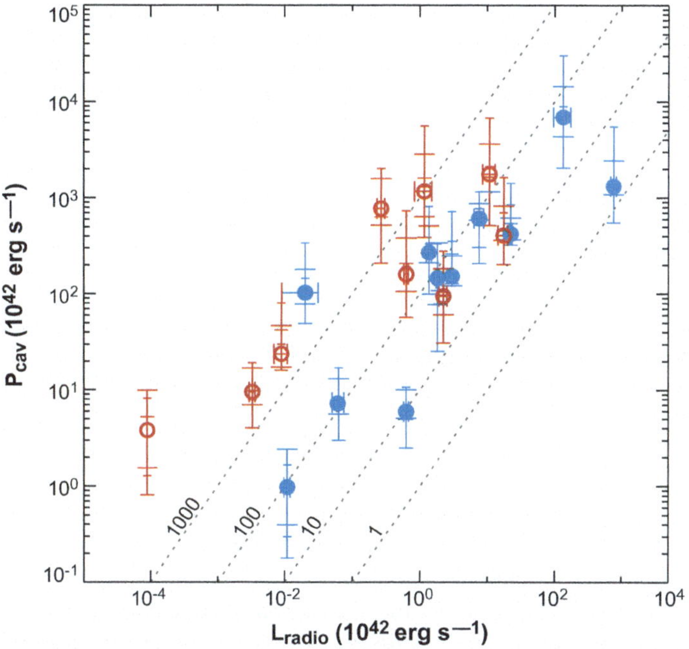

Fig. 7.12 Total radio luminosity between 10 MHz and 10 GHz plotted against jet power ($4pV/t_{buoy}$). Open red symbols represent ghost cavities. Solid blue symbols represent radio filled cavities. The diagonal lines represent ratios of constant jet power to radio synchrotron power. Jet power correlates with synchrotron power but with a large scatter in their ratio. Radio sources in cooling flows are dominated by mechanical power. The radio measurements were made with the Very Large Array telescope (from McNamara and Nulsen 2007)

M87 (Forman et al. 2005) to diameters approaching 200 kpc in the MS0735.6+7421 and Hydra A clusters (McNamara et al. 2005; Nulsen et al. 2005; Wise et al. 2007). They are often surrounded by a series of complex structures, such as belts (Smith et al. 2002), arms (Young et al. 2002; Forman et al. 2005), sheets (Fabian et al. 2006), shock fronts, and filaments, perhaps, by magnetic fields threaded along their lengths (Nipoti and Binney 2004). At present, this wealth of structure is not well understood. The cavities in these interaction systems occupy between 5% and 10% of the volume within 300 kpc, giving the hot ICM a Swiss cheese-like topology (Wise et al. 2007). Evidently, the cavities are close to being in pressure balance with the surrounding ICM or halo. The cool rims are probably composed of displaced gas dragged outward from the center by the buoyant cavities (Blanton et al. 2001; Churazov et al. 2001; Reynolds et al. 2001).

The work required to inflate the cavities against the surrounding pressure is roughly $pV \sim 10^{55}$ erg in giant elliptic galaxies (Finoguenov and Jones 2001) and upward of $pV \sim 10^{61}$ erg in rich clusters (Rafferty et al. 2006). The total energy needed to create a cavity is the sum of its internal (thermal) energy, E, and the work required to inflate it, i.e., its enthalpy,

$$H = E + pV = \frac{\gamma}{1-\gamma} pV, \tag{7.4}$$

where the ratio of specific heats (i.e. the adiabatic exponent) $\gamma = 4/3$ for a relativistic gas, $5/3$ for a non-relativistic gas, and 2 for the case dominated by magnetic field, and the corresponding enthalpy $H = 4pV$, $2.5pV$, and $2pV$, respectively. While the radio luminosity, L_{radio}, seems to be correlative with the cavity power, $P_{cav} = H/t_{buoy}$, but with a large scatter that is poorly understood, where t_{buoy} is the bouyant time of the cavity from the center of the cluster to its present location, a reasonable estimate for the cavity age (Bîrzan et al. 2004; Dunn and Fabian 2006). Figure 7.12 shows that the total radio luminosity between 10 MHz and 10 GHz plotted against cavity power provided by the driven jets ($4pV/t_{buoy}$), where the open red symbols represent ghost cavities, the solid blue symbols represent radio filled cavities, and the diagonal lines represent ratios of constant jet power to radio synchrotron power. From Fig. 7.12, jet power correlates with synchrotron power but with a large scatter in their ratio (McNamara and Nulsen 2007).

Meanwhile, Fig. 7.12 also shows the total power of the cavity, in general, is much larger than the loss rate of its radio emission, implying that the majority of its carrying energy contributes to the interaction with the surrounding ICM or halo, including to heat the surrounding hot ICM or halo and to replenish their energy loss due to the X-ray emission. These interaction processes and active phenomena of hot atmospheres are strikingly similar with very familiar solar active and eruptive phenomena frequently occurring in the solar atmosphere, such as solar flares and coronal mass ejections, as well as the well-known solar coronal heating, but in much larger scales. However, our understanding of the physical process and mechanism that transforms the cavity energies into the kinetic energies of ICM particles is very poor. One of the most promising candidate mechanisms, perhaps, also is the heating caused by KAWs or KAW turbulence in the ICM, like the heating in the solar atmosphere and the solar wind. These KAWs and KAW turbulence may be excited or created by the cavities inflating into the ICM.

In fact, not only large-scale structures observed in ICM have their appearances similar with the active structures in the solar corona and the solar wind, but in larger scales, but also the ICM turbulence has an inertial spectrum similar to that measured in the solar wind turbulence as well as in the ISM turbulence. Such ICM turbulence has been observed to present in the Coma cluster of galaxies by Schuecker et al. (2004). By means of a spectral reduction of deep XMM-Newton observations, they investigated the stochastic turbulence and the corresponding turbulent spectrum based on the projected pressure map of the ICM in the central region of the Coma galaxy cluster (Schuecker et al. 2004). The Coma cluster has a very flat appearance, characterized by a very large core radius of the X-ray surface brightness of $r_c \sim 420$ kpc (Briel et al. 1992). This enables us to treat the configuration of the central region of the Coma cluster in the first approximation as a slab geometry, with corrections to

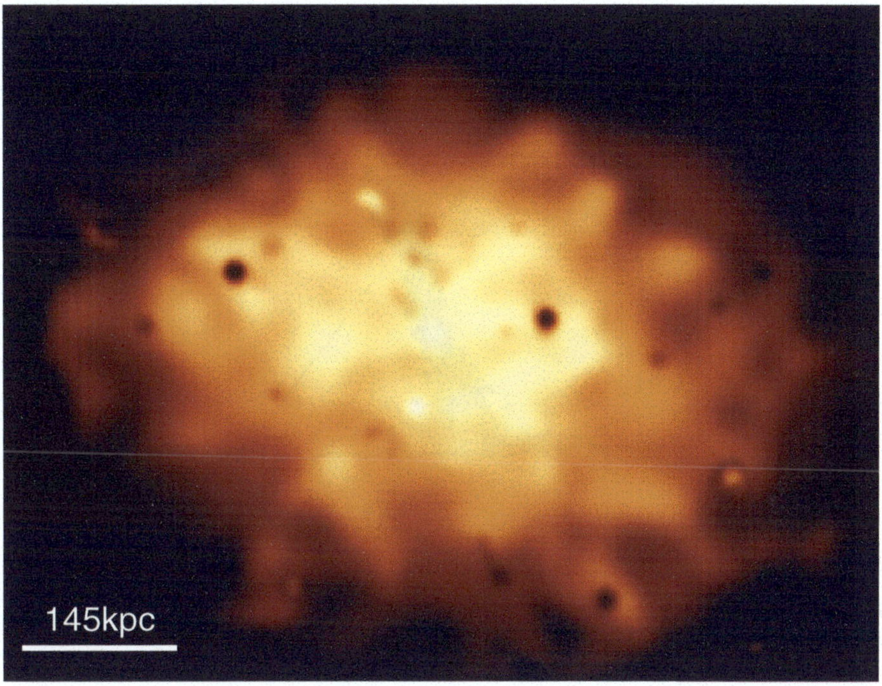

Fig. 7.13 Detailed overview of the projected pressure distribution of the ICM in the central region of the Coma cluster. The 145 kpc scale corresponds to the largest size of the turbulent eddies indicated by the pressure spectrum. The smallest turbulent eddies have scales of around 20 kpc. On smaller scales the number of photons used for the spectral analysis is too low for reliable pressure measurements (from Schuecker et al. 2004)

the power spectrum applied later. Moreover, more importantly, the core region of the Coma cluster has sufficiently high signal-to-noise X-ray spectra at a comparatively small angular resolution.

Figure 7.13 shows the map of the projected pressure distribution of the ICM in the central region of the Coma cluster, which was obtained by calculating the plasma density from the X-ray surface brightness (with an assumed depth of the ICM in the line of sight), then deriving the temperature by a spectral analysis of the data, and finally yielding the pressure by means of the ideal gas equation of state (Schuecker et al. 2004). The analysis of the fluctuation spectrum testing for a turbulent power law spectrum was performed by using the pressure rather than density or temperature fluctuations in order to avoid confusion with the static entropy fluctuation in pressure equilibrium, which may be produced by contact discontinuities. In Fig. 7.13, the turbulent structure of pressure fluctuation can be displayed clearly, in which the largest turbulent eddies associated with the power spectrum of pressure fluctuation rather than large-scale structures such as discontinuities or shocks, have a scale size $\sim$145 kpc, about a fraction of the size of the radius of the ICM core of the Coma

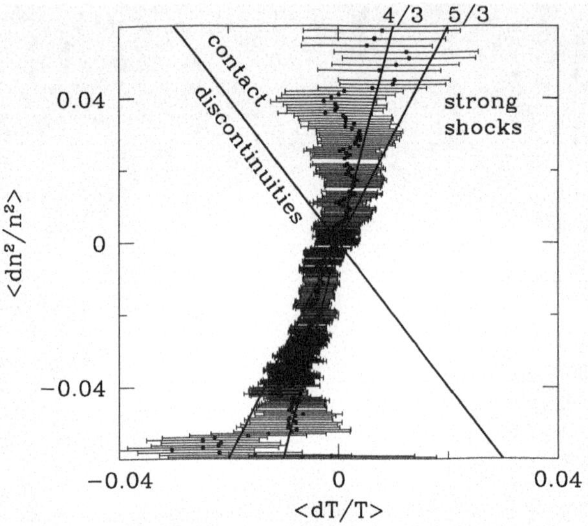

Fig. 7.14 Correlation between relative fluctuations of density squared (n^2) and temperature (T) with the 1-σ error bars. The two thick lines represent the adiabatic exponent $\gamma = 5/3$ (for an ideal gas) and $\gamma = 4/3$ (for the observed data of the Coma cluster), respectively, and another thick line shows the anticorrelation of density and pressure expected by contact discontinuities (from Schuecker et al. 2004)

cluster. While the sizes of the smallest turbulent eddies are about 20 kpc, which is restricted by the observational limit, because on smaller scales the number of detected photons used for the spectral analysis is too low for reliable pressure measurements (Schuecker et al. 2004).

Figure 7.14 presents the correlation of the relative fluctuation in squared density (n^2) and temperature (T) with the 1-σ error bars, where the thick line denoted by "4/3" represents the observed correlation for the ICM fluctuations in the central region of the Coma cluster, which has an adiabatic exponent $\gamma = 4/3$ and, as a comparison, the line denoted by "5/3" gives the correlation for an idea gas with the adiabatic exponent $\gamma = 5/3$. While, as a contrast, another thick line shows the anticorrelation of density and pressure, which is expected by contact discontinuities (Schuecker et al. 2004). The results presented in Fig. 7.14 show that the observed density fluctuation is dominated by the turbulent pressure fluctuation rather than contact discontinuities because the observed fluctuation obviously is inconsistent with the expectation of the anticorrelation between density and pressure fluctuation by contact discontinuities. By contrary, the positive correlation exponent ($\gamma = 4/3$) given by the observation is rather close to the adiabatic exponent of $\gamma = 5/3$ for an ideal gas, implying that adiabatic pressure fluctuations as expected for the ram pressure effects of turbulent motions dominate also the density and temperature fluctuation spectrum. Therefore, the small-scale structures in the density and temperature fluctuations with a positive correlation occupy different regions than the large-scale distribution of the ICM, such as contact discontinuities and strong shocks.

In order to find out whether or not such fluctuations are organized as in a turbulent regime, Schuecker et al. (2004) further investigated the power spectrum of the spatial pressure fluctuations. Figure 7.15 presents the shot-noise subtracted power spectrum on scales between 40 and 90 kpc with 1-σ error bars, from the Fourier analysis of the pressure fluctuation of the ICM in the central region of the Coma cluster. In Fig. 7.15, as a comparison, the dashed lines give the power-law spectra projected in an analogous way as the observed spectrum, which are labeled with the exponents $-7/3$, $-5/3$, $-3/3$, and $-1/3$, respectively, of the non-projected three-dimensional power spectra (Schuecker et al. 2004). From Fig. 7.15, the observations lay in between a power law exponent of $-5/3$ and $-7/3$. An exponent of $-7/3$ is the one expected for the pressure turbulence by the Kolmogorov-Obuchov inertial spectrum (Kolmogorov 1941; Oboukhov 1941). This indicates that an inertial-regime turbulent spectrum between the driving and dissipation scales can be established in the ICM of the Coma cluster, at least in its central region (Schuecker et al. 2004). The presence of this inertial-regime turbulent spectrum in cosmological scales range also raises some challenging problems. For example, what is the driving source for the cosmological turbulence? What is its association with the cosmic evolution? What is the role of magnetic fields in this turbulence? Is this ICM turbulence observed in the central region in the Coma cluster universal in other galaxy clusters and can extend to larger and smaller scales? Finally, is this ICM turbulence a part of the universal turbulence presented in Fig. 7.1, or its straight extension in larger scales?

In smaller scale ranges, the important information on the ICM turbulence can also be obtained by measuring the level of resonant scattering in emission lines observed in the ICM. The ICM is generally assumed to be optically thin. As pointed out by Gilfanov et al. (1987), measurements of the optical depth may give important information about the turbulent velocities in the hot plasma because the optical depth of the resonance line depends on the characteristic velocity of small-scale motions of plasmas in the ICM. Werner et al. (2009) used deprojected density and temperature profile for a giant elliptical galaxy, which is obtained by the Chandra satellite, to model the radial intensity profile of the strongest resonance lines (e.g., the Fe XVII line at 15.01 Å), accounting for the effects of resonant scattering, for different values of the characteristic turbulent velocity. Comparing the model to the data they found that the isotropic turbulent velocities on spatial scales smaller than $\sim$1 kpc are less than 100 km/s in the center of the galaxy (Werner et al. 2009).

In the previous section Sect. 7.1, by combining the in situ measurements of the satellites from the solar wind to the local ISM and the observations of interstellar scintillations of radio waves in ISM, a large power spectrum of cosmic plasma turbulence has been found from the kinetic scales of plasma particles to cosmological scales of 100 pc. Now this cosmic plasma turbulent spectrum further may extend to larger scales of 100 kpc in extragalactic ICM. Certainly, our knowledge about plasma physical processes and active phenomena occurring on the Sun, the astrophysical laboratory, can greatly help us further to understand physical mechanisms

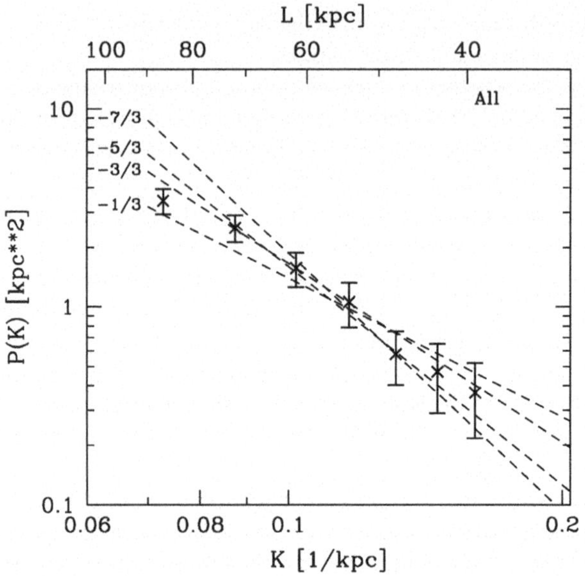

Fig. 7.15 Observed projected shot-noise subtracted power spectral densities of the pressure fluctuations of the ICM in the central region of the Coma cluster, with the 1-σ error bars, obtained by an average of the results from the 20×20 arcsec^2 (corresponding to 13.5×13.5 kpc^2) and 40×40 arcsec^2 (corresponding to 27×27 kpc^2) grids. As a comparison, dashed lines represent the power-law spectra with different spectra exponents of $-7/3$, $-5/3$, $-3/3$, and $-1/3$, projected in an analogous way as the observed spectrum (from Schuecker et al. 2004)

of plasma phenomena in other astrophysical objects, such as normal stellar coronae, accretion coronae around dense objects, galactic halos, ISM and ICM.

As shown in the Chaps. 5 and 6, KAWs or KAW turbulence are the most promising candidate responsible for the plasma heating in the solar corona and the solar wind. For the case of hot ICM or galactic halos, the results presented in Figs. 7.13, 7.14, and 7.15 (Schuecker et al. 2004) and discussions by Werner et al. (2009) show that a wide turbulent spectrum over a scale range from $\sim$100 kpc to $\sim$1 kpc can exist in ICM and has a power-law behavior similar to the big ISM turbulent spectrum covering the scale range from $\sim$0.1 kpc to the particle kinetic scales of $\sim$50 m presented in Fig. 7.1, which has been identified well to exist in the ISM of our galaxy (Lee and Lee 2019). Therefore, we can in reason expect that the ICM turbulent spectrum may extend into sub-kpc scale range, although the sub-kpc scale turbulence in ICM is still below resolvable observations. In particular, KAW turbulence is the natural result of the turbulent cascade of the AW turbulence towards small scales into the kinetic scales in ICM. Like the case of in the solar atmosphere and the solar wind, KAW turbulence also may be one of the most promising candidates responsible for the plasma heating of hot ICM and hot halos.

7.4 KAWs and Their Roles in Extended Radio Sources

Besides the ICM heating, another important problem in the plasma physics of extragalaxies and cluster of galaxies is the reacceleration of synchrotron-emitting relativistic electrons in powerful extragalactic extended radio sources (ERS), especially in extragalactic radio jets. In general, powerful extragalactic ERS consist of three parts: a central compact radio source located in near the central engine of the central dominant galaxy of the associated cluster, extended radio lobes with a size $\sim$ tens or hundreds of kpc at a large distance from the central engine $\sim$ Mpc, and narrow radio jets linking the radio lobes to the central engine. It has been widely believed that the central engine consists of a massive rotating black hole (called Kerr black hole) with a strong gravitational field and a steady accretion disk nearly at the Kepler motion around its rotating axis. A large amount of gravitational energy released by the accretion process can drive two oppositely-directed relativistic plasma beams along the rotating axis, although the specific driven mechanism still is an open problem. It is the relativistic plasma beams that supply the observed double radio jets and lobes with synchrotron-emitting relativistic electrons, in which the basic emission mechanism is the incoherent synchrotron emission produced by a nonthermal population of relativistic electrons gyrating in a magnetic field, based on a common fundamental postulate proposed by Alfvén and Herlofson (1950).

The early theories of ERS were to some extent analogies drawn with solar flares and supernova remnants, involving the ejection of magnetized plasmas associated with mysterious explosions driven by the central engine (van der Laan 1963; De Young and Axford 1967; Burbidge 1967; Sturrock and Feldman 1968; Mills and Sturrock 1970). One immediate difficulty with explosion models is the so-called adiabatic loss problem. In dynamical expansion, the internal energy associated with the relativistic electrons is converted into bulk kinetic energy, and the electron energies scale in proportion to the cube root of the background density. The extremely relativistic electrons with energies of GeV orders in the extended sources would need to have been unacceptably energetic in the initial central region if they had been transported adiabatically. This energy shortage problem was put on a firmer footing with the discovery of the hot spots in the strong ERS, Cygnus A, in which the radiative lifetimes of the emitting electrons are much shorter than the transferring times (Blandford and Rees 1974; Begelman et al. 1984). Therefore, it is very clear that some of the bulk kinetic energy must be channeled back into particle energy through some in situ acceleration mechanisms, such as shocks or turbulence, in the ERS.

It is a widely adopted view that the existent narrow radio jets link the powerful extended radio lobes at distances from the central engine directly to the central engine and channel energetic relativistic plasma beams from the central engine into the distant radio lobes, although, sometimes, these narrow radio jets may be invisible. Thus, the other basic problem in the physics of ERS is how to explain the highly collimated degree of the radio jets from their origin $\sim$0.1 pc to distances $\sim$100 kpc, which is a dynamic range $>$ 6 orders of magnitude, since it is these narrow radio jets that power the distant powerful extended radio lobes. Hydrodynamic models of narrow radio jets confined by external ICM pressure usually may be valid only in dynamic ranges of the order of one order of magnitude (Norman et al. 1982; Bicknell 1984). Further analyses show that the minimum pressure in a number of jets, including the famous M87 jet associated with the radio source Virgo A, appears to exceed the pressure in the surrounding ICM calculated from the X-ray emission by an order of magnitude, and while the continuous energy input required to produce the jet emission is incompatible with a freely expanding jet (Potash and Wardle 1980; Burns et al. 1983; Hardee 1985).

For a current-carrying jet that is very likely case, in fact, a natural explanation for the jet collimation is that the jet is self-confined by the Lorentz force $\mathbf{J} \times \mathbf{B}$, where $\mathbf{J}$ and $\mathbf{B}$ are the electric current density and magnetic field in the jet. This indicates that the confinement mechanism for the narrow radio jets is the magnetic pinch, that is, the classic Bennett pinch, in which the plasma pressure within a jet, p, may be balanced by the tension of an azimuthal magnetic field, B_ϕ, that is, $p = B_\phi^2/2\mu_0$. However, this requires the jet has an axial self-collimated current, I_C, as follows:

$$j_z = \frac{2B_\phi}{\mu_0 r} \Rightarrow I_C = 2\pi r\sqrt{\frac{2p}{\mu_0}} \simeq 7.7 \times 10^{16} r_{kpc}\sqrt{p_{-12}}\ (\text{A}), \tag{7.5}$$

where r_{kpc} is the jet radius in the unit of kpc, p_{-12} is the jet pressure in the unit of 10^{-12} dyn/cm^2. Therefore, a jet may be magnetically confined if it carries a net axial current $I_C \sim 10^{17} r_{kpc}\sqrt{p_{-12}}$ A (Benford 1978; Begelman et al. 1984). However, as pointed out by Eilek (1985), the strong electrostatic potential produced by the charge accumulation caused by the axial current in the end of the jet or the radio lobes would be large enough to stop the jet in only 10 to 10^3 years. This implies that a complete circuit as a electrodynamically coupling beam-return system must exist in the jet-lobe system and its surrounding environment, including the jets, lobes, the central engine, and the surrounding ICM.

In fact, the presence or injection of a relativistic fast electron beam emanating from the central engine in the jet inevitably causes a charge- and current-neutralizing return current in the jet-lobe system and sets up a beam-return current system, in which the return current can significantly compensate the charge and current carried by the fast electron beam (Hammer and Rostoker 1970; van den Oord 1990). The electrodynamics of the beam-return current system have been extensively investigated in the solar atmosphere (Spicer and Sudan 1984; Brown and Bingham 1984; van den Oord 1990; Xu et al. 2013; Chen et al. 2014, 2017). In the case of jet-lobe systems, the return current possibly presents on the boundary layer of the surrounding jet-lobe system (like the magnetopause current described in Sect. 4.1) or in between the axial current channels within the jet (like the auroral field-aligned current described in Sect. 3.2), and hence has a very diffuse distribution over a much larger area than the intense central axial current (Benford 1978, 1984; Elphic 1985). Whatever the specific details of the electrodynamics are, the beam-return current system can effectively excite AWs by the current instability, called the self-generated AWs (Wu et al. 2014; Chen et al. 2017). For a low-β plasma with the kinetic to magnetic pressure ratio (i.e., the plasma pressure parameter) $\beta \ll 1$, the saturation strength of the self-generated AW, B_{AW}, may be given approximately by (Wu et al. 2014, Chen et al. 2017)

$$B_{AW} \simeq \sqrt{\beta_0} B_0, \tag{7.6}$$

where the β_0 and B_0 are the background plasma pressure parameter and the background magnetic field strength. The local observations from solar and space plasmas show that this relation for the saturation level of AW turbulence seems to be a universal in a wide range, although it is obtained from the case of the self-generated AW excited by the current instability (Wu et al. 2014; Chen et al. 2017).

However, based on the simple model of cosmic ray propagation, the current intensity carried by a beam of charged particles can not exceed the so-called Alfvén limit, $I_A \simeq 17\beta_J\gamma_J$ kA, where $\beta_J \equiv v_J/c$ is the jet velocity in the unit of the light velocity c and $\gamma_J \equiv 1/\sqrt{1-\beta_J^2}$ is the relativistic factor (Alfvén 1939). In general, this Alfvén limit I_A is much less than the self-collimated current I_C in Eq. (7.5), unless the relativistic factor of the jet reaches $\gamma_J > 10^{13}$, an unreasonably high value. On the other hand, the in situ observations of magnetic structures of plasma ejections or flows by satellites in the solar wind and the planetary magnetospheres show that they usually have a rope-like structure with $B \sim B_z$ near the axis and $B \sim B_\phi$ far from the axis (Elphic 1985; Russell 1985), where B_z and B_ϕ are, respectively, the axial and the azimuthal components of the magnetic field. If taking account of the effect of an axial magnetic field B_z, the Alfvén limit above can be largely upgraded to (Hammer and Rostoker 1970; Lee and Sudan 1971; Brown and Bingham 1984; Spicer and Sudan 1984; Melrose 1990; van den Oord 1990)

$$I_A = \frac{2\pi r B_z}{\mu_0} = 1.5 r_{kpc} B_{nT} \times 10^{17} \text{ (A)}, \tag{7.7}$$

where B_{nT} is the axial magnetic field in the unit of nT $= 10\ \mu$G. Comparing Eqs. (7.5) and (7.7), one has $I_A > I_C \Rightarrow B_{nT} > 0.5\sqrt{p_{-12}}$, implying that a sufficiently strong axial magnetic field of $B_{nT} > 0.5\sqrt{p_{-12}}$ can allow the jet to carry a current enough to generate a self-confining azimuthal field B_ϕ. As for the source of the axial magnetic field B_z, the most possible case is that it originates from the central engine, such as the direct extension or the portion dragged by the jet of the magnetic field of the accretion disk around the Kerr black hole (Lovelace 1976; Blandford 1976).

Jafelice and Opher (1987a, b)) proposed that KAWs can be responsible for the generation of the self-collimated current leading to the self-confined magnetic field in the jets as well as the reacceleration of relativistic synchrotron-emitting electrons. As noted by Rees (1982), the plasma conditions in ERS are similar to those found in the local ISM, where the plasma may be concentrated in filaments or clouds. In fact, as revealed by the X-ray observations presented in previous sections, the ICM has complex magnetized plasma structures (see e.g. Figs. 7.9, 7.10, and 7.11). Like the ICM case, there is much observational and theoretical evidence suggesting that the

ERS probably do not have a homogeneous internal structure, even on relatively small scales, and hence the existence of more or less intense gradients of density, magnetic fields, or both, are expected inside ERS (Lacombe 1977; Rees 1982; Begelman et al. 1984). For instance, a number of filamentary structures have been observed in the radio lobes of Cygnus A (Perley et al. 1984). These ubiquitous inhomogeneities in ERS can be very plentiful and efficient sources of exciting KAWs through the density striation instability (Wu and Chen 2013; Chen et al. 2015) or the resonant mode conversion of AWs (Hasegawa 1985; Xiang et al. 2019).

Besides the current instability, in fact, the majority of the macroscopic MHD instabilities, such as the Kelvin-Helmholtz and Rayleigh-Taylor instabilities, all can produce efficiently AWs (Hasegawa 1976). The fact that these instabilities are very likely to be generally present in ERS (Blake 1972; Begelman et al. 1984) suggests that the intense AW turbulence can prevail in ERS, just like they exist ubiquitously in the solar atmosphere, solar wind, and ISM. In a weakly inhomogeneous plasma such as ERS (radio lobes or jets), as described in Sect. 1.5, one of the most effective mechanisms for generating KAWs is the resonant mode conversion of AWs (Hasegawa and Chen 1976). Like the resonant mode conversion process occurring in the magnetopause and magnetotail described in Chap. 4, in the inhomogeneous ERS (in density, magnetic field, or both), the resonant mode conversion process of AW with a frequency ω_0 into KAW also can occur at the Alfvén resonant surface of $\omega_0 = \omega_A(x) \equiv k_\parallel v_A(x)$, where $v_A(x) = B(x)/\sqrt{\mu_0 m_i n(x)}$ is the local Alfvén velocity in the inhomogeneous ERS with the inhomogeneous magnetic field strength $B(x)$ and plasma density $n(x)$, and x is the distance along the inhomogeneity with a characteristic scale size $a = \kappa^{-1} = |\partial \ln n(x)/\partial x|^{-1}$. In consideration of the fully collisionless nature of plasmas in the ERS, it should be a very reasonable assumption that the AW energy damping is primarily due to the mode conversion to KAW in ERS, implying a conversion rate of AW to KAW of $\sim$100%, and that these AW energy ultimately all will be transferred to the plasma, field-aligned accelerating electrons and cross-field heating ions (Hasegawa and Uberoi 1982; Ross et al. 1982; Heyvaerts and Priest 1983; Steinolfson 1985; Lee and Roberts 1986; Steinolfson et al. 1986).

Assuming that the inhomogeneity of ERS, especially jets, can be characterized by a wealth of filamentary structures and these filaments have the inhomogeneity scale size a and average distance d, implying that the fraction of the space occupied by the AWs is $a/(d+a) \sim a/d$ for $a \ll d$ in the ERS, the inhomogeneity of an individual filament may be approximately described by a linear variation as follows (Lacombe 1977; Rees 1982; Begelman et al. 1984; Perley et al. 1984):

$$\omega_A(x)^2 \equiv k_\parallel^2 v_A^2(x) = \omega_+^2 + \omega_-^2 \kappa x, \tag{7.8}$$

where $\omega_\pm^2 \equiv \left(\omega_M^2 \pm \omega_m^2\right)/2$, $\omega_{M(m)} = k_\parallel v_{AM(m)}$, and $v_{AM(m)}$ is the maximal (minimal) Alfvén velocity in the filament. The damping time of the resonant AW with the resonant frequency $\omega_0 \simeq \omega_+$ (i.e., the resonant mode conversion time of the AW into the KAW), τ_r, can be given by

$$\tau_r = \frac{4\omega_+}{\pi k_\parallel a\omega_-^2} \simeq \frac{4\sqrt{2}\omega_0^{-1}}{\pi k_\parallel a} \simeq \frac{1.8\tau_A}{k_\parallel a}, \tag{7.9}$$

where the approximation $\omega_M^2 \gg \omega_m^2$ has been used and the Alfvén time (the wave period) $\tau_A = \omega_0^{-1} = 1/k_\parallel v_A$ (Lee and Roberts 1986).

At appreciable distances from the central engine, the nonrelativistic MHD is still valid and the Alfvén velocity v_A is much less than the light velocity c ($v_A \ll c$). For an intermediate-β of $1 \gg \beta \gg Q \equiv m_e/m_i$, in the resonant mode conversion process of the AW into the KAW, the magnetic field strength of the KAW, B_{KAW}, can be intensified significantly in comparison to the AW strength, B_{AW}, (Hasegawa 1976),

$$B_{KAW} \simeq \frac{1 + (k_x\bar{\rho}_i)^2}{\sqrt{\kappa\bar{\rho}_i}} B_{AW} \simeq \frac{2}{\sqrt{\kappa\bar{\rho}_i}} B_{AW}, \tag{7.10}$$

where k_x is the perpendicular wave number of the KAW along the inhomogeneity, $\bar{\rho}_i$ is the ion gyroradius at an average energy $\bar{E}_i$ and $k_x\bar{\rho}_i \sim 1$ for the KAW. The corresponding field-aligned electric potential of the KAW, ψ, is proportional to $\bar{\rho}_i v_A B_{KAW}$, and hence the potential energy can be written as (Hasegawa 1976)

$$e\psi \simeq \frac{2}{\sqrt{\kappa\bar{\rho}_i}} \frac{B_{AW}}{B_0} \bar{p}_i v_A \sim 2\sqrt{\frac{\beta_0}{\kappa\bar{\rho}_i}} \bar{p}_i v_A, \tag{7.11}$$

where $\bar{p}_i \equiv eB_0\bar{\rho}_i = \sqrt{2m_i\bar{E}_i}$ is the average momentum of ions with energy $\bar{E}_i$. In particular, if the average energy density of relativistic beam particles in the jet, $n_i\bar{E}_i$, is comparable to the ambient magnetic field energy density, $B_0^2/2\mu_0$, that is, $B_0 \sim \sqrt{2\mu_0 n_i\bar{E}_i}$, Eq. (7.11) leads to

$$e\psi \sim 4\sqrt{\frac{\beta_0}{\kappa\bar{\rho}_i}} \bar{E}_i. \tag{7.12}$$

The electric potential energy of the KAW in Eq. (7.12) can give an estimation for the electron energy accelerated by the KAW (Jafelice and Opher 1987a). For the case in ERS, which is a low density cavity surrounded by the hot ICM, it is proper to take the ambient magnetic field $B_0 \sim 5$ nT (i.e., 50 μG), temperature $T \sim 10^7$ K, and density $n_0 \sim 10^{-2}$–10^{-3} cm^{-3}. This leads to the plasma pressure parameter $\beta_0 \sim 10^{-1}$–10^{-2} in the ERS. For the scale size of the filamentary inhomogeneity in the ERS, $a = \kappa^{-1} \sim 10^2 - 10^3\bar{\rho}_i$ possibly is a reasonable scale range. The result indicates $\beta_0/\kappa\bar{\rho}_i \sim 10$, that is, the acceleration potential energy of the KAW is higher than the beam electron energy by an order, $e\psi \sim 10\bar{E}_i$. The acceleration time (i.e., the damping time of the KAW), τ_a, may by estimated by (Lee and Roberts 1986; Jafelice and Opher 1987a)

$$\tau_a \sim 10^2 \tau_A \frac{k_\parallel}{k_x} \sim \kappa \bar{\rho}_i \tau_r \ll \tau_r, \tag{7.13}$$

where $k_\parallel a \sim 0.1$ and $k_x \bar{\rho}_i \sim 1$ have been used for the resonant AW and KAW, respectively. This indicates that the transferring time of the AW energy into the energy of the beam electrons, that is, so-called the reacceleration time, $\tau_{ra} = \tau_r + \tau_a \simeq \tau_r$. On the other hand, if the expansion time is larger than the synchrotron radiation time τ_s (i.e., the average synchrotron lifetime of the beam electrons), the cooling time of the beam electrons $\tau_L \sim \tau_s$. Therefore, the energy balance by this reacceleration process results in

$$\frac{\tau_L}{\tau_{ra}} \beta_0 f = 1 \Rightarrow \tau_r \sim \beta_0 f \tau_s \Rightarrow \tau_A \sim \frac{\beta_0 f}{18} \tau_s, \tag{7.14}$$

where $f \equiv a^2/(d+a)^2 \sim a^2/d^2$ is the filling factor of filaments in the ERS and Eq. (7.9) has been used for τ_r. For a typical synchrotron-radiating lifetime of relativistic electrons $\tau_s \sim 3 \times 10^{11}$ s (i.e., $\sim 10^4$ years) and $\beta_0 \sim 10^{-2}$, the required Alfvén time $\tau_A \sim 2 \times 10^8 f$ s $\sim 2 \times 10^5$ s for the filling factor $f \sim 10^{-3}$ and the corresponding wavelength of the AW $\lambda_\parallel \sim v_A \tau_A \sim 4$ AU for $v_A \sim 0.01c$. For the typical parameters of the ERS above given and the typical energy $\bar{E}_i \sim 10$ MeV, one has $\bar{\rho}_i \sim 6 \times 10^4$ km (i.e., $a = \kappa^{-1} \sim 0.4$ AU $\sim 0.1\lambda_\parallel$) and hence the acceleration potential energy $e\psi \sim 10\bar{E}_i \sim 100$ MeV.

For extremely relativistic electrons at higher energies, which are expected to present in jets and lobes, the cyclotron-resonant acceleration (CRA) by ubiquitous AWs can provide a more effective acceleration mechanism (Lacombe 1977; Begelman et al. 1984). In fact, the physical nature of the above acceleration by the field-aligned electric potential energy of KAWs, $e\psi$, is the Landau damping acceleration of KAWs. For the case of AWs the Landau damping acceleration can not work, and while the CRA by AW requires the following cyclotron resonance condition (Lacombe 1977; Begelman et al. 1984; Jafelice and Opher 1987a):

$$\omega + \frac{s\omega_{ce}}{\gamma_e} + k_z v_z = 0, \tag{7.15}$$

where s is the harmonic number and γ_e is the relativistic factor of electrons. For the low-frequency AW the cyclotron resonant condition of Eq. (7.15) can be reduced approximately as

$$\gamma_e \simeq \frac{\omega_{ce}}{\omega} \frac{v_A}{c} \sim 10^{-2} \frac{\omega_{ce}}{\omega}, \tag{7.16}$$

where the harmonic number $s = 1$ has been used. For a typical plasma environment in ERS the AW turbulence can have a wide spectrum ranging between the resonant mode conversion frequency ω_+ and the ion gyrofrequency ω_{ci} in the frequency space, that is, the AWs with frequency $\omega_{ci} > \omega > \omega_A = 1/\tau_A$ are available for the CRA of electrons. For above the typical parameters in ERS, we have $\omega_{ce} \sim 140$ Hz, $\omega_{ci} \sim 7.6 \times 10^{-2}$ Hz, and $\omega_A \sim 5 \times 10^{-6}$ Hz, implying that the resonant electrons are required to have a relativistic factor $20 < \gamma_e < 2.8 \times 10^5$, that is, to have an energy between 10 MeV and 140 GeV.

In fact, the above arguments and results suggest a competitive and cooperative scenario for the reacceleration of relativistic electrons by AWs and KAWs in the ERS (Lacombe 1977; Begelman et al. 1984; Jafelice and Opher 1987a). Firstly, the cooling nonrelativistic electrons due to the Landau damping acceleration by KAWs at the resonant mode conversion frequency ω_A, which is much less than the ion gyrofrequency become moderately relativistic electrons with energies of MeV orders. Then, these moderately relativistic electrons further are accelerated to extremely relativistic energies of GeV orders through the CRA by AWs with frequencies above ω_A but below ω_{ci}.

The CRA by AWs increases mainly the cross-field energy of the accelerated electrons, which can efficiently compensate the synchrotron-emitting loss of the relativistic electrons. While the Landau damping acceleration by KAWs can greatly increase the field-aligned energy of the accelerated electrons, which may contribute significantly to the self-collimated current in the radio jets. Jafelice and Opher (1987b) further calculated the field-aligned current caused due to the Landau damping acceleration of electrons by KAWs for the Maxwellian and power-law distributions of the electrons in momentum. In principle, the field-aligned current $J_z = -en_e u_{ez}$ carried by the accelerated electrons with the distribution $f(\mathbf{p})$ in the momentum space can be obtained by the field-aligned integrating velocity u_{ez} as follows:

$$u_{ez} = \int_{\mathbf{p}} v_{ez}(\mathbf{p}) f(\mathbf{p}) d^3\mathbf{p} = 2\pi \int_{p_m}^{+\infty} \int_{-\infty}^{+\infty} \frac{p_z}{\gamma_e m_e} f(p_r, p_z) p_r dp_z dp_r, \qquad (7.17)$$

where the axial symmetry has been assumed for the distribution function $f(\mathbf{p})$, $\mathbf{p} = \gamma_e m_e \mathbf{v}$, v_{ez} is the z-component of the electron velocity with the momentum p_z, and p_m is the low-momentum cut-off $p_m \sim 0.1 T_e/c$ in the distribution.

Under the simplest approximation that the perturbation of KAW alters the relativistic electrons distribution just in the neighborhood of the resonant velocity v_A (i.e., momentum p_{zA}), it results in a plateau in the distribution function at the value $f(p; p_z = p_{zA})$ (where $p_{zA} \equiv \gamma_e m_e v_A$) with a characteristic width $2\Delta p_z = 2\delta e\psi / v_A$ centered at p_{zA}. The factor $\delta \leq 1$ determines the magnitude of the perturbation of electron distribution. Neglecting terms of the order of $(v_A/c)^2$ and using $\gamma_e = p/m_e c$ with $p = \sqrt{p_r^2 + p_z^2}$, then $p_{zA} \simeq (v_A/c)\, p_r$ and the primary contribution to u_{ez} in Eq. (7.17) comes from the momentum $p_r \gg p_z$. Thus, Eq. (7.17) may be reduced to

$$u_{ez} = 2\pi c \int_{p_m}^{+\infty} g(p_r) p_r dp_r, \qquad (7.18)$$

where

$$g(p_r) \equiv \int_{-\infty}^{+\infty} \frac{f(p_r, p_z)}{\sqrt{p_r^2 + p_z^2}} p_z dp_z \qquad (7.19)$$

can be specified for the Maxwellian and power-law distributions, respectively.

On the other hand, the current $J_z = -en_e u_{ez}$ caused by KAWs exists only in the thin layer where the resonant mode conversion of AWs into KAWs occurs. Taking the typical parameter values used above for a jet, for a relativistic Maxwellian distribution function of the electrons as follows:

$$f(p) = \frac{c^3}{8\pi T_e^3} \exp\left(-\frac{pc}{T_e}\right), \tag{7.20}$$

Jafelice and Opher (1987b) obtained the field-aligned current density caused by KAWs J_z is

$$J_z \simeq 10^{-10}\ \mathrm{A/m^2} \tag{7.21}$$

and the current intensity

$$I_J \simeq 2 \times 10^{28} f\ \mathrm{A} \simeq 2 \times 10^{25}\ \mathrm{A}, \tag{7.22}$$

where the filling factor $f = 10^{-3}$ has been used. By comparing with the Alfvén limit I_A in Eq. (7.7), the field-aligned current caused by KAWs I_J evidently exceeds greatly the Alfvén limit I_A, implying that a strongly electrodynamic coupling beam-return current system must be established in the jet-lobe system, in which the majority of the KAW I_J is compensated by a return current I_R and the net current $I_J - I_R \lesssim I_A \sim I_C$, sufficiently to confined the jet.

For a power-law distribution function of nonthermal electrons as follows:

$$f(p) = \frac{s-4}{4\pi c} p_m^{s-4} \bar{E}_e p^{-s}, \tag{7.23}$$

where $s = 2\alpha_s + 3$ and α_s is the synchrotron spectral index, Jafelice and Opher (1987b) gave the electric current density generated by KAW as follows:

$$J_z = 3 \times 10^{-7}\ \mathrm{A/m^2}, \tag{7.24}$$

where $\bar{E}_e = 2^{1/(s-4)} p_m c = 10$ MeV, $\alpha_s = 0.65$, and hence $s = 4.3$ has been used. The total electric current intensity generated by KAWs is

$$I_J = 7 \times 10^{31} f\ \mathrm{A} \sim 7 \times 10^{28}\ \mathrm{A} \tag{7.25}$$

for the filling factor $f \sim 10^{-3}$.

From above analyses, in both two cases of relativistic electron distributions: the Maxwellian and power-law distribution, the field-aligned currents caused by KAWs I_J are much larger than both the Alfvén limit current I_A and the collimated current I_C. Moreover, for the typical parameters of extragalactic radio jets, we can have $I_A \gtrsim I_C$. The condition $I_J \gg I_A$ indicates that the return current and a strongly coupling beam-return current system must exist in extragalactic jet-lobe system, and while the condition $I_A \gtrsim I_C$ implies that the self-collimated mechanism for the relativistic jets by the field-aligned current caused by KAWs is plausible. Therefore, KAWs can not only play an important role in the reacceleration of synchrotron-emitting electrons, but also their field-aligned currents can provide an effective self-collimated mechanism for the jets.

References

Alcalá, J. M., Krautter, J., Covino, E., et al. (1997). A study of the Chamaeleon star-forming region from the ROSAT All-Sky Survey. II. The pre-main sequence population. *Astronomy & Astrophysics*, *319*, 184–200.

Alexandrova, O., Carbone, V., Veltri, P., & Sorriso-Valvo, L. (2008). Small-scale energy cascade of the solar wind turbulence. *The Astrophysical Journal*, *674*, 1153–1157.

Alfvén, H. (1939). On the motion of cosmic rays in interstellar space. *Physical Review*, *55*(425), 429.

Alfvén, H., & Herlofson, N. (1950). Cosmic radiation and radio stars. *Physical Review*, *78*, 616–616.

Armstrong, J. W., Cordes, J. M., & Rickett, B. J. (1981). Density power spectrum in the local interstellar medium. *Nature*, *291*, 561–564.

Armstrong, J. W., Rickett, B. J., & Spangler, S. R. (1995). Electron density power spectrum in the local interstellar medium. *The Astrophysical Journal*, *443*, 209.

Bîrzan, L., Rafferty, D. A., Mcnamara, B. R., et al. (2004). A systematic study of radio induced x-ray cavities in clusters, groups, and galaxies. *The Astrophysical Journal*, *607*, 800–809.

Baganoff, F. K., Bautz, M. W., Brandt, W. N., et al. (2001). Rapid X-ray flaring from the direction of the supermassive black hole at the Galactic Centre. *Nature*, *413*, 45–48.

Beck, R. (2001). Galactic and extragalactic magnetic fields, Space Sci. *Review*, *99*, 243–260.

Begelman, M. C., Blandford, R. D., & Rees, M. J. (1984). Theory of extragalactic radio sources. *Reviews of Modern Physics*, *56*, 255–351.

Belsole, E., Sauvageot, J. L., Böhringer, H., et al. (2001). An XMM-Newton study of the sub-structure in M 87's halo. *Astronomy & Astrophysics*, *365*, L188–L194.

Benevolenskaya, E. E., Kosovichev, A. G., Lemen, J. R., et al. (2002). Large-scale solar coronal structures in soft x-rays and their relationship to the magnetic flux. *The Astrophysical Journal*, *571*, L181.

Benford, G. (1978). Current-carrying beams in astrophysics: models for double radio sources and jets. *Monthly Notices of the Royal Astronomical Society*, *183*, 29–48.

Benford, G. (1984). Magnetically ordered jets from pulsars. *The Astrophysical Journal*, *282*, 154–160.

Benz, A. O., & Güdel, M. (1994). X-ray/microwave ratio of flares and coronae. *Astronomy & Astrophysics*, *285*, 621–630.

Berghöfer, T. W., Schmitt, J. H. M. M., & Cassinelli, J. P. (1996). The ROSAT all-sky survey catalogue of optically bright OB-type stars. *Astronomy & Astrophysics Supplement*, *118*, 481–494.

Bicknell, G. V. (1984). A model for the surface brightness of a turbulent low mach number jet. I. Theoretical development and application to 3C 31. *The Astrophysical Journal*, *286*, 68–87.

Birkinshaw, M. (1999). *The effects of nearby clusters of galaxies on the microwave background radiation*. Smithsonian Astrophysical Observatory Cambridge, MA United States: Technical Report.

Blake, G. M. (1972). Fluid dynamic stability of double radio sources. *Monthly Notices of the Royal Astronomical Society*, *181*, 465.

Blandford, R. D. (1976). Accretion disc electrodynamics-a model for double radio sources. *Monthly Notices of the Royal Astronomical Society*, *176*, 465–481.

Blandford, R. D., & Rees, M. J. (1974). A "twin-exhaust" model for double radio sources. *Monthly Notices of the Royal Astronomical Society*, *169*, 395–415.
Blanton, E. L., Sarazin, C. L., McNamara, B. R., & Wise, M. W. (2001). Chandra observations of the radio source/X-ray gas interaction in the cooling flow cluster Abell 2052. *The Astrophysical Journal Letters*, *558*, L15–L19.
Böhringer, H., & Werner, N. (2010). X-ray spectroscopy of galaxy clusters: studying astrophysical processes in the largest celestial laboratories. *Astronomy & Astrophysics Review*, *18*, 127–196.
Böhringer, H., Belsole, E., Kennea, J., et al. (2001). XMM-Newton observations of M 87 and its X-ray halo. *Astronomy & Astrophysics*, *365*, L181–L187.
Böhringer, H., Nulsen, P. E. J., Braun, R., & Fabian, A. C. (1995). The interaction of the radio halo of M87 with the cooling intracluster medium of the Virgo cluster. *Monthly Notices of the Royal Astronomical Society*, *274*, L67–L71.
Braginskii, S. I. (1965). Transport processes in a plasma. *Reviews of Modern Plasma Physics*, *1*, 205–311.
Branduardi-Raymont, G., Fabricant, D., Feigelson, E., et al. (1981). Soft X-ray images of the central region of the Perseus cluster. *The Astrophysical Journal*, *248*, 55–60.
Briel, U. G., Henry, J. P., & Boehringer, H. (1992). Observation of the Coma cluster of galaxies with ROSAT during the all-sky-survey. *Astronomy & Astrophysics*, *259*, L31–L34.
Brown, J. C., & Bingham, R. (1984). Electrodynamics effects in beam-return current systems and their implications for solar impulsive bursts. *Astronomy & Astrophysics*, *131*, L11.
Burbidge, G. (1967). Generation of radio sources. *Nature*, *216*, 1287–1289.
Burns, J. O., Feigelson, E. D., & Schreier, E. J. (1983). The inner radio structure of Centaurus A: Clues to the origin of the jet X-ray emission. *The Astrophysical Journal*, *273*, 128–153.
Carilli, C. L., & Taylor, G. B. (2002). Cluster magnetic fields. *Annual Reviews Astronomy & Astrophysics*, *40*, 319–348.
Catura, R. C., Acton, L. W., & Johnson, H. M. (1975). Evidence for x-ray emission from Capella. *The Astrophysical Journal*, *196*, L47–L49.
Cavaliere, A., & Fusco-Femiano, R. (1976). X-rays from hot plasma in clusters of galaxies. *Astronomy & Astrophysics*, *49*, 137–144.
Chandran, B. D. G., Quataert, E., Howes, G. G., et al. (2009). Constraining low-frequency Alfvénic turbulence in the solar wind using density-fluctuation measurements. *The Astrophysical Journal*, *707*, 1668–1675.
Chen, C. H. K., Salem, C. S., Bonnell, J. W., et al. (2012). Density fluctuation spectrum of solar wind turbulence between ion and electron scales. *Physical Review Letters*, *109*, 035001.
Chen, L., Wu, D. J., Zhao, G. Q., & Tang, J. F. (2017). A self-consistent mechanism for electron cyclotron maser emission and its application to type III solar radio bursts. *Journal of Geophysical Research*, *122*, 35–49.
Chen, L., Wu, D. J., Zhao, G. Q., et al. (2014). Excitation of kinetic Alfvén waves by fast electron beams. *The Astrophysical Journal*, *793*, 13.
Chen, L., Wu, D. J., Zhao, G. Q., et al. (2015). A possible mechanism for the formation of filamentous structures in magnetoplasmas by kinetic Alfvén waves. *Journal of Geophysical Research*, *120*, 61–69.
Chepurnov, A., & Lazarian, A. (2010). Extending the big power law in the sky with turbulence spectra from Wisconsin H mapper data. *The Astrophysical Journal*, *710*, 853–858.
Chiaberge, M., Capetti, A., & Celotti, A. (1999). The HST view of FR I radio galaxies: Evidence for non-thermal nuclear sources. *Astronomy & Astrophysics*, *349*, 77–87.
Cho, J., & Vishniac, E. T. (2000). The anisotropy of MHD Alfvénic turbulence. *The Astrophysical Journal*, *539*, 273–282.
Churazov, E., Bruggen, M., Kaiser, C. R., et al. (2001). Evolution of Buoyant bubbles in M87. *The Astrophysical Journal*, *554*, 261–273.
Clarke, T. E., Kronberg, P. P., & Böhringer, H. (2001). A new radio-X-ray probe of galaxy cluster magnetic fields. *The Astrophysical Journal*, *547*, L111–L114.

Clarke, T. E., Sarazin, C. L., Blanton, E. L., et al. (2005). Low-frequency radio observations of X-ray ghost bubbles in A2597: A history of radio activity in the core. *The Astrophysical Journal, 625*, 748–753.
Corbel, S., Nowak, M. A., Fender, R. P., et al. (2003). Radio/X-ray correlation in the low/hard state of GX 339–4. *Astronomy & Astrophysics, 400*, 1007–1012.
Cordes, J. M., Weisberg, J. M., & Boriakoff, V. (1985). Small-scale electron density turbulence in the interstellar medium. *The Astrophysical Journal, 288*, 221–247.
Cowie, L. L. (1981). Theoretical models of X-ray emission from rich clusters of galaxies. In X-ray Astronomy with the Einstein Satellite (Proceedings of the Meeting, Cambridge M. A., January 28–30, 1980, ed. R. Giacconi), Dordrecht, Reidel, 227–240.
Cowie, L. L., & Binney, J. (1977). Radiative regulation of gas flow within clusters of galaxies: A model for cluster X-ray sources. *The Astrophysical Journal, 215*, 723–732.
Crovisier, J., & Dickey, J. M. (1983). The spatial power spectrum of galactic neutral hydrogen from observations of the 21-cm emission line. *Astronomy & Astrophysics, 122*, 282–296.
Dai, Z. G., Wang, X. Y., Wu, X. F., & Zhang, B. (2006). X-ray flares from postmerger, millisecond pulsars. *Science, 311*, 1127.
Dame, T. M., Elmegreen, B. G., Cohen, R. S., & Thaddeus, P. (1986). The largest molecular cloud complexes in the first galactic quadrant. *The Astrophysical Journal, 305*, 892–908.
David, L. P., Nulsen, P. E. J., McNamara, B. R., et al. (2001). A high-resolution study of the Hydra A cluster with Chandra: Comparison of the core mass distribution with theoretical predictions and evidence for feedback in the cooling flow. *The Astrophysical Journal, 557*, 546–559.
De Young, D. S., & Axford, W. I. (1967). Inertial confinement of extended radio sources. *Nature, 216*, 129–131.
Di Matteo, T., Celotti, A., & Fabian, A. C. (1999). Magnetic flares in accretion disc coronae and the spectral states of black hole candidates: The case of GX339-4. *Monthly Notices of the Royal Astronomical Society, 304*, 809.
Dunn, R. J. H., & Fabian, A. C. (2006). Investigating AGN heating in a sample of nearby clusters. *Monthly Notices of the Royal Astronomical Society, 373*, 959–971.
Eilek, J. A. (1985). Current systems in radio jets. In M. R. Kundu & G. D. Holman (eds.) Unstable Current Systems and Plasma Instabilities in Astrophysics *IAU Symposia, 107*, 433–437.
Elphic, R. C. (1985). Magnetic flux ropes of Venus-A paradigm for helical magnetic structures in astrophysical systems. In M. R. Kundu & G. D. Holman (eds.) *Unstable Current Systems and Plasma Instabilities in Astrophysics. IAU Symposia, 107*, 43–46.
Ettori, S. (2000). β-model and cooling flows in X-ray clusters of galaxies. *Monthly Notices of the Royal Astronomical Society, 318*, 1041–1046.
Fabian, A. C. (1994). Cooling flows in clusters of galaxies. *Annual Reviews Astronomy & Astrophysics, 32*, 277–318.
Fabian, A. C., & Nulsen, P. E. J. (1977). Subsonic accretion of cooling gas in clusters of galaxies. *Monthly Notices of the Royal Astronomical Society, 180*, 479–484.
Fabian, A. C., Celotti, A., Blundell, K. M., et al. (2002). The properties of the X-ray holes in the intracluster medium of the Perseus cluster. *Monthly Notices of the Royal Astronomical Society, 331*, 369–375.
Fabian, A. C., Hu, E. M., Cowie, L. L., & Grindlay, J. (1981). The distribution and morphology of x-ray-emitting gas in the core of the perseus cluster. *The Astrophysical Journal, 248*, 47–54.
Fabian, A. C., Reynolds, C. S., Taylor, G. B., & Dunn, R. J. H. (2005). On viscosity, conduction and sound waves in the intracluster medium. *Monthly Notices of the Royal Astronomical Society, 363*, 891–896.
Fabian, A. C., Sanders, J. S., Taylor, G. B., et al. (2006). A very deep Chandra observation of the Perseus cluster: Shocks, ripples and conduction. *Monthly Notices of the Royal Astronomical Society, 366*, 417–428.
Fabricant, D., & Gorenstein, P. (1983). Further evidence for M87's massive, dark halo. *The Astrophysical Journal, 267*, 535–546.

Falcke, H., & Biermann, P. L. (1995). The jet-disk symbiosis. I. Radio to X-ray emission models for quasars. *Astronomy & Astrophysics, 293*, 665–682.
Falcke, H., Kording, E., & Markoff, S. (2004). A scheme to unify low-power accreting black holes, Jet-dominated accretion flows and the radio/X-ray correlation. *Astronomy & Astrophysics, 414*, 895–903.
Feretti, L., Gioia, I. M. & Giovannini, G. (2002). Merging processes in galaxy clusters, Edited by L. Feretti, Istituto di Radioastronomia CNR, Bologna, Italy; I.M. Gioia, Istituto di Radioastronomia CNR, Bologna, Italy; G. Giovannini, Physics Department, University of Bologna, Italy. Astrophysics and Space Science Library, Vol. 272. Kluwer Academic Publishers, Dordrecht.
Finoguenov, A., & Jones, C. (2001). Chandra observation of M84, a radio lobe elliptical galaxy in the Virgo cluster. *The Astrophysical Journal, 547*, L107–L110.
Ford, H. C., Harms, R. J., Tsvetanov, Z. I., et al. (1994). Narrowband HST images of M87: Evidence for a disk of ionized gas around a massive black hole. *Astrophysical Journal Letters, 435*, L27.
Forman, W., & Jones, C. (1982). X-ray-imaging observations of clusters of galaxies. *Annual Reviews Astronomy & Astrophysics, 20*, 547–585.
Forman, W., Jones, C., Churazov, E., et al. (2007). Filaments, bubbles, and weak shocks in the gaseous atmosphere of M87. *The Astrophysical Journal, 665*, 1057–1066.
Forman, W., Kellogg, E., Gursky, H., et al. (1972). Observations of the extended X-ray sources in the Perseus and Coma clusters from UHURU. *The Astrophysical Journal, 178*, 309–316.
Forman, W., Nulsen, P., Heinz, S., et al. (2005). Reflections of active galactic nucleus outbursts in the gaseous atmosphere of M87. *The Astrophysical Journal, 635*, 894–906.
Furlanetto, S. R., & Loeb, A. (2001). Intergalactic magnetic fields from quasar outflows. *The Astrophysical Journal, 556*, 619–634.
Gallo, E., Fender, R. P., & Pooley, G. G. (2003). A universal radio-X-ray correlation in low/hard state black hole binaries. *Monthly Notices of the Royal Astronomical Society, 344*, 60–72.
Ghisellini, G., Padovani, P., Celotti, A., & Maraschi, L. (1993). Relativistic bulk motion in active galactic nuclei. *The Astrophysical Journal, 407*, 65.
Giacconi, R., Kellogg, E., Gorenstein, P., et al. (1971). An X-ray scan of the galactic plane from UHURU. *The Astrophysical Journal, 165*, L27.
Gilfanov, M. R., Syunyaev, R. A., & Churazov, E. M. (1987). The X-ray surface brightness distribution of clusters of galaxies in resonance lines. *Pisma v Astronomicheskii Zhurnal, 13*, 7–18.
Gillmon, K., Sanders, J. S., & Fabian, A. C. (2004). An X-ray absorption analysis of the high-velocity system in NGC 1275. *Monthly Notices of the Royal Astronomical Society, 348*, 159–164.
Goldreich, P., & Sridhar, S. (1995). Toward a theory of interstellar turbulence. 2: Strong Alfvénic turbulence. *The Astrophysical Journal, 438*, 763–775.
Goldreich, P., & Sridhar, S. (1997). Magnetohydrodynamic turbulence revisited. *The Astrophysical Journal, 485*, 680–688.
Gorenstein, P., Fabricant, D., Topka, K., & ... (1977). Structure of the X-ray source in the Virgo cluster of galaxies. *The Astrophysical Journal, 216*, L95–L99.
Govoni, F. (2006). Observations of magnetic fields in regular and irregular clusters. *Astronomische Nachrichten, 327*, 539.
Govoni, F., & Feretti, L. (2004). Magnetic fields in clusters of galaxies. *International Journal of Modern Physics D, 13*, 1549–1594.
Gronenschild, E. H. B. M., & Mewe, R. (1978). Calculated X-radiation from optically thin plasmas. III. Abundance effects on continuum emission. *Astronomy and Astrophysics Supplement, 32*, 283–305.
Güdel, M. (2004). X-ray astronomy of stellar coronae. *Astronomy & Astrophysics Review, 12*, 71–237.
Güdel, M., & Benz, A. O. (1993). X-Ray/microwave relation of different types of active stars. *Astrophysical Journal Letters, 405*, L63.
Gurnett, D. A., Kurth, W. S., Burlaga, L. F., & Ness, N. F. (2013). In situ observations of interstellar plasma with Voyager 1. *Science, 341*, 1489–1492.

Gurnett, D. A., Kurth, W. S., Stone, E. C., et al. (2015). Precursors to interstellar shocks of solar origin. *The Astrophysical Journal*, *809*, 121.

Gursky, H., & Schwartz, D. A. (1977). Extragalactic X-ray sources. *Annual Reviews Astronomy & Astrophysics*, *15*, 541–568.

Gursky, H., Kellogg, E., Murray, S., et al. (1971). A strong X-ray source in the coma cluster observed by UHURU. *The Astrophysical Journal*, *167*, L81.

Haardt, F., Maraschi, L., & Ghisellini, G. (1997). X-Ray variability and correlations in the two-Phase disk-corona model for Seyfert galaxies. *The Astrophysical Journal*, *476*, 620–631.

Haisch, B., Strong, K. T., & Rodono, M. (1991). Flares on the Sun and other stars, Annual Rev. *Astronomy & Astrophysics*, *29*, 275–324.

Hammer, D. A., & Rostoker, N. (1970). Propagation of high current relativistic electron beams. *Physics of Fluids*, *13*, 1831–1850.

Hardee, P. E. (1985). Is the jet in M87 magnetically confined? In M. R. Kundu & G. D. Holman (eds.)*Unstable Current Systems and Plasma Instabilities in Astrophysics*, IAU Synmp. 107, 442, 443.

Harms, R. J., Ford, H. C., Tsvetanov, Z. I., et al. (1994). HST FOS spectroscopy of M87: Evidence for a disk of ionized gas around a massive black hole. *Astrophysical Journal Letters*, *435*, L35.

Harris, D. E., Biretta, J. A., Junor, W., et al. (2003). Flaring X-ray emission from HST-1, a knot in the M87 jet. *The Astrophysical Journal*, *586*, L41–L44.

Hasegawa, A. (1976). Kinetic theory of MHD instabilities in a nonuniform plasma. *Solar Physics*, *47*, 325–330.

Hasegawa, A. (1985). Plasma heating by Alfvén waves-kinetic properties of magnetohydrodynamic disturbances. In M. R. Kundu & G. D. Holman (eds.) *Unstable Current Systems and Plasma Instabilities in Astrophysics IAU Symposia*, *107*, 381–388.

Hasegawa, A., & Chen, L. (1976). Kinetic process of plasma heating by resonant mode conversion of Alfvén wave. *Physics of Fluids*, *19*, 1924–1934.

Hasegawa, A., & Uberoi, C. (1982). *The Alfvén Waves*. Tech: Inf. Center, US Dept. of Energy, Oak Ridge.

Hattori, M., Kneib, J., & Makino, N. (1999). Gravitational lensing in clusters of galaxies. *Progress of Theoretical Physics Supplements*, *133*, 1–51.

Hawley, J. F., Balbus, S. A., & Winters, W. F. (1999). Local hydrodynamic stability of accretion disks. *The Astrophysical Journal*, *518*, 394–404.

Heise, J., Brinkman, A. C., Schrijver, J., et al. (1975). Evidence for x-ray emission from flare stars observed by ANS. *The Astrophysical Journal*, *202*, L73–L76.

Heyvaerts, J., & Priest, E. R. (1983). Coronal heating by phase-mixed shear Alfvén waves. *Astronomy & Astrophysics*, *117*, 220–234.

Higdon, J. C. (1984). Density fluctuations in the interstellar medium: Evidence for anisotropic magnetogasdynamic turbulence. I-Model and astrophysical sites. *The Astrophysical Journal*, *285*, 109–123.

Horbury, T. S., Forman, M., & Oughton, S. (2008). Anisotropic scaling of magnetohydrodynamic turbulence. *Physical Review Letters*, *101*, 175005.

Howes, G. G., Cowley, S. C., Dorland, W., et al. (2006). Astrophysical gyrokinetics: basic equations and linear theory. *The Astrophysical Journal*, *651*, 590–614.

Hünsch, M., Schmitt, J. H. M. M., & Voges, W. (1998a). The ROSAT all-sky survey catalogue of optically bright late-type giants and supergiants. *Astronomy and Astrophysics Supplement*, *127*, 251–255.

Hünsch, M., Schmitt, J. H. M. M., & Voges, W. (1998b). The ROSAT all-sky survey catalogue of optically bright main-sequence stars and subgiant stars. *Astronomy and Astrophysics Supplement*, *132*, 155–171.

Hünsch, M., Schmitt, J. H. M. M., Sterzik, M. F., & Voges, W. (1999). The ROSAT all-sky survey catalogue of the nearby stars. *Astronomy and Astrophysics Supplement*, *135*, 319–338.

Jafelice, L. C., & Opher, R. (1987a). Kinetic Alfvén waves in extended radio sources, I. Reacceleration. *Astrophysics and Space Science*, *137*, 303–315.

Jafelice, L. C., & Opher, R. (1987b). Kinetic Alfvén waves in extended radio sources, II. Electric Currents, Collimated Jets, and Inhomogeneities. *Astrophysics and Space Science*, *138*, 23–39.
Kellogg, P. J., & Horbury, T. S. (2005). Rapid density fluctuations in the solar wind. *Annales Geophysicae*, *23*, 3765–3773.
Kim, T. K., Pogorelov, N. V., & Burlaga, L. F. (2017). Modeling shocks detected by Voyager 1 in the local interstellar medium. *The Astrophysical Journal*, *843*, L32.
Kolmogorov, A. N. (1941). The local structure turbulence in incompressible viscous fluids for very large Reynolds numbers, Dokl. Akad. Nauk. SSSR 30, 301–305. Reprinted in 1991. *Proceedings of the Royal Society London A*, *434*, 9–13.
Kronberg, P. P. (2003). Galaxies and the magnetization of intergalactic space. *Physics of Plasmas*, *10*, 1985–1991.
Kronberg, P. P., Dufton, Q. W., Li, H., & Colgate, S. A. (2001). Magnetic energy of the intergalactic medium from galactic black holes. *The Astrophysical Journal*, *560*, 178–186.
Lacombe, C. (1977). Acceleration of particles and plasma heating by turbulent Alfvén waves in a radiogalaxy. *Astronomy & Astrophysics*, *54*, 1–16.
Larson, R. B. (1979). Stellar kinematics and interstellar turbulence. *Monthly Notices of the Royal Astronomical Society*, *186*, 479–490.
Larson, R. B. (1981). Turbulence and star formation in molecular clouds. *Monthly Notices of the Royal Astronomical Society*, *194*, 809–826.
Lawson, W. A., Feigelson, E. D., & Huenemoerder, D. P. (1996). An improved HR diagram for Chamaeleon I pre-main-sequence stars. *Monthly Notices of the Royal Astronomical Society*, *280*, 1071–1088.
Lee, K. H. (2017). Generation of parallel and quasi-perpendicular EMIC waves and mirror waves by fast magnetosonic shocks in the solar wind. *Journal of Geophysical Research*, *122*, 7307–7322.
Lee, K. H., & Lee, L. C. (2019). Interstellar turbulence spectrum from in situ observations of Voyager 1. *Nature Astronomy*, *3*, 154–159.
Lee, L. C., & Jokipii, J. R. (1976). The irregularity spectrum in interstellar space. *The Astrophysical Journal*, *206*, 735–743.
Lee, M. A., & Roberts, B. (1986). On the behavior of hydromagnetic surface waves. *The Astrophysical Journal*, *301*, 430–439.
Lee, R., & Sudan, R. N. (1971). Return current induced by a relativistic beam propagating in a magnetized plasma. *Physics of Fluids*, *14*, 1213–1225.
Liang, E. P. (1998). Multi-wavelength signatures of galactic black holes: observation and theory. *Physics Reports*, *302*, 67–142.
Little, L. T., & Matheson, D. N. (1973). Radio scintillations due to plasma irregularities with power law spectra: the interplanetary medium. *Monthly Notices of the Royal Astronomical Society*, *162*, 329–338.
Lovelace, R. V. E. (1976). Dynamo model of double radio sources. *Nature*, *262*, 649–652.
Luo, Q. Y., & Wu, D. J. (2010). Observations of anisotropic scaling of solar wind turbulence. *Astrophysical Journal Letters*, *714*, L138–L141.
Maccarone, T. J. (2003). Do X-ray binary spectral state transition luminosities vary? *Astronomy & Astrophysics*, *409*, 697–706.
Macchetto, F., Marconi, A., Axon, D. J., et al. (1997). The super massive black hole of M87 and the kinematics of its associated gaseous disk. *The Astrophysical Journal*, *489*, 579–600.
Maron, J., & Goldreich, P. (2001). Simulations of incompressible magnetohydrodynamic turbulence. *The Astrophysical Journal*, *554*, 1175.
Marshall, H. L., Miller, B. P., Davis, D. S., et al. (2002). A high-resolution X-ray image of the jet in M87. *The Astrophysical Journal*, *564*, 683–687.
Mathews, W. G., & Bregman, J. N. (1978). Radiative accretion flow onto giant galaxies in clusters. *The Astrophysical Journal*, *224*, 308–319.
Matsushita, K., Belsole, E., Finoguenov, A., & Böhringer, H. (2002). XMM-Newton observation of M 87. I. Single-phase temperature structure of intracluster medium. *Astronomy & Astrophysics*, *386*, 77–96.

McNamara, B. R., & Nulsen, P. E. J. (2007). Heating hot atmospheres with active galactic nuclei. *Annual Review Astronomy & Astrophysics*, *45*, 117–175.

McNamara, B. R., Wise, M. W., Nulsen, P. E. J., et al. (2001). Discovery of ghost cavities in the X-ray atmosphere of Abell 2597. *The Astrophysical Journal*, *562*, L149–L152.

McNamara, B., Donahue, M., Nulsen, P., et al. (2005), Star bursts and super cavities in clusters of galaxies, *Spitzer Proposal ID 20345*.

Melrose, D. B. (1990). Particle beams in the solar atmosphere: General overview. *Solar Physics*, *130*, 3–18.

Mewe, R., Heise, J., Gronenschild, E. H. B. M., et al. (1975). Detection of x-ray emission from stellar coronae with ANS. *The Astrophysical Journal*, *202*, L67–L71.

Mills, D. M., & Sturrock, P. A. (1970). A model of extragalactic radio sources. *Astrophysical Journal Letters*, *5*, 105.

Mitchell, R. J., Culhane, J. L., Davison, P. J. N., & Ives, J. C. (1976). Ariel 5 observations of the X-ray spectrum of the Perseus cluster. *Monthly Notices of the Royal Astronomical Society*, *175*, 29–34.

Molendi, S. (2002). On the temperature structure of M87. *The Astrophysical Journal*, *580*, 815–823.

Montgomery, D., Brown, M. R., & Matthaeus, W. H. (1987). Density fluctuation spectra in magnetohydrodynamic turbulence. *Journal of Geophysical Research*, *92*, 282–284.

Müller, W. C., Biskamp, D., & Grappin, R. (2003). Statistical anisotropy of magnetohydrodynamic turbulence. *Physical Review E*, *67*, 066302.

Myers, P. C. (1983). Dense cores in dark clouds III-subsonic turbulence. *The Astrophysical Journal*, *270*, 105–118.

Nagar, N. M., Falcke, H., Wilson, A. S., & Ho, L. C. (2000). Radio sources in low-luminosity active galactic nuclei. I. VLA detections of compact flat-spectrum cores. *The Astrophysical Journal*, *542*, 186–196.

Narayan, R. (1992). The physics of pulsar scintillation. *Philosophical Transactions of the Royal Society A*, *341*, 151–165.

Neugebauer, M. (1975). The enhancement of solar wind fluctuations at the proton thermal gyroradius. *Journal of Geophysical Research*, *80*, 998–1002.

Nipoti, C., & Binney, J. (2004). Cold filaments in galaxy clusters: Effects of heat conduction. *Monthly Notices of the Royal Astronomical Society*, *349*, 1509–1515.

Norman, M. L., Winkler, K. H. A., Smarr, L., & Smith, M. D. (1982). Structure and dynamics of supersonic jets. *Astronomy & Astrophysics*, *113*, 285–302.

Nulsen, P. E. J., David, L. P., McNamara, B. R., et al. (2002). Interaction of radio lobes with the hot intracluster medium: Driving convective outflow in Hydra A. *The Astrophysical Journal*, *568*, 163–173.

Nulsen, P. E. J., McNamara, B. R., Wise, M. W., & David, L. P. (2005). The cluster-scale AGN outburst in Hydra A. *The Astrophysical Journal*, *628*, 629–636.

Oboukhov, A. (1941). On the distribution of energy in the spectrum of turbulent fow. *Dokl. Akad. Nauk SSSR*, *32*, 22.

Pallavicini, R., Tagliaferri, G., & Stella, L. (1990). X-ray emission from solar neighborhood flare stars: A comprehensive survey of EXOSAT results. *Astronomy & Astrophysics*, *228*, 403–425.

Perley, R. A., Dreher, J. W., & Cowan, J. J. (1984). The jet and filaments in Cygnus A. *The Astrophysical Journal*, *285*, L35–L38.

Perlman, E. S., Biretta, J. A., Sparks, W. B., et al. (2001). The optical-near-infrared spectrum of the M87 jet from Hubble space telescope observations. *The Astrophysical Journal*, *551*, 206–222.

Perryman, M. A. C., & the Hipparcos Science Team,. (1997). *The Hipparcos and Tycho catalogues, ESA report SP-1200*. Noordwijk: ESA.

Peterson, J. R., & Fabian, A. C. (2006). X-ray spectroscopy of cooling clusters. *Physics Reports*, *427*, 1–39.

Peterson, J. R., Kahn, S. M., Paerels, F. B. S., et al. (2003). High-resolution X-ray spectroscopic constraints on cooling-flow models for clusters of galaxies. *The Astrophysical Journal*, *590*, 207–224.

Peterson, J. R., Paerels, F. B. S., Kaastra, J. S., et al. (2001). X-ray imaging-spectroscopy of Abell 1835. *Astronomy & Astrophysics*, *365*, L104–L109.

Pevtsov, A. A., Fisher, G. H., Acton, L. W., et al. (2003). The relationship between X-ray radiance and magnetic flux. *The Astrophysical Journal*, *598*, 1387–1391.

Phillips, J. A., & Wolszczan, A. (1991). Time variability of pulsar dispersion measures. *The Astrophysical Journal*, *382*, L27–L30.

Podesta, J. J. (2009). Dependence of solar-wind power spectra on the direction of the local mean magnetic field. *The Astrophysical Journal*, *698*, 986–999.

Pointecouteau, E., Arnaud, M., Kaastra, J., & de Plaa, J. (2004). XMM-Newton observation of the relaxed cluster A478: Gas and dark matter distribution from 0.01R200 to 0.5R200. *Astronomy & Astrophysics*, *423*, 33–47.

Potash, R. I., & Wardle, J. F. C. (1980). 4C 32.69: a quasar with a radio jet. *The Astrophysical Journal*, *239*, 42–49.

Pratt, G. W., Böhringer, H., Croston, J. H., et al. (2007). Temperature profiles of a representative sample of nearby X-ray galaxy clusters. *Astronomy & Astrophysics*, *461*, 71–80.

Rafferty, D. A., McNamara, B. R., Nulsen, P. E. J., & Wise, M. W. (2006). The feedback-regulated growth of black holes and bulges through gas accretion and star bursts in cluster central dominant galaxies. *The Astrophysical Journal*, *652*, 216–231.

Rees, M. J. (1982), in D. S. Heesehen & C. M. Wade (eds.), 'Extragalactic radio sources', IAU Syrup. 97, 21.

Reynolds, C. S., Heinz, S., & Begelman, M. C. (2001). Shocks and sonic booms in the intracluster medium: X-ray shells and radio galaxy activity. *The Astrophysical Journal*, *549*, L179–L182.

Rickett, B. J. (1970). Interstellar scintillation and pulsar intensity variations. *Monthly Notices of the Royal Astronomical Society*, *150*, 67–91.

Rickett, B. J., Coles, W. A., & Bourgois, G. (1984). Slow scintillation in the interstellar medium. *Astronomy & Astrophysics*, *134*, 390–395.

Rickett, B. J. (1990). Radio propagation through the turbulent interstellar plasma. *Annual Review Astronomy & Astrophysics*, *28*, 561–605.

Ross, D. W., Chen, G. L., & Mahajan, S. M. (1982). Kinetic description of Alfvén wave heating. *Physics of Fluids*, *25*, 652–667.

Russell, C. T. (1985). Patchy reconnection and magnetic ropes in astrophysical plasmas. In M. R. Kundu & G. D. Holman (eds.) *Unstable Current Systems and Plasma Instabilities in Astrophysics IAU Symposia*, *107*, 25–42.

Šafránková, J., Němeček, Z., Němec, F., et al. (2015). Solar wind density spectra around the ion spectral break. *The Astrophysical Journal*, *803*, 107.

Sahraoui, F., Goldstein, M. L., Robert, P., & Khotyaintsev, Y. V. (2009). Evidence of a cascade and dissipation of solar-wind turbulence at the electron gyroscale. *Physical Review Letters*, *102*, 231102.

Salpeter, E. E. (1969). Pulsar amplitude variations. *Nature*, *221*, 31–33.

Sambruna, R. M., Maraschi, L., & Urry, C. M. (1996). On the spectral energy distributions of Blazars. *The Astrophysical Journal*, *463*, 444.

Sarazin, C. L. (1988). *X-ray emission from clusters of galaxies*. Cambridge Astrophysics Series, Cambridge: Cambridge University Press.

Schekochihin, A. A., Cowley, S. C., Dorland, W., et al. (2009). Astrophysical gyrokinetics: Kinetic and fluid turbulent cascades in magnetized weakly collisional plasmas. *The Astrophysical Journal Supplement*, *182*, 310–377.

Scheuer, P. A. G. (1968). Amplitude variations in pulsed radio sources. *Nature*, *218*, 920–922.

Schödel, R., Ott, T., Genzel, R., et al. (2002). A star in a 15.2-year orbit around the supermassive black hole at the center of the Milky Way. *Nature*, *419*, 694–696.

Schuecker, P., Finoguenov, A., Miniati, F., et al. (2004). Probing turbulence in the Coma galaxy cluster. *Astronomy & Astrophysics*, *426*, 387–397.

Serlemitsos, P. J., Smith, B. W., Boldt, E. A., et al. (1977). X-radiation from clusters of galaxies: Spectral evidence for a hot evolved gas. *The Astrophysical Journal*, *211*, L63–L66.

Shakura, N. I. & Sunyaev, R. A. (1973). Black holes in binary systems: Observational appearance, Edited by H. Bradt and Riccardo Giacconi, IAU Symposium 55, Dordrecht, Holland, Boston, D. Reidel, 155.

Silk, J. (1976). Accretion by galaxy clusters and the relationship between X-ray luminosity and velocity dispersion. *The Astrophysical Journal*, *208*, 646–649.

Simionescu, A., Böhringer, H., Bruggen, M., & Finoguenov, A. (2007). The gaseous atmosphere of M 87 seen with XMM-Newton. *Astronomy & Astrophysics*, *465*, 749–758.

Simionescu, A., Roediger, E., Nulsen, P. E. J., et al. (2009). The large-scale shock in the cluster of galaxies Hydra A. *Astronomy & Astrophysics*, *495*, 721–732.

Simionescu, A., Werner, N., Finoguenov, A., et al. (2008). Metal-rich multi-phase gas in M 87. AGN-driven metal transport, magnetic-field supported multi-temperature gas, and constraints on non-thermal emission observed with XMM-Newton. *Astronomy & Astrophysics*, *482*, 97–112.

Smith, D. A., Wilson, A. S., Arnaud, K. A., et al. (2002). A Chandra X-ray study of Cygnus A. III. The cluster of galaxies. *The Astrophysical Journal*, *565*, 195–207.

Solomon, P. M., Rivolo, A. R., Barrett, J., & Yahil, A. (1987). Mass, luminosity, and line width relations of galactic molecular clouds. *The Astrophysical Journal*, *319*, 730–741.

Sparks, W. B., Biretta, J. A., & Macchetto, F. (1996). The jet of M87 at Tenth-Arcsecond resolution: Optical, ultraviolet, and radio observations. *The Astrophysical Journal*, *473*, 254.

Spicer, D. S., & Sudan, R. N. (1984). Beam-return current systems in solar flares. *The Astrophysical Journal*, *280*, 448–456.

Spinrad, H., Djorgovski, S., Marr, J., & Aguilar, L. (1985). A third update of the status of the 3 CR sources: further new redshifts and new identifications of distant galaxies. *Publications of the Astronomical Society of the Pacific*, *97*, 932–961.

Sridhar, S., & Goldreich, P. (1994). Toward a theory of interstellar turbulence 1: weak Alfvénic turbulence. *The Astrophysical Journal*, *432*, 612–621.

Steinolfson, R. S. (1985). Resistive wave dissipation on magnetic inhomogeneities Normal modes and phase mixing. *The Astrophysical Journal*, *295*, 213–219.

Steinolfson, R. S., Priest, E. R., Poedts, S., et al. (1986). Viscous normal modes on coronal inhomogeneities and their role as a heating mechanism. *The Astrophysical Journal*, *304*, 526.

Sturrock, P. A., & Feldman, P. A. (1968). A mechanism for continuum radiation from quasi-stellar radio sources with application to 3c 273B. *The Astrophysical Journal*, *152*, L39.

Tamura, T., Kaastra, J. S., Peterson, J. R., et al. (2001). X-ray spectroscopy of the cluster of galaxies Abell 1795 with XMM-Newton. *Astronomy & Astrophysics*, *365*, L87–L92.

Tanaka, Y., & Lewin, W. H. G. (1995). Black-hole binaries. In W. H. G. Lewin, J. van Paradijs, & E. P. J. van den Heuvel (Eds.), *x-ray Binaries* (pp. 126–174). Cambridge: Cambridge University Press.

Terashima, Y., & Wilson, A. S. (2003). Chandra snapshot observations of low-luminosity active galactic nuclei with a compact radio source. *The Astrophysical Journal*, *583*, 145–158.

Tribble, P. C. (1989). The reduction of thermal conductivity by magnetic fields in clusters of galaxies. *Monthly Notices of the Royal Astronomical Society*, *238*, 1247–1260.

Tsurutani, B. T., Smith, E. J., Anderson, R. R., et al. (1982). Lion roars and nonoscillatory drift mirror waves in the magnetosheath. *Journal of Geophysical Research*, *87*, 6060–6072.

Vaiana, G. S., Cassinelli, J. P., Fabbiano, G., et al. (1981). Results from an extensive Einstein stellar survey. *The Astrophysical Journal*, *244*, 163–182.

Vallée, J. P. (2004). Cosmic magnetic fields-as observed in the universe, in galactic dynamos, and in the Milky Way, New Astron. *Review*, *48*, 763–841.

van den Oord, G. H. J. (1990). The electrodynamics of beam/return current systems in the solar corona. *Astronomy & Astrophysics*, *234*, 496–518.

van der Laan, H. (1963). Radio galaxies, II. *Monthly Notices of the Royal Astronomical Society*, *126*, 535.

Vikhlinin, A., Kravtsov, A., Forman, W., et al. (2006). Chandra sample of nearby relaxed galaxy clusters: mass, gas fraction, and mass-temperature relation. *The Astrophysical Journal*, *640*, 691–709.

Vikhlinin, A., Markevitch, M., Forman, W., & Jones, C. (2001). Zooming in on the coma cluster with Chandra: compressed warm gas in the brightest cluster galaxies. *The Astrophysical Journal*, *555*, L87–L90.

von Hoerner, S. (1951). Eine methode zur untersuchung der turbulenz der interstellaren materie. mit 10 textabbildungen. *Z. Astrophysics*, *30*, 17–64.

von Weizsäcker, C. F. (1951). The evolution of galaxies and stars. *The Astrophysical Journal*, *114*, 165.

Walter, F. M., Linsky, J. L., Bowyer, S., & Garmire, G. (1980). HEAO 1 observations of active coronae in main-sequence and subgiant stars. *The Astrophysical Journal*, *236*, L137–L141.

Wentzel, D. G. (1968). Hydromagnetic waves excited by slowly streaming cosmic rays. *The Astrophysical Journal*, *152*, 987.

Werner, N., Zhuravleva, I., Churazov, E., et al. (2009). Constraints on turbulent pressure in the X-ray haloes of giant elliptical galaxies from resonant scattering. *Monthly Notices of the Royal Astronomical Society*, *398*, 23–32.

White, N. E., Sanford, P. W., & Weiler, E. J. (1978). An X-ray outburst from the RS CVn binary HR1099. *Nature*, *274*, 569.

Wicks, R. T., Horbury, T. S., Chen, C. H. K., & Schekochihin, A. A. (2010). Power and spectral index anisotropy of the entire inertial range of turbulence in the fast solar wind. *Monthly Notices of the Royal Astronomical Society*, *407*, L31–L35.

Wise, M. W., McNamara, B. R., Nulsen, P. E. J., et al. (2007). X-ray super cavities in the Hydra A cluster and the outburst history of the central galaxy's active nucleus. *The Astrophysical Journal*, *659*, 1153–1158.

Wu, D. J., Chen, L., Zhao, G. Q., & Tang, J. F. (2014). A novel mechanism for electron-cyclotron maser. *Astronomy & Astrophysics*, *566*, A138.

Wu, D. J., & Chen, L. (2013). Excitation of kinetic Alfvén waves by density striation in magnetoplasmas. *The Astrophysical Journal*, *771*, 3.

Xiang, L., Chen, L., & Wu, D. J. (2019). Resonant Mode Conversion of Alfvén Waves to Kinetic Alfvén Waves in an Inhomogeneous Plasma. *The Astrophysical Journal*, *881*, 61.

Xu, L., Chen, L., & Wu, D. J. (2013). Anomalous resistivity in beam-return currents and hard-X ray spectra of solar flares. *Astronomy & Astrophysics*, *550*, A63.

Young, A. J., Wilson, A. S., & Mundell, C. G. (2002). Chandra imaging of the X-ray core of the Virgo cluster. *The Astrophysical Journal*, *579*, 560–570.

Zhang, S. N., Cui, W., Chen, W., et al. (2000). Three-layered atmospheric structure in accretion disks around stellar-mass black holes. *Science*, *287*, 1239–1241.

The manufacturer's authorised representative in the EU is Springer
Nature Customer Service Centre GmbH, Europaplatz 3, 69115 Heidelberg,
Germany. If you have any concerns regarding our products, please
contact ProductSafety@springernature.com

Printed and bound by CPI Group (UK) Ltd, Croydon, CR0 4YY
30/09/2026
02225449-0013